Linear Discrete-Time Systems

Linear Discrete-Time Systems

Zoran M. Buchevats • Lyubomir T. Gruyitch

CRC Press
Taylor & Francis Group
Boca Raton London New York

CRC Press is an imprint of the
Taylor & Francis Group, an **informa** business

CRC Press
Taylor & Francis Group
6000 Broken Sound Parkway NW, Suite 300
Boca Raton, FL 33487-2742

First issued in paperback 2022

ISBN-13: 978-1-138-03959-9 (hbk)
ISBN-13: 978-1-03-233938-2 (pbk)
DOI: 10.1201/9781138039629

Visit the Taylor & Francis Web site at
http://www.taylorandfrancis.com

and the CRC Press Web site at
http://www.crcpress.com

Contents

List of Figures

Preface

On the state of the art

Linear Discrete-Time Systems discovers and fills the fundamental missing parts of the theory and practice of the discrete-*time time*-invariant linear dynamical Multiple Input-Multiple Output (*MIMO*) systems, short for the systems in the sequel.

Initial (input, state and output) conditions contain all relevant information on the system history, external actions and their consequences upon the system until the initial moment. The initial conditions are unforeseeable, unknown a priori and untouchable. It is well known that the $Z-$transform of shifts of a variable contains the corresponding initial conditions. They do exist in the complex domain. They appear in the $Z-$transform of the system state and the system output. Their appearance for the systems of the order two or higher is under double sums that contain products of the system parameters and initial conditions. Such double sums seemed nontreatable effectively mathematically. In order to avoid the original mathematical problem of how to study effectively the system described completely, in the forced regime under arbitrary initial conditions, the superposition principle has been exploited to justify the assumption on the zero initial conditions, which has been unconditionally accepted and has governed the courses and the research on the systems.

The fact is that the system transfers simultaneously the influence of the input variables, i.e., the influence of the input vector, and the influence of the initial conditions on its state and its output.

The system, which is in reality dynamical, possesses its internal dynamics that determines its state and state variables, i.e., its state vector. This fact has been ignored by the unjustified accepted condition that the concept of state has a sense if and only if the shifts of the input vector do not influence the system internal dynamics. This condition has no physical meaning. However, it appeared to be very useful to develop a mathematical machinery for the related studies. This explains the reason why the theory and practice have been well developed only for one class of the physical systems. The treatment of other classes of the physical systems requires formal mathematical transformation of their mathematical models with the complete loss of the physical sense of the new mathematical model.

A fundamental lacuna is that the concept of the state, state variables, state vector and state space is well defined and widely effectively used only in the framework of the systems described by the first-order vector linear discrete state equation and by the algebraic vector linear discrete output equation, which are referred to as *Input-State-Output* (ISO) (or *state-space*) systems. This fundamental lacuna is the result of the nonexistence of the clear, well-defined concept of the state also for the much less theoretically treated systems described by an arbitrary order vector linear input-output discrete-time equation, which are called *Input-Output* (IO) systems, nor is there an unambiguous concept of the state for a new, more general class of the systems described by an arbitrary order vector linear discrete-time state equation and by a vector linear discrete-time output equation, which are called *Input-Internal and Output dynamical* (IIO) systems.

The second fundamental lacuna is related to the use of the system transfer function matrix $\boldsymbol{G}(z)$, defined and valid only for all zero initial conditions and a nonzero input vector, to test Lyapunov stability properties defined for arbitrary initial conditions and for the zero input vector.

The third fundamental lacuna refers to the restriction of the bounded input stability concept imposed by the condition that all initial conditions are equal to zero.

Last but not least, a fundamental lacuna arises from the nonexistence of a complex domain method for effective treatment based on the $Z-$transform of the systems subjected simultaneously to both input actions and initial conditions. It is an outcome of the unreal and unjustifiable assumption on all zero initial conditions, which is accepted for the definition and determination of the system transfer function (transfer function matrix $\boldsymbol{G}(z)$ in general) $G(z)$ and for the system matrix $\boldsymbol{P}(z)$. One of its consequences is the inconsistent link between the stability definitions and the stability criteria determined in terms of the system transfer function matrix $\boldsymbol{G}(z)$.

The inconsistence results from the definition of the former exclusively for the free regime under arbitrary initial conditions and the definition of the latter exclusively for the forced regime under zero initial conditions.

Another consequence is due to the unclear criterion for the pole-zero cancellation. The treatment of mainly the system behavior in the forced regime under zero initial conditions is the third consequence.

The conceptual drawback of the widely accepted existing theory is the nonexistence of the state space theory for the *IO* dynamical systems. Moreover, the discrete-time *IIO* systems have not been studied. These drawbacks reflect a profound scientific and engineering vagueness.

Among many excellent books on the linear dynamical systems theory, in writing this book, the authors consulted in particular the books by the following writers (mentioned in alphabetical order): J. Ackermann [2], B. D. O. Anderson and J. B. Moore [4], P. J. Antsaklis and A. N. Michel [5], K. J. Astrom and B. J. Wittenmark [6], A. B. Bishop [8], J. G. Bollinger and N. A. Duffie [9], P. Borne et al. [10], W. L. Brogan [12], G. S. Brown and D. P. Campbell [13], J. A. Cadzow and H. R. Martens [14], F. M. Callier and C. A. Desoer [15] and [16], C.-T. Chen [18], H. Chestnut and R. W. Mayer [19], R. I. Damper [20], J. J. D'Azzo and C. H. Houpis [21], C. A. Desoer [22], C. A. Desoer and M. Vidyasagar [23], Lj. T. Grujić [41], Ly. T. Gruyitch [68], Ly. T. Gruyitch [69], Y. Hasegawa [78], C. H. Houpis and G.B. Lamont [80], R. Iserman [81], W. Hahn [83], T. Kailath [84], N. N. Krasovskii [86] and [87], B. C. Kuo [88] and [89], [90], H. Kwakernaak and R. Sivan [91], A. M. Lyapunov [93], L. A. MacColl [94], J. M. Maciejowski [95], M. Mandal and A. Asif [96], J. L. Melsa and D. G. Schultz [97], B. R. Milojković and Lj. T. Grujić[1] [99], K. M. Moudgalya [101], K. Ogata [105] and [106], D. H. Owens [107], C.L. Phillips and H.T. Nagle [108], H. M. Power and R. J. Simpson [109], C. A. Rabbath and N. Léchevin [110], H. H. Rosenbrock [111], E. N. Sanchez, F. Ornelas-Tellez [112], R. E. Skelton [117], H.F. Vanlandingham [122], M. Vidyasagar [123], J. C. West [124], D. M. Wiberg [125], W. A. Wolovich [126], and W. M. Wonham [127].

On the book

Linearized continuous-*time* and then discretized in *time* mathematical descriptions of physical (biological, economical, technical) systems, or natu-

[1] Author's name is written in Serbo-Croatian as Ljubomir T. Grujić using the Croatian Roman alphabet, and Lyubomir T. Gruyitch written in English and sufficiently correctly in French (in which it fully correctly reads: Lyoubomir T. Grouyitch [35]). Since it is not possible to preserve the original Cyrillic writing of Serb names in other languages, and since it is important to preserve personal identification, in order to save at least the correct pronunciation, then the adequate transcription should be applied, for which, in this case, the author's name is transcribed in other languages as shown herein.

rally occurring discrete-*time* systems, can be studied in the *time* domain or in the *complex* domain. In the time domain, all system variables are expressed in terms of the *discrete time* k. In the complex domain, all system variables are represented by their $Z-$transforms in terms of *the complex variable* z. In this book, the linear *time*-invariant discrete-*time* mathematical models of physical (biological, economical, technical) dynamical systems will be referred to as *the systems.* The book treats some general aspects of them. This book is the full discrete-time counter part to the books [68] and [69]. The former fully relies on, follows and presents a discrete-time analogy of [68] and [69]. These two books are completely literally mutually analogous and constitute an entity.

The authors will treat the three classes of the discrete-*time* *time*-invariant dynamical systems, for short: **systems**.

- *The Input-Output* (*IO*) *systems* described by the ν*-th* order linear vector discrete-time equation in the output vector $\mathbf{Y} \in \mathcal{R}^N$.

- *The (first-order) Input-State-Output* (*ISO*) *systems* determined by the first-order linear vector discrete-time equation in the state vector $\mathbf{X}$, *the state equation,* $\mathbf{X} \in \mathcal{R}^n$, and by the discrete-time algebraic vector equation in the output $\mathbf{Y}$, *the output equation.* They are well known as *the state-space systems.*

- *The Input-Internal and Output dynamical* (*IIO*) *systems* characterized by the α*-th* order linear vector discrete-time equation in the internal dynamics vector $\mathbf{R}$, *the internal dynamics equation,* $\mathbf{R} \in \mathcal{R}^\rho$, and by the ν*-th* order linear vector discrete-time (if $\nu > 0$), or linear vector discrete-time algebraic (if $\nu = 0$), equation in the output vector $\mathbf{Y}$, *the output equation.* This class of the systems has not been studied so far.

The *time* domain mathematical models and their studies allow for the direct insight into the physical properties and dynamical behavior of the systems. Physical phenomena of and processes in dynamical systems can be adequately explained in the *time* domain. It is also appropriate for defining dynamical properties of the systems. However, their direct mathematical treatment in the *time* domain is not very effective in the case of the analysis of their qualitative dynamical properties.

The complex domain mathematical descriptions of the systems enable the effective mathematical treatment of their qualitative properties. There are several basic tools it employs such as the $Z-$transform and, induced by it, the fundamental dynamic system characteristic that is well known, in general, as the transfer function matrix $\boldsymbol{G}(z)$ of a *Multiple-Input Multiple-Output* (*MIMO*) system, or, in a simpler case, the transfer function $G(z)$ of a *Single-Input Single-Output* (*SISO*) system. The usage of $G(z)$ and $\boldsymbol{G}(z)$ allows us to study effectively many system qualitative dynamical properties [e.g., the system response *under all zero initial conditions*, system controllability and observability, Lyapunov stability of a completely controllable and observable system, and *Bounded-Input Bounded-Output* (*BIBO*) stability under *zero initial conditions*]. It has a number of advantages. It represents the powerful mathematical system characteristic to treat dynamical properties of the system, by definition, under *all zero initial conditions*. Its substantial feature is its full independence of external actions (i.e., of the system inputs), and, naturally, of initial conditions since it is defined *for all zero initial conditions*. It is completely determined exclusively by the system itself so that it describes in the complex domain how the system transfers in the course of *time* the influence of the external action on the system to its output variables (i.e., to its output vector). It is *only the system input-output transfer function (matrix), nothing more*. It is inapplicable in the cases when the initial conditions are not all equal to zero. It does not and cannot express in the complex domain how the initial conditions influence the system output behavior in the course of *time*.

This poses the following basic question:

Does a linear *time*-invariant system have a complex domain characteristic mathematically expressed in the form of a matrix such that it satisfies the following two conditions:

- it describes in the complex domain how the system transfers in the course of *time* the influences of *both* the input vector action and of *all initial conditions* on the system output response, and

- it is completely determined by the system itself, meaning its full independence of *both the input vector and the vector of all initial conditions*?

The reply is affirmative.

Such system characteristic is its **full (complete) transfer function matrix** $\boldsymbol{F}(z)$. It will be precisely defined and determined for all three classes of the systems. Its usage permits us to refine studies of system dynamical properties (e.g., the complete system response, Lyapunov stability, Bounded Input (*BI*) stability properties, system minimal realization, system tracking and its trackability [70]). It shows exactly when poles and zeros may (not) be cancelled.

Even the simplest first-order linear *SISO* dynamical system induces this characteristic in the matrix form rather than the scalar form.

It seems that the problem which impeded the determination of the full transfer function matrix refers mainly to the application of the $Z-$transform to the *IO* (*Input-Output*) (and to *the Input-Internal and Output dynamical, i.e., IIO*) systems descriptions, which are in the form of the high-order vector linear discrete equations with constant matrix coefficients.

The $Z-$transform of the *IO* and *IIO* systems introduces double sums that contain the products of initial conditions and system parameters [(3.10), (3.11), (3.14a), (3.14b), (3.15a), and (3.15b), in Subsection 3.3.3]. It will be shown how those double sums can be represented, e.g., for the *IO* system, as a product of a matrix $\boldsymbol{G}_{IO_0}(z)$ and of the vector $\mathbf{C}_{IO_0}$ composed of all (input and output) initial conditions,

$$\mathbf{C}_{IO_0} = \begin{bmatrix} \mathbf{I}_0 \\ E^1\mathbf{I}_0 \\ \vdots \\ E^{\mu-1}\mathbf{I}_0 \\ \mathbf{Y}_0 \\ E^1\mathbf{Y}_0 \\ \vdots \\ E^{\nu-1}\mathbf{Y}_0 \end{bmatrix} \in \mathcal{R}^{\mu M+\nu N}.$$

The vector $\mathbf{C}_{IO_0}$ is a subvector of the vector $\mathbf{V}_{IO}(z)$ composed of the $Z-$transform $\mathbf{I}(z)$ of the input vector $\mathbf{I}(k)$ and of all (input and output) initial conditions composing the vector $\mathbf{C}_{IO_0}$,

$$\mathbf{V}_{IO}(z) = \begin{bmatrix} \mathbf{I}(z) \\ \mathbf{C}_{IO_0} \end{bmatrix} \in \mathcal{R}^{(\mu+1)M+\nu N}.$$

A complete list of the notation is in Appendix A.

The matrix $\boldsymbol{G}_{IO_0}(z)$ is determined exclusively by the system parameters. It is completely independent of the initial conditions and of the input vector. It is *the system transfer function matrix relative to the initial conditions,* which along the system transfer function matrix $\boldsymbol{G}_{IO}(z)$, compose the *full transfer function matrix* $\boldsymbol{F}_{IO}(z)$ of the *Input-Output* (*IO*) *systems,*

$$\boldsymbol{F}_{IO}(z) = \begin{bmatrix} \boldsymbol{G}_{IO}(z) & \boldsymbol{G}_{IO_0}(z) \end{bmatrix} \Longleftrightarrow \mathbf{Y}(z) = \boldsymbol{F}_{IO}(z)\mathbf{V}_{IO}(z).$$

The analogy also holds for the *Input-State-Output* (*ISO*) *systems*,

$$\boldsymbol{F}_{ISO}(z) = \begin{bmatrix} \boldsymbol{G}_{ISO}(z) & \boldsymbol{G}_{ISO_0}(z) \end{bmatrix} \Longleftrightarrow \mathbf{Y}(s) = \boldsymbol{F}_{ISO}(z)\mathbf{V}_{ISO}(z),$$

and for the *Input-Internal dynamics-Output* (*IIO*) *systems*,

$$\boldsymbol{F}_{IIO}(z) = \begin{bmatrix} \boldsymbol{G}_{IIO}(z) & \boldsymbol{G}_{IIO_0}(z) \end{bmatrix} \Longleftrightarrow \mathbf{Y}(z) = \boldsymbol{F}_{IIO}(z)\mathbf{V}_{IIO}(z).$$

The use of the system full transfer function matrix $\boldsymbol{F}(z)$ requires the same knowledge of mathematics as for the application of the system transfer function matrix $\boldsymbol{G}(z)$. Nothing more. However, it requires some new notation that is presented in the simple symbolic vector or matrix form analogous to the scalar form.

The introduction of both the system full transfer function matrix $\boldsymbol{F}(z)$ and the $Z-$transform $\mathbf{V}(z)$ of the complete action vector $\mathbf{v}(k)$ makes it possible to generalize the block diagram technique to the **full block diagram technique**.

The definition and the determination of the full transfer function matrix $\boldsymbol{F}(z)$ is unified herein for all three classes of the systems. The new compact, simple, scalar-like, vector-matrix notation permits us to integrate the treatment of the IO systems and the IIO systems in order to study them in the same manner as the ISO systems. Lyapunov stability theory (definitions, method and theorems) becomes directly applicable to the IO and IIO systems in the same manner as to the ISO systems. It will be found herein that the properties of the system full transfer function matrix $\boldsymbol{F}(z)$ and of the system transfer function matrix $\boldsymbol{G}(z)$ show that $\boldsymbol{F}(z)$ is adequate for Lyapunov stability study and investigations of BI stability properties under arbitrary bounded initial conditions while the application of $\boldsymbol{G}(z)$ can yield essentially the wrong result (e.g., that the equilibrium state is asymptotically stable although it is really unstable). The crucial generalization of the BI stability concept has been achieved herein. Additionally, Lyapunov stability and BI stability criteria are inherently refined by proving new ones in the complex domain. This will be possible only due to the use of the system full transfer function matrix $\boldsymbol{F}(z)$.

The purpose of this book is threefold: to contribute to the advancement of the linear dynamical systems theory, improvement of the corresponding university courses, and to open up new directions for the research in this theory and its engineering applications. It represents a further development of the existing linear systems theory, which will not be repeated herein.

The system full transfer function matrix $\boldsymbol{F}(s)$ (of continuous-*time* systems) was introduced by the author of the book Lyubomir T. Gruyitch,[2] to seniors through the undergraduate courses on linear dynamical systems and on control systems first at the Department of Electrical Engineering, University of Natal in Durban (UND), R. South Africa, 1993, and at the National Engineering School (Ecole Nationale d'Ingénieurs de Belfort, $ENIB$) in Belfort, France, 1994, and continued until 1999. This was a topic of lectures held by the author of the book Lyubomir T. Gruyitch, to freshmen or juniors of the new University of Technology Belfort, Montbéliard ($UTBM$), which was created as the union of the $ENIB$ and the Polytechnic Institute of Sevenans. The UND students had the lecture notes [34] available during the course. The lecture notes [29], [30], [35], [36] [59], [60], [61] were available to the students, immediately after the classes, in the copy center of $ENIB$ / $UTBM$.

The system full transfer function matrix $\boldsymbol{F}(z)$ (of discrete-*time* systems) was also introduced by the author of the book Lyubomir T. Gruyitch through the course on Control of industrial processes (2002, 2003) [61] at the University of Technology Belfort-Montbéliard, Belfort, France.

Hopefully, *the 21st-century linear systems courses and linear control courses:*

– **will incorporate**:

- *all three classes of the systems:* IO, ISO and IIO systems;
- *the system full (complete) transfer function matrix* $\boldsymbol{F}(z)$ [62] as the basic system dynamical characteristic in the complex domain, as well as its applications to various issues of dynamical systems in general, and control systems in particular, e.g., to the system complete response, pole-zero cancellation, stability, observability, controllability, tracking, trackability ([37]-[40], [42]-[50], [52]-[54], [56]-[58], [70]-[76], [102]-[104]), and optimality, which are the basic issues of the system theory in general, and the control theory in particular;

  and
- *the tracking and trackability theory*, as the fundamentals of the control science and control engineering, which express the pri-

---

[2] It is the common to transcribe the Serb Cyrillic letter "ћ" in English into "ch." However, "ch" is always used as an English transcription of another Slavic (Russian, Serb, $\cdots$) Cyrillic letter "ч." In order to avoid the confusion, the author has accepted to use "tch" for "ћ" because the linguists have not implemented another solution that might be more adequate.

mary control goal (presented in the accompanying books [70], [71]);

- **will refine the study in the complex domain** *of the qualitative system properties* ***by using the system full transfer function matrix*** $\boldsymbol{F}(z)$ *instead of the system transfer function matrix* $\boldsymbol{G}(z)$;
- **will devote more attention to the basic system characteristics** *such as*
  - *system regimes,*
  - *system stationary vectors,*
  - *system equilibrium vectors;*

  and
- **will pay attention to the differences between**
  - *the transfer function matrix realization* and *the system realization,*
  - *the irreducible complex rational matrix functions* and *the degenerate complex rational matrix functions.*

Note 0.1 ***On the proofs presentation***
Since this book provides novel results, all the details of the derivations of formulae and proofs of the main results are preserved and exposed in the text, even those that might appear very simple (or even trivial) to some readers, but with which some students or young researchers might be unfamiliar. Such presentation gives the reader a full understanding and an easy verification of the statements and results.

In gratitude

The authors are indebted to the publisher, the editor and their team, in particular to:

Ms. Nora Konopka, Editor, Engineering, for leading and organizing the publication process elegantly and effectively

Ms. Kyra Lindholm, Editor, Engineering, for leading effectively the administrative process

Ms. Karen Simon, Production Editor, for effectively leading the editing process

Ms. Michele Smith, Editor, Engineering, for her very useful assistance

Zoran M. Buchevats,[3] author
Lyubomir T. Gruyitch, author
Belgrade

[3]The Serb Cyrillic letter "ц" is transcribed in English into "ts."

Part I

BASIC TOPICS

Chapter 1

Introduction

1.1 *Discrete time*, physical variables, and systems

All processes, motions and movements, all behaviors of systems and their responses, as well as all external actions on the systems, occur and propagate in *time*. It is natural from the physical point of view to study systems directly in the temporal domain. This requires to be clear about how we understand what is *time* and what are its properties. It was explained in brief in [68], [69] and for the more complete analysis can be seen in [64], [65] and [66].

Currently, the most actual approach in studying dynamical systems is to observe their phenomena, properties, behaviors, only at some *moments* (*instants*), called *discrete moments* (*discrete instants*).

Definition 1.1 *The truncated time called* ***the discrete time*** *and denoted by t_d is the physical and mathematical variable, the values of which belong only to the set of all discrete instants (i.e., of all discrete moments).*

Since the discrete instants are temporal values of time t then they flow monotonously, permanently, independently of everybody and everything with constant and invariant time speed v_t, the numerical value of which is equal to one (1) everywhere and always [64], [65], [66]. This means that the speed v_{t_d} of the discrete instants flow is constant, invariant with the numerical value one,

$$v_{t_d} = v_t = 1 \left\langle 1_t 1_t^{-1} \right\rangle \left[\mathrm{TT}^{-1} \right]. \tag{1.1}$$

In (1.1) [T] means the physical dimension of *time* where T stands for *time*.

Let Int_t be an arbitrarily chosen and fixed *time* interval represented as $\text{Int}_t = [0, T]$ or $\text{Int}_t = [t, t+T]$ in general. Its duration is usually very short: $T - 0 = T$ or $t + T - t = T$ in general. Then, only some moments (instants) defined by integer multiples kT of T, $k \in \mathbb{Z}$, where $\mathbb{Z}$ is the set of integers, are observed. Obviously, instants defined in such a way are discrete regardless of whether T is dimensionless or expressed in terms of any *time* unity 1_t [68]. These discrete instants are instants of the *discrete time* t_d. As adopted, the *time* interval Int_t is known. The discrete instants are well determined by the integers k so that they are usually only used to represent the instants of the *discrete time*, but, it is clear, only with respect to the accepted and fixed *time* interval Int_t.

There is exactly one real number that can be assigned to every moment (instant), and vice versa, while $\text{num}\, t$ means the numerical value of the moment t which is exactly that real number, $\text{num}\, t \in \mathcal{R}$.

The *discrete* instants speed is the same time speed also over the *time* interval Int_t because it is the variance of the time values from 0 to T, or t to $t+T$, in general, during the time interval $\text{Int}_t = [0, T]$, or $\text{Int}_t = [t, t+T]$, in general, respectively,

$$v_{\text{Int}_t} = TT^{-1} = TT^{-1}\left\langle 1_t 1_t^{-1} \right\rangle = v_{t_d} = v_t, \text{num}\, v_{\text{Int}_t} = \text{num}\, v_{t_d} = 1 = \text{num}\, v_t.$$

This illustrates Equation (1.1). It is natural, because *discrete* instants are constituents of the main *time* stream.

Let $\exists!$ mean "there is exactly one," i.e., "there exists exactly one."

Discrete time set $\mathcal{T}_d$ is the set of all discrete moments in the previously defined sense. It is in the biunique (one-to-one) correspondence with the set $\mathcal{R}_d \subset \mathcal{R}$ of some real numbers, $\mathcal{R}_d = \{x : x = k\, \text{num}\, T \in \mathcal{R}, k \in \mathbb{Z}\}$,

$$\mathcal{T}_d = \left\{ \begin{array}{c} t_d : \text{num}\, t_d = k\, \text{num}\, T \in \mathcal{R}_d \subset \mathcal{R}, \\ \Delta t_d = (k+1)T - kT = T > 0, v_{t_d} \equiv 1 \end{array} \right\} \tag{1.2a}$$

$$\forall t_d \in \mathcal{T}_d \Longleftrightarrow \exists! i \in \mathbb{Z} \Longrightarrow i = \text{num}\, t_d \,(\text{num}\, T)^{-1}, \text{ or} \tag{1.2b}$$

$$\forall t_d \in \mathcal{T}_d \Longleftrightarrow \exists! y \in \mathcal{R}_d \Longrightarrow y = \text{num}\, t_d \tag{1.2c}$$

and

$$\forall i \in \mathbb{Z} \Longleftrightarrow \exists! t_d \in \mathcal{T}_d \Longrightarrow \text{num}\, t_d \,(\text{num}\, T)^{-1} = i, \text{ or} \tag{1.2d}$$

$$\forall y \in \mathcal{R}_d \Longleftrightarrow \exists! t_d \in \mathcal{T}_d \Longrightarrow \text{num}\, t_d = y. \tag{1.2e}$$

Obviously, the set $\mathcal{T}_d$ is left and right unbounded. The set $\mathcal{T}_d$ can be extended by $\{-\infty, +\infty\}$ in set $\mathcal{T}_d^*$, $\mathcal{T}_d^* = \mathcal{T}_d \cup \{-\infty, +\infty\}$. $\mathcal{T}_d$ as a subset of $\mathcal{T}_d^*$ has got in $\mathcal{T}_d^*$ its infimum and supremum, $\text{num}\inf T_d = \text{num}\, t_{d\inf} =$

$-\infty \in T_d^*$ and num sup $\mathcal{T}_d = \text{num}\, t_{d\sup} = \infty \in \mathcal{T}_d^*$ so that inf $\mathcal{T}_d \notin \mathcal{T}_d$ and sup $\mathcal{T}_d \notin \mathcal{T}_d$.

In the case of the ideal sampling process, it is clear that the duration T of the interval Int_t corresponds to the constant sampling period, which is in literature usually denoted also by T [2], [9], [18], [41], [80], [81], [89], [108], [122]. We will denote in sequel, the sampling period also by T. Since the sampling period T is numerically integrated in a discrete time value during the sampling process, the only visible representative of *discrete time* remains the integer k. It is in the full agreement with the already accepted presentation of *discrete time* only by integers k.

Let int_t be an another arbitrarily chosen and fixed *time* interval, $\text{int}_t = [0, \psi]$ or $\text{int}_t = [t, t+\psi]$, of the duration ψ, $\psi = \psi - 0$ or $\psi = t + \psi - t$ such that $0 < \frac{\text{num}\,\psi}{\text{num}\,T} \ll 1$. Other moments (instants) defined by integer multiples kT of T increased by ψ, $kT + \psi$, can be also observed. These moments (instants) are obviously discrete, too.

In view of a nonideal sampling process, ψ corresponds to the sampling interval, in literature denoted in different ways, by, ψ [41], γ [80], h [2], [81], p [89], τ [122]. Nonideal sampling process corresponds to the happenings in reality, i.e., to the real sampling process. Then, no matter how small sampling interval ψ is, it is finite and has got finite numerical value.

It is accepted herein for the relative *discrete time* zero moment t_{dzero} to be the same as the relative zero moment t_{zero} [65], [66], [68], [69] i.e., that it has the zero numerical value, $\text{num}\, t_{dzero} = \text{num}\, t_{zero} = 0$. As $\text{num}\, T \neq 0$, consequently, for the adopted t_{dzero} the *discrete time representative* k is equal to zero, $k = 0$. Also, this moment is adopted herein to be the initial moment t_{d0}, $t_{d0} = t_{dzero}$, $\text{num}\, t_{d0} = 0$, in view of the *time*-invariance of the system to be studied. Consequently, $k_0 = 0$, too. This determines the subset $\mathcal{T}_{d0}$ of $\mathcal{T}_d$,

$$\mathcal{T}_{d0} = \{t_d : t_d \in \mathcal{T}_d, \text{num}\, t_d \in [0, \infty[\}\,.$$

Obviously, the *discrete time* set $\mathcal{T}_d$ is not **continuum** as the *time* set $\mathcal{T}$ is [68], [69]. Consequently, **Physical Continuity and Uniqueness Principle (*PCUP*)** and ***Time* Continuity and Uniqueness Principle (*TCUP*) in all their forms: either scalar, matrix, vector, or system form** [68], [69] are not valid with respect to *discrete time* t_d only in the ideal case when $\psi = 0$ or in case that $\psi \to 0^+$. This illustrates the unrealizability of $\psi = 0$.

However, all these Principles rest valid inside real sampling intervals,

$$t \in [k\,\text{num}\, T,\, k\,\text{num}\, T + \text{num}\,\psi]\,, \forall k \in \mathbb{Z},\, 0 < \psi \ll T,$$

in case of a nonideal sampling process: $\psi > 0$, i.e., in the case of a real sampling process.

However, the mathematical modeling of physical variables and physical systems in terms of the *discrete time* t_d is very rare by using a nonideal, real, sampling process. It is for the mathematical description and treatment of variables and systems very complicate and cumbersome.

Mostly, mathematical modeling of variables and systems with respect to the *discrete time* t_d is by means of an ideal sampling process. To do that, some conditions should be satisfied, under which a real sampling process is approximately close to an ideal sampling process. One of the conditions is dominant: the sampling interval is to be much less than the sampling period, $0 < \text{num}\,\psi \ll \text{num}\,T$, as previously stated.

Furthermore, in this text, variables and systems will be mathematically modeled with respect to *discrete time* t_d by means of an ideal sampling process, whereby T is numerically integrated in variables values. The only visible representative of *discrete instants (moments)* is integer k.

Hence, the statement $t_d \in \mathcal{T}_d$ is equivalent to $k \in \mathbb{Z}$, but having in mind, that it is with respect to the known $\text{Int}_t = [0, T]$, i.e., relative to the known T.

Similarly, $t_d \in \mathcal{T}_{d0} \subset \mathcal{T}_d$ is equivalent to $k \in \mathcal{N}_0$, where $\mathcal{N}_0$ is the extended set of natural numbers by zero:

$$\mathcal{N}_0 = \{k : k \in \mathbb{Z} \wedge k \in [0, \infty[\},$$

again having in mind, that it is with respect to the known Int_t. $\mathcal{N}$ is the set of all natural numbers, $\mathcal{N} = \{1, 2, \cdots, n, \cdots\}$, so that $\mathcal{N}_0 = \{0\} \cup \mathcal{N}$.

There exist variables and systems which are available at specific instants only. The instants are equidistant, which is again expressed by $k\,\text{Int}_t$. They are, naturally occurring, *discrete time* variables and systems (see examples in [101]). They are much less present than the previous ones. Formally, models of the variables and systems are identical as those previously considered ones. Int_t is numerically integrated in variables values and only visible representative of the specific time instants is k.

1.2 *Discrete time* and system dynamics

Regardless of whether a physical system behavior is observed with respect to *discrete time* t_d or a system is naturally *discrete-time* system, we call the system *discrete-time* system. Any variation of the discrete-time system behavior occurs in the course of *discrete time.* The values of all discrete-time system variables depend on *discrete time.*

There exist three main characteristic groups of the variables related to or linked with the dynamical system. Their definitions follow [18, Definition 3-6, p. 83], [41, p. 239], [71], [88, p. 105], [99, 2. Definition, p. 380], [100, 2. Definition, p. 380], [105, p. 4], [106, p. 664].

Definition 1.2 ***Input variables, input vector and input space***

A variable that acts on the system and its influence is essential for the system behavior, is the ***system input variable*** *denoted by* $I \in \mathcal{R}$. *The system can be under the action of several mutually independent input variables* $I_1, I_2, \cdots, I_M$. *They compose* ***the system input vector*** *(for short,* ***input****)*

$$\mathbf{I} = \begin{bmatrix} I_1 & I_2 & \cdots & I_M \end{bmatrix}^T \in \mathcal{R}^M, \tag{1.3}$$

which spans the input space $\mathcal{R}^M$.

The capital letters I *and* $\mathbf{I}$ *denote the total (scalar, vector) values of the variable* I *and the vector* $\mathbf{I}$ *relative to their total zero (scalar, vector) value, if it exists, or relative to their accepted zero (scalar, vector) value, respectively.*

The crucial characteristics of discrete-time dynamical systems is the existence of their *internal dynamics* and/or *output dynamics.* The system internal dynamics represents its *state,* which is determined by the values at discrete instants of the corresponding variables called *state variables,* and the output dynamics is determined by the *output variables and their shiftings,* in the sense of the following definitions:

Definition 1.3 ***Output variables, output vector, output space and response***

A variable $Y \in \mathcal{R}$ *is an* ***output variable*** *of the system if and only if its values result from the system behavior, they are (directly or indirectly) measurable, and we are interested in them.*

The number N *is the maximal number of the mutually independent output variables* $Y_1, Y_2, \cdots, Y_N$ *of the system. They form the* ***output vector*** $\mathbf{Y}$ *of the system, which spans* ***the output space*** $\mathcal{R}^N$:

$$\mathbf{Y} = \begin{bmatrix} Y_1 & Y_2 & \cdots & Y_N \end{bmatrix}^T \in \mathcal{R}^N. \tag{1.4}$$

The discrete time variation of the system output vector $\mathbf{Y}$ *is the* ***system (output) response.***

The capital letters Y and $\mathbf{Y}$ denote the total (scalar, vector) values of the variable Y and the vector $\mathbf{Y}$ relative to their total zero (scalar, vector) value, if it exists, or relative to their accepted zero (scalar, vector) value, respectively.

In what follows, the term *mathematical system* denotes the accepted mathematical model (i.e., the description) with respect to *discrete time* t_d of the corresponding continuous-time or discrete-time physical system.

Definition 1.4 ***State of a discrete-time dynamical system***

The (internal, output, full) state of a discrete-time dynamical physical system at a moment $k_\tau \in \mathbb{Z}$ *is the system (internal, output, internal and output) dynamical situation at the moment* k_τ *that, together with the input vector and its shiftings acting on the system at every moment* $(k \geq k_\tau) \in \mathbb{Z}$, *determines uniquely the system (internal, output, internal and output) behavior [i.e., the system (internal, output, full) state and the system output response] for all* $(k \geq k_\tau) \in \mathbb{Z}$, *respectively.*

The (internal, output, full) state of a discrete-time dynamical mathematical system at a moment $k_\tau \in \mathbb{Z}$ *is the minimal amount of information about the system (internal, output, internal and output) dynamics at the moment* k_τ *that, together with information about the action on the system (the system input and its shiftings acting on the system) at every moment* $(k \geq k_\tau) \in \mathbb{Z}$, *determines uniquely the system (internal, output, internal and output) behavior [i.e., the system (internal, output, full) state and the system output response] for all* $(k \geq k_\tau) \in \mathbb{Z}$, *respectively.*

The minimal number n_I, n_O *or* n *of mutually independent variables* S_j, $j = 1, 2, \cdots, K$, $K \in \{n_I, n_O, n\}$, *the values* $S_j(k_\tau)$ *of which at any moment* $k_\tau \in \mathbb{Z}$ *are in the biunivoque correspondence with the system (internal, output, full) state at the same moment* k_τ, *are* ***the (internal, output, full) state variables of the system, respectively****. They compose* ***the (internal, output or full) state vector*** $\mathbf{S}_I$, $\mathbf{S}_O$ *or* $\mathbf{S}$ ***of the system,***

$$\mathbf{S}_I = \begin{bmatrix} S_{I1} & S_{I2} & \cdots & S_{In_I} \end{bmatrix}^T \in \mathcal{R}^{n_I}, \tag{1.5}$$

$$\mathbf{S}_O = \begin{bmatrix} S_{O1} & S_{O2} & \cdots & S_{On_O} \end{bmatrix}^T \in \mathcal{R}^{n_O}. \tag{1.6}$$

$$\mathbf{S} = \begin{bmatrix} S_1 & S_2 & \cdots & S_n \end{bmatrix}^T \in \mathcal{R}^{n}, \; n \geq n_I + n_O, \tag{1.7}$$

respectively.

Definition 1.5 *The space* $\mathcal{R}^K$, $K \in \{n_I, n_O, n\}$, *is* ***the (internal, output, full) state space of the system, respectively.***

This definition broadens and generalizes the well-known and commonly accepted definition of the state of the *ISO* systems.

Definition 1.6 ***State and motion***

The system (internal, output, full) state vector $\mathbf{S}_{(\cdot)}(k)$ *at a moment* $k \in \mathbb{Z}$ *is the vector value of the system (internal, output, full) motion* $\mathcal{S}_{(\cdot)}(\cdot; k_0; \mathbf{S}_0, \mathbf{I})$ *at the same moment* k*:*

$$\mathbf{S}_{(\cdot)}(k) \equiv \mathcal{S}_{(\cdot)}(k; k_0; \mathbf{S}_0, \mathbf{I}) \Longrightarrow \mathbf{S}_{(\cdot)}(k_0) \equiv \mathcal{S}_{(\cdot)}(k_0; k_0; \mathbf{S}_0, \mathbf{I}) \equiv \mathbf{S}_{(\cdot)0},$$

$(\cdot) \in \{I, O, \emptyset\}$, *where* $\emptyset$ *denotes the empty space.*

1.3 Discrete-time systems and complex domain

Considerations of dynamical properties of *time*-invariant linear discrete-*time* dynamical systems, in further text denoted only by ***systems***, directly in the temporal domain can be mathematically more difficult and ineffective than in the complex domain.

The effectiveness of analysis and synthesis of *time*-invariant linear systems in the complex domain relies largely on the notion and on the properties of *the system transfer function for the SISO system*, and on its generalization — *the transfer function matrix for the MIMO system.* They represent input-output dynamical characteristics of the systems in the domain of the complex variable z,

$$z = (\sigma_z + j\omega_z) \in \mathcal{C},\ \sigma_z \in \mathcal{R},\ \omega_z \in \mathcal{R}, \tag{1.8}$$

where $\mathcal{C}$ is the set of all complex numbers z, σ_z and ω_z are real numbers, or real-valued scalar variables, and $j = \sqrt{-1}$ is the imaginary unit. They describe in the complex domain $\mathcal{C}$ how the system transfers, in course of *discrete time*, influences of the input variables $I_{(\cdot)}$ (of the input vector $\mathbf{I}$) on the output variables $Y_{(\cdot)}$ (on the output vector $\mathbf{Y}$) *under all zero initial conditions.*

For the *SISO* systems, we will write I instead of I_1, $\mathbf{I} = (I_1) = (I) \in \mathcal{R}^1$, and Y instead of Y_1, $\mathbf{Y} = (Y_1) = (Y) \in \mathcal{R}^1$, or simply $I \in \mathcal{R}$ and $Y \in \mathcal{R}$, respectively.

Here, $\mathcal{R}^1$ is the one-dimensional real vector space, the elements of which are one-dimensional real-valued vectors, while the elements of $\mathcal{R}$ are scalars (real numbers). Division is not defined in $\mathcal{R}^1$, but it is defined in $\mathcal{R}$.

The superposition principle, i.e., the system linearity, permits us to treat, mathematically, separately the influence of the input vector and the influences of all initial conditions on the system dynamical behavior. We have treated the problems of the pole-zero cancellation, the system realization, *BIBO* stability, and, most often, tracking by assuming all zero initial conditions; hence, by using the related system transfer function (matrix) and the block diagram technique. We use it also to investigate Lyapunov stability properties of the systems although they are defined exclusively for zero input vector and nonzero initial conditions. Besides, we study often the system response to the input vector action under all zero initial conditions.

The past (i.e., the history) and the present of a dynamical system influence its future behavior. If the system is without a memory and without a *discrete time* delay, then the initial conditions express and transfer, in the very condensed form, the permanent influence of the system past on its future behavior. Evidently, the past is untouchable; the initial conditions cannot be chosen, they are even most often unpredictable. The influence of the system's past is unavoidable; the initial conditions are imposed by the system history regardless of forms and intensities of current actions of the input variables.

When we wish to study the real system dynamical behavior, we may not avoid considering the influence of the initial conditions. O. I. Elgerd [24], H. M. Power and R. J. Simpson [109], and R. E. Skelton [117] observed this well, but related to the continuous-time system, by introducing various system transfer function matrices. Skelton defined the transfer function matrix relative to the initial state of the *ISO* system. Besides, he introduced the block diagram of the state-space system description with the initial state vector [117]. But, returning to the main stream of the system and control theories, he continued to use only the system transfer function matrix and the classical block diagram technique valid exclusively under all zero initial conditions.

The following examples of trivial *Single-Input Single-Output* (*SISO*) systems explain simply the crucial difference between the system transfer function $G(z)$ and the system *full (complete) matrix transfer function* $\boldsymbol{F}(z)$. They show that the use of the latter is indispensable.

Example 1.1 *Let us consider the simplest systems,*

$$a)\ aY(k+1) = bI(k) \tag{1.9a}$$

and

$$b)\ aY(k+1)+bY(k)=cI(k+1)+dI(k),$$
$$\left(abcd\neq 0, \frac{b}{a}=\frac{d}{c}\right)\in\mathcal{R}, \tag{1.9b}$$

Y is the output variable, I is the input variable

Their values are measured in their total scales.

The Z−transform $\mathcal{Z}\{\cdot\}$ (Appendix B), of the preceding equations read, respectively,

$$a)\ Y(z)=\frac{b}{az}I(z)+\frac{az}{az}Y(0) \tag{1.10a}$$

and

$$b)\ Y(z)=\frac{c\left(z+\frac{d}{c}\right)}{a\left(z+\frac{b}{a}\right)}I(z)-\frac{cz}{az+b}I(0)+\frac{az}{az+b}Y(0). \tag{1.10b}$$

We can set these equations in the more compact vector-matrix form,

$$a)\ Y(z)=\underbrace{\left[\begin{array}{cc}\overbrace{\frac{b}{az}}^{G_a(z)} & \overbrace{\frac{az}{az}}^{G_{ay0}(z)=G_{a0}(z)}\end{array}\right]}_{\boldsymbol{F}_a(z)}\underbrace{\left[\begin{array}{c}I(z)\\ Y(0)\end{array}\right]}_{\mathbf{V}_a(z)}=\boldsymbol{F}_a(z)\mathbf{V}_a(z), \tag{1.11a}$$

$$\boldsymbol{F}_a(z)=\left[\begin{array}{cc}\frac{b}{az} & \frac{az}{az}\end{array}\right]=\left[\begin{array}{cc}G_a(z) & G_{a0}(z)\end{array}\right],$$
$$\mathbf{V}_a(z)=\left[\begin{array}{c}I(z)\\ C_{0a}\end{array}\right],\ C_{0a}=Y(0), \tag{1.11b}$$

$$b)\ Y(z)=\underbrace{\left[\begin{array}{cc}\overbrace{\frac{c\left(z+\frac{d}{c}\right)}{a\left(z+\frac{b}{a}\right)}}^{G_b(z)} & \underbrace{\overbrace{-\frac{cz}{az+b}}^{G_{bi0}(z)}\ \overbrace{\frac{az}{az+b}}^{G_{by0}(z)}}_{G_{b0}(z)}\end{array}\right]}_{\boldsymbol{F}_b(z)}\underbrace{\left[\begin{array}{c}I(z)\\ I(0)\\ Y(0)\end{array}\right]}_{\mathbf{V}_b(z)}=$$
$$=\boldsymbol{F}_b(z)\mathbf{V}_b(z), \tag{1.12a}$$

$$\boldsymbol{F}_b(z)=\left[\begin{array}{ccc}\frac{c\left(z+\frac{d}{c}\right)}{a\left(z+\frac{b}{a}\right)} & -\frac{cz}{az+b} & \frac{az}{az+b}\end{array}\right]=\left[\begin{array}{ccc}G_b(z) & G_{bi0}(z) & G_{by0}(z)\end{array}\right]=$$
$$=\left[\begin{array}{cc}G_b(z) & \mathbf{G}_{b0}(z)\end{array}\right],\ \mathbf{V}_b(z)=\left[\begin{array}{c}I(z)\\ \mathbf{C}_{0b}\end{array}\right],\ \mathbf{C}_{0b}=\left[\begin{array}{c}I(0)\\ Y(0)\end{array}\right]. \tag{1.12b}$$

The complex function $\boldsymbol{F}_{(..)}(\cdot) : \mathcal{C} \to \mathcal{C}^{1\times q}$ *describes fully (completely) in the complex domain how the system transfers in the temporal domain influences of all actions: of the history through the initial conditions and the input variable, on its output behavior, where* $q = 2$ *in case* a), *and* $q = 3$ *in case* b). *The function* $\boldsymbol{F}_{(..)}(\cdot)$ *is a matrix function despite the dynamical systems are the simplest ones, scalar of the first order. We will call the function* $\boldsymbol{F}_{(..)}(\cdot)$ *the full (complete) matrix transfer function of the system. We can use it by following Skelton [117] to extend the notion of the system block; see Fig. 1.1, Fig. 1.2, Fig. 1.3, and Fig. 1.4, respectively.*

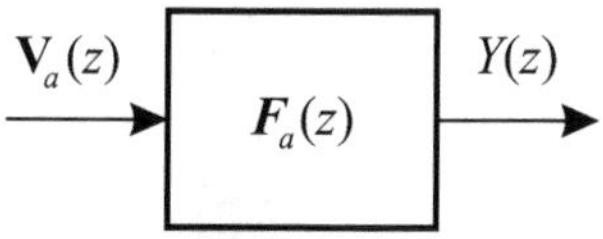

Figure 1.1: The full block of system a) in general for the nonzero initial output, $Y_0 \neq 0$.

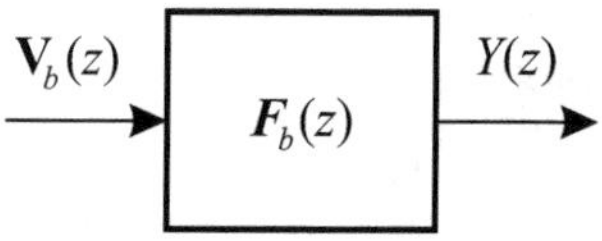

Figure 1.2: The full block of system b) in general for the nonzero initial input, $I_0 \neq 0$, and the initial output, $Y_0 \neq 0$.

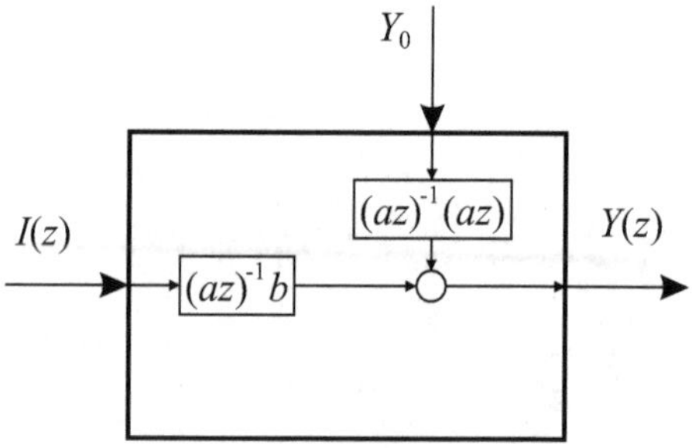

Figure 1.3: The full block diagram of system a).

We can recognize transmissions and transformations of different influences through the system to its output. They are expressed and described in the complex domain by the corresponding transfer functions $G_{(..)}(\cdot) : \mathcal{C} \to \mathcal{C}$ *that are scalar entries in these examples of* $\boldsymbol{F}_{(..)}(z)$, *Fig. 1.3 through Fig. 1.8. In the case of*

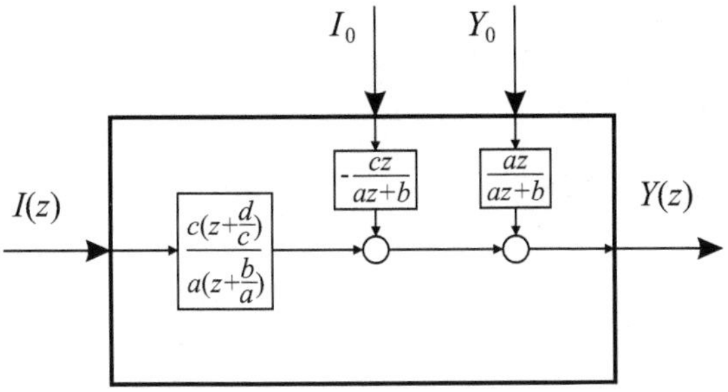

Figure 1.4: The full block diagram of system b).

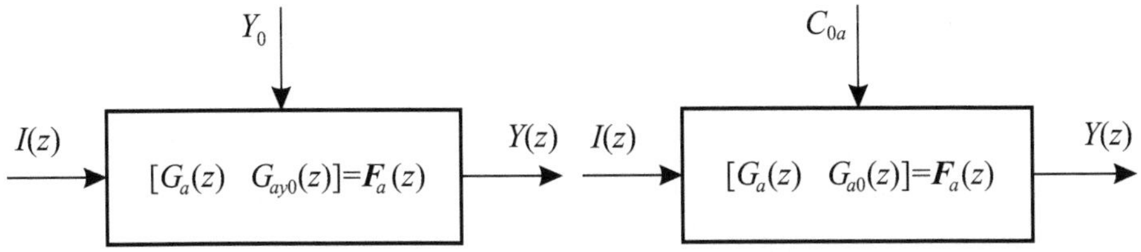

Figure 1.5: The system full block with the submatrices of the full system transfer function matrix $\boldsymbol{F}(z)$ of system a).

the system under b) above, we find its irreducible transfer function to be constant, $G_b(z) = a^{-1}c = \varkappa$, $(\varkappa \neq 0) \in \mathcal{R}$. *It yields the minimal system realization* $Y(k) = \varkappa I(k)$ *under zero initial condition, i.e., the transfer function realization. The same result for the system output response follows from* $Y(z) = G_b(z)I(z)$ *regardless of the form of* $G_b(z)$ *(either reducible,* $G_b(z) = \left[a\left(z+\frac{b}{a}\right)\right]^{-1} c\left(z+\frac{d}{c}\right)$, *or irreducible,* $G_b(z) = \varkappa$*). We know that this is incorrect in general because it is valid only for the zero initial conditions. The equation* $Y(k) = \varkappa I(k)$ *corresponds to a static system, the behavior of which does not depend on initial conditions. Hence, dynamical (e.g., controllability, observability, stability) problems do not exist for such a system. However, the correct relationship between output and input in general is* $Y(k) = \left(-\frac{b}{a}\right)^k Y(0) + \sum_{j=1}^{k} \left(-\frac{b}{a}\right)^{k-j} \varkappa \left[I(j) + \frac{d}{c}I(j-1)\right]$. *It results from the system IO discrete equation after its solving. It describes the output response of a dynamical system. Its equilibrium state* $X = Y = 0$ *is stable for* $\left|-a^{-1}b\right| \leq 1$, *and attractive for* $\left|-a^{-1}b\right| < 1$ *so as well asymptotically stable, i.e., the system is critically stable, or stable, respectively. The same result follows if we use the full transfer function matrix* $\boldsymbol{F}_b(z)$ *and the vector* $\mathbf{C}_{0b}$ *of all initial*

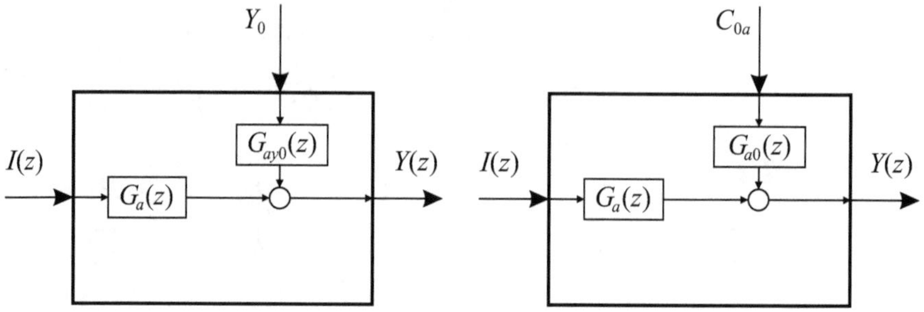

Figure 1.6: The system full block diagram with the submatrices of the full system transfer function matrix $\boldsymbol{F}(z)$ of system a).

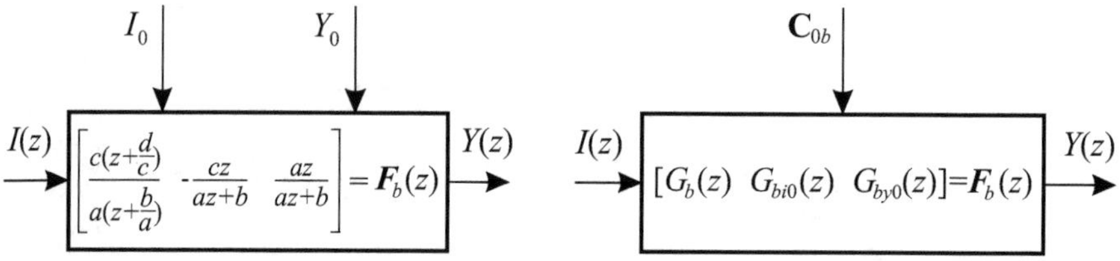

Figure 1.7: The system full block with the submatrices of the full system transfer function matrix $\boldsymbol{F}(z)$ of system b).

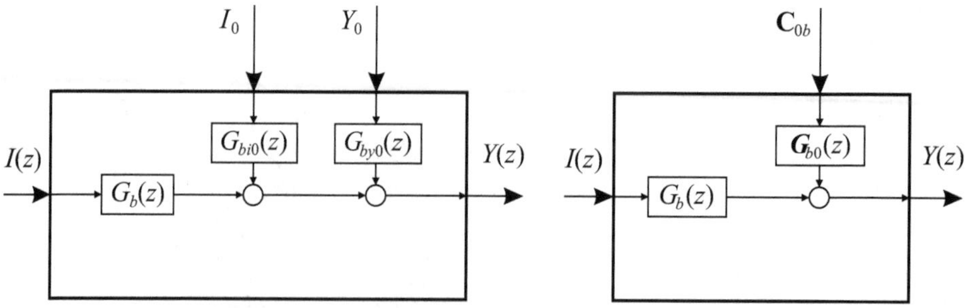

Figure 1.8: The system full block diagram with the submatrices of the full system transfer function matrix $\boldsymbol{F}(z)$ of system b).

conditions, or simply if we use $\mathbf{V}_b(z)$, in the expression for $Y(z)$,

$$Y(z) = \boldsymbol{F}_b(z)\mathbf{V}_b(z), \ \mathbf{V}_b(z) = \begin{bmatrix} I(z) \\ \mathbf{C}_{0b} \end{bmatrix}, \ \mathbf{C}_{0b} = \begin{bmatrix} I(0) \\ Y(0) \end{bmatrix},$$

and when we apply the inverse of the $Z-$transform to this equation. The denominator polynomial of $\boldsymbol{F}_b(z)$ is its characteristic polynomial $\Delta(z)$ and, in this case, its minimal polynomial $m(s)$, $\Delta(z) = m(z) = z + a^{-1}b$. The cancellation of the zero z_1^0 and the pole z_1^*, $z_1^0 = -c^{-1}d = z_1^* = -a^{-1}b$, of $G_b(z)$ is not possible in $\boldsymbol{F}_b(z)$, although it is possible in the transfer function $G_b(z)$,

$$G_b(z) = \frac{c\left(z + \frac{d}{c}\right)}{a\left(z + \frac{b}{a}\right)}, \ \boldsymbol{F}_b(z) = \begin{bmatrix} \frac{c\left(z+\frac{d}{c}\right)}{a\left(z+\frac{b}{a}\right)} & -\frac{cz}{az+b} & \frac{az}{az+b} \end{bmatrix}.$$

The use of $G_b(z)$ for the pole-zero cancellation or for Lyapunov stability test is wrong. We should use instead the full transfer function matrix $\boldsymbol{F}_b(z)$.

Example 1.2 *The ignorance of the initial conditions in the complex domain becomes more severe if the reducible form of the system transfer function (matrix) has an unstable pole that is cancelled with the corresponding zero. Let*

$$Y(k+2) - 1.5Y(k+1) - Y(k) = -2I(k) + I(k+1),$$

so that

$$Y(z) = \underbrace{\begin{bmatrix} \overbrace{\frac{z-2}{z^2-1.5z-1}}^{G(z)} & \overbrace{\frac{-z}{z^2-1.5z-1}}^{G_{I_0}(z)} & \overbrace{\frac{z(z-1.5)}{z^2-1.5z-1}}^{G_{Y_0}(z)} & \overbrace{\frac{z}{z^2-1.5z-1}}^{G_{Y_1}(z)} \end{bmatrix}}_{\boldsymbol{F}(z)} \cdot \underbrace{\begin{bmatrix} I(z) \\ I(0) \\ Y(0) \\ Y(1) \end{bmatrix}}_{\mathbf{V}(z)},$$

$$Y(z) = \boldsymbol{F}(z)\mathbf{V}(z), \ \mathbf{V}(z) = \begin{bmatrix} I(z) \\ \mathbf{C}_0 \end{bmatrix}, \ \mathbf{C}_0 = \begin{bmatrix} I(0) \\ Y(0) \\ Y(1) \end{bmatrix}.$$

The system transfer function $G(z) = \left(z^2 - 1.5z - 1\right)^{-1}(z-2)$ is reducible. Its irreducible form $G(z) = (z+0.5)^{-1}$ yields its minimal realization

$$Y(k+1) + 0.5Y(k) = I(k),$$

which is not the system minimal realization. Obviously, the irreducible form $(z+0.5)^{-1}$ *of* $G(z)$ *may not be used to test Lyapunov stability properties of the system, to test system BIBO stability under bounded nonzero initial conditions, or to determine the system output response under nonzero initial conditions. The cancellation of the zero* $z^0 = 2$ *and the unstable pole* $z^* = 2$ *in* $G(z) = \left(z^2 - 1.5z - 1\right)^{-1}(z-2)$ *is not possible in* $\boldsymbol{F}(z)$.

The use of $\boldsymbol{F}(z)$ enables us to get all correct results on the general validity of the pole-zero cancellation, the system minimal polynomial, the (minimal) system realization, the system complete output response, Lyapunov stability properties, *BIBO* stability under bounded nonzero initial conditions (also under zero initial conditions), and on tracking (under nonzero or zero initial conditions [70], [71]). Moreover, the properties of $\boldsymbol{F}(z)$ lead to the generalization of the block diagram technique; i.e., they imply the full (complete) block diagram technique (established in Chapter 9).

The full transfer function matrix $\boldsymbol{F}(z)$ was introduced, defined and determined for *time*-invariant continuous-*time* linear systems in [34], and for *time*-invariant discrete-*time* linear systems in [61]. It was used in these references, as well as in [29], [30], [35], [36], [59], [60], [61], [62], [67], for the analysis of the complete system output response. We will show in the sequel how it enables us to obtain new results (in the complex domain) on system minimal realization, the zero-pole cancellation, and stability analysis. It can be effectively exploited also for stabilizing, tracking [70], and/or optimal stabilizing control synthesis [32], [51], [55], [113], [114]. However, these issues exceed the scope of this work that establishes a new basis for them.

1.4 Notational preliminaries

Lowercase and capital, ordinary letters denote scalars; bold, lowercase and capital, upright, Greek and Roman, letters signify vectors; capital bold italic letters stand for matrices; and capital $\mathcal{CALLIGRAPHIC}$ letters indicate sets and spaces.

Note 1.1 *On the new notation*

In order to define and use effectively the system full transfer function matrix $\boldsymbol{F}(z)$, *we need new, simple and elegant notation. For example, in-*

stead of using

$$\mathbf{Y}(z) = \boldsymbol{F}(z) \begin{bmatrix} I(z) \\ I(0) \\ Y(0) \\ Y(1) \end{bmatrix}$$

we can use, [34],

$$\mathbf{Y}(z) = \boldsymbol{F}(z)\mathbf{V}(z), \ \mathbf{V}(z) = \begin{bmatrix} I(z) \\ \mathbf{C}_0 \end{bmatrix},$$

$$\mathbf{C}_0 = \begin{bmatrix} I(0) \\ \mathbf{Y}^1(0) \end{bmatrix}, \ \mathbf{Y}^1(0) = \begin{bmatrix} Y(0) \\ Y(1) \end{bmatrix},$$

by introducing the general compact vector notation

$$\mathbf{Y}^j = \begin{bmatrix} E^0\mathbf{Y} \\ E^1\mathbf{Y} \\ \vdots \\ E^j\mathbf{Y} \end{bmatrix} = \begin{bmatrix} \mathbf{Y}(k) \\ \mathbf{Y}(k+1) \\ \vdots \\ \mathbf{Y}(k+j) \end{bmatrix} \in \mathcal{R}^{(j+1)N}, \ j \in \{0, 1, \cdots\}, \ E^0\mathbf{Y} = \mathbf{Y},$$

where E is shifting operator

$$E\mathbf{Y}(k) = \mathbf{Y}(k+1),$$
$$E^2\mathbf{Y}(k) = E[E\mathbf{Y}(k)] = E[\mathbf{Y}(k+1)] = \mathbf{Y}(k+2) \in \mathcal{R}^N,$$
$$E^j\mathbf{Y}(k) = \underbrace{E[E[\cdots[E\mathbf{Y}(k)]]]}_{j-times} = \mathbf{Y}(k+j), \ j \in \{0, 1, \cdots\}.$$

The system matrices $\boldsymbol{A}_i \in \mathcal{R}^{N\times N}, i \in \{0, 1, \cdots, \nu\}$, induce the extended system matrix $\boldsymbol{A}^{(\nu)}$,

$$\boldsymbol{A}^{(\nu)} = \begin{bmatrix} \boldsymbol{A}_0 & \boldsymbol{A}_1 & \cdots & \boldsymbol{A}_\nu \end{bmatrix} \in \mathcal{R}^{N\times(\nu+1)N}, \ \boldsymbol{A}^{(\nu)} \neq \boldsymbol{A}^\nu = \underbrace{\boldsymbol{A}\boldsymbol{A}\cdots\boldsymbol{A}}_{\nu-times}.$$

We use the matrix function $\boldsymbol{S}_i^{(j)}(\cdot) : \mathcal{C} \longrightarrow \mathcal{C}^{\ i(j+1)\times i}$ of z,

$$\boldsymbol{S}_i^{(j)}(z) = \begin{bmatrix} z^0\boldsymbol{I}_i & z^1\boldsymbol{I}_i & z^2\boldsymbol{I}_i & \cdots & z^j\boldsymbol{I}_i \end{bmatrix}^T \in \mathcal{C}^{\ i(j+1)\times i},$$
$$(j, i) \in \{(\mu, M), \ (\nu, N)\}, \tag{1.13}$$

in order to set

$$\sum_{i=0}^{i=\nu} \boldsymbol{A}_i z^i$$

into the compact form $\boldsymbol{A}^{(\nu)}\boldsymbol{S}_N^{(\nu)}(z)$,

$$\sum_{i=0}^{i=\nu} \boldsymbol{A}_i z^i = \boldsymbol{A}^{(\nu)}\boldsymbol{S}_N^{(\nu)}(z).$$

Note 1.2 *If a system is of higher order and/or dimension, the new notation is more advantageous.*

We will use the symbolic vector notation and operations in the elementwise sense as follows:

- *the zero and unit vectors,*

$$\mathbf{0}_N = \begin{bmatrix} 0 & 0 & \cdots & 0 \end{bmatrix}^T \in \mathcal{R}^N, \ \mathbf{1}_N = \begin{bmatrix} 1 & 1 & \cdots & 1 \end{bmatrix}^T \in \mathcal{R}^N,$$

- *the matrix* $\boldsymbol{E}$ *associated elementwise with the vector* $\boldsymbol{\varepsilon}$,

$$\begin{gathered} \boldsymbol{\varepsilon} = \begin{bmatrix} \varepsilon_1 & \varepsilon_2 & \cdots & \varepsilon_N \end{bmatrix}^T \in \mathcal{R}^N \Longrightarrow \\ \boldsymbol{E} = \operatorname{diag}\left\{ \varepsilon_1 \quad \varepsilon_2 \quad \cdots \quad \varepsilon_N \right\} \in \mathcal{R}^{N\times N}, \end{gathered}$$

- *the vector and matrix absolute values,*

$$|\boldsymbol{\varepsilon}| = \begin{bmatrix} |\varepsilon_1| & |\varepsilon_2| & \cdots & |\varepsilon_N| \end{bmatrix}^T, |\boldsymbol{E}| = \operatorname{diag}\left\{ |\varepsilon_1| \quad |\varepsilon_2| \quad \cdots \quad |\varepsilon_N| \right\},$$

- *the elementwise vector inequality,*

$$\begin{aligned} \mathbf{w} &= \begin{bmatrix} w_1 & w_2 & \cdots & w_N \end{bmatrix}^T, \\ \mathbf{w} \neq \boldsymbol{\varepsilon} &\Longleftrightarrow w_i \neq \varepsilon_i, \ \forall i = 1, 2, \cdots, N. \end{aligned}$$

We define the scalar sign function:

- $\operatorname{sign}(\cdot) : \mathcal{R} \to \{-1, 0, 1\}$ *the signum scalar function.*

Other notation is defined at its first use and in Appendix A.

Chapter 2

Classes of discrete-time linear systems

2.1 *IO* systems

The classes of the systems studied in this book are the following discrete-time time-invariant linear dynamical systems: the known but not sufficiently explored *IO* systems, the very well known and studied but still not completely explored *ISO* systems, and new, more general, *IIO* systems.

We start with the *Input-Output* (*IO*) systems determined by (2.1):

$$\sum_{r=0}^{r=\nu} \boldsymbol{A}_r E^r \mathbf{Y}(k) = \sum_{r=0}^{r=\mu} \boldsymbol{B}_r E^r \mathbf{I}(k), \det \boldsymbol{A}_\nu \neq 0, \forall k \in \mathcal{N}_0,\ \nu \geq 1,\ 0 \leq \mu \leq \nu,$$
$$E^r \mathbf{Y}(k) = \mathbf{Y}(k+r),\ \boldsymbol{A}_r \in \mathcal{R}^{N\times N},\ \boldsymbol{B}_r \in \mathcal{R}^{N\times M}, r = 0, 1, \cdots, \nu,$$
$$\mu < \nu \Longrightarrow \boldsymbol{B}_i = \boldsymbol{O}_{N,M},\ i = \mu+1, \mu+2, \cdots, \nu. \tag{2.1}$$

$\mathcal{R}^r$ represents the r-dimensional real vector space, $r \in \{M, N, n\}$, $\mathcal{R}_+$ is the set of all nonnegative real numbers, and $\mathcal{C}^r$ stands for the r-dimensional complex vector space. $\boldsymbol{O}_{M\times N}$ designates the zero matrix in $\mathcal{R}^{M\times N}$, and $\boldsymbol{O}_N$ represents the zero matrix in $\mathcal{R}^{N\times N}$, $\boldsymbol{O}_N = \boldsymbol{O}_{N\times N}$. Furthermore, $\mathbf{0}_r \in \mathcal{R}^r$ is the zero vector in $\mathcal{R}^r$. The notation for the system total *input vector* is $\mathbf{I}$,

$$\mathbf{I} = \begin{bmatrix} I_1 & I_2 & \cdots & I_M \end{bmatrix}^T \in \mathcal{R}^M,$$

(Definition 1.2), and for the total *output vector* is

$$\mathbf{Y} = \begin{bmatrix} Y_1 & Y_2 & \cdots & Y_N \end{bmatrix}^T \in \mathcal{R}^N.$$

The values I_i and Y_j are the total values of the input and the output variables $I_i(\cdot)$ and $Y_j(\cdot)$ (Definition 1.3). *The total value* of a variable is its value measured with respect to the total zero value of the variable, if it has the total zero value; and if it does not have the total zero value, then an appropriate value is accepted to represent the (relative) total zero value. Kelvin zero is the total zero value of temperature. Position does not have a total zero value. We proclaim a point for the origin in the space and with respect to it we measure the positions of all other points, things, and objects.

The following compact notation for the extended matrices, which were proposed in [34] (see Note 1.1 in *Notational preliminaries,* Section 1.4), simplifies enormously the study of nontrivial (third- and higher-order) systems up to complex $MIMO$ systems:

$$\boldsymbol{A}^{(\nu)} = \begin{bmatrix} \boldsymbol{A}_0 & \boldsymbol{A}_1 & \cdots & \boldsymbol{A}_\nu \end{bmatrix} \in \mathcal{R}^{N\times(\nu+1)N}, \tag{2.2a}$$

$$\boldsymbol{B}^{(\mu)} = \begin{bmatrix} \boldsymbol{B}_0 & \boldsymbol{B}_1 & \cdots & \boldsymbol{B}_\mu \end{bmatrix} \in \mathcal{R}^{N\times(\mu+1)M}. \tag{2.2b}$$

$$\mathbf{I}^\mu(k) = \begin{bmatrix} E^0\mathbf{I}^T(k) & E^1\mathbf{I}^T(k) & \cdots & E^\mu\mathbf{I}^T(k) \end{bmatrix}^T \in \mathcal{R}^{(\mu+1)M}, \tag{2.3a}$$

$$\mathbf{Y}^\nu(k) = \begin{bmatrix} E^0\mathbf{Y}^T(k) & E^1\mathbf{Y}^T(k) & \cdots & E^\nu\mathbf{Y}^T(k) \end{bmatrix}^T \in \mathcal{R}^{(\nu+1)N}. \tag{2.3b}$$

Their initial vectors are

$$\mathbf{I}_0^{\mu-1} = \mathbf{I}^{\mu-1}(0) = \begin{bmatrix} E^0\mathbf{I}_0^T & E^1\mathbf{I}_0^T & \cdots & E^{\mu-1}\mathbf{I}_0^T \end{bmatrix}^T \in \mathcal{R}^{\mu M}, \tag{2.4a}$$

$$\mathbf{Y}_0^{\nu-1} = \mathbf{Y}^{\nu-1}(0) = \begin{bmatrix} E^0\mathbf{Y}_0^T & E^1\mathbf{Y}_0^T & \cdots & E^{\nu-1}\mathbf{Y}_0^T \end{bmatrix}^T \in \mathcal{R}^{\nu N}. \tag{2.4b}$$

Notice that superscript ν in the parentheses in $\boldsymbol{A}^{(\nu)}$ denotes the extended matrix of the matrix $\boldsymbol{A}$. Be careful to distinguish $\boldsymbol{A}^{(\nu)}$ from the $\nu - th$ power $\boldsymbol{A}^\nu$ of $\boldsymbol{A}$,

$$\boldsymbol{A}^{(\nu)} = \begin{bmatrix} \boldsymbol{A}_0 & \boldsymbol{A}_1 & \cdots & \boldsymbol{A}_\nu \end{bmatrix} \neq \boldsymbol{A}^\nu = \underbrace{\boldsymbol{A}\boldsymbol{A}\cdots\boldsymbol{A}}_{\nu-times}.$$

Different meaning has the superscript μ of a vector. It is not in the parentheses in $\mathbf{I}^\mu(k)$ because when it is in the parentheses as in $\mathbf{I}^{(\mu)}(k)$ then it designates the $\mu - th$ derivative $d^\mu\mathbf{I}(k)/dt^\mu$ of $\mathbf{I}(k)$, what has no sense, i.e., it is not defined of the discrete-time variable (at discrete moments k),

$$\mathbf{I}^\mu(k) = \begin{bmatrix} \mathbf{I}^T(k) & E^1\mathbf{I}^T(k) & \cdots & E^\mu\mathbf{I}^T(k) \end{bmatrix}^T \neq \mathbf{I}^{(\mu)}(k) = \frac{d^\mu\mathbf{I}(k)}{dt^\mu}.$$

As the significant system modeling result of the usage of the above compact notation is a very simple, clear and elegant mathematical description (2.5) of the IO system 2.1):

$$A^{(\nu)}\mathbf{Y}^{\nu}(k) = B^{(\mu)}\mathbf{I}^{\mu}(k), \ \forall k \in \mathcal{N}_0. \tag{2.5}$$

We continue to use mostly (2.5) instead of (2.1).

The left-hand side of the Equation (2.1), equivalently to (2.5), describes simultaneously the internal dynamics, i.e., *the internal state* $\mathbf{S}_I$ (Definition 1.4), the output dynamics, i.e., *the output state* $\mathbf{S}_O$, and the full dynamics, i.e., *the state* $\mathbf{S}$, of the IO system, where

$$\mathbf{S} = \mathbf{S}_I = \mathbf{S}_O = \mathbf{Y}^{\nu-1}. \tag{2.6}$$

The output vector $\mathbf{Y}(k)$ and its shiftings $E^1\mathbf{Y}(k)$, $E^2\mathbf{Y}(k)$, $\cdots$, $E^{\nu-1}\mathbf{Y}(k)$ determine them at any $k \in \mathbb{Z}$. They form the extended output vector $\mathbf{Y}^{\nu-1}$,

$$\begin{gathered}\mathbf{S}(k) = \begin{bmatrix} S_1(k) & S_2(k) & \cdots & S_{\nu N} \end{bmatrix}^T = \\ = \mathbf{Y}^{\nu-1}(k) = \begin{bmatrix} \mathbf{Y}^T(k) & E^1\mathbf{Y}^T(k) & \cdots & E^{\nu-1}\mathbf{Y}^T(k) \end{bmatrix}^T = \\ = \begin{bmatrix} E^0\mathbf{Y}^T(k) & E^1\mathbf{Y}^T(k) & \cdots & E^{\nu-1}\mathbf{Y}^T(k) \end{bmatrix}^T \in \mathcal{R}^{\nu N},\end{gathered} \tag{2.7}$$

that is *the state vector* $\mathbf{S}$ *at the moment* k of the IO system (Definition 1.4), $\mathbf{S} = \mathbf{Y}^{\nu-1}$. The system motion $\mathcal{S}(\cdot; k_0; \mathbf{S}_0; \mathbf{I})$ is $\mathcal{Y}^{\nu-1}(\cdot; k_0; \mathbf{Y}_0^{\nu-1}; \mathbf{I})$,

$$\begin{gathered}\mathcal{S}(\cdot; k_0; \mathbf{S}_0; \mathbf{I}) = \mathcal{Y}^{\nu-1}(\cdot; k_0; \mathbf{Y}_0^{\nu-1}; \mathbf{I}) \Longrightarrow \\ \mathcal{S}(k; k_0; \mathbf{S}_0; \mathbf{I}) = \mathcal{Y}^{\nu-1}(k; k_0; \mathbf{Y}_0^{\nu-1}; \mathbf{I}) = \mathbf{Y}^{\nu-1}(k; k_0; \mathbf{Y}_0^{\nu-1}; \mathbf{I}) \Longrightarrow \\ \mathcal{S}(k_0; k_0; \mathbf{S}_0; \mathbf{I}) = \mathcal{Y}^{\nu-1}(k_0; k_0; \mathbf{Y}_0^{\nu-1}; \mathbf{I}) = \\ = \mathbf{Y}^{\nu-1}(k_0; k_0; \mathbf{Y}_0^{\nu-1}; \mathbf{I}) \equiv \mathbf{Y}_0^{\nu-1}.\end{gathered} \tag{2.8}$$

The system output response $\mathcal{Y}(\cdot; k_0; \mathbf{Y}_0^{\nu-1}; \mathbf{I})$ is the first component of the system motion $\mathcal{Y}^{\nu-1}(\cdot; k_0; \mathbf{Y}_0^{\nu-1}; \mathbf{I})$,

$$\mathcal{Y}(k_0; k_0; \mathbf{Y}_0^{\nu-1}; \mathbf{I}) \equiv \mathbf{Y}_0. \tag{2.9}$$

We assume that input vector functions $\mathbf{I}(\cdot) : \mathcal{N}_0 \to \mathcal{R}^M$ belong to the class $\mathcal{I}$ of *discrete time* dependent bounded functions such that their

$Z-$transforms are either proper or strictly proper real rational vector functions of the complex variable z,

$$\mathcal{I}=\left\{\begin{array}{l} \mathbf{I}(\cdot):\exists\gamma(\mathbf{I})\in\mathcal{R}^{+}\Longrightarrow\|\mathbf{I}(k)\|<\gamma(\mathbf{I}),\ \forall k\in\mathcal{N}_{0}, \\ \mathcal{Z}\{\mathbf{I}(k)\}=\mathbf{I}(z)=\left[\begin{array}{llll} I_{1}(z) & I_{2}(z) & \cdots & I_{M}(z) \end{array}\right]^{T}, \\ I_{r}(z)=\dfrac{\sum_{j=0}^{j=\zeta_{r}} a_{rj}z^{j}}{\sum_{j=0}^{j=\psi_{r}} b_{rj}z^{j}}, 0\leq\zeta_{r}\leq\psi_{r},\ \forall r=1,2,\cdots,M. \end{array}\right. \tag{2.10}$$

The $Z-$transform $\mathbf{I}(z)$ of the input vector function $\mathbf{I}(\cdot)\in\mathcal{I}$ may be either proper or strictly proper so that the original $\mathbf{I}(k)$ contains or does not contain an discrete impulse component which is certainly bounded.

Notice that the zero input vector function $\mathbf{I}(\cdot)$, $\mathbf{I}(k)\equiv\mathbf{0}_{M}$, belongs to $\mathcal{I}$.

$\mathcal{D}^{i}=\mathcal{D}(\mathcal{R}^{i})$ is *the family of all functions defined on* $\mathcal{R}^{i}$, and

$\mathcal{D}=\mathcal{D}(\mathcal{T}_{d0})$ is *the family of all functions defined and with a finite number of the first-order discontinuities on* $\mathcal{T}_{d0}$,

$\mathcal{J}$ is a given, or to be determined, family of all bounded permitted input vector functions $\mathbf{I}(\cdot)\in\mathcal{D}\cap\mathcal{I}$,

$$\mathcal{J}\subset\mathcal{D}\cap\mathcal{I}. \tag{2.11}$$

$\mathcal{J}_{-}$ is a subfamily of $\mathcal{J}$, $\mathcal{J}_{-}\subset\mathcal{J}$, such that the modulus of every pole of the $Z-$transform $\mathbf{I}(z)$ of every $\mathbf{I}(\cdot)\in\mathcal{J}_{-}$ is less than one.

Note 2.1 *The imposed condition* $\det\boldsymbol{A}_{\nu}\neq 0$ *is sufficient, but not necessary, for all the output variables of system (2.1) to have the same order* ν *of their highest shifting. If the order* r *of the highest shifting of one output variable* Y_{i} *were lower than the highest shifting order* ν *of some other output variable* Y_{j}*, then all entries of the* $i-th$ *column of* $\boldsymbol{A}_{r+1},\cdots,\boldsymbol{A}_{\nu}$ *would be equal to zero implying their singularity:* $\det\boldsymbol{A}_{m}=0,\ m=r+1,...,\nu$. *Such systems are* ***degenerate****, called also* ***singular****. The system that satisfy the condition* $\det\boldsymbol{A}_{\nu}\neq 0$ *are the* ***regular systems****. We concentrate our study only on the regular systems.*

Condition 2.1 *The matrix* $\boldsymbol{A}_{\nu}$ *of the IO system (2.1) obeys*

$$\det\boldsymbol{A}_{\nu}\neq 0. \tag{2.12}$$

The condition (2.12), imposed already in (2.1), ensures

$$\exists z\in\mathcal{C}\Longrightarrow\det\left(\sum_{r=0}^{r=\nu}\boldsymbol{A}_{r}z^{r}\right)\neq 0,$$

and it permits the solvability of the $Z-$transform of (2.1) in $\mathbf{Y}(z)$, see (8.15) in proof of Theorem 8.1.

Example 2.1 *Let*

$$M = 1,\ N = 2,\ \nu = 1,\ \mu = 0,$$

$$\underbrace{\begin{bmatrix} 2 & 2 \\ 4 & 4 \end{bmatrix}}_{\boldsymbol{A}_1} E^1\mathbf{Y}(k) + \underbrace{\begin{bmatrix} \frac{1}{2} & 1 \\ 1 & 2 \end{bmatrix}}_{\boldsymbol{A}_0} E^0\mathbf{Y}(k) = \underbrace{\begin{bmatrix} 4 \\ 2 \end{bmatrix}}_{\boldsymbol{B}_0} E^0 I(k) \Longrightarrow$$

$$\det \boldsymbol{A}_1 = \begin{vmatrix} 2 & 2 \\ 4 & 4 \end{vmatrix} = 0,\ \det\left(\boldsymbol{A}_1 z + \boldsymbol{A}_0\right) = \begin{vmatrix} 2z+\frac{1}{2} & 2z+1 \\ 4z+1 & 4z+2 \end{vmatrix} = 0.$$

Both Y_1 and Y_2 have the same highest shifting that is the first order, but $\boldsymbol{A}_1$ is singular. This illustrates that the condition $\det \boldsymbol{A}_\nu \neq 0$ *is not necessary (although it is sufficient) for all the output variables of system (2.1) to have the same order of their highest shifting. There does not exist a solution to this vector discrete equation because* $\det\left(\boldsymbol{A}_1 z + \boldsymbol{A}_0\right) \equiv 0$. *We explain this first by considering the scalar form of the mathematical model of the system,*

$$2E^1Y_1(k) + 2E^1Y_2(k) + \frac{1}{2}E^0Y_1(k) + E^0Y_2(k) = 4E^0I(k),$$
$$4E^1Y_1(k) + 4E^1Y_2(k) + E^0Y_1(k) + 2E^0Y_2(k) = 2E^0I(k).$$

We multiply the first equation by 2. The result is

$$4E^1Y_1(k) + 4E^1Y_2(k) + E^0Y_1(k) + 2E^0Y_2(k) = 8E^0I(k),$$
$$4E^1Y_1(k) + 4E^1Y_2(k) + E^0Y_1(k) + 2E^0Y_2(k) = 2E^0I(k).$$

The left-hand sides of these equations are the same. Their right-hand sides are different. They do not have a solution for $I(k) \neq 0$.

The $Z-$transform $\mathcal{Z}\{\cdot\}$ of the system mathematical model for all zero initial conditions reads

$$\left(\boldsymbol{A}_1 z + \boldsymbol{A}_0\right)\mathbf{Y}(z) = \begin{bmatrix} 2z+\frac{1}{2} & 2z+1 \\ 4z+1 & 4z+2 \end{bmatrix} \mathbf{Y}(z) = \begin{bmatrix} 4 \\ 2 \end{bmatrix} I(z),$$

where $\mathbf{Y}(z)$ and $I(z)$ are the $Z-$transforms $\mathcal{Z}\{\cdot\}$ of $\mathbf{Y}(k)$ and of $I(k)$, $\mathbf{Y}(z) = \mathcal{Z}\{\mathbf{Y}(k)\}$ and $I(z) = \mathcal{Z}\{I(k)\}$. Since $\det\left(\boldsymbol{A}_1 z + \boldsymbol{A}_0\right) \equiv 0$, *then the preceding vector equation is not solvable in $\mathbf{Y}(k)$.*

Note 2.2 ***We accept the Condition 2.1.***

The dimensions of the system matrices $\boldsymbol{A}_r \in \mathcal{R}^{N\times N}$, $\boldsymbol{B}_r \in \mathcal{R}^{N\times M}$, $r = 0, 1, .., \nu$, and Condition 2.1 imply

$$\deg\left(\sum_{r=0}^{r=\nu} \boldsymbol{A}_r z^r\right) = \nu, \tag{2.13a}$$

$$\deg\left[\det\left(\sum_{r=0}^{r=\nu} \boldsymbol{A}_r z^r\right)\right] = \eta,\ \eta = \nu N, \tag{2.13b}$$

$$\deg\left[\operatorname{adj}\left(\sum_{r=0}^{r=\nu} \boldsymbol{A}_r z^r\right)\right] = \sigma,\ \sigma = \nu\left(N-1\right), \tag{2.13c}$$

$$\deg\left(\sum_{r=0}^{r=\mu} \boldsymbol{B}_r z^r\right) = \mu, \tag{2.13d}$$

where

$$\deg\left(\sum_{r=0}^{r=\nu} \boldsymbol{A}_r z^r\right),\ \deg\left[\operatorname{adj}\left(\sum_{r=0}^{r=\nu} \boldsymbol{A}_r z^r\right)\right] \text{ and } \deg\left(\sum_{r=0}^{r=\mu} \boldsymbol{B}_r z^r\right)$$

denote the greatest power of z over all elements of

$$\sum_{r=0}^{r=\nu} \boldsymbol{A}_r z^r,\ \operatorname{adj}\left(\sum_{r=0}^{r=\nu} \boldsymbol{A}_r z^r\right) \text{ and } \sum_{r=0}^{r=\mu} \boldsymbol{B}_r z^r,$$

respectively.

Notice that in general

$$\deg\left[\det\left(\sum_{r=0}^{r=\nu} \boldsymbol{A}_r z^r\right)\right] = \eta,\ 0 \le \eta \le \nu N,$$

$$\deg\left[\operatorname{adj}\left(\sum_{r=0}^{r=\nu} \boldsymbol{A}_r z^r\right)\right] = \sigma,\ 0 \le \sigma \le \nu\left(N-1\right).$$

Definition 2.1 ***A realization of the IO system (2.1); i.e., of (2.5), for an arbitrary input vector function and for arbitrary input and output initial conditions*** *is the quadruple* $(\nu,\ \mu,\ \mathbf{A}^{(\nu)},\ \mathbf{B}^{(\mu)})$.

Comment 2.1 ***The IO realization*** *of the IO system (2.1), i.e., of (2.5) is the* $(\nu,\ \mu,\ \mathbf{A}^{(\nu)},\ \mathbf{B}^{(\mu)})$. *For the ISO realization of the IO system (2.1), i.e., of (2.5), see Subsection 2.4.1.*

It is indispensable to distinguish the order of a system, from its dimension and its dynamical dimension (see in the sequel Definition 12.3 in Section 12.1).

Definition 2.2 *The number ν of the highest shifting of the output vector function $\mathbf{Y}(\cdot)$ in (2.1) is* ***the order of the IO system (2.1)****.*

Definition 2.3 *The dimension N of the system output vector $\mathbf{Y}$ in (2.1),* $\dim_{IO} = \dim \mathbf{Y} = N$, *is* ***the dimension,*** *denoted by* $\dim_{IO}$, ***of the IO system (2.1)****.*

Example 2.2 *The IO system*

$$\underbrace{\begin{bmatrix} 3 & 0 & 0 \\ 0 & 0 & 0 \\ 0 & 0 & 0 \end{bmatrix}}_{\boldsymbol{A}_2} E^2\mathbf{Y}(k) + \underbrace{\begin{bmatrix} 0 & 0 & 0 \\ 0 & 1 & 0 \\ 0 & 0 & 0 \end{bmatrix}}_{\boldsymbol{A}_1} E^1\mathbf{Y}(k) + \underbrace{\begin{bmatrix} 0 & 0 & 0 \\ 0 & 0 & 0 \\ 0 & 0 & 1 \end{bmatrix}}_{\boldsymbol{A}_0} E^0\mathbf{Y}(k) =$$

$$= \underbrace{\begin{bmatrix} 2 & 0 \\ 0 & 1 \\ 1 & 0 \end{bmatrix}}_{\boldsymbol{B}_0} E^0\mathbf{I}(k) + \underbrace{\begin{bmatrix} 1 & 0 \\ 0 & 1 \\ 1 & 1 \end{bmatrix}}_{\boldsymbol{B}_2} E^2\mathbf{I}(k)$$

yields

$$\nu = 2,\ \mu = 2,\ N = 3,\ M = 2,\ \det \boldsymbol{A}_\nu = \det \boldsymbol{A}_2 = \det \begin{bmatrix} 3 & 0 & 0 \\ 0 & 0 & 0 \\ 0 & 0 & 0 \end{bmatrix} = 0,$$

$$\deg \left[\det \left(\sum_{r=0}^{r=\nu=2} \boldsymbol{A}_r z^r \right) \right] = \deg \begin{vmatrix} 3z^2 & 0 & 0 \\ 0 & z & 0 \\ 0 & 0 & 1 \end{vmatrix} = \deg \left(3z^3\right) = 3 = \eta > \nu = 2,$$

$$\deg \left[\operatorname{adj} \left(\sum_{r=0}^{r=\nu=2} \boldsymbol{A}_r z^r \right) \right] = \deg \begin{bmatrix} z & 0 & 0 \\ 0 & 3z^2 & 0 \\ 0 & 0 & 3z^3 \end{bmatrix} = \deg \left(3z^3\right) = 3 = \sigma,$$

$$\deg \left(\sum_{r=0}^{r=\mu=2} \boldsymbol{B}_r z^r \right) = \deg \begin{bmatrix} 2+z^2 & 0 \\ 0 & 1+z^2 \\ 1+z^2 & z^2 \end{bmatrix} = \deg \left(2+z^2\right) = 2 = \mu,$$

$$\sigma = \ \eta = 3 > \nu = 2,\ \eta < \nu N = 2 \times 3 = 6,\ \sigma = \ 3 < \nu\left(N-1\right) = 2 \times 2 = 4.$$

This is the second-order system, $\nu = 2$. Its dimension equals $N = 3$, $\dim \mathbf{Y} = N = 3$. Their product is bigger than the degree η of the system characteristic polynomial, $\nu N = 6 \geq 3 = \eta$. The nonzero entries of the matrices $\boldsymbol{A}_2$ through $\boldsymbol{A}_0$ show that the second shifting exists only of the first output variable Y_1, the first shifting exists only of the second output variable Y_2, and there is not any shifting of the third output variable Y_3 in the system mathematical model.

Example 2.3 *The IO system*

$$\begin{bmatrix} 3 & 0 & 1 \\ 2 & 0 & 0 \\ 0 & 1 & 1 \end{bmatrix} E^2\mathbf{Y}(k) + \begin{bmatrix} 0 & 0 & 0 \\ 0 & 1 & 0 \\ 0 & 0 & 0 \end{bmatrix} E^1\mathbf{Y}(k) + \begin{bmatrix} 0 & 0 & 0 \\ 0 & 0 & 0 \\ 0 & 0 & 1 \end{bmatrix} E^0\mathbf{Y}(k) =$$

$$= \begin{bmatrix} 2 & 0 \\ 0 & 1 \\ 1 & 0 \end{bmatrix} E^0\mathbf{I}(k) + \begin{bmatrix} 1 & 0 \\ 0 & 1 \\ 1 & 1 \end{bmatrix} E^2\mathbf{I}(k)$$

induces

$$\nu = 2,\ \mu = 2,\ N = 3,\ \ M = 2,$$

$$\det \boldsymbol{A}_\nu = \det \boldsymbol{A}_2 = \det \begin{bmatrix} 3 & 0 & 1 \\ 2 & 0 & 0 \\ 0 & 1 & 1 \end{bmatrix} = 2 \neq 0,$$

$$\deg\left(\sum_{r=0}^{r=\nu=2} \boldsymbol{A}_r z^r\right) = \deg \begin{bmatrix} 3z^2 & 0 & z^2 \\ 2z^2 & z & 0 \\ 0 & z^2 & z^2+1 \end{bmatrix} = 2 = \nu,$$

$$\deg\left[\det\left(\sum_{r=0}^{r=\nu=2} \boldsymbol{A}_r z^r\right)\right] = \deg \begin{vmatrix} 3z^2 & 0 & z^2 \\ 2z^2 & z & 0 \\ 0 & z^2 & z^2+1 \end{vmatrix} =$$

$$= \deg\left(2z^6 + 3z^5 + 3z^3\right) = 6 = \eta = 2 \times 3 = \nu N,$$

$$\deg\left[\operatorname{adj}\left(\sum_{r=0}^{r=\nu=2} \boldsymbol{A}_r z^r\right)\right] = \deg \begin{bmatrix} z^3+z & z^4 & -z^3 \\ -2z^4-2z^2 & 3z^4+3z^2 & 2z^4 \\ 2z^4 & -3z^4 & 3z^3 \end{bmatrix} =$$

$$= \deg\left(z^4\right) = 4 = \sigma = \nu(N-1) = 2(3-1),$$

$$\deg\left(\sum_{r=0}^{r=\mu=2} \boldsymbol{B}_r z^r\right) = \deg \begin{bmatrix} 2+z^2 & 0 \\ 0 & 1+z^2 \\ 1+z^2 & z^2 \end{bmatrix} = \deg\left(2+z^2\right) = 2 = \mu,$$

$$\sigma = 4 = 2 \times 2 = 2 \times (3-1) = \nu(N-1) < \eta = \nu N = 6.$$

This is also the second-order system of dimension three. In this case, the product of the system order ($\nu = 2$) and of the system dimension ($N = 3$) equals the degree ($\eta = 6$) of the system characteristic polynomial, $\nu N = 6 = \eta$.

2.2 *ISO systems*

The *Input-State-Output* (*ISO*) *systems* have been the most frequently treated systems and the best explored systems. They are described by *the state Equation* (2.14a) and by *the output Equation* (2.14b),

$$\mathbf{X}(k+1) = \boldsymbol{A}\mathbf{X}(k) + \boldsymbol{B}\mathbf{I}(k),\ \forall k \in \mathcal{N}_0, \tag{2.14a}$$
$$\mathbf{Y}(k) = \boldsymbol{C}\mathbf{X}(k) + \boldsymbol{D}\mathbf{I}(k),\ \forall k \in \mathcal{N}_0, \tag{2.14b}$$
$$\boldsymbol{A} \in \mathcal{R}^{n\times n}, \boldsymbol{B} \in \mathcal{R}^{n\times M}, \boldsymbol{C} \in \mathcal{R}^{N\times n}, \boldsymbol{C} \neq \boldsymbol{O}_{N,n}, \boldsymbol{D} \in \mathcal{R}^{N\times M}, n \geq N.$$

The vector $\mathbf{X}$, the system *internal state vector* $\mathbf{S}_I$, and *the state vector* $\mathbf{S}$, are all the same vector and commonly denoted by $\mathbf{X}$,

$$\mathbf{S} = \mathbf{S}_I = \mathbf{X} \in \mathcal{R}^n. \tag{2.15}$$

(Definition 1.4). The system motion $\mathcal{S}(\cdot; k_0; \mathbf{S}_0; \mathbf{I})$ is $\mathcal{X}(\cdot; k_0; \mathbf{X}_0; \mathbf{I})$,

$$\begin{gathered} \mathcal{S}(\cdot; k_0; \mathbf{S}_0; \mathbf{I}) = \mathcal{X}(\cdot; k_0; \mathbf{X}_0; \mathbf{I}) \Longrightarrow \\ \mathbf{S}(k; k_0; \mathbf{S}_0; \mathbf{I}) = \mathbf{X}(k; k_0; \mathbf{X}_0; \mathbf{I}), \\ \mathbf{S}(k_0; k_0; \mathbf{S}_0; \mathbf{I}) = \mathbf{X}(k_0; k_0; \mathbf{X}_0; \mathbf{I}) \equiv \mathbf{X}_0. \end{gathered} \tag{2.16}$$

The system output response $\mathcal{Y}(\cdot; k_0; \mathbf{X}_0; \mathbf{I})$ obeys

$$\mathbf{Y}(k_0; k_0; \mathbf{X}_0; \mathbf{I}) \equiv \mathbf{Y}_0. \tag{2.17}$$

We often refer to the *ISO* system (2.14a) and (2.14b) as *the state-space system.*

Note 2.3 ***Equivalence between the*** ISO ***system and the*** IO ***system***
Subsection 2.4.2 demonstrates how the ISO system (2.14a) and (2.14b) can be formally mathematically transformed into the IO system (2.1), Section 2.1.

Subsection 2.4.1 presents the formal mathematical transformation, without any physical sense in general, of the IO system (2.1), Section 2.1, into the equivalent ISO system (2.14a) and (2.14b).

In the dynamical systems theory it is well known that the definition of the realization of the transfer function matrix $\boldsymbol{G}_{ISO}(z)$ of the ISO system is valid if and only if, all initial conditions are equal to zero (vector). We generalize it to the systems with arbitrary input vector function $\mathbf{I}(\cdot)$ and for any initial state vector $\mathbf{X}_0$:

Definition 2.4 *The quadruple* $(\boldsymbol{A},\ \boldsymbol{B},\ \boldsymbol{C},\ \boldsymbol{D})$ *is* ***a realization of the ISO system*** *(2.14a) and (2.14b) for an arbitrary input vector function* $\mathbf{I}(\cdot)$ *and for an arbitrary initial state vector* $\mathbf{X}_0$.

Comment 2.2 *The realization* $(\boldsymbol{A},\ \boldsymbol{B},\ \boldsymbol{C},\ \boldsymbol{D})$ *of the ISO system (2.14a), (2.14b) is* ***its ISO realization.*** *For its IO realization, see Subsection 2.4.2.*

Definition 2.5 ***The order of the ISO system (realization)*** *(2.14a) and (2.14b) is the order of the highest shifting of the state vector function* $\mathbf{X}(\cdot)$ *in (2.14a); i.e., it is one* (1).

This definition characterizes all ISO systems as the first-order systems. Their state Equation (2.14a) is in *the Cauchy (i.e., normal) form.* Only shifting of the state vector $\mathbf{X}(k)$ in (2.14a) is the first shifting $E\mathbf{X}(k)$, i.e., $\mathbf{X}(k+1)$.

Definition 2.6 ***The dimension****, denoted by* $\dim_{ISO}$, ***of the ISO system*** *(2.14a) and (2.14b) is the dimension n of its state vector* $\mathbf{X}$, $\dim_{ISO} = \dim \mathbf{X} = n$.

2.3 *IIO* systems

This introduces a new general class of (time-invariant discrete-time linear dynamical) systems. Their mathematical model expressed in terms of the total vector variables, and which is consisted of two discrete-time equations of an arbitrary order each, reads:

$$\sum_{r=0}^{r=\alpha} \boldsymbol{Q}_r E^r \mathbf{R}(k) = \sum_{r=0}^{r=\beta} \boldsymbol{P}_r E^r \mathbf{I}(k), \ \ \alpha \geq 1, \alpha \geq \beta \geq 0, \tag{2.18a}$$

$$\sum_{r=0}^{r=\nu} \boldsymbol{E}_r E^r \mathbf{Y}(k) = \sum_{r=0}^{r=\alpha} \boldsymbol{R}_r E^r \mathbf{R}(k) + \sum_{r=0}^{r=\mu \leq \nu} \boldsymbol{T}_r E^r \mathbf{I}(k), \tag{2.18b}$$

$$\nu, \mu \in \mathcal{R}_+, \ \forall k \in \mathcal{N}_0.$$

The characteristic of the systems is that they possess both the internal and output dynamics, which determines their name: *Input-Internal and Output dynamics systems,* for short *IIO systems.* In their mathematical model (2.18a) and (2.18b), which we continue to call also *the IIO system (2.18a) and (2.18b),* $\boldsymbol{Q}_r \in \mathcal{R}^{\rho\times\rho}$, $\mathcal{R}_+$ is the set of all nonnegative real numbers, $\mathbf{R} \in \mathcal{R}^{\rho}$, $\boldsymbol{P}_r \in \mathcal{R}^{\rho\times M}$, $\boldsymbol{E}_r \in \mathcal{R}^{N\times N}$, $\boldsymbol{R}_r \in \mathcal{R}^{N\times\rho}$, $\boldsymbol{R}_\alpha = \boldsymbol{O}_{N,\rho}$, and $\boldsymbol{T}_r \in \mathcal{R}^{N\times M}$.

Condition 2.2 *The system matrices $\boldsymbol{Q}_r$ and $\boldsymbol{E}_r$ obey:*

$$\det \boldsymbol{Q}_\alpha \neq 0, \text{ which implies } \exists z \in \mathcal{C} \Longrightarrow \det\left[\sum_{r=0}^{r=\alpha} z^r \boldsymbol{Q}_r\right] \neq 0, \tag{2.19a}$$

$$\det \boldsymbol{E}_\nu \neq 0, \text{ which implies } \exists z \in \mathcal{C} \Longrightarrow \det\left[\sum_{r=0}^{r=\nu} z^r \boldsymbol{E}_r\right] \neq 0. \tag{2.19b}$$

Note 2.4 *We adopt Condition 2.2 in this book for the same reasons for which we accepted Condition 2.1 (for details see Example 2.1), (Section 2.1).*

Let, in order to get the compact form of (2.18a), (2.18b), $\boldsymbol{Q}^{(\alpha)}$, $\boldsymbol{P}^{(\beta)}$, $\boldsymbol{E}^{(\nu)}$, $\boldsymbol{R}^{(\alpha)}$, and $\boldsymbol{T}^{(\mu)}$ be defined in the sense of (2.2a), (2.2b) (Section 2.1) by:

$$\boldsymbol{Q}^{(\alpha)} = \begin{bmatrix} \boldsymbol{Q}_0 & \boldsymbol{Q}_1 & \cdots & \boldsymbol{Q}_\alpha \end{bmatrix} \in \mathcal{R}^{\rho\times(\alpha+1)\rho}, \tag{2.20a}$$

$$\boldsymbol{P}^{(\beta)} = \begin{bmatrix} \boldsymbol{P}_0 & \boldsymbol{P}_1 & \cdots & \boldsymbol{P}_\beta \end{bmatrix} \in \mathcal{R}^{\rho\times(\beta+1)M}, \tag{2.20b}$$

$$\boldsymbol{E}^{(\nu)} = \begin{bmatrix} \boldsymbol{E}_0 & \boldsymbol{E}_1 & \cdots & \boldsymbol{E}_\nu \end{bmatrix} \in \mathcal{R}^{N\times(\nu+1)N}, \tag{2.20c}$$

$$\boldsymbol{R}^{(\alpha)} = \begin{bmatrix} \boldsymbol{R}_0 & \boldsymbol{R}_1 & \cdots & \boldsymbol{R}_{\alpha-1} & \boldsymbol{O}_{N,\rho} \end{bmatrix} \in \mathcal{R}^{N\times(\alpha+1)\rho}, \tag{2.20d}$$

$$\boldsymbol{T}^{(\mu)} = \begin{bmatrix} \boldsymbol{T}_0 & \boldsymbol{T}_1 & \cdots & \boldsymbol{T}_\mu \end{bmatrix} \in \mathcal{R}^{N\times(\mu+1)M}, \tag{2.20e}$$

and $\mathbf{R}^{\alpha-1}$ be defined in the sense of (2.3a) and (2.3b) (in Section 2.1) by:

$$\mathbf{R}^{\alpha-1}(k) = \begin{bmatrix} E^0\mathbf{R}^T(k) & E^1\mathbf{R}^T(k) & \cdots & E^{\alpha-1}\mathbf{R}^T(k) \end{bmatrix}^T \in \mathcal{R}^{\alpha\rho}. \tag{2.21}$$

The left-hand side of the Equation (2.18a) describes *the internal dynamics of the system* (called for short *the internal dynamics*), i.e., *the internal state* $\mathbf{S}_I$ *of the system* (Definition 1.4),

$$\mathbf{S}_I = \mathbf{R}^{\alpha-1}. \tag{2.22}$$

The left-hand side of the Equation (2.18b) describes *the output dynamics of the system* (called for shot *the output dynamics*) if and only if $\nu > 0$, i.e., *the output state* $\mathbf{S}_O$ *of the system* (Definition 1.4).

The output vector $\mathbf{Y}(k)$ and its shiftings $E^1\mathbf{Y}(k)$, $E^2\mathbf{Y}(k)$, $\cdots$, $E^{\nu-1}\mathbf{Y}(k)$ determine *the output dynamics, i.e., the output state* of the system at any $k \in \mathbb{Z}$. We call the extended output vector $\mathbf{Y}^{\nu-1}$,

$$\mathbf{Y}^{\nu-1}(k) = \begin{bmatrix} \mathbf{Y}^T(k) & E^1\mathbf{Y}^T(k) & \cdots & E^{\nu-1}\mathbf{Y}^T(k) \end{bmatrix}^T \in \mathcal{R}^{\nu N}, \quad (2.23)$$

also *the output state vector* $\mathbf{S}_O$ of the *IIO* system if and only if $\nu \geq 1$,

$$\mathbf{S}_O = \mathbf{Y}^{\nu-1}, \ \nu \geq 1. \quad (2.24)$$

If and only if $\nu = 0$ then the system does not have the output dynamics.

The left-hand sides of both Equations (2.18a) and (2.18b), determine *the system full dynamics* (the internal dynamics, i.e., the internal state, and the output dynamics, i.e., the output state) of the *IIO* system (2.18a) and (2.18b) which is also called *the system complete dynamics,* i.e., *the system full state* $\mathbf{S}$ *of the IIO* system (2.18a), (2.18b),

$$\mathbf{S} = \begin{cases} \begin{bmatrix} \mathbf{S}_I \\ \mathbf{S}_O \end{bmatrix} = \begin{bmatrix} \mathbf{R}^{\alpha-1} \\ \mathbf{Y}^{\nu-1} \end{bmatrix}, \ \nu \geq 1, \\ \mathbf{S}_I = \mathbf{R}^{\alpha-1} = \mathbf{S}_I, \ \nu = 0. \end{cases} \quad (2.25)$$

The system internal motion $\mathrm{S}_I(\cdot; k_0; \mathbf{S}_0; \mathbf{I})$ is $\mathcal{R}^{\alpha-1}(\cdot; k_0; \mathbf{R}_0^{\alpha-1}; \mathbf{I})$,

$$\begin{gathered} \mathcal{S}_I(\cdot; k_0; \mathbf{S}_{I0}; \mathbf{I}) = \mathcal{R}^{\alpha-1}(\cdot; k_0; \mathbf{R}_0^{\alpha-1}; \mathbf{I}) \Longrightarrow \\ \mathbf{S}_I(k; k_0; \mathbf{S}_{I0}; \mathbf{I}) = \mathcal{R}^{\alpha-1}(k; k_0; \mathbf{R}_0^{\alpha-1}; \mathbf{I}) \Longrightarrow \\ \mathbf{S}_I(k_0; k_0; \mathbf{S}_{I0}; \mathbf{I}) = \mathbf{R}_0^{\alpha-1} = \mathbf{S}_{I0}. \end{gathered} \quad (2.26)$$

The system full output response is *the system full output motion* $\mathcal{Y}^{\nu-1}(\cdot; k_0; \mathbf{Y}_0^{\nu-1}; \mathbf{I})$,

$$\begin{gathered} \mathcal{S}_O(\cdot; k_0; \mathbf{S}_0; \mathbf{I}) = \mathbf{Y}^{\nu-1}(\cdot; k_0; \mathbf{R}_0^{\alpha-1}; \mathbf{Y}_0^{\nu-1}; \mathbf{I}) \Longrightarrow \\ \mathbf{S}_O(k; k_0; \mathbf{S}_0; \mathbf{I}) = \mathbf{Y}^{\nu-1}(k; k_0; \mathbf{R}_0^{\alpha-1}; \mathbf{Y}_0^{\nu-1}; \mathbf{I}) \Longrightarrow \\ \mathbf{S}_O(k_0; k_0; \mathbf{S}_0; \mathbf{I}) = \mathbf{Y}_0^{\nu-1} = \mathbf{S}_{O0}. \end{gathered} \quad (2.27)$$

The IIO system output response $\mathcal{Y}(\cdot; k_0; \mathbf{R}_0^{\alpha-1}; \mathbf{Y}_0^{\nu-1}; \mathbf{I})$ is the first component of the system full output response $\mathcal{Y}^{\nu-1}(\cdot; k_0; \mathbf{R}_0^{\alpha-1}; \mathbf{Y}_0^{\nu-1}; \mathbf{I})$,

$$\mathcal{Y}(k_0; k_0; \mathbf{R}_0^{\alpha-1}; \mathbf{Y}_0^{\nu-1}; \mathbf{I}) \equiv \mathbf{Y}(k_0; k_0; \mathbf{R}_0^{\alpha-1}; \mathbf{Y}_0^{\nu-1}; \mathbf{I}) \equiv \mathbf{Y}_0. \quad (2.28)$$

The *time* evolution $\mathcal{S}(\cdot;k_0;\mathbf{S}_0;\mathbf{I})$ of the *IIO* system *state vector* $\mathbf{S}$,

$$\mathbf{S}=\begin{cases}\begin{bmatrix}\mathbf{S}_I\\ \mathbf{S}_O\end{bmatrix}\in\mathcal{R}^{\alpha\rho+\nu N}, & \nu\geq 1,\\ \mathbf{S}_I\in\mathcal{R}^{\alpha\rho}, & \nu=0.\end{cases}\tag{2.29}$$

is the system *(full) motion*. It is the vector function

$$\mathcal{S}(\cdot;k_0;\mathbf{S}_0;\mathbf{I})=\begin{cases}\begin{bmatrix}\mathcal{S}_I(\cdot;k_0;\mathbf{S}_{I0};\mathbf{I})\\ \mathcal{S}_O(\cdot;k_0;\mathbf{S}_{O0};\mathbf{I})\end{bmatrix}, & \nu\geq 1,\\ \mathcal{S}_I(\cdot;k_0;\mathbf{S}_{I0};\mathbf{I}), & \nu=0.\end{cases}\tag{2.30}$$

The preceding simple and elegant vector notation (2.21) and (2.23), allows us to put (2.18a) and (2.18b) into its following compact form

$$\boldsymbol{Q}^{(\alpha)}\mathbf{R}^{\alpha}(k)=\boldsymbol{P}^{(\beta)}\mathbf{I}^{\beta}(k),\ \forall k\in\mathcal{N}_0,\tag{2.31a}$$

$$\boldsymbol{E}^{(\nu)}\mathbf{Y}^{\nu}(k)=\boldsymbol{R}^{(\alpha)}\mathbf{R}^{\alpha}(k)+\boldsymbol{T}^{(\mu)}\mathbf{I}^{\mu}(k),\ \forall k\in\mathcal{N}_0.\tag{2.31b}$$

Definition 2.7 ***The order of the IIO system (2.18a), (2.18b)*** *is the number equal to the sum* $\alpha+\nu$ *of the highest shiftings of the substate vector function* $\mathbf{R}(\cdot)$ *and of the output vector function* $\mathbf{Y}(\cdot)$ *in (2.18a), (2.18b).*

Definition 2.8 ***The dimension****, denoted by* $\dim_{IIO}$, ***of the IIO system (2.18a), (2.18b)*** *is the sum of the dimension* ρ *of its internal substate vector* $\mathbf{R}$ *and of the dimension* N *of its output vector* $\mathbf{Y}$*; i.e.,* $\rho+N$, $\dim_{IIO}=\dim_{IIO}\begin{bmatrix}\mathbf{R}^T & \mathbf{Y}^T\end{bmatrix}^T=\rho+N$.

The generality of the *IIO* systems (2.18a) and (2.18b) comes out from the following facts: the results valid for the *IIO* systems are valid also for the *IO* systems, the *ISO* systems, and the systems described by (2.32a) and (2.32b)

$$\boldsymbol{Q}^{(\alpha)}\mathbf{R}^{\alpha}(k)=\boldsymbol{P}^{(\beta)}\mathbf{I}^{\beta}(k),\ \forall k\in\mathcal{N}_0,\tag{2.32a}$$

$$\mathbf{Y}(k)=\boldsymbol{R}^{(\alpha)}\mathbf{R}^{\alpha}(k)+\boldsymbol{T}^{(\mu)}\mathbf{I}^{\mu}(k),\ \forall k\in\mathcal{N}_0.\tag{2.32b}$$

These systems are discrete-time counterpart to *the Polynomial Matrix Description* of the continuous-time systems (for short: *the PMD systems*) due to P. J. Antsaklis and A. N. Michel [5, p. 553].

The *IIO* systems determined by $\alpha=1$, $\beta=0$, $\boldsymbol{Q}_1=\boldsymbol{I}_{\rho,\rho}$, $\boldsymbol{Q}_0=-\boldsymbol{A}$, $\boldsymbol{P}_0=\boldsymbol{B}$, $\boldsymbol{R}_0=\boldsymbol{R}=\boldsymbol{C}$, $\boldsymbol{R}_1=\boldsymbol{O}_{N,\rho}$, $\mathbf{R}=\mathbf{X}$, $\nu=0$, and $\mu\geq 1$, i.e.,

$$E^1\mathbf{X}(k)=\boldsymbol{A}\mathbf{X}(k)+\boldsymbol{B}\mathbf{I}(k),\ \forall k\in\mathcal{N}_0,\tag{2.33a}$$

$$\mathbf{Y}(k)=\boldsymbol{C}\mathbf{X}(k)+\boldsymbol{T}^{(\mu)}\mathbf{I}^{\mu}(k),\ \forall k\in\mathcal{N}_0.\tag{2.33b}$$

is discrete-time counterpart to continuous-time *Rosenbrock systems* (RS) after H. H. Rosenbrock [111].

Another subclass of the IIO systems is determined by $\nu = 0$, $\boldsymbol{E}_0 = \boldsymbol{I}_N$, $\boldsymbol{R}_\alpha = \boldsymbol{O}_{N,\rho}$, and $\boldsymbol{T}_0 = \boldsymbol{T}$, $\boldsymbol{T}_r = \boldsymbol{O}_{N\times M}$, $r = 1, 2, \cdots, \mu$, $\mu = 0$,

$$\sum_{r=0}^{r=\alpha} \boldsymbol{Q}_r E^r \mathbf{R}(k) = \sum_{r=0}^{r=\beta} \boldsymbol{P}_r E^r \mathbf{I}(k),\ \forall k \in \mathcal{N}_0, \tag{2.34a}$$

$$\mathbf{Y}(k) = \sum_{r=0}^{r=\alpha} \boldsymbol{R}_r E^r \mathbf{R}(k) + \boldsymbol{T}\mathbf{I}(k),\ \forall k \in \mathcal{N}_0, \tag{2.34b}$$

or, equivalently, in the compact form:

$$\boldsymbol{Q}^{(\alpha)}\mathbf{R}^{\alpha}(k) = \boldsymbol{P}^{(\beta)}\mathbf{I}^{\beta}(k),\ \forall k \in \mathcal{N}_0,\ \alpha \geq 1, \alpha \geq \beta \geq 0, \tag{2.35a}$$

$$\mathbf{Y}(k) = \boldsymbol{R}^{(\alpha)}\mathbf{R}^{\alpha}(k) + \boldsymbol{T}\mathbf{I}(k) = \boldsymbol{R}^{(\alpha-1)}\mathbf{R}^{\alpha-1}(k) + \boldsymbol{T}\mathbf{I}(k), \tag{2.35b}$$

$$\boldsymbol{R}^{(\alpha)} \in \mathcal{R}^{N\times\rho(\alpha+1)}, \boldsymbol{R}^{(\alpha-1)} \in \mathcal{R}^{N\times\rho\alpha}, \mathbf{R}^{\alpha-1}(k) \in \mathcal{R}^{\rho\alpha}, \boldsymbol{T} \in \mathcal{R}^{N\times M}, \forall k \in \mathcal{N}_0.$$

We decide to call the systems described by (2.35a) and (2.35b) *General Input-State-Output Systems* (for short: *GISO systems*). Their characteristics are the existence of their state, but the nonexistence of the output dynamics. The ISO systems compose a special class of the IIO systems (2.18a) and (2.18b).

The analysis of the relationships among different classes of systems is in Subsection 2.4.3.

Let in the sequel

$$\gamma = \max\{\beta, \mu\}. \tag{2.36}$$

Definition 2.9 *The quintuple* ($\boldsymbol{E}^{(\nu)}$, $\boldsymbol{P}^{(\beta)}$, $\boldsymbol{Q}^{(\alpha)}$, $\boldsymbol{R}^{(\alpha-1)}$, $\boldsymbol{T}^{(\mu)}$) *is* ***an IIO realization of the IIO system (2.18a) and (2.18b) equivalently of (2.31a) and (2.31b) for an arbitrary input vector function and arbitrary input, internal (dynamics) and output initial conditions.***

2.4 System forms

2.4.1 From IO system to ISO system

In order to transform the IO system (2.1) (in Section 2.1),

$$\sum_{r=0}^{r=\nu} \boldsymbol{A}_r E^r \mathbf{Y}(k) = \sum_{r=0}^{r=\mu} \boldsymbol{B}_r E^r \mathbf{I}(k), \tag{2.37}$$

$$\det \boldsymbol{A}_\nu \neq 0, \forall k \in \mathcal{N},\ \nu \geq 1,\ 0 \leq \mu \leq \nu,$$

into the equivalent *ISO* system (2.14a) and (2.14b) (in Section 2.2), i.e., into

$$\mathbf{X}(k+1) = \boldsymbol{A}\mathbf{X}(k) + \boldsymbol{B}\mathbf{I}(k),\ \forall k \in \mathcal{N} \tag{2.38a}$$

$$\mathbf{Y}(k) = \boldsymbol{C}\mathbf{X}(k) + \boldsymbol{D}\mathbf{I}(k),\ \forall k \in \mathcal{N}, \tag{2.38b}$$

we define purely mathematically, without any physical justification or meaning, subsidiary vector variables $\mathbf{X}_1$, $\mathbf{X}_2$, ... $\mathbf{X}_\nu$ by

$$\mathbf{X}_1 = \mathbf{Y} - \boldsymbol{B}_\nu\mathbf{I}, \tag{2.39a}$$

$$\mathbf{X}_2 = E^1\mathbf{X}_1 + \boldsymbol{A}_{\nu-1}\mathbf{Y} - \boldsymbol{B}_{\nu-1}\mathbf{I}, \tag{2.39b}$$

$$\mathbf{X}_3 = E^1\mathbf{X}_2 + \boldsymbol{A}_{\nu-2}\mathbf{Y} - \boldsymbol{B}_{\nu-2}\mathbf{I}, \tag{2.39c}$$

$$\vdots$$

$$\mathbf{X}_{\nu-2} = E^1\mathbf{X}_{\nu-3} + \boldsymbol{A}_3\mathbf{Y} - \boldsymbol{B}_3\mathbf{I}, \tag{2.39d}$$

$$\mathbf{X}_{\nu-1} = E^1\mathbf{X}_{\nu-2} + \boldsymbol{A}_2\mathbf{Y} - \boldsymbol{B}_2\mathbf{I}, \tag{2.39e}$$

$$\mathbf{X}_\nu = E^1\mathbf{X}_{\nu-1} + \boldsymbol{A}_1\mathbf{Y} - \boldsymbol{B}_1\mathbf{I} \tag{2.39f}$$

which we can rewrite as

$$\mathbf{X}_1 = \mathbf{Y} - \boldsymbol{B}_\nu\mathbf{I}, \tag{2.40a}$$

$$E^1\mathbf{X}_1 = \mathbf{X}_2 - \boldsymbol{A}_{\nu-1}\mathbf{Y} + \boldsymbol{B}_{\nu-1}\mathbf{I}, \tag{2.40b}$$

$$E^1\mathbf{X}_2 = \mathbf{X}_3 - \boldsymbol{A}_{\nu-2}\mathbf{Y} + \boldsymbol{B}_{\nu-2}\mathbf{I}, \tag{2.40c}$$

$$\vdots$$

$$E^1\mathbf{X}_{\nu-3} = \mathbf{X}_{\nu-2} - \boldsymbol{A}_3\mathbf{Y} + \boldsymbol{B}_3\mathbf{I}, \tag{2.40d}$$

$$E^1\mathbf{X}_{\nu-2} = \mathbf{X}_{\nu-1} - \boldsymbol{A}_2\mathbf{Y} + \boldsymbol{B}_2\mathbf{I}, \tag{2.40e}$$

$$E^1\mathbf{X}_{\nu-1} = \mathbf{X}_\nu - \boldsymbol{A}_1\mathbf{Y} + \boldsymbol{B}_1\mathbf{I}. \tag{2.40f}$$

We solve the Equation (2.40a) for $\mathbf{Y}$,

$$\mathbf{Y} = \mathbf{X}_1 + \boldsymbol{B}_\nu\mathbf{I} \tag{2.41}$$

and then we replace $\mathbf{Y}$ by $\mathbf{X}_1 + \boldsymbol{B}_\nu \mathbf{I}$ in all other Equations (2.40b) to (2.40f):

$$E^1\mathbf{X}_1 = \mathbf{X}_2 - \boldsymbol{A}_{\nu-1}(\mathbf{X}_1 + \boldsymbol{B}_\nu \mathbf{I}) + \boldsymbol{B}_{\nu-1}\mathbf{I}, \tag{2.42a}$$
$$E^1\mathbf{X}_2 = \mathbf{X}_3 - \boldsymbol{A}_{\nu-2}(\mathbf{X}_1 + \boldsymbol{B}_\nu \mathbf{I}) + \boldsymbol{B}_{\nu-2}\mathbf{I}, \tag{2.42b}$$
$$\vdots$$
$$E^1\mathbf{X}_{\nu-3} = \mathbf{X}_{\nu-2} - \boldsymbol{A}_3(\mathbf{X}_1 + \boldsymbol{B}_\nu \mathbf{I}) + \boldsymbol{B}_3\mathbf{I}, \tag{2.42c}$$
$$E^1\mathbf{X}_{\nu-2} = \mathbf{X}_{\nu-1} - \boldsymbol{A}_2(\mathbf{X}_1 + \boldsymbol{B}_\nu \mathbf{I}) + \boldsymbol{B}_2\mathbf{I}, \tag{2.42d}$$
$$E^1\mathbf{X}_{\nu-1} = \mathbf{X}_\nu - \boldsymbol{A}_1(\mathbf{X}_1 + \boldsymbol{B}_\nu \mathbf{I}) + \boldsymbol{B}_1\mathbf{I} \tag{2.42e}$$

which we set in the following form

$$E^1\mathbf{X}_1 = -\boldsymbol{A}_{\nu-1}\mathbf{X}_1 + \mathbf{X}_2 + (\boldsymbol{B}_{\nu-1} - \boldsymbol{A}_{\nu-1}\boldsymbol{B}_\nu)\mathbf{I}, \tag{2.43a}$$
$$E^1\mathbf{X}_2 = -\boldsymbol{A}_{\nu-2}\mathbf{X}_1 + \mathbf{X}_3 + (\boldsymbol{B}_{\nu-2} - \boldsymbol{A}_{\nu-2}\boldsymbol{B}_\nu)\mathbf{I}, \tag{2.43b}$$
$$\vdots$$
$$E^1\mathbf{X}_{\nu-3} = -\boldsymbol{A}_3\mathbf{X}_1 + \mathbf{X}_{\nu-2} + (\boldsymbol{B}_3 - \boldsymbol{A}_3\boldsymbol{B}_\nu)\mathbf{I}, \tag{2.43c}$$
$$E^1\mathbf{X}_{\nu-2} = -\boldsymbol{A}_2\mathbf{X}_1 + \mathbf{X}_{\nu-1} + (\boldsymbol{B}_2 - \boldsymbol{A}_2\boldsymbol{B}_\nu)\mathbf{I}, \tag{2.43d}$$
$$E^1\mathbf{X}_{\nu-1} = -\boldsymbol{A}_1\mathbf{X}_1 + \mathbf{X}_\nu + (\boldsymbol{B}_1 - \boldsymbol{A}_1\boldsymbol{B}_\nu)\mathbf{I}. \tag{2.43e}$$

We replace the first shifts $E^1\mathbf{X}_1$, $E^1\mathbf{X}_2, \cdots, E^1\mathbf{X}_{\nu-1}$ into (2.39b) through (2.39f):

$$\mathbf{X}_1 = \mathbf{Y} - \boldsymbol{B}_\nu \mathbf{I},$$
$$\mathbf{X}_2 = E^1\mathbf{Y} - \boldsymbol{B}_\nu E^1\mathbf{I} + \boldsymbol{A}_{\nu-1}\mathbf{Y} - \boldsymbol{B}_{\nu-1}\mathbf{I},$$
$$\mathbf{X}_3 = E^2\mathbf{Y} - \boldsymbol{B}_\nu E^2\mathbf{I} + \boldsymbol{A}_{\nu-1}E^1\mathbf{Y} - \boldsymbol{B}_{\nu-1}E^1\mathbf{I} + \boldsymbol{A}_{\nu-2}\mathbf{Y} - \boldsymbol{B}_{\nu-2}\mathbf{I},$$
$$\vdots$$
$$\mathbf{X}_{\nu-2} = \begin{cases} E^{\nu-3}\mathbf{Y} - \boldsymbol{B}_\nu E^{\nu-3}\mathbf{I} + \boldsymbol{A}_{\nu-1}E^{\nu-4}\mathbf{Y} - \boldsymbol{B}_{\nu-1}E^{\nu-4}\mathbf{I} + \\ + \boldsymbol{A}_{\nu-2}E^{\nu-5}\mathbf{Y} - \boldsymbol{B}_{\nu-2}E^{\nu-5}\mathbf{I} + \cdots + \boldsymbol{A}_4E^1\mathbf{Y} - \boldsymbol{B}_4E^1\mathbf{I} + \\ + \boldsymbol{A}_3\mathbf{Y} - \boldsymbol{B}_3\mathbf{I}, \end{cases}$$
$$\mathbf{X}_{\nu-1} = \begin{cases} E^{\nu-2}\mathbf{Y} - \boldsymbol{B}_\nu E^{\nu-2}\mathbf{I} + \boldsymbol{A}_{\nu-1}E^{\nu-3}\mathbf{Y} - \boldsymbol{B}_{\nu-1}E^{\nu-3}\mathbf{I} + \\ + \boldsymbol{A}_{\nu-2}E^{\nu-4}\mathbf{Y} - \boldsymbol{B}_{\nu-2}E^{\nu-4}\mathbf{I} + \cdots + \boldsymbol{A}_3E^1\mathbf{Y} - \boldsymbol{B}_3E^1\mathbf{I} + \\ + \boldsymbol{A}_2\mathbf{Y} - \boldsymbol{B}_2\mathbf{I}, \end{cases}$$
$$\mathbf{X}_\nu = \begin{cases} E^{\nu-1}\mathbf{Y} - \boldsymbol{B}_\nu E^{\nu-1}\mathbf{I} + \boldsymbol{A}_{\nu-1}E^{\nu-2}\mathbf{Y} - \boldsymbol{B}_{\nu-1}E^{\nu-2}\mathbf{I} + \\ + \boldsymbol{A}_{\nu-2}E^{\nu-3}\mathbf{Y} - \boldsymbol{B}_{\nu-2}E^{\nu-3}\mathbf{I} + \cdots + \boldsymbol{A}_2E^1\mathbf{Y} - \boldsymbol{B}_2E^1\mathbf{I} + \\ + \boldsymbol{A}_1\mathbf{Y} - \boldsymbol{B}_1\mathbf{I}. \end{cases}$$

We shift once the last equation

$$E^1\mathbf{X}_\nu = E^\nu\mathbf{Y} + \boldsymbol{A}_{\nu-1}E^{\nu-1}\mathbf{Y} + \boldsymbol{A}_{\nu-2}E^{\nu-2}\mathbf{Y} + \cdots + \boldsymbol{A}_2E^2\mathbf{Y} + \boldsymbol{A}_1E^1\mathbf{Y} - \\ -\boldsymbol{B}_\nu E^\nu\mathbf{I} - \boldsymbol{B}_{\nu-1}E^{\nu-1}\mathbf{I} - \boldsymbol{B}_{\nu-2}E^{\nu-2}\mathbf{I} - \cdots - \boldsymbol{B}_2E^2\mathbf{I} - \boldsymbol{B}_1E^1\mathbf{I} = \\ = \sum_{r=1}^{r=\nu} \boldsymbol{A}_rE^r\mathbf{Y}(k) - \sum_{r=1}^{r=\mu} \boldsymbol{B}_rE^r\mathbf{I}(k)$$

and we use (2.37)

$$\sum_{r=1}^{r=\nu} \boldsymbol{A}_rE^r\mathbf{Y}(k) - \sum_{r=1}^{r=\mu} \boldsymbol{B}_rE^r\mathbf{I}(k) = \boldsymbol{B}_0\mathbf{I}(k) - \boldsymbol{A}_0\mathbf{Y}(k) = \\ = \boldsymbol{B}_0\mathbf{I}(k) - \boldsymbol{A}_0\left[\mathbf{X}_1(k) + \boldsymbol{B}_\nu\mathbf{I}(k)\right] = -\boldsymbol{A}_0\mathbf{X}_1(k) + (\boldsymbol{B}_0 - \boldsymbol{A}_0\boldsymbol{B}_\nu)\mathbf{I}(k).$$

Hence,

$$E^1\mathbf{X}_\nu = -\boldsymbol{A}_0\mathbf{X}_1 + (\boldsymbol{B}_0 - \boldsymbol{A}_0\boldsymbol{B}_\nu)\mathbf{I}.$$

We gather this and (2.43a) through (2.43e),

$$\underbrace{\begin{bmatrix} E^1\mathbf{X}_1 \\ E^1\mathbf{X}_2 \\ \vdots \\ E^1\mathbf{X}_{\nu-3} \\ E^1\mathbf{X}_{\nu-2} \\ E^1\mathbf{X}_{\nu-1} \\ E^1\mathbf{X}_\nu \end{bmatrix}}_{E^1\mathbf{X}} = \begin{bmatrix} -\boldsymbol{A}_{\nu-1}\mathbf{X}_1 + \mathbf{X}_2 + (\boldsymbol{B}_{\nu-1} - \boldsymbol{A}_{\nu-1}\boldsymbol{B}_\nu)\mathbf{I} \\ -\boldsymbol{A}_{\nu-2}\mathbf{X}_1 + \mathbf{X}_3 + (\boldsymbol{B}_{\nu-2} - \boldsymbol{A}_{\nu-2}\boldsymbol{B}_\nu)\mathbf{I} \\ \vdots \\ -\boldsymbol{A}_3\mathbf{X}_1 + \mathbf{X}_{\nu-2} + (\boldsymbol{B}_3 - \boldsymbol{A}_3\boldsymbol{B}_\nu)\mathbf{I} \\ -\boldsymbol{A}_2\mathbf{X}_1 + \mathbf{X}_{\nu-1} + (\boldsymbol{B}_2 - \boldsymbol{A}_2\boldsymbol{B}_\nu)\mathbf{I} \\ -\boldsymbol{A}_1\mathbf{X}_1 + \mathbf{X}_\nu + (\boldsymbol{B}_1 - \boldsymbol{A}_1\boldsymbol{B}_\nu)\mathbf{I} \\ -\boldsymbol{A}_0\mathbf{X}_1 + (\boldsymbol{B}_0 - \boldsymbol{A}_0\boldsymbol{B}_\nu)\mathbf{I} \end{bmatrix} =$$

$$= \underbrace{\begin{bmatrix} -\boldsymbol{A}_{\nu-1} & \boldsymbol{I}_N & \boldsymbol{O}_N & \cdots & \boldsymbol{O}_N & \boldsymbol{O}_N & \boldsymbol{O}_N \\ -\boldsymbol{A}_{\nu-2} & \boldsymbol{O}_N & \boldsymbol{I}_N & \cdots & \boldsymbol{O}_N & \boldsymbol{O}_N & \boldsymbol{O}_N \\ \vdots & \vdots & \vdots & \vdots & \vdots & \vdots & \vdots \\ -\boldsymbol{A}_3 & \boldsymbol{O}_N & \boldsymbol{O}_N & \cdots & \boldsymbol{I}_N & \boldsymbol{O}_N & \boldsymbol{O}_N \\ -\boldsymbol{A}_2 & \boldsymbol{O}_N & \boldsymbol{O}_N & \cdots & \boldsymbol{O}_N & \boldsymbol{I}_N & \boldsymbol{O}_N \\ -\boldsymbol{A}_1 & \boldsymbol{O}_N & \boldsymbol{O}_N & \cdots & \boldsymbol{O}_N & \boldsymbol{O}_N & \boldsymbol{I}_N \\ -\boldsymbol{A}_0 & \boldsymbol{O}_N & \boldsymbol{O}_N & \cdots & \boldsymbol{O}_N & \boldsymbol{O}_N & \boldsymbol{O}_N \end{bmatrix}}_{\boldsymbol{A}} \underbrace{\begin{bmatrix} \mathbf{X}_1 \\ \mathbf{X}_2 \\ \vdots \\ \mathbf{X}_{\nu-3} \\ \mathbf{X}_{\nu-2} \\ \mathbf{X}_{\nu-1} \\ \mathbf{X}_\nu \end{bmatrix}}_{\mathbf{X}} +$$

$$+\underbrace{\begin{bmatrix} \boldsymbol{B}_{\nu-1}-\boldsymbol{A}_{\nu-1}\boldsymbol{B}_{\nu} \\ \boldsymbol{B}_{\nu-2}-\boldsymbol{A}_{\nu-2}\boldsymbol{B}_{\nu} \\ \\ \boldsymbol{B}_{3}-\boldsymbol{A}_{3}\boldsymbol{B}_{\nu} \\ \boldsymbol{B}_{2}-\boldsymbol{A}_{2}\boldsymbol{B}_{\nu} \\ \boldsymbol{B}_{1}-\boldsymbol{A}_{1}\boldsymbol{B}_{\nu} \\ \boldsymbol{B}_{0}-\boldsymbol{A}_{0}\boldsymbol{B}_{\nu} \end{bmatrix}}_{\boldsymbol{B}}\mathbf{I}. \tag{2.44}$$

We set (2.41) into the following form:

$$\mathbf{Y}=\mathbf{X}_1+\boldsymbol{B}_{\nu}\mathbf{I}=\underbrace{[\ \boldsymbol{I}_N \quad \boldsymbol{O}_N \quad \boldsymbol{O}_N \quad \cdots \quad \boldsymbol{O}_N \quad \boldsymbol{O}_N \quad \boldsymbol{O}_N\ ]}_{\boldsymbol{C}}\mathbf{X}+\underbrace{[\boldsymbol{B}_{\nu}]}_{\boldsymbol{D}}\mathbf{I},$$

which, together with (2.44), yields

$$E^1\mathbf{X}(k)=\boldsymbol{A}\mathbf{X}(k)+\boldsymbol{B}\mathbf{I}(k),\ \mathbf{Y}(k)=\boldsymbol{C}\mathbf{X}(k)+\boldsymbol{D}\mathbf{I}(k),\ \forall k\in\mathcal{N}. \tag{2.45}$$

The quadruple $(\boldsymbol{A},\ \boldsymbol{B},\ \boldsymbol{C},\ \boldsymbol{D})$ is the *ISO* realization of the *IO* system (2.5).

2.4.2 From *ISO* system to *IO* system

We can transform the *ISO* system (2.14a) and (2.14b) (in Section 2.2),

$$\mathbf{X}(k+1)=\boldsymbol{A}\mathbf{X}(k)+\boldsymbol{B}\mathbf{I}(k),\ \forall k\in\mathcal{N} \tag{2.46a}$$

$$\mathbf{Y}(k)=\boldsymbol{C}\mathbf{X}(k)+\boldsymbol{D}\mathbf{I}(k),\ \forall k\in\mathcal{N}, \tag{2.46b}$$

into the *IO* system by applying first the $Z-$transform to (2.46a) and (2.46b) for all zero initial conditions,

$$(z\boldsymbol{I}-\boldsymbol{A})\mathbf{X}(z)=\boldsymbol{B}\mathbf{I}(z)\Longrightarrow$$
$$\mathbf{X}(z)=(z\boldsymbol{I}-\boldsymbol{A})^{-1}\boldsymbol{B}\mathbf{I}(z),$$
$$\mathbf{Y}(z)=\left[\boldsymbol{C}(z\boldsymbol{I}-\boldsymbol{A})^{-1}\boldsymbol{B}+\boldsymbol{D}\right]\mathbf{I}(z)\Longrightarrow$$
$$[\det(z\boldsymbol{I}-\boldsymbol{A})]\,\mathbf{Y}(z)=[\boldsymbol{C}\,\mathrm{adj}\,(z\boldsymbol{I}-\boldsymbol{A})\,\boldsymbol{B}+\boldsymbol{D}\det(z\boldsymbol{I}-\boldsymbol{A})]\,\mathbf{I}(z).$$

Let

$$f(z)=\det(z\boldsymbol{I}-\boldsymbol{A})=\sum_{i=0}^{i=n}c_i z^i,\ c_n=1,$$

$$\boldsymbol{C}\,\mathrm{adj}\,(z\boldsymbol{I}-\boldsymbol{A})\,\boldsymbol{B}+\boldsymbol{D}\det(z\boldsymbol{I}-\boldsymbol{A})=\sum_{i=0}^{i=n}\boldsymbol{K}_i z^i,\ \boldsymbol{K}_i\in\mathcal{R}^{N\times M},$$

so that

$$\sum_{i=0}^{i=n} c_i \left[z^i \mathbf{Y}(z) \right] = \sum_{i=0}^{i=n} \boldsymbol{K}_i \left[z^i \mathbf{I}(z) \right].$$

The application of the Inverse $Z-$transform for all zero initial conditions results into

$$\sum_{i=0}^{i=n} c_i E^i \mathbf{Y}(k) = \sum_{i=0}^{i=n} \boldsymbol{K}_i E^i \mathbf{I}(k), \ \forall k \in \mathcal{N}.$$

This is the IO description of the ISO system (2.46a) and (2.46b).

Let

$$\begin{aligned} \boldsymbol{C}^{(n)} &= \begin{bmatrix} c_0 \boldsymbol{I} & c_1 \boldsymbol{I} & \cdots & c_n \boldsymbol{I} \end{bmatrix} \in \mathcal{R}^{N \times (n+1)N}, \\ \boldsymbol{K}^{(n)} &= \begin{bmatrix} \boldsymbol{K}_0 & \boldsymbol{K}_1 & \cdots & \boldsymbol{K}_n \end{bmatrix} \in \mathcal{R}^{N \times (n+1)M}. \end{aligned}$$

The quadruple $(n, n, \boldsymbol{C}^{(n)}, \boldsymbol{K}^{(n)})$ is the IO realization of the ISO system (2.46a) and (2.46b).

2.4.3 Relationships among system descriptions

If $\boldsymbol{Q}_r = \boldsymbol{A}_r$, for $r = 0, 1, 2, ..., \alpha$, $\boldsymbol{P}_r = \boldsymbol{B}_r$, for $r = 0, 1, 2, ..., \beta$, $\boldsymbol{E}_r = \boldsymbol{O}_N$, for $r = 1, 2, ..., \nu$, i.e., $\nu = 0$, $\boldsymbol{E}_0 = \boldsymbol{I}_N$, and $\boldsymbol{T}_r = \boldsymbol{O}_{N,M}$, for $r = 0, 1, 2, ..., \mu$, then (2.18a) and (2.18b) (in Section 2.3) reduces to (2.1) (in Section 2.1), and (3.63a) and (3.63b) (in Subsection 3.5.4), reduces to (3.55) (in Subsection 3.5.2), i.e., the IIO system becomes the IO system. Then, the IIO systems incorporate the IO systems. The IO system (2.5),

$$\boldsymbol{A}^{(\nu_{IO})} \mathbf{Y}^{\nu_{IO}}(k) = \boldsymbol{B}^{(\mu)} \mathbf{I}^{\mu}(k), \ \forall k \in \mathcal{N}_0$$

can be set in the special IIO system (2.31a) and (2.31b) form, i.e., in the *GISO system form* (2.35a) and (2.35b) by formally defining $\mathbf{R} = \mathbf{Y}$, $\alpha = \nu_{IO}$, $\beta = \mu$, and $\rho = N$,

$$\begin{aligned} \boldsymbol{A}^{(\alpha)} \mathbf{R}^{\alpha}(k) &= \boldsymbol{B}^{(\beta)} \mathbf{I}^{\beta}(k), \ \forall k \in \mathcal{N}_0, \\ \mathbf{Y}(k) &= \mathbf{R}(k), \ \forall k \in \mathcal{N}_0. \end{aligned}$$

We can formally consider the IO systems as a subclass of the IIO systems. These two classes of the systems are equivalent if $\nu = 0$. However, if $\nu > 0$, then the *IIO* system (2.31a) and (2.31b) can be put into the following form:

$$\underbrace{\begin{bmatrix} \boldsymbol{Q}^{(\alpha)} & \boldsymbol{O}_{\rho,(\nu+1)N} \\ -\boldsymbol{R}^{(\alpha)} & \boldsymbol{E}^{(\nu)} \end{bmatrix}}_{\boldsymbol{A}^{(\alpha+\nu)}} \underbrace{\begin{bmatrix} \mathbf{R}^{\alpha}(k) \\ \mathbf{Y}^{\nu}(k) \end{bmatrix}}_{\mathbf{Y}^{*\alpha+\nu}(k)} =$$

$$= \underbrace{\begin{bmatrix} \boldsymbol{P}^{(\beta)} & \boldsymbol{O}_{\rho,(\beta+1)M} \\ \boldsymbol{O}_{N,(\mu+1)M} & \boldsymbol{T}^{(\mu)} \end{bmatrix}}_{\boldsymbol{B}^{(\beta+\mu)}} \underbrace{\begin{bmatrix} \mathbf{I}^{\beta}(k) \\ \mathbf{I}^{\mu}(k) \end{bmatrix}}_{\mathbf{I}^{*\gamma}(k)}, \forall k \in \mathcal{N}_0. \tag{2.47}$$

so that

$$\boldsymbol{A}^{(\alpha+\nu)}\mathbf{Y}^{*^{\alpha+\nu}}(k) = \boldsymbol{B}^{(\beta+\mu)}\mathbf{I}^{*^{\gamma}}(k),\ \forall k \in \mathcal{N}_0, \tag{2.48a}$$

$$\mathbf{Y}(k) = \begin{bmatrix} \boldsymbol{O}_{N,(\alpha+1)\rho} & \boldsymbol{I}_N & \boldsymbol{O}_{N,\nu N} \end{bmatrix} \mathbf{Y}^{*^{\alpha+\nu}}(k),\ \forall k \in \mathcal{N}_0. \tag{2.48b}$$

This is the *IO* form relative to the vector $\mathbf{Y}^*$, but it rests the *IIO* form relative to the real output vector $\mathbf{Y}$. The *IO* system (2.5) (in Section 2.1), and the *IIO* system (2.31a) and (2.31b), are different systems if $\nu > 0$. These two classes of the systems are not equivalent if $\nu > 0$.

If $\rho = n$, $\mathbf{R} = \mathbf{X}$ ($\mathbf{r} = \mathbf{x}$), $\alpha = 1$, $\boldsymbol{Q}_1 = \boldsymbol{I}_n$, $\boldsymbol{Q}_0 = -\boldsymbol{A}$, $\beta = 0$, $\boldsymbol{P}_0 = \boldsymbol{B}$, $\nu = 0$, $\boldsymbol{E}_0 = \boldsymbol{I}_N$, $\boldsymbol{R}_0 = \boldsymbol{C}$, $\boldsymbol{R}_1 = \boldsymbol{O}_{N,\rho}$ and $\mu = 0$, $\boldsymbol{T}_0 = \boldsymbol{D}$, then (2.18a) and (2.18b) reduces to (2.14a) and (2.14b) (in Section 2.2) and (3.63a) and (3.63b) reduces to (3.60a) and (3.60b) (in Subsection 3.5.3). The *IIO* becomes the *ISO* system. The latter is a special case of the former.

If $\nu = 0$, $\boldsymbol{E}_0 = \boldsymbol{I}_N$, then (2.18a) and (2.18b) (in Section 2.3) corresponds to the *PMD* system (2.32a) and (2.32b) (in Section 2.3), and (3.63a) and (3.63b) reduces to (3.65a) and (3.65b) (in Subsection 3.5.4). The *IIO* is then a *PMD* system. The family of the *IIO* systems incorporates the family of the *PMD* systems described by (2.32a) and (2.32a) (in Section 2.3).

If $\rho = n$, $\mathbf{r} = \mathbf{x}$, $\alpha = 1$, $\boldsymbol{Q}_1 = \boldsymbol{I}_n$, $\boldsymbol{Q}_0 = -\boldsymbol{A}$, $\beta = 0$, $\boldsymbol{P}_0 = \boldsymbol{B}$, $\nu = 0$, $\boldsymbol{E}_0 = \boldsymbol{I}_N$, $\boldsymbol{R}_0 = \boldsymbol{C}$, and $\boldsymbol{R}_1 = \boldsymbol{O}_{N\times\rho}$, then the *IIO* system (2.18a) and (2.18b) (in Section 2.3) reduces to the Rosenbrock system (2.33a) and (2.33b) (in Section 2.3), and (3.64a) and (3.64b) (in Subsection 3.5.4), represents the Rosenbrock system *RS* (3.66a) and (3.66b) (in Subsection 3.5.4), which is the special case of the former and the *PMD* system (2.32a) and (2.32b).

If $\nu = 0$, $\boldsymbol{E}_0 = \boldsymbol{I}_N$, $\boldsymbol{R}_0 = \boldsymbol{R}$, $\boldsymbol{R}_r = \boldsymbol{O}_{N\times\rho}$, $r = 1, 2, ..., \alpha > 0$, and $\boldsymbol{T}_0 = \boldsymbol{T}$, $\boldsymbol{T}_r = \boldsymbol{O}_{N\times M}$, $r = 1, 2, ..., \mu$, then (2.18a) and (2.18b) reduces to the *GISO* system (2.35a) and (2.35b) (in Section 2.3), and (3.63a) and (3.63b), i.e., (3.64a) and (3.64b) becomes (3.68a) and (3.68b) (in Subsection 3.5.4). The *GISO* systems are a special subclass of the *PMD* systems, which incorporates Rosenbrock systems.

Subsection 2.4.1 presents the transformation of the *IO* system (2.1) (in Section 2.1), into the equivalent *ISO* system (2.14a) and (2.14b). Subsection 2.4.2 shows how the *ISO* system (2.14a) and (2.14b) can be transformed into the *IO* system (2.1).

We can summarize this by using notation $\{\cdot\}$ for a class of systems, e.g., $\{IIO\}$ is the family of all *IIO* systems (3.63a) and (3.63b),

$$\{IO\} \simeq \{ISO\},\ \{ISO\} \subset \{GISO\} \subset \{RS\} \subset \{PMD\} \subset \{IIO\},$$
$$\nu{=}\mu{=}0,\ \boldsymbol{R}_r{=}\boldsymbol{O}_{N\times\rho}, r{=}1,2,...,\alpha>0,\ \boldsymbol{T}_0{=}\boldsymbol{O}_{N\times M} \Rightarrow\ \{IO\}{\simeq}\{IIO\},$$
$$\nu > 0 \Longrightarrow \{IO\} \subset \{IIO\},$$

where $\simeq$ means *equivalent.*

Chapter 3

System regimes

3.1 System regime meaning

The system dynamical behavior (*the system behavior*) reflects the law of the evolution of a process, work, and movement or of the output response of the system in the course of time. The system behavior is governed by both:

a) *the system properties,* and

b) *the actions upon the system.*

We distinguish two main groups of the system properties: *quantitative* or *qualitative* system properties. *A quantitative system property* is, for example, the settling *time* of the system response. Quantitative system properties characterize the system behavior under specific external and internal conditions. *Qualitative properties of the system* are, for example, its controllability, observability, stability, and trackability ([31], [34]-[40], [42]-[50], [52]-[54], [56], [57], [70]-[76], [102]-[104]). Their qualitative system properties determine the system behavior for a set of external and/or internal conditions. Their set can be finite or infinite, bounded or unbounded, but not a singleton that is typical for quantitative system properties, such as a settling time.

There exist two different principal *actions upon the system* depending on the time of their observation:

- *The system history and the system past* fully determine the *impact of the external actions on the system* and of *the system state until the initial moment* k_0. ***Initial conditions*** (initial conditions of input, state, and output variables) express that impact of the past on the system future behavior after the initial moment k_0. We will treat the impact of *arbitrary initial conditions*.

- *Actions upon the system since the initial moment* $t_{d0} = 0$ $(k_0 = 0)$ *on.* These actions are usually ***the external actions***. They are called ***the input variables*** if and only if their influence on the system is essential for the system behavior.

The behavior of the system can be:

- *independent of time* (invariable in *time*, i.e., *time-invariant*, i.e., *time-independent*), which is a constant behavior also called a *stationary behavior*,

  or

- *dependent on time* (variable in *time*, i.e., *time-dependent*, i.e., *time-varying*). In this case, the system behavior is either *periodic* or *aperiodic* (also called *transient*). The system behavior is periodic if and only if it repeats itself every σ discrete instants (moments), where σ is a positive integer. Such a minimal number σ is *the period* T_p of the system behavior. Otherwise, the system behavior is *aperiodic (transient)*.

Definition 3.1 ***A system regime*** *represents the set of all (initial and exterior) conditions under which the system operates, and the type of its behavior (i.e., the type of the temporal evolution: of a process, work, movement of the system and/or its response).*

This definition discovers that system regimes can be different and that they can be classified with respect to the following criteria:

- *The existence (the nonexistence) of the initial conditions.* Their existence (nonexistence) signifies that their values are different from (equal to) zero, respectively.
- *The existence (the nonexistence) of the exterior actions.*
- *The realization of the system demanded/required behavior*. If the system is an object/a plant, its demanded behavior is called its *desired behavior* and it is defined by its *desired response* (or, more precisely, by its *desired output response*) denoted by $\mathbf{Y}_d(\cdot;\cdot;\cdot) = \mathbf{Y}_d(\cdot)$.

- *The type of the system behavior.*

We will consider different system regimes classified by these criteria.

3.2 System regimes and initial conditions

With respect to *the existence of the initial conditions* a system can be in a regime during which:

- all initial conditions are *absent*, i.e., *equal to zero*,

  or

- there are *nonzero initial conditions.*

These regimes do not have special names.

Important qualitative system dynamical properties (e.g., Lyapunov stability properties, controllability, trackability and observability) reflect the system behavior under nonzero initial conditions. However, Bounded-Input Bounded-Output ($BIBO$) stability has been restricted so far by the demand that all initial conditions must be equal to zero.

Initial conditions are most often unpredictable and different from zero.

3.3 Forced and free regimes

3.3.1 Introduction

Let

$$(k_\sigma, \infty[\in \{]k_\sigma, \infty[, \ [k_\sigma, \infty[\}, \ k_\sigma \in \mathbb{Z}.$$

Let the system regimes be classified according to the following criterion: *the existence of the exterior actions.*

Definition 3.2 (*a*) *A system is in* ***a forced regime*** *on* $(k_\sigma, \infty[$ *if and only if there exists a moment* k *in* $(k_\sigma, \infty[$*, when the input vector different from the zero vector acts on the system:*

$$\exists k \in (k_\sigma, \infty[\Longrightarrow \mathbf{I}(k) \neq \mathbf{0}_M.$$

(*b*) *A system is in* ***a free regime*** *on* $(k_\sigma, \infty[$ *if and only if its input vector is equal to the zero vector always on* $(k_\sigma, \infty[$ *:*

$$\mathbf{I}(k) = \mathbf{0}_M, \ \forall k \in (k_\sigma, \infty[.$$

The expression "on $(k_\sigma, \infty[$" is to be omitted if and only if $k_\sigma = 0$, i.e., $(k_\sigma, \infty[= \mathcal{N}_0$.

Definition 3.3 *A system behavior is* ***trivial*** *if and only if the system movement is equal to the zero vector all the time. Otherwise, it is* ***non-trivial.***

The linearity of the system implies that for a system behavior to be trivial in a free regime, it is necessary and sufficient that all initial conditions are equal to zero. Physically this means that the physical behavior of the system in a free regime can be nontrivial if and only if there is an accumulated non-nominal energy in the system at the initial moment. The nominal system accumulated energy is equal to zero if and only if the system is described by total valued variables.

Lyapunov stability properties, and essentially linear system observability, deal with the system behavior in a free regime. *BIBO* stability, practical stability, controllability, trackability, and tracking treat the system behavior in a forced regime.

We analyze now in details the following descriptions of an arbitrary system, the forms of which depend on the domain of the independent variable that can be *discrete time* $t_d \in \mathcal{T}_{d0}$ $(k \in \mathcal{N}_0)$, or a complex variable $z \in \mathcal{C}$.

3.3.2 The temporal domain descriptions. The independent variable is *discrete time* $t_d \in \mathcal{T}_{d0}$ $(k \in \mathcal{N}_0)$

The description of the IO system in terms of the total coordinates

- in a *forced regime* is determined by (2.1) (in Section 2.1),

$$\sum_{r=0}^{r=\nu} \boldsymbol{A}_r E^r \mathbf{Y}(k) = \sum_{r=0}^{r=\mu} \boldsymbol{B}_r E^r \mathbf{I}(k), \ \forall k \in \mathcal{N}_0, \tag{3.1}$$

  or, equivalently by (2.5) (in Section 2.1),

$$\boldsymbol{A}^{(\nu)} \mathbf{Y}^{\nu}(k) = \boldsymbol{B}^{(\mu)} \mathbf{I}^{\mu}(k), \ \forall k \in \mathcal{N}_0, \tag{3.2}$$

- in a *free regime* has the form

$$\sum_{r=0}^{r=\nu} \boldsymbol{A}_r E^r \mathbf{Y}(k) = \mathbf{0}_N, \ \forall k \in \mathcal{N}_0; \tag{3.3}$$

  or, equivalently by

$$\boldsymbol{A}^{(\nu)} \mathbf{Y}^{\nu}(k) = \mathbf{0}_N, \ \forall k \in \mathcal{N}_0. \tag{3.4}$$

The description of the ISO system in terms of the total coordinates

- in a *forced regime* is determined by (2.14a) and (2.14b) (in Section 2.2),

$$\mathbf{X}(k+1) = \boldsymbol{A}\mathbf{X}(k) + \boldsymbol{B}\mathbf{I}(k),\ \forall k \in \mathcal{N}_0, \tag{3.5a}$$

$$\mathbf{Y}(k) = \boldsymbol{C}\mathbf{X}(k) + \boldsymbol{D}\mathbf{I}(k),\ \forall k \in \mathcal{N}_0, \tag{3.5b}$$

- in a *free regime* takes the following form, respectively:

$$\mathbf{X}(k+1) = \boldsymbol{A}\mathbf{X}(k),\ \forall k \in \mathcal{N}_0, \tag{3.6a}$$

$$\mathbf{Y}(k) = \boldsymbol{C}\mathbf{X}(k),\ \forall k \in \mathcal{N}_0. \tag{3.6b}$$

The description of the IIO system in terms of the total coordinates

- in a *forced regime* is determined by (2.18a) and (2.18b) (in Section 2.3),

$$\sum_{r=0}^{r=\alpha} \boldsymbol{Q}_r E^r \mathbf{R}(k) = \sum_{r=0}^{r=\beta\leq\alpha} \boldsymbol{P}_r E^r \mathbf{I}(k),\ \forall k \in \mathcal{N}_0, \tag{3.7a}$$

$$\sum_{r=0}^{r=\nu} \boldsymbol{E}_r E^r \mathbf{Y}(k) = \sum_{r=0}^{r=\alpha} \boldsymbol{R}_r E^r \mathbf{R}(k) + \sum_{r=0}^{r=\mu} \boldsymbol{T}_r E^r \mathbf{I}(k),\ \forall k \in \mathcal{N}_0, \tag{3.7b}$$

or, equivalently by (2.31a) and (2.31b) (in Section 2.3),

$$\boldsymbol{Q}^{(\alpha)} \mathbf{R}^{\alpha}(k) = \boldsymbol{P}^{(\beta)} \mathbf{I}^{\beta}(k),\ \forall k \in \mathcal{N}_0, \tag{3.8a}$$

$$\boldsymbol{E}^{(\nu)} \mathbf{Y}^{\nu}(k) = \boldsymbol{R}^{(\alpha)} \mathbf{R}^{\alpha}(k) + \boldsymbol{T}^{(\mu)} \mathbf{I}^{\mu}(k),\ \forall k \in \mathcal{N}_0; \tag{3.8b}$$

- in a *free regime* becomes:

$$\boldsymbol{Q}^{(\alpha)} \mathbf{R}^{\alpha}(k) = \mathbf{0}_{\rho},\ \forall k \in \mathcal{N}_0, \tag{3.9a}$$

$$\boldsymbol{E}^{(\nu)} \mathbf{Y}^{\nu}(k) = \boldsymbol{R}^{(\alpha)} \mathbf{R}^{\alpha}(k),\ \forall k \in \mathcal{N}_0. \tag{3.9b}$$

The mathematical models simplify essentially if the system is in a free regime.

The temporal domain mathematical descriptions of the systems do not show explicitly the influence of initial conditions. It appears only in the solutions of the mathematical models.

3.3.3 The complex domain system descriptions. The independent variable is the complex variable $z \in \mathcal{C}$.

The $Z-$transform (Appendix B.1) is the basic tool to obtain the complex domain description of the system.

The complex domain description of the IO system

is determined by the $Z-$transform of (3.1), when we apply the properties of the $Z-$transform,

- in a *forced regime* by

$$\sum_{r=0}^{r=\nu} \boldsymbol{A}_r \left[z^r \mathbf{Y}(z) - \sum_{k=0}^{k=r-1} z^{r-k} \mathbf{Y}(k) \right] = \\ = \sum_{r=0}^{r=\mu} \boldsymbol{B}_r \left[z^r \mathbf{I}(z) - \sum_{k=0}^{k=r-1} z^{r-k} \mathbf{I}(k) \right], \tag{3.10}$$

- and in a *free regime* by

$$\sum_{r=0}^{r=\nu} \boldsymbol{A}_r \left[z^r \mathbf{Y}(z) - \sum_{k=0}^{k=r-1} z^{r-k} \mathbf{Y}(k) \right] = \mathbf{0}_N. \tag{3.11}$$

The complex domain description of the ISO system

is determined by the $Z-$transform of (3.5a) and (3.5b), when we apply the properties of the $Z-$transform,

- in a *forced regime* by

$$(z\boldsymbol{I} - \boldsymbol{A})\,\mathbf{X}(z) - z\mathbf{X}(0) = \boldsymbol{B}\mathbf{I}(z), \tag{3.12a}$$

$$\mathbf{Y}(z) = \boldsymbol{C}\mathbf{X}(z) + \boldsymbol{D}\mathbf{I}(z), \tag{3.12b}$$

- which in a *free regime* becomes

$$(z\boldsymbol{I} - \boldsymbol{A})\,\mathbf{X}(z) - z\mathbf{X}(0) = \mathbf{0}_N, \tag{3.13a}$$

$$\mathbf{Y}(z) = \boldsymbol{C}\mathbf{X}(z). \tag{3.13b}$$

The complex domain description of the IIO system

follows from the application of the $Z-$transform and its properties to (3.7a) and (3.7b)

- in a *forced regime*

$$\sum_{r=0}^{r=\alpha} \boldsymbol{Q}_r \left[z^r \mathbf{R}(z) - \sum_{k=0}^{k=r-1} z^{r-k} \mathbf{R}(k) \right] =$$

$$= \sum_{r=0}^{r=\beta\leq\alpha} \boldsymbol{P}_r \left[z^r \mathbf{I}(z) - \sum_{k=0}^{k=r-1} z^{r-k} \mathbf{I}(k) \right], \tag{3.14a}$$

$$\sum_{r=0}^{r=\nu} \boldsymbol{E}_r \left[z^r \mathbf{Y}(z) - \sum_{k=0}^{k=r-1} z^{r-k} \mathbf{Y}(k) \right] =$$

$$= \sum_{r=0}^{r=\alpha} \boldsymbol{R}_r \left[z^r \mathbf{R}(z) - \sum_{k=0}^{k=r-1} z^{r-k} \mathbf{R}(k) \right] +$$

$$+ \sum_{r=0}^{r=\mu} \boldsymbol{T}_r \left[z^r \mathbf{I}(z) - \sum_{k=0}^{k=r-1} z^{r-k} \mathbf{I}(k) \right]; \tag{3.14b}$$

- in a *free regime*

$$\sum_{r=0}^{r=\alpha} \boldsymbol{Q}_r \left[z^r \mathbf{R}(z) - \sum_{k=0}^{k=r-1} z^{r-k} \mathbf{R}(k) \right] = \mathbf{0}_\rho, \tag{3.15a}$$

$$\sum_{r=0}^{r=\nu} \boldsymbol{E}_r \left[z^r \mathbf{Y}(z) - \sum_{k=0}^{k=r-1} z^{r-k} \mathbf{Y}(k) \right] =$$

$$= \sum_{r=0}^{r=\alpha} \boldsymbol{R}_r \left[z^r \mathbf{R}(z) - \sum_{k=0}^{k=r-1} z^{r-k} \mathbf{R}(k) \right]. \tag{3.15b}$$

The complex domain descriptions in free regimes are essentially simpler from those related to forced regimes. However, both contain explicitly initial conditions. We resolve them in terms of the $Z-$transforms of unknown system variables.

The solution of the complex domain description (3.10) of the IO system

- in a *forced regime* under **nonzero initial conditions**

$$\mathbf{Y}(z) = \left(\sum_{r=0}^{r=\nu} \boldsymbol{A}_r z^r\right)^{-1} \left\{\sum_{r=0}^{r=\nu} \boldsymbol{A}_r \left[\sum_{k=0}^{k=r-1} z^{r-k}\mathbf{Y}(k)\right] + \right.$$
$$\left. + \sum_{r=0}^{r=\mu} \boldsymbol{B}_r \left[z^r \mathbf{I}(z) - \sum_{k=0}^{k=r-1} z^{r-k}\mathbf{I}(k)\right]\right\}, \tag{3.16}$$

- in a *free regime* under **nonzero initial conditions**

$$\mathbf{Y}(z) = \left(\sum_{r=0}^{r=\nu} \boldsymbol{A}_r z^r\right)^{-1} \cdot \sum_{r=0}^{r=\nu} \boldsymbol{A}_r \left[\sum_{k=0}^{k=r-1} z^{r-k}\mathbf{Y}(k)\right]. \tag{3.17}$$

The solution of the complex domain description of the ISO system

follows from (3.12a) and (3.12b)

- in a *forced regime* under **nonzero initial conditions**

$$\mathbf{X}(z) = (z\boldsymbol{I} - \boldsymbol{A})^{-1}\left[\boldsymbol{B}\mathbf{I}(z) + z\mathbf{X}(0)\right], \tag{3.18a}$$
$$\mathbf{Y}(z) = \boldsymbol{C}\mathbf{X}(z) + \boldsymbol{D}\mathbf{I}(z), \tag{3.18b}$$

- in a *free regime* under **nonzero initial conditions**

$$\mathbf{X}(z) = z(z\boldsymbol{I} - \boldsymbol{A})^{-1}\mathbf{X}(0), \tag{3.19a}$$
$$\mathbf{Y}(z) = \boldsymbol{C}\mathbf{X}(z). \tag{3.19b}$$

The solution of the complex domain description of the IIO system

results from (3.14a) and (3.14b)

- in a *forced regime* under **nonzero initial conditions**

$$\mathbf{R}(z) = \left(\sum_{r=0}^{r=\alpha} \boldsymbol{Q}_r z^r\right)^{-1} \left\{\sum_{r=0}^{r=\alpha} \boldsymbol{Q}_r \left[\sum_{k=0}^{k=r-1} z^{r-k}\mathbf{R}(k)\right] + \right.$$
$$\left. + \sum_{r=0}^{r=\beta\leq\alpha} \boldsymbol{P}_r \left[z^r\mathbf{I}(z) - \sum_{k=0}^{k=r-1} z^{r-k}\mathbf{I}(k)\right]\right\}, \tag{3.20a}$$

$$\mathbf{Y}(z) = \left(\sum_{r=0}^{r=\nu} \boldsymbol{E}_r z^r\right)^{-1} \left\{\sum_{r=0}^{r=\nu} \boldsymbol{E}_r \left[\sum_{k=0}^{k=r-1} z^{r-k}\mathbf{Y}(k)\right] + \right.$$
$$+ \sum_{r=0}^{r=\alpha} \boldsymbol{R}_r \left[z^r\mathbf{R}(z) - \sum_{k=0}^{k=r-1} z^{r-k}\mathbf{R}(k)\right] +$$
$$\left. + \sum_{r=0}^{r=\mu} \boldsymbol{T}_r \left[z^r\mathbf{I}(z) - \sum_{k=0}^{k=r-1} z^{r-k}\mathbf{I}(k)\right]\right\}, \tag{3.20b}$$

- in a *free regime* under **nonzero initial conditions**

$$\mathbf{R}(z) = \left(\sum_{r=0}^{r=\alpha} \boldsymbol{Q}_r z^r\right)^{-1} \left\{\sum_{r=0}^{r=\alpha} \boldsymbol{Q}_r \left[\sum_{k=0}^{k=r-1} z^{r-k}\mathbf{R}(k)\right]\right\} \tag{3.21a}$$

$$\mathbf{Y}(z) = \left(\sum_{r=0}^{r=\nu} \boldsymbol{E}_r z^r\right)^{-1} \left\{\sum_{r=0}^{r=\nu} \boldsymbol{E}_r \left[\sum_{k=0}^{k=r-1} z^{r-k}\mathbf{Y}(k)\right] + \right.$$
$$\left. + \sum_{r=0}^{r=\alpha} \boldsymbol{R}_r \left[z^r\mathbf{R}(z) - \sum_{k=0}^{k=r-1} z^{r-k}\mathbf{R}(k)\right]\right\}. \tag{3.21b}$$

Although the complex domain mathematical models of the systems simplify in free regimes, they still have complex forms because of double sums of the terms containing initial conditions.

Let us accept that **all initial conditions are equal to zero.**

The complex domain description for zero initial conditions

from (3.10) of the *IO* system

- in a *forced regime* under **zero initial conditions**

$$\mathbf{Y}(z) = \left[\left(\sum_{r=0}^{r=\nu} \boldsymbol{A}_r z^r\right)^{-1} \left(\sum_{r=0}^{r=\mu} \boldsymbol{B}_r z^r\right)\right] \mathbf{I}(z), \tag{3.22}$$

- in a *free regime* under **zero initial conditions**

$$\mathbf{Y}(z) \equiv \mathbf{0}, \tag{3.23}$$

from (3.12a) and (3.12b) of the *ISO* system

- in a *forced regime* under **zero initial conditions**

$$\mathbf{X}(z) = \left[(z\boldsymbol{I} - \boldsymbol{A})^{-1} \boldsymbol{B}\right] \mathbf{I}(z), \tag{3.24a}$$

$$\mathbf{Y}(z) = \boldsymbol{C}\mathbf{X}(z) + \boldsymbol{D}\mathbf{I}(z), \tag{3.24b}$$

- in a *free regime* under **zero initial conditions**

$$\mathbf{X}(z) \equiv \mathbf{0}, \tag{3.25a}$$

$$\mathbf{Y}(z) \equiv \mathbf{0}, \tag{3.25b}$$

from (3.14a) and (3.14b) of the *IIO* system

- in a *forced regime* under **zero initial conditions**

$$\mathbf{R}(z) = \left[\left(\sum_{r=0}^{r=\alpha} \boldsymbol{Q}_r z^r\right)^{-1} \left(\sum_{r=0}^{r=\beta\leq\alpha} \boldsymbol{P}_r z^r\right)\right] \mathbf{I}(z), \tag{3.26a}$$

$$\mathbf{Y}(z) = \left(\sum_{r=0}^{r=\nu} \boldsymbol{E}_r z^r\right)^{-1} \left[\left(\sum_{r=0}^{r=\alpha} \boldsymbol{R}_r z^r\right) \mathbf{R}(z) + \right.$$
$$\left. + \left(\sum_{r=0}^{r=\mu} \boldsymbol{T}_r z^r\right) \mathbf{I}(z)\right], \tag{3.26b}$$

- in a *free regime* under **zero initial conditions**

$$\mathbf{R}(z) \equiv \mathbf{0}, \tag{3.27a}$$

$$\mathbf{Y}(z) \equiv \mathbf{0}. \tag{3.27b}$$

3.3.4 Basic problem

We can conclude that for a system in a forced regime all the preceding cases, in the forced regime, the relationships between $Z-$transforms of the output and input vectors are determined only by system parameters, whatever the form of the input vector function is.

This raises the following basic problem to be solved in this book:

Problem 3.1 ***The basic problem***

What is the complex domain description of the system such that the relationship between the $Z-$transforms of the output and input vectors is determined only by system parameters, regardless of both the form of the input vector function, and initial conditions?

The solutions to this problem for different classes of the systems form the core of this book. They discover the existence of (the undiscovered until 1993 [34]) the dynamical system characteristic that generalizes the system transfer function matrix $\boldsymbol{G}(z)$. It is ***the system full (complete) transfer function matrix*** denoted by $\boldsymbol{F}(z)$ ([34], [35], [59], [60], [61], [68], [69]). Its use permits us to treat fully and correctly many qualitative dynamical properties as well as quantitative dynamical characteristics of the systems in the complex domain.

3.4 Desired regime

3.4.1 Introduction

What follows deals with system regimes determined according to the following criterion: *the realization of the demanded system behavior.* The criterion has the complete meaning for a system that is an object O (a plant P), which is to be controlled or which is controlled. A desired regime of a system (object/plant) is defined by its demanded, i.e., desired (output) response $\mathbf{Y}_d(\cdot)$.

Definition 3.4 ***Desired regime***

A system is in ***a desired*** *(also called:* ***nominal*** *or* ***nonperturbed****)* ***regime on*** $\mathcal{T}_{d0}$ *i.e.,* $\mathcal{N}_0$ *(for short: in* ***a desired regime****) if and only if it realizes its desired (output) response* $\mathbf{Y}_d(k)$ *all the time,*

$$\mathbf{Y}(k) = \mathbf{Y}_d(k),\ \forall k \in \mathcal{N}_0. \tag{3.28}$$

This definition directly determines a necessary (but not a sufficient) condition for a system to be in a desired (nominal, nonperturbed) regime.

Proposition 3.1 *In order for the plant to be in a desired (nominal, nonperturbed) regime, i.e.,*

$$\mathbf{Y}(k) = \mathbf{Y}_d(k),\ \forall k \in \mathcal{N}_0 \Longrightarrow \mathbf{Y}_0 = \mathbf{Y}_{d0},$$

it is necessary that the initial real output vector is equal to the initial desired output vector,

$$\mathbf{Y}_0 = \mathbf{Y}_{d0}.$$

The system cannot be in a nominal regime (on $\mathcal{N}_0$) if its initial real output vector is different from the initial desired output vector:

$$\mathbf{Y}_0 \neq \mathbf{Y}_{d0} \Longrightarrow \exists k_\sigma \in \mathcal{N}_0 \Longrightarrow \mathbf{Y}(k_\sigma) \neq \mathbf{Y}_d(k_\sigma).$$

The real initial output vector $\mathbf{Y}(0) = \mathbf{Y}_0$ is most often different from the desired initial output vector $\mathbf{Y}_d(0) = \mathbf{Y}_{d0}$. Therefore, the system is most often in a *nondesired* (*nonnominal, perturbed, disturbed*) regime.

Definition 3.5 ***Nominal input***

An input vector function $\mathbf{I}^*(\cdot)$ of a system is ***nominal relative to the desired response*** $\mathbf{Y}_d(\cdot)$, *which is denoted by $\mathbf{I}_N(\cdot)$, if and only if, $\mathbf{I}(\cdot) = \mathbf{I}^*(\cdot)$ ensures that the corresponding real response $\mathbf{Y}(\cdot) = \mathbf{Y}^*(\cdot)$ of the system obeys $\mathbf{Y}^*(k) = \mathbf{Y}_d(k)$ all the time as soon as all the internal and the output system initial conditions are desired (nominal, nonperturbed).*

This is general definition. It specifies forms for different classes of systems, which depend on the internal and output system initial conditions.

Note 3.1 *An input vector function $\mathbf{I}^*(\cdot)$ can be nominal relative to the desired response $\mathbf{Y}_{d1}(\cdot)$ of a system, but it need not be nominal with respect to another desired response $\mathbf{Y}_{d2}(\cdot)$ of the system. This explains the relative sense of the notion "nominal relative to the desired response $\mathbf{Y}_d(\cdot)$."*

Definition 3.6 $\mathbf{Y}_d(\cdot)$ ***realizable in*** $\mathcal{J}$

The desired response $\mathbf{Y}_d(\cdot)$ of the system is ***realizable in*** $\mathcal{J}$ *if and only if there exist $\mathbf{I}^*(\cdot) \in \mathcal{J}$ and the initial conditions $E^0\mathbf{Y}_0^*,\ E^1\mathbf{Y}_0^*, \cdots$ of the output vector function $\mathbf{Y}(\cdot)$ and of its shiftings such that $\mathbf{I}^*(\cdot)$ is nominal relative to $\mathbf{Y}_d(\cdot)$,*

$$\exists \mathbf{I}^*(\cdot) \in \mathcal{J},\ \exists E^0\mathbf{Y}_0^*, E^1\mathbf{Y}_0^*, \cdots \in \mathcal{R}^N \Longrightarrow$$
$$\Longrightarrow \mathbf{Y}\left(k; \mathbf{I}^*; E^0\mathbf{Y}_0^*, E^1\mathbf{Y}_0^*, \cdots\right) = \mathbf{Y}_d(k),\ \forall k \in \mathcal{N}_0.$$

Definition 3.7 $\mathbf{Y}_d(.)$ *realizable on* $\mathcal{J}$

The desired response $\mathbf{Y}_d(\cdot)$ of the system is **realizable on** $\mathcal{J}$ *if and only if for every $\mathbf{I}(\cdot) \in \mathcal{J}$ there exist the initial conditions $E^0\mathbf{Y}_0^*, E^1\mathbf{Y}_0^*, \cdots$ of the output vector function $\mathbf{Y}(\cdot)$ such that $\mathbf{Y}(\cdot;\mathbf{I};E^0\mathbf{Y}_0^*, E^1\mathbf{Y}_0^*, \cdots)$ is equal to the desired output vector function $\mathbf{Y}_d(\cdot)$,*

$$\forall \mathbf{I}^*(\cdot) \in \mathcal{J}, \exists E^0\mathbf{Y}_0^*, E^1\mathbf{Y}_0^*, \cdots \in \mathcal{R}^N \Longrightarrow$$
$$\Longrightarrow \mathbf{Y}\left(k;\mathbf{I};E^0\mathbf{Y}_0^*, E^1\mathbf{Y}_0^*, \cdots\right) = \mathbf{Y}_d(k), \ \forall k \in \mathcal{N}_0.$$

Comment 3.1 *The realizability of $\mathbf{Y}_d(\cdot)$ in $\mathcal{J}$ is necessary, but not sufficient, for the realizability of $\mathbf{Y}_d(\cdot)$ on $\mathcal{J}$. The realizability of $\mathbf{Y}_d(\cdot)$ on $\mathcal{J}$ is sufficient, but not necessary, for the realizability of $\mathbf{Y}_d(\cdot)$ in $\mathcal{J}$.*

Problem 3.2 *Under what conditions there exists a nominal vector function $\mathbf{I}_N(\cdot)$ relative to the system desired (nominal) output response, or equivalently, under what conditions is the system desired output response realizable in $\mathcal{J}$ and/or realizable on $\mathcal{J}$?*

There are qualitative system properties that have a sense if and only if there exists an affirmative solution to the preceding problem. Such properties are Lyapunov stability properties of the desired motion.

3.4.2 *IO* systems

We will accommodate Definition 3.4 and Definition 3.5 (in Subsection 3.4.1) to the *IO* system (plant) (2.1) (in Section 2.1).

In order to present effectively, simply and clearly the complex domain condition for an input vector function to be nominal for the system relative to its desired output vector response, we introduce the following complex matrix functions. The first one is $\boldsymbol{S}_i^{(r)}(\cdot) : \mathcal{C} \longrightarrow \mathcal{C}^{i(r+1)\times i}$, defined by

$$\boldsymbol{S}_i^{(r)}(z) = \left[\begin{array}{ccccc} z^0\boldsymbol{I}_i & z^1\boldsymbol{I}_i & z^2\boldsymbol{I}_i & \cdots & z^r\boldsymbol{I}_i \end{array}\right]^T \in \mathcal{C}^{i(r+1)\times i},$$
$$(r,i) \in \{(\mu, M), \ (\nu, N)\}, \tag{3.29}$$

with $\boldsymbol{I}_i$ being the *i-th* order identity matrix, $\boldsymbol{I}_i \in \mathcal{R}^{i\times i}$. The second one is

$\boldsymbol{Z}_r^{(\varsigma)}(\cdot): \mathcal{C} \rightarrow \mathcal{C}^{(\varsigma+1)r \times \varsigma r}$,

$$\boldsymbol{Z}_r^{(\varsigma)}(z) = \begin{bmatrix} \boldsymbol{O}_r & \boldsymbol{O}_r & \boldsymbol{O}_r & \cdots & \boldsymbol{O}_r \\ z^{1-0}\boldsymbol{I}_r & \boldsymbol{O}_r & \boldsymbol{O}_r & \cdots & \boldsymbol{O}_r \\ z^{2-0}\boldsymbol{I}_r & z^{2-1}\boldsymbol{I}_r & \boldsymbol{O}_r & \cdots & \boldsymbol{O}_r \\ \vdots & \vdots & \vdots & \vdots & \vdots \\ z^{\varsigma-0}\boldsymbol{I}_r & z^{\varsigma-1}\boldsymbol{I}_r & z^{\varsigma-2}\boldsymbol{I}_r & \cdots & z^{\varsigma-(\varsigma-1)}\boldsymbol{I}_r \end{bmatrix}, \varsigma \geq 1,$$
$$\boldsymbol{Z}_r^{(\varsigma)}(z) \in \mathcal{C}^{(\varsigma+1)r \times \varsigma r}, \quad (\varsigma, r) \in \{(\mu, M), \ (\nu, N)\}. \tag{3.30}$$

These complex matrix functions enable us also to resolve effectively the basic Problem 3.1 (in Subsection 3.3.4). They also permit us to solve also Problem 5.1 (Chapter 5).

Note 3.2 *If $\varsigma = 0$, then the matrix $\boldsymbol{Z}_r^{(\varsigma)}(z) = \boldsymbol{Z}_r^{(0)}(z)$ is not defined, it does not exist and it should be omitted rather than replaced by the zero matrix. The matrix $\boldsymbol{Z}_r^{(\varsigma)}(z)$ is not defined for $\zeta \leq 0$ and should be treated as a nonexisting one. Derivatives exist only for natural numbers, i.e., $\mathbf{Y}^{(\varsigma)}(t)$ can exist only for $\varsigma \geq 1$. As shifts $E^{\varsigma}\mathbf{Y}(k)$ are originating from derivatives $\mathbf{Y}^{(\varsigma)}(t)$, this type of shifts can exist only for $\varsigma \geq 1$, too. Matrix function $\boldsymbol{Z}_r^{(\varsigma)}(\cdot)$ is related to the $Z-$transform of shifts originating from derivatives only (see [68], [69]).*

Theorem 3.1 *In order for an input vector function $\mathbf{I}^*(\cdot)$ to be nominal for the IO system (2.1), i.e., for (2.5), relative to its desired response $\mathbf{Y}_d(\cdot): \mathbf{I}^*(\cdot) = \mathbf{I}_N(\cdot)$, it is necessary and sufficient that 1) and 2) hold:*

1) rank $\boldsymbol{B}^{(\mu)} = N \leq M$, equivalently

$$\text{rank} \sum_{r=0}^{r=\mu} \boldsymbol{B}_r z^r = \text{rank}\, \boldsymbol{B}^{(\mu)} \boldsymbol{S}_M^{(\mu)}(z) = N \leq M, \text{ and}$$

2) anyone of the following equations is valid:

$$\sum_{r=0}^{r=\mu} \boldsymbol{B}_r E^r \mathbf{I}^*(k) = \sum_{r=0}^{r=\nu} \boldsymbol{A}_r E^r \mathbf{Y}_d(k), \ \forall k \in \mathcal{N}_0, \tag{3.31a}$$
$$\boldsymbol{B}^{(\mu)} \mathbf{I}^{*\mu}(k) = \boldsymbol{A}^{(\nu)} \mathbf{Y}_d^{\nu}(k), \ \forall k \in \mathcal{N}_0, \tag{3.31b}$$

or equivalently in the complex domain:

$$\mathbf{I}^*(z) = \left(\sum_{r=0}^{r=\mu} \boldsymbol{B}_r z^r\right)^T \left[\left(\sum_{r=0}^{r=\mu} \boldsymbol{B}_r z^r\right) \cdot \left(\sum_{r=0}^{r=\mu} \boldsymbol{B}_r z^r\right)^T\right]^{-1} \cdot$$
$$\cdot \left\{\sum_{r=0}^{r=\mu} \boldsymbol{B}_r \left[\sum_{k=0}^{k=r-1} z^{r-k}\mathbf{I}^*(k)\right] + \right.$$
$$\left. + \sum_{r=0}^{r=\nu} \boldsymbol{A}_r \left[z^r \mathbf{Y}_d(z) - \sum_{k=0}^{k=r-1} z^{r-k}\mathbf{Y}_d(k)\right]\right\}, \tag{3.32a}$$

i.e.,

$$\mathbf{I}^*(z) = \left(\boldsymbol{B}^{(\mu)} \boldsymbol{S}_M^{(\mu)}(z)\right)^T \cdot$$
$$\cdot \left[\left(\boldsymbol{B}^{(\mu)} \boldsymbol{S}_M^{(\mu)}(z)\right) \cdot \left(\boldsymbol{B}^{(\mu)} \boldsymbol{S}_M^{(\mu)}(z)\right)^T\right]^{-1} \left\{\boldsymbol{B}^{(\mu)} \boldsymbol{Z}_M^{(\mu)}(z) \mathbf{I}^{*^{\mu-1}}(0) + \right.$$
$$\left. + \boldsymbol{A}^{(\nu)} \left[\boldsymbol{S}_N^{(\nu)}(z)\mathbf{Y}_d(z) - \boldsymbol{Z}_N^{(\nu)}(z)\mathbf{Y}_d^{\nu-1}(0)\right]\right\}. \tag{3.32b}$$

Proof. *Necessity*: Let a vector function $\mathbf{I}^*(\cdot)$ be nominal for the *IO* system (2.1), i.e., for (2.5), relative to its desired response $\mathbf{Y}_d(\cdot)$. Definition 3.5 (in Subsection 3.4.1) holds. Holding the definition 3.5 and (2.1), i.e., (2.5), imply

$$\begin{aligned} \sum_{r=0}^{r=\nu} \boldsymbol{A}_r E^r \mathbf{Y}_d(k) &= \sum_{r=0}^{r=\mu} \boldsymbol{B}_r E^r \mathbf{I}^*(k), \ \forall k \in \mathcal{N}_0, \\ \boldsymbol{A}^{(\nu)} \mathbf{Y}_d^{\nu}(k) &= \boldsymbol{B}^{(\mu)} \mathbf{I}^{*^{\mu}}(k), \ \forall k \in \mathcal{N}_0, \end{aligned}$$

These equations are (3.31a) and (3.31b) in another forms, respectively. Their $Z-$transforms solved in $\mathbf{I}^*(z)$ are given in (3.32a) and (3.32b), respectively. Since they are solvable in $\mathbf{I}^*(z)$ it follows that the conditions 1) and 2) hold.

Sufficiency: Let the conditions 1) and 2) hold. The input vector function $\mathbf{I}^*(\cdot)$ to the *IO* system (2.1), i.e., (2.5),

$$\boldsymbol{A}^{(\nu)} \mathbf{Y}^{\nu}(k) = \boldsymbol{B}^{(\mu)} \mathbf{I}^{*^{\mu}}(k), \ \forall k \in \mathcal{N}_0,$$

satisfies (3.31a), hence (3.31b):

$$\boldsymbol{A}^{(\nu)} \mathbf{Y}_d^{\nu}(k) = \boldsymbol{B}^{(\mu)} \mathbf{I}^{*^{\mu}}(k), \ \forall k \in \mathcal{N}_0. \tag{3.33}$$

These equations and

$$\boldsymbol{\varepsilon} = \mathbf{Y}_d - \mathbf{Y} \tag{3.34}$$

yield

$$\boldsymbol{A}^{(\nu)}\boldsymbol{\varepsilon}^{\nu}(k) = \mathbf{0}_N,\ \forall k \in \mathcal{N}_0. \tag{3.35}$$

Definition 3.5 requires $\boldsymbol{\varepsilon}^{\nu}(0) = \mathbf{0}_{N(\nu+1)}$, which implies the trivial solution $\boldsymbol{\varepsilon}(k) = \mathbf{0}_N$, $\forall k \in \mathcal{N}_0$, of (3.35). This and $\boldsymbol{\varepsilon} = \mathbf{Y}_d - \mathbf{Y}$ prove $\mathbf{Y}(k) = \mathbf{Y}_d(k)$, $\forall k \in \mathcal{N}_0$.

Let the input vector function $\mathbf{I}^*(\cdot)$ to the *IO* system (2.1), i.e., to (2.5), obeys (3.32a), equivalently (3.32b). The $Z-$transforms of (2.1) and (2.5) read for the input vector function $\mathbf{I}(\cdot) = \mathbf{I}^*(\cdot)$:

$$\mathbf{I}^*(z) = \left(\sum_{r=0}^{r=\mu}\boldsymbol{B}_r z^r\right)^T \left[\left(\sum_{r=0}^{r=\mu}\boldsymbol{B}_r z^r\right)\cdot\left(\sum_{r=0}^{r=\mu}\boldsymbol{B}_r z^r\right)^T\right]^{-1}\cdot$$
$$\cdot\left\{\sum_{r=0}^{r=\mu}\boldsymbol{B}_r\left[\sum_{k=0}^{k=r-1} z^{r-k}\mathbf{I}^*(k)\right] + \right.$$
$$\left. +\sum_{r=0}^{r=\nu}\boldsymbol{A}_r\left[z^r\mathbf{Y}(z) - \sum_{k=0}^{k=r-1} z^{r-k}\mathbf{Y}(k)\right]\right\},$$

i.e.,

$$\mathbf{I}^*(z) = \left(\boldsymbol{B}^{(\mu)}\boldsymbol{S}_M^{(\mu)}(z)\right)^T\left[\left(\boldsymbol{B}^{(\mu)}\boldsymbol{S}_M^{(\mu)}(z)\right)\cdot\left(\boldsymbol{B}^{(\mu)}\boldsymbol{S}_M^{(\mu)}(z)\right)^T\right]^{-1}\cdot$$
$$\cdot\left\{\boldsymbol{B}^{(\mu)}\boldsymbol{Z}_M^{(\mu)}(z)\mathbf{I}^{*\mu-1}(0) + \boldsymbol{A}^{(\nu)}\left[\boldsymbol{S}_N^{(\nu)}(z)\mathbf{Y}(z) - \boldsymbol{Z}_N^{(\nu)}(z)\mathbf{Y}^{\nu-1}(0)\right]\right\}.$$

These equations multiplied on the left by $\left(\sum_{r=0}^{r=\mu}\boldsymbol{B}_r z^r\right)$, i.e., by $\left(\boldsymbol{B}^{(\mu)}\boldsymbol{S}_M^{(\mu)}(z)\right)$, respectively, and (3.32a) and (3.32b) imply, respectively,

$$\sum_{r=0}^{r=\nu}\boldsymbol{A}_r\left[z^r\mathbf{Y}_d(z) - \sum_{k=0}^{k=r-1} z^{r-k}\mathbf{Y}_d(k)\right] =$$
$$= \sum_{r=0}^{r=\nu}\boldsymbol{A}_r\left[z^r\mathbf{Y}(z) - \sum_{k=0}^{k=r-1} z^{r-k}\mathbf{Y}(k)\right]$$

and

$$\boldsymbol{A}^{(\nu)}\left[\boldsymbol{S}_N^{(\nu)}(z)\mathbf{Y}_d(z) - \boldsymbol{Z}_N^{(\nu)}(z)\mathbf{Y}_d^{\nu-1}(0)\right] =$$
$$= \boldsymbol{A}^{(\nu)}\left[\boldsymbol{S}_N^{(\nu)}(z)\mathbf{Y}(z) - \boldsymbol{Z}_N^{(\nu)}(z)\mathbf{Y}^{\nu-1}(0)\right].$$

Definition 3.5 requires $\mathbf{Y}^{\nu-1}(0) = \mathbf{Y}_d^{\nu-1}(0)$ that reduces the preceding equations to

$$\sum_{r=0}^{r=\nu} \boldsymbol{A}_r z^r \left[\mathbf{Y}_d(z) - \mathbf{Y}(z)\right] = \sum_{r=0}^{r=\nu} \boldsymbol{A}_r z^r \boldsymbol{\varepsilon}(z) = \mathbf{0}_N,$$

$$\boldsymbol{A}^{(\nu)} \boldsymbol{S}_N^{(\nu)}(z) \left[\mathbf{Y}_d(z) - \mathbf{Y}(z)\right] = \boldsymbol{A}^{(\nu)} \boldsymbol{S}_N^{(\nu)}(z) \boldsymbol{\varepsilon}(z) = \mathbf{0}_N.$$

These equations imply $\boldsymbol{\varepsilon}(z) = \mathbf{0}_N$ due to Condition 2.1 (in Section 2.1), hence $\boldsymbol{\varepsilon}(z) = \mathbf{0}_N$, $\forall z \in \mathcal{C}$, which is equivalent to $\boldsymbol{\varepsilon}(k) = \mathbf{0}_N$, $\forall k \in \mathcal{N}_0$, i.e., $\mathbf{Y}(k) = \mathbf{Y}_d(k)$, $\forall k \in \mathcal{N}_0$. ■

This is general theorem for all *IO* systems (2.1), i.e., (2.5).

Note 3.3 *We accept that $N \leq M$ holds in the sequel.*

3.4.3 *ISO* systems

Definition 3.5 (in Subsection 3.4.1) takes the following form in the framework of the *ISO* systems (2.14a) and (2.14b) (in Section 2.2).

Definition 3.8 *A functional vector pair $[\mathbf{I}^*(\cdot), \mathbf{X}^*(\cdot)]$ is* ***nominal for the ISO system (2.14a) and (2.14b) relative to its desired response*** *$\mathbf{Y}_d(\cdot)$, which is denoted by $[\mathbf{I}_N(\cdot), \mathbf{X}_N(\cdot)]$, if and only if $[\mathbf{I}(\cdot), \mathbf{X}(\cdot)] = [\mathbf{I}^*(\cdot), \mathbf{X}^*(\cdot)]$ ensures that the corresponding real response $\mathbf{Y}(\cdot) = \mathbf{Y}^*(\cdot)$ of the system obeys $\mathbf{Y}^*(k) = \mathbf{Y}_d(k)$ all the time,*

$$[\mathbf{I}^*(\cdot), \mathbf{X}^*(\cdot)] = [\mathbf{I}_N(\cdot), \mathbf{X}_N(\cdot)] \Longleftrightarrow \langle \mathbf{Y}^*(k) = \mathbf{Y}_d(k), \ \forall k \in \mathcal{N}_0 \rangle .$$

The time evolution $\mathbf{X}_N(k; \mathbf{X}_{N0}; \mathbf{I}_N)$, $\mathbf{X}_N(0; \mathbf{X}_{N0}; \mathbf{I}_N) \equiv \mathbf{X}_{N0}$, of the nominal state vector $\mathbf{X}_N$ is ***the desired motion*** *$\mathbf{X}_d(\cdot; \mathbf{X}_{d0}; \mathbf{I}_N)$* ***of the ISO system (2.14a) and (2.14b) relative to its desired response*** *$\mathbf{Y}_d(\cdot)$, for short,* ***the desired motion****,*

$$\mathbf{X}_d(k; \mathbf{X}_{d0}; \mathbf{I}_N) \equiv \mathbf{X}_N(k; \mathbf{X}_{N0}; \mathbf{I}_N),$$
$$\mathbf{X}_d(0; \mathbf{X}_{d0}; \mathbf{I}_N) \equiv \mathbf{X}_{d0} \equiv \mathbf{X}_{N0}. \tag{3.36}$$

Let us remind that $\boldsymbol{I}$ is the identity matrix of the dimension $n : \boldsymbol{I}_n = \boldsymbol{I}$. The system (2.14a), (2.14b) suggests the introduction of the following matrix:

$$\begin{bmatrix} -\boldsymbol{B} & z\boldsymbol{I} - \boldsymbol{A} \\ \boldsymbol{D} & \boldsymbol{C} \end{bmatrix} \in \mathcal{C}^{(N+n)\times(M+n)}.$$

Theorem 3.2 *In order for a functional vector pair* $[\mathbf{I}^*(\cdot), \mathbf{X}^*(\cdot)]$ *to be nominal for the ISO system (2.14a) and (2.14b) relative to its desired response* $\mathbf{Y}_d(\cdot)$,

$$[\mathbf{I}^*(\cdot), \mathbf{X}^*(\cdot)] = [\mathbf{I}_N(\cdot), \mathbf{X}_N(\cdot)],$$

it is necessary and sufficient that it obeys the following equations:

$$-\boldsymbol{B}\mathbf{I}^*(k) + \mathbf{X}^*(k+1) - \boldsymbol{A}\mathbf{X}^*(k) = \mathbf{0}_n,\ \forall k \in \mathcal{N}_0, \tag{3.37a}$$

$$\boldsymbol{D}\mathbf{I}^*(k) + \boldsymbol{C}\mathbf{X}^*(k) = \mathbf{Y}_d(k),\ \forall k \in \mathcal{N}_0, \tag{3.37b}$$

or equivalently,

$$\begin{bmatrix} -\boldsymbol{B} & z\boldsymbol{I} - \boldsymbol{A} \\ \boldsymbol{D} & \boldsymbol{C} \end{bmatrix} \begin{bmatrix} \mathbf{I}^*(z) \\ \mathbf{X}^*(z) \end{bmatrix} = \begin{bmatrix} z\mathbf{X}_0^* \\ \mathbf{Y}_d(z) \end{bmatrix}. \tag{3.38}$$

Proof. *Necessity*: Let $[\mathbf{I}^*(\cdot), \mathbf{X}^*(\cdot)]$ be a nominal functional (input and state) vector pair for the *ISO* system (2.14a) and (2.14b) relative to its desired response $\mathbf{Y}_d(\cdot)$. Hence, the system is in its desired regime relative to $\mathbf{Y}_d(\cdot)$. Definition 3.8 shows that $[\mathbf{I}(\cdot), \mathbf{X}(\cdot)] = [\mathbf{I}^*(\cdot), \mathbf{X}^*(\cdot)]$ implies $\mathbf{Y}(\cdot) = \mathbf{Y}_d(\cdot)$. This and the *ISO* model (2.14a) and (2.14b) yield the following equations:

$$\mathbf{X}^*(k+1) = \boldsymbol{A}\mathbf{X}^*(k) + \boldsymbol{B}\mathbf{I}^*(k),\ \forall k \in \mathcal{N}_0, \tag{3.39a}$$

$$\mathbf{Y}(k) = \mathbf{Y}_d(k) = \boldsymbol{C}\mathbf{X}^*(k) + \boldsymbol{D}\mathbf{I}^*(k),\ \forall k \in \mathcal{N}_0, \tag{3.39b}$$

which can be easily set in the form of Equations (3.37a) and (3.37b). Application of the $Z-$transform together with its properties transforms Equations (3.37a) and (3.37b) into (3.38).

Sufficiency. We accept that all the conditions of the theorem are valid. We chose $[\mathbf{I}(\cdot), \mathbf{X}(\cdot)] = [\mathbf{I}^*(\cdot), \mathbf{X}^*(\cdot)]$. Equation (3.37a) written in the normal state form,

$$\mathbf{X}^*(k+1) = \boldsymbol{A}\mathbf{X}^*(k) + \boldsymbol{B}\mathbf{I}^*(k),\ \forall k \in \mathcal{N}_0,$$

shows that the pair $[\mathbf{I}^*(\cdot), \mathbf{X}^*(\cdot)]$ satisfies (2.14a). Furthermore, Equation (2.14b) takes the following form:

$$\mathbf{Y}(k) = \boldsymbol{C}\mathbf{X}^*(k) + \boldsymbol{D}\mathbf{I}^*(k),\ \forall k \in \mathcal{N}_0.$$

It, subtracted from (3.37b), yields

$$\mathbf{Y}(k) - \mathbf{Y}_d(k) = \mathbf{0},\ \forall k \in \mathcal{N}_0,$$

i.e.,

$$\mathbf{Y}(k) = \mathbf{Y}_d(k), \ \forall k \in \mathcal{N}_0.$$

The $Z-$transform of Equations (3.37a) and (3.37b) yield (3.38). This completes the proof. ■

We can freely choose the initial state vector $\mathbf{X}^*(0)$. Let us consider the conditions for the existence of the solutions of Equations (3.37a) and (3.37b), or equivalently (3.38). Notice that there exist $(M+n)$ unknown variables in $(N+n)$ Equations (3.37a), (3.37b). The unknown variables are the entries of $\mathbf{I}^*(z) \in \mathcal{C}^M$ and $\mathbf{X}^*(z) \in \mathcal{C}^n$. The following three different cases of the relationship between M and N exist:

Case 3.1 $N > M$

If $N > M$, then Equations (3.37a) and (3.37b), or equivalently (3.38), do not have a solution. The number of the unknown variables is less than the number of the available equations.

Let us consider the case that seems contrary to this assertion: rank $\boldsymbol{D} = M$, *which implies* $\det\left(\boldsymbol{D}^T\boldsymbol{D}\right) \neq 0$. *This creates an impression that we can find* $\mathbf{I}^*(z)$ *and* $\mathbf{X}^*(z)$ *from (3.38):*

$$\mathbf{I}^*(z) = \left(\boldsymbol{D}^T\boldsymbol{D}\right)^{-1}\left[-\boldsymbol{D}^T\boldsymbol{C}\mathbf{X}^*(z) + \boldsymbol{D}^T\mathbf{Y}_d(z)\right],$$
$$\left[z\boldsymbol{I} - \boldsymbol{A} + \boldsymbol{B}\left(\boldsymbol{D}^T\boldsymbol{D}\right)^{-1}\boldsymbol{D}^T\boldsymbol{C}\right]\mathbf{X}^*(z) = \boldsymbol{B}\left(\boldsymbol{D}^T\boldsymbol{D}\right)^{-1}\boldsymbol{D}^T\mathbf{Y}_d(z) + z\mathbf{X}_0^*.$$

In order for these solutions to satisfy

$$\mathbf{Y}^*(z) = \boldsymbol{C}\mathbf{X}^*(z) + \boldsymbol{D}\mathbf{I}^*(z) = \mathbf{Y}_d(z)$$

the following condition should hold:

$$\boldsymbol{D}\left(\boldsymbol{D}^T\boldsymbol{D}\right)^{-1}\boldsymbol{D}^T = \boldsymbol{I}_N.$$

In view of rank$\boldsymbol{D} = M < N$, *we accept* $\boldsymbol{D} \in \mathcal{R}^{N\times M}$ *in the following form:*

$$\boldsymbol{D} = \begin{bmatrix} \widetilde{\boldsymbol{D}} \\ \boldsymbol{O} \end{bmatrix} \in \mathcal{R}^{N\times M}, \widetilde{\boldsymbol{D}} \in \mathcal{R}^{M\times M}, \ \text{rank}\,\widetilde{\boldsymbol{D}} = M, \ \boldsymbol{O} = \boldsymbol{O}_{(N-M),M},$$

so that the condition $\mathbf{D}(\mathbf{D}^T\mathbf{D})^{-1}\mathbf{D}^T = \mathbf{I}_N$ *reads*

$$\mathbf{D}\left(\mathbf{D}^T\mathbf{D}\right)^{-1}\mathbf{D}^T = \begin{bmatrix} \widetilde{\mathbf{D}} \\ \mathbf{O} \end{bmatrix} \left\{ \left(\begin{bmatrix} \widetilde{\mathbf{D}} \\ \mathbf{O} \end{bmatrix}^T \begin{bmatrix} \widetilde{\mathbf{D}} \\ \mathbf{O} \end{bmatrix} \right)^{-1} \right\} \begin{bmatrix} \widetilde{\mathbf{D}} \\ \mathbf{O} \end{bmatrix}^T =$$

$$= \begin{bmatrix} \widetilde{\mathbf{D}} \\ \mathbf{O} \end{bmatrix} \left(\widetilde{\mathbf{D}}^T \widetilde{\mathbf{D}} \right)^{-1} \begin{bmatrix} \widetilde{\mathbf{D}} \\ \mathbf{O} \end{bmatrix}^T =$$

$$= \begin{bmatrix} \widetilde{\mathbf{D}}\left(\widetilde{\mathbf{D}}^T\widetilde{\mathbf{D}}\right)^{-1}\widetilde{\mathbf{D}}^T & \mathbf{O}_{M,(N-M)} \\ \mathbf{O}_{(N-M),M} & \mathbf{O}_{(N-M),(N-M)} \end{bmatrix} \neq \mathbf{I}_N.$$

The condition $\mathbf{D}(\mathbf{D}^T\mathbf{D})^{-1}\mathbf{D}^T = \mathbf{I}_N$ *is unrealizable. Hence, if* $N > M$ *then Equations (3.37a) and (3.37b), or equivalently (3.38), do not have a solution. If the number* N *of the output variables, i.e., the dimension* N *of the output vector, is bigger than the number* M *of the input variables, i.e., the dimension* M *of the input vector, then there is not a nominal functional vector pair* $[\mathbf{I}^*(\cdot), \mathbf{X}^*(\cdot)]$ *for the ISO system (2.14a) and (2.14b) relative to its desired response* $\mathbf{Y}_d(\cdot)$.

In general, the following should hold if $N > M$*:*

$$\underbrace{\begin{bmatrix} -\mathbf{B} & z\mathbf{I}-\mathbf{A} \\ \mathbf{D} & \mathbf{C} \end{bmatrix}}_{(N+n)\times(M+n)} \underbrace{\left\{ \begin{bmatrix} -\mathbf{B} & z\mathbf{I}-\mathbf{A} \\ \mathbf{D} & \mathbf{C} \end{bmatrix}^T \begin{bmatrix} -\mathbf{B} & z\mathbf{I}-\mathbf{A} \\ \mathbf{D} & \mathbf{C} \end{bmatrix} \right\}^{-1}}_{(M+n)\times(M+n)} \cdot$$

$$\cdot \underbrace{\begin{bmatrix} -\mathbf{B} & z\mathbf{I}-\mathbf{A} \\ \mathbf{D} & \mathbf{C} \end{bmatrix}^T}_{(M+n)\times(N+n)} = \mathbf{I}_{N+n}, \ \forall z \in \mathcal{C},$$

for the existence of a solution $[\mathbf{I}^*(k), \mathbf{X}^*(k)]$ *to (3.37a) and (3.37b), i.e., to (3.38), which is impossible.*

Case 3.2 $N \leq M$

If $N \leq M$ *and*

$$\text{rank} \begin{bmatrix} -\mathbf{B} & z\mathbf{I}-\mathbf{A} \\ \mathbf{D} & \mathbf{C} \end{bmatrix} = N + n \leq M + n$$

for all complex numbers z *for which* $\det(zI - \mathbf{A}) \neq 0$,

then

$$\det \left\{ \begin{bmatrix} -\mathbf{B} & z\mathbf{I}-\mathbf{A} \\ \mathbf{D} & \mathbf{C} \end{bmatrix} \begin{bmatrix} -\mathbf{B} & z\mathbf{I}-\mathbf{A} \\ \mathbf{D} & \mathbf{C} \end{bmatrix}^T \right\} \neq 0.$$

Equations (3.37a) and (3.37b), or equivalently (3.38), have the solution determined by

$$\begin{bmatrix} \mathbf{I}^*(z) \\ \mathbf{X}^*(z) \end{bmatrix} = \begin{bmatrix} -\boldsymbol{B} & z\boldsymbol{I}-\boldsymbol{A} \\ \boldsymbol{D} & \boldsymbol{C} \end{bmatrix}^T \cdot$$

$$\cdot \left\{ \begin{bmatrix} -\boldsymbol{B} & z\boldsymbol{I}-\boldsymbol{A} \\ \boldsymbol{D} & \boldsymbol{C} \end{bmatrix} \begin{bmatrix} -\boldsymbol{B} & z\boldsymbol{I}-\boldsymbol{A} \\ \boldsymbol{D} & \boldsymbol{C} \end{bmatrix}^T \right\}^{-1} \begin{bmatrix} z\mathbf{X}_0^* \\ \mathbf{Y}_d(z) \end{bmatrix}.$$

Case 3.3 $N = M$

If $N = M$ *and*

$$\text{rank} \begin{bmatrix} -\boldsymbol{B} & z\boldsymbol{I}-\boldsymbol{A} \\ \boldsymbol{D} & \boldsymbol{C} \end{bmatrix} = N + n = M + n$$

for all complex numbers z *for which* $\det(zI - \boldsymbol{A}) \neq 0$,

then Equations (3.37a) and (3.37b), or equivalently (3.38), have the unique solution determined by

$$\begin{bmatrix} \mathbf{I}^*(z) \\ \mathbf{X}^*(z) \end{bmatrix} = \begin{bmatrix} -\boldsymbol{B} & z\boldsymbol{I}-\boldsymbol{A} \\ \boldsymbol{D} & \boldsymbol{C} \end{bmatrix}^{-1} \begin{bmatrix} z\mathbf{X}_0^* \\ \mathbf{Y}_d(z) \end{bmatrix}.$$

Conclusion 3.1 *In order to exist a nominal functional vector pair* $[\mathbf{I}_N(\cdot), \mathbf{X}_N(\cdot)]$ *for system (2.14a) and (2.14b) relative to its desired response* $\mathbf{Y}_d(\cdot)$ *it is necessary and sufficient that the conditions of Case 3.2 hold. The functional vector pair* $[\mathbf{I}_N(\cdot), \mathbf{X}_N(\cdot)]$ *is nominal relative to the desired response* $\mathbf{Y}_d(\cdot)$ *of system (2.14a) and (2.14b).*

We have resolved completely both the problem of the existence of a nominal functional vector pair $[\mathbf{I}_N(\cdot), \mathbf{X}_N(\cdot)]$ for the *ISO* system (2.14a) and (2.14b) relative to its desired response $\mathbf{Y}_d(\cdot)$, and the problem of the realizability of $\mathbf{Y}_d(\cdot)$. The realizability of $\mathbf{Y}_d(\cdot)$ is independent of the characteristics of $\mathbf{Y}_d(\cdot)$, and it is determined only by the system parameters, i.e., by the system realization $(\boldsymbol{A}, \boldsymbol{B}, \boldsymbol{C}, \boldsymbol{D})$.

Note 3.4 *We accept that the conditions of Case 3.2 for the realizability of their desired output vector functions* $\mathbf{Y}_d(\cdot)$ *hold in the sequel.*

3.4.4 *IIO* systems

Definition 3.5 (in Subsection 3.4.1) to the *IIO* system (2.18a) and (2.18b) (in Section 2.3) imply the following:

Definition 3.9 *A functional vector pair* $[\mathbf{I}^*(\cdot), \mathbf{R}^*(\cdot)]$ *is* ***nominal for the*** *IIO system (2.18a) and (2.18b)* ***relative to its desired response*** $\mathbf{Y}_d(\cdot)$*, which is denoted by* $[\mathbf{I}_N(\cdot), \mathbf{R}_N(\cdot)]$*, if and only if,* $[\mathbf{I}(\cdot), \mathbf{R}(\cdot)] = [\mathbf{I}^*(\cdot), \mathbf{R}^*(\cdot)]$ *ensures that the corresponding real response* $\mathbf{Y}(\cdot) = \mathbf{Y}^*(\cdot)$ *of the system obeys* $\mathbf{Y}^*(k) = \mathbf{Y}_d(k)$ *all the time as soon as* $\mathbf{Y}_0^{\nu-1} = \mathbf{Y}_{d0}^{\nu-1}$,

$$[\mathbf{I}^*(\cdot), \mathbf{R}^*(\cdot)] = [\mathbf{I}_N(\cdot), \mathbf{R}_N(\cdot)] \Longleftrightarrow$$
$$\Longleftrightarrow \left\langle \mathbf{Y}_0^{\nu-1} = \mathbf{Y}_{d0}^{\nu-1} \Longrightarrow \mathbf{Y}^*(k) = \mathbf{Y}_d(k), \forall k \in \mathcal{N}_0 \right\rangle.$$

Let

$$\mathbf{w}_1(z) = \sum_{r=0}^{r=\beta\leq\alpha} \boldsymbol{P}_r \left[\sum_{k=0}^{k=r-1} z^{r-k} E^k \mathbf{I}_0^* \right] - $$
$$-\sum_{r=0}^{r=\alpha} \boldsymbol{Q}_r \left[\sum_{k=0}^{k=r-1} z^{r-k} E^k \mathbf{R}_0^* \right], \tag{3.40a}$$

$$\mathbf{w}_2(z) = \sum_{r=0}^{r=\nu} \boldsymbol{E}_r \left[z^r \mathbf{Y}_d(z) - \sum_{k=0}^{k=r-1} z^{r-k} E^k \mathbf{Y}_{d0} \right] +$$
$$+\sum_{r=0}^{r=\alpha} \boldsymbol{R}_r \left[\sum_{k=0}^{k=r-1} z^{r-k} E^k \mathbf{R}_0^* \right] + \sum_{r=0}^{r=\mu} \boldsymbol{T}_r \left[\sum_{k=0}^{k=r-1} z^{r-k} E^k \mathbf{I}_0^* \right]. \tag{3.40b}$$

Theorem 3.3 *In order for a functional vector pair* $[\mathbf{I}^*(\cdot), \mathbf{R}^*(\cdot)]$ *to be nominal for the IIO system (2.18a) and (2.18b) relative to its desired response* $\mathbf{Y}_d(\cdot)$,

$$[\mathbf{I}^*(\cdot), \mathbf{R}^*(\cdot)] = [\mathbf{I}_N(\cdot), \mathbf{R}_N(\cdot)],$$

it is necessary and sufficient that it obeys the following equations:

$$\sum_{r=0}^{r=\beta\leq\alpha} \boldsymbol{P}_r E^r \mathbf{I}^*(k) - \sum_{r=0}^{r=\alpha} \boldsymbol{Q}_r E^r \mathbf{R}^*(k) =$$
$$= \boldsymbol{P}^{(\beta)} \mathbf{I}^{*\beta}(k) - \boldsymbol{Q}^{(\alpha)} \mathbf{R}^{*\alpha}(k) = \mathbf{0}, \forall k \in \mathcal{N}_0, \tag{3.41a}$$

$$\sum_{r=0}^{r=\alpha} \boldsymbol{R}_r E^r \mathbf{R}^*(k) + \sum_{r=0}^{r=\mu} \boldsymbol{T}_r E^r \mathbf{I}^*(k) = \boldsymbol{R}^{(\alpha)} \mathbf{R}^{*\alpha}(k) + \boldsymbol{T}^{(\mu)} \mathbf{I}^{*\mu}(k) =$$

$$= \sum_{r=0}^{r=\nu} \boldsymbol{E}_r E^r \mathbf{Y}_d(k) = \boldsymbol{E}^{(\nu)} \mathbf{Y}_d^{\nu}(k), \forall k \in \mathcal{N}_0, \tag{3.41b}$$

or equivalently,

$$\begin{bmatrix} \boldsymbol{P}^{(\beta)}\boldsymbol{S}_M^{(\beta)}(z) & -\boldsymbol{Q}^{(\alpha)}\boldsymbol{S}_\rho^{(\alpha)}(z) \\ \boldsymbol{T}^{(\mu)}\boldsymbol{S}_M^{(\mu)}(z) & \boldsymbol{R}^{(\alpha)}\boldsymbol{S}_\rho^{(\alpha)}(z) \end{bmatrix} \begin{bmatrix} \mathbf{I}^*(z) \\ \mathbf{R}^*(z) \end{bmatrix} = \begin{bmatrix} \mathbf{w}_1(z) \\ \mathbf{w}_2(z) \end{bmatrix}. \tag{3.42}$$

Proof. *Necessity*: Let $[\mathbf{I}^*(\cdot), \mathbf{R}^*(\cdot)]$ be a nominal functional (input and internal substate) vector pair for the *IIO* system (2.18a) and (2.18b) relative to its desired response $\mathbf{Y}_d(\cdot)$. Hence, the system is in its desired regime relative to $\mathbf{Y}_d(\cdot)$. Definition 3.9 shows that $[\mathbf{I}(\cdot), \mathbf{R}(\cdot)] = [\mathbf{I}^*(\cdot), \mathbf{R}^*(\cdot)]$ implies $\mathbf{Y}(\cdot) = \mathbf{Y}_d(\cdot)$. This and the *IIO* system model (2.18a) and (2.18b) yield the following equations:

$$\boldsymbol{Q}^{(\alpha)}\mathbf{R}^{*\alpha}(k) = \boldsymbol{P}^{(\beta)}\mathbf{I}^{*\beta}(k), \forall k \in \mathcal{N}_0, \tag{3.43a}$$

$$\boldsymbol{E}^{(\nu)}\mathbf{Y}_d^\nu(k) = \boldsymbol{R}^{(\alpha)}\mathbf{R}^{*\alpha}(k) + \boldsymbol{T}^{(\mu)}\mathbf{I}^{*\mu}(k), \forall k \in \mathcal{N}_0, \tag{3.43b}$$

which can be easily set in the form of Equations (3.41a) and (3.41b). When we apply the $Z-$transform together with its properties to Equations (3.41a) and (3.41b) then they become

$$\left[\boldsymbol{P}^{(\beta)}\boldsymbol{S}_M^{(\beta)}(z)\right]\mathbf{I}^*(z) - \left[\boldsymbol{Q}^{(\alpha)}\boldsymbol{S}_\rho^{(\alpha)}(z)\right]\mathbf{R}^*(z) =$$
$$= \underbrace{\sum_{r=0}^{r=\beta\leq\alpha} \boldsymbol{P}_r \left[\sum_{k=0}^{k=r-1} z^{r-k}E^k\mathbf{I}_0^*\right] - \sum_{r=0}^{r=\alpha} \boldsymbol{Q}_r \left[\sum_{k=0}^{k=r-1} z^{r-k}E^k\mathbf{R}_0^*\right]}_{\mathbf{w}_1(z)\ (3.40\text{a})}, \tag{3.44a}$$

$$\left[\boldsymbol{T}^{(\mu)}\boldsymbol{S}_M^{(\beta)}(z)\right]\mathbf{I}^*(z) + \left[\boldsymbol{R}^{(\alpha)}\boldsymbol{S}_\rho^{(\alpha)}(z)\right]\mathbf{R}^*(z) =$$
$$= \underbrace{\left\{ \begin{array}{l} \sum_{r=0}^{r=\nu} \boldsymbol{E}_r \left[z^r\mathbf{Y}_d(z) - \sum_{k=0}^{k=r-1} z^{r-k}E^k\mathbf{Y}_{d0}\right] + \\ \quad + \sum_{r=0}^{r=\alpha} \boldsymbol{R}_r \left[\sum_{k=0}^{k=r-1} z^{r-k}E^k\mathbf{R}_0^*\right] + \\ \quad + \sum_{r=0}^{r=\mu} \boldsymbol{T}_r \left[\sum_{k=0}^{k=r-1} z^{r-k}E^k\mathbf{I}_0^*\right] \end{array} \right.}_{\mathbf{w}_2(z)\ (3.40\text{b})}, \tag{3.44b}$$

or equivalently (3.42).

Sufficiency. We accept that all the conditions of the theorem hold. We choose $[\mathbf{I}(\cdot), \mathbf{R}(\cdot)] = [\mathbf{I}^*(\cdot), \mathbf{R}^*(\cdot)]$. Equation (3.41a) written in the form (3.43a) shows that the pair $[\mathbf{I}^*(\cdot), \mathbf{R}^*(\cdot)]$ satisfies Equation (2.18a). Besides, Equation (2.18b) takes the following form:

$$\boldsymbol{E}^{(\nu)}\mathbf{Y}^{\nu}(k) = \boldsymbol{R}^{(\alpha)}\mathbf{R}^{*\alpha}(k) + \boldsymbol{T}^{(\mu)}\mathbf{I}^{*\mu}(k), \forall k \in \mathcal{N}_0.$$

It, subtracted from (3.41b), implies

$$\boldsymbol{E}^{(\nu)}\left[\mathbf{Y}_d^{\nu}(k) - \mathbf{Y}^{\nu}(k)\right] = \mathbf{0}_N, \forall k \in \mathcal{N}_0,$$

i.e.,

$$\boldsymbol{E}^{(\nu)}\boldsymbol{\varepsilon}^{\nu}(k) = \mathbf{0}_N, \forall k \in \mathcal{N}_0.$$

The solution of this time-invariant linear discrete equation is trivial, $\boldsymbol{\varepsilon}^{\nu}(k) \equiv \mathbf{Y}_d^{\nu}(k) - \mathbf{Y}^{\nu}(k) \equiv \mathbf{0}_{(\nu+1)N}$, because all initial conditions are equal to zero, $\boldsymbol{\varepsilon}_0^{\nu} = \mathbf{Y}_{d0}^{\nu-1} - \mathbf{Y}_0^{\nu-1} = \mathbf{0}_{(\nu+1)N}$, due to the conditions of the theorem and Definition 3.9, $\mathbf{Y}_0^{\nu-1} = \mathbf{Y}_{d0}^{\nu-1}$. This completes the proof. ■

Let us reply to the question: what are conditions for the existence of the solutions of Equations (3.44a) and (3.44b), i.e., of (3.42). There are $(M+\rho)$ unknown variables and $(N+\rho)$ equations of the system (3.41a) and (3.41b), equivalently of (3.42). The unknown variables are the entries of $\mathbf{I}^*(z) \in \mathcal{C}^M$ and of $\mathbf{R}^*(z) \in \mathcal{C}^{\rho}$.

Case 3.4 $N > M$

If $N > M$, then Equations (3.44a) and (3.44b), or equivalently (3.42), do not have a solution for the same reasons as for the ISO systems explained in Case 3.2.

Case 3.5 $N \leq M$

If $N \leq M$ and

$$\text{rank}\begin{bmatrix} \boldsymbol{P}^{(\beta)}\boldsymbol{S}_M^{(\beta)}(z) & -\boldsymbol{Q}^{(\alpha)}\boldsymbol{S}_{\rho}^{(\alpha)}(z) \\ \boldsymbol{T}^{(\mu)}\boldsymbol{S}_M^{(\mu)}(z) & \boldsymbol{R}^{(\alpha)}\boldsymbol{S}_{\rho}^{(\alpha)}(z) \end{bmatrix} = N + \rho \leq M + \rho$$

for almost all complex numbers z, then

$$\det\begin{bmatrix} \boldsymbol{P}^{(\beta)}\boldsymbol{S}_M^{(\beta)}(z) & -\boldsymbol{Q}^{(\alpha)}\boldsymbol{S}_{\rho}^{(\alpha)}(z) \\ \boldsymbol{T}^{(\mu)}\boldsymbol{S}_M^{(\mu)}(z) & \boldsymbol{R}^{(\alpha)}\boldsymbol{S}_{\rho}^{(\alpha)}(z) \end{bmatrix}\begin{bmatrix} \boldsymbol{P}^{(\beta)}\boldsymbol{S}_M^{(\beta)}(z) & -\boldsymbol{Q}^{(\alpha)}\boldsymbol{S}_{\rho}^{(\alpha)}(z) \\ \boldsymbol{T}^{(\mu)}\boldsymbol{S}_M^{(\mu)}(z) & \boldsymbol{R}^{(\alpha)}\boldsymbol{S}_{\rho}^{(\alpha)}(z) \end{bmatrix}^T \neq 0$$

for almost all complex numbers z.

There is a solution determined by

$$\begin{bmatrix} \mathbf{I}^*(z) \\ \mathbf{R}^*(z) \end{bmatrix} = \begin{bmatrix} \boldsymbol{P}^{(\beta)} \boldsymbol{S}_M^{(\beta)}(z) & -\boldsymbol{Q}^{(\alpha)} \boldsymbol{S}_\rho^{(\alpha)}(z) \\ \boldsymbol{T}^{(\mu)} \boldsymbol{S}_M^{(\mu)}(z) & \boldsymbol{R}^{(\alpha)} \boldsymbol{S}_\rho^{(\alpha)}(z) \end{bmatrix}^T \cdot$$

$$\cdot \left\{ \begin{bmatrix} \boldsymbol{P}^{(\beta)} \boldsymbol{S}_M^{(\beta)}(z) & -\boldsymbol{Q}^{(\alpha)} \boldsymbol{S}_\rho^{(\alpha)}(z) \\ \boldsymbol{T}^{(\mu)} \boldsymbol{S}_M^{(\mu)}(z) & \boldsymbol{R}^{(\alpha)} \boldsymbol{S}_\rho^{(\alpha)}(z) \end{bmatrix} \begin{bmatrix} \boldsymbol{P}^{(\beta)} \boldsymbol{S}_M^{(\beta)}(z) & -\boldsymbol{Q}^{(\alpha)} \boldsymbol{S}_\rho^{(\alpha)}(z) \\ \boldsymbol{T}^{(\mu)} \boldsymbol{S}_M^{(\mu)}(z) & \boldsymbol{R}^{(\alpha)} \boldsymbol{S}_\rho^{(\alpha)}(z) \end{bmatrix}^T \right\}^{-1} \cdot$$

$$\cdot \begin{bmatrix} \mathbf{w}_1(z) \\ \mathbf{w}_2(z) \end{bmatrix}.$$

Case 3.6 $N = M$

If $N = M$ *and*

$$\operatorname{rank} \begin{bmatrix} \boldsymbol{P}^{(\beta)} \boldsymbol{S}_M^{(\beta)}(z) & -\boldsymbol{Q}^{(\alpha)} \boldsymbol{S}_\rho^{(\alpha)}(z) \\ \boldsymbol{T}^{(\mu)} \boldsymbol{S}_M^{(\mu)}(z) & \boldsymbol{R}^{(\alpha)} \boldsymbol{S}_\rho^{(\alpha)}(z) \end{bmatrix} = N + \rho = M + \rho$$

for almost all complex numbers z,

then Equations (3.44a) and (3.44b), or equivalently (3.42), have the unique solution determined by

$$\begin{bmatrix} \mathbf{I}^*(z) \\ \mathbf{R}^*(z) \end{bmatrix} = \begin{bmatrix} \boldsymbol{P}^{(\beta)} \boldsymbol{S}_M^{(\beta)}(z) & -\boldsymbol{Q}^{(\alpha)} \boldsymbol{S}_\rho^{(\alpha)}(z) \\ \boldsymbol{T}^{(\mu)} \boldsymbol{S}_M^{(\mu)}(z) & \boldsymbol{R}^{(\alpha)} \boldsymbol{S}_\rho^{(\alpha)}(z) \end{bmatrix}^{-1} \begin{bmatrix} \mathbf{w}_1(z) \\ \mathbf{w}_2(z) \end{bmatrix}.$$

We have found the solution to both the problem of the existence of a nominal functional vector pair $[\mathbf{I}_N(\cdot), \mathbf{R}_N(\cdot)]$ for the *IIO* system (2.18a), (2.18b) relative to its desired response $\mathbf{Y}_d(\cdot)$, $[\mathbf{I}^*(\cdot), \mathbf{R}^*(\cdot)] = [\mathbf{I}_N(\cdot), \mathbf{R}_N(\cdot)]$, and the problem of the realizability of $\mathbf{Y}_d(\cdot)$. The realizability of $\mathbf{Y}_d(\cdot)$ is independent of the characteristics of $\mathbf{Y}_d(\cdot)$. It is determined only by the system parameters.

Note 3.5 *We accept in the sequel the conditions of Case 3.5 for the realizability of their desired output vector functions* $\mathbf{Y}_d(\cdot)$.

3.5 Deviations and mathematical models

3.5.1 Introduction

The fulfillment of the special conditions is necessary for the system to be in its nominal regime. If and only if such conditions are not satisfied then the system operate in *a nonnominal regime* that is called also *a perturbed regime*. The nonnominal regime is more probable than the nominal one. This justifies to call a nonnominal regime also *a real regime*. Since the nonnominal regime is a perturbed regime so that if deviates from the desired regime, then it opens the need for the study of the relationship between behaviors of the system in real regimes and its behavior in the nominal regime.

For that purpose we use *the output deviation (vector)* $\mathbf{y}$ **(3.45),** and *the output error (vector)* $\boldsymbol{\varepsilon}$ (3.46) [already defined in (3.34), Subsection 3.4.2], of the real behavior from the nominal one,

$$\text{the output deviation vector: } \mathbf{y} = \mathbf{Y} - \mathbf{Y}_d, \tag{3.45}$$

$$\text{the output error vector: } \boldsymbol{\varepsilon} = \mathbf{Y}_d - \mathbf{Y}. \tag{3.46}$$

These equations imply

$$\boldsymbol{\varepsilon} = -\mathbf{y}. \tag{3.47}$$

Equations (3.45) and (3.46) represent *Lyapunov coordinate transformations* after A. M. Lyapunov [93]. He was the first to relate the coordinates of a real *motion* to the coordinates of the nominal (i.e., desired) *motion* of the system; see Fig. 3.1.

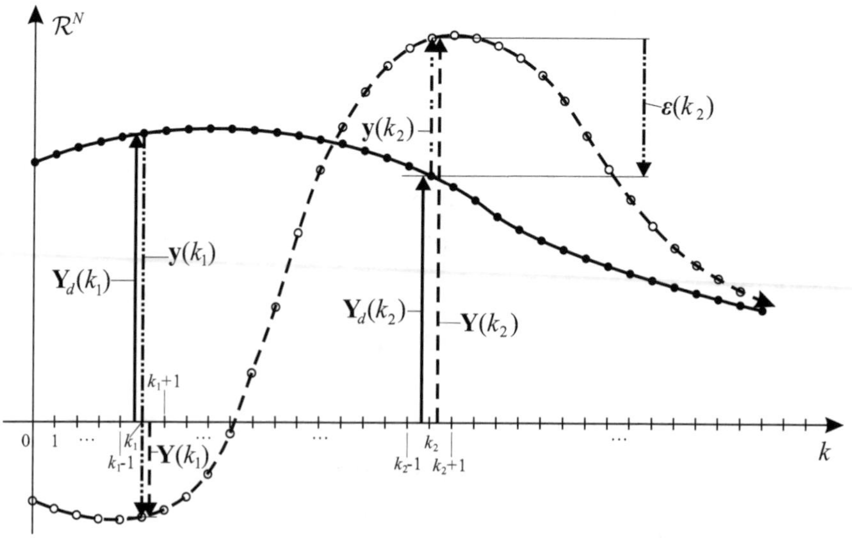

Figure 3.1: Lyapunov coordinate transformations.

Comment 3.2 *The preceding equations prove the equivalence among*

$$\boldsymbol{\varepsilon} = \mathbf{0}_N, \ \mathbf{y} = \mathbf{0}_N \text{ and } \mathbf{Y} = \mathbf{Y}_d,$$

$$\mathbf{Y} = \mathbf{Y}_d \Longleftrightarrow \boldsymbol{\varepsilon} = \mathbf{0}_N \Longleftrightarrow \mathbf{y} = \mathbf{0}_N.$$

The zero output error vector $\boldsymbol{\varepsilon} = \mathbf{0}_N$ *and the zero output deviation vector* $\mathbf{y} = \mathbf{0}_N$ *correspond to the total desired output vector* $\mathbf{Y}_d$.

Conclusion 3.2 *We can reduce the study of the properties of the desired (nominal) total response [of the desired (nominal) total movement] to the study of the same properties of the zero deviation (of the zero error) vector.*

Lyapunov coordinate transformations can be extended to other variables,

$$\mathbf{d} = \mathbf{D} - \mathbf{D}_N, \tag{3.48}$$

$$\mathbf{i} = \mathbf{I} - \mathbf{I}_N, \tag{3.49}$$

$$\mathbf{r} = \mathbf{R} - \mathbf{R}_N, \tag{3.50}$$

$$\mathbf{x} = \mathbf{X} - \mathbf{X}_N. \tag{3.51}$$

Problem 3.3 ***How to determine mathematical models of the systems in terms of the deviations, what are their forms, and what are their relationships to the original models expressed in terms of total coordinates.***

The solution to this problem will be derived separately for *IO*, *ISO* and *IIO* systems in what follows.

3.5.2 *IO* systems

This book treats only well designed *IO* systems, Note 3.3 (in Subsection 3.4.2). This means that the dimension M of the input vector $\mathbf{I}$ is not less than the dimension N of the output vector $\mathbf{Y}$.

The *IO* system (2.1) (in Section 2.1) is determined in terms of total valued variables of the variables and holds for an arbitrary regime, i.e., for a real regime:

$$\sum_{r=0}^{r=\nu} \boldsymbol{A}_r E^r \mathbf{Y}(k) = \sum_{r=0}^{r=\mu} \boldsymbol{B}_r E^r \mathbf{I}(k),$$
$$\det \boldsymbol{A}_\nu \neq 0, \forall k \in \mathcal{N}_0, \ \nu \geq 1, \ 0 \leq \mu \leq \nu. \tag{3.52}$$

If the *IO* system (2.1) is in its desired (nominal) regime, then (3.52) becomes (3.53) due to (3.31a) (Theorem 3.1, in Subsection 3.4.2):

$$\sum_{r=0}^{r=\nu} \boldsymbol{A}_r E^r \mathbf{Y}_d(k) = \sum_{r=0}^{r=\mu} \boldsymbol{B}_r E^r \mathbf{I}_N(k), \forall k \in \mathcal{N}_0. \tag{3.53}$$

Assumption 3.1 *The desired output response of the IO system (2.1), i.e., (3.52), is realizable.*

The validity of the assumption makes reasonable to subtract (3.53) from (3.52),

$$\sum_{r=0}^{r=\nu} \boldsymbol{A}_r E^r \left[\mathbf{Y}(k) - \mathbf{Y}_d(k)\right] = \sum_{r=0}^{r=\mu} \boldsymbol{B}_r E^r \left[\mathbf{I}(k) - \mathbf{I}_N(k)\right], \forall k \in \mathcal{N}_0. \tag{3.54}$$

The usage of the deviations $\mathbf{y} = \mathbf{Y} - \mathbf{Y}_d$ (3.45) and $\mathbf{i} = \mathbf{I} - \mathbf{I}_N$ (3.49) (in Subsection 3.5.1), to (3.54) simplifies the preceding equation to

$$\sum_{r=0}^{r=\nu} \boldsymbol{A}_r E^r \mathbf{y}(k) = \sum_{r=0}^{r=\mu} \boldsymbol{B}_r E^r \mathbf{i}(k), \forall k \in \mathcal{N}_0, \ \nu \geq 1, \ 0 \leq \mu \leq \nu. \tag{3.55}$$

Note 3.6 *Equation (3.55) represents the IO system expressed in terms of the deviations of all variables. Its form, order, and matrices are the same as those of the system model expressed in total values of the variables (2.1), i.e., (3.52). The IO systems (2.1) and (3.55) possess the same characteristics and properties by noting once more that* $\mathbf{y} = \mathbf{0}$ *represents* $\mathbf{Y} = \mathbf{Y}_d$. *For example, they have the same transfer function matrices, and the stability properties of* $\mathbf{y} = \mathbf{0}$ *of (3.55) are simultaneously the same stability properties of* $\mathbf{Y}_d(k)$ *of (2.1), i.e., of (3.52). Therefore, we will continue with the IO system description in terms of the deviations (3.55).*

The compact form of (3.55) reads, in view of (2.2a), (2.2b), (2.3a), and (2.3b) (in Section 2.1):

$$\boldsymbol{A}^{(\nu)} \mathbf{y}^{\nu}(k) = \boldsymbol{B}^{(\mu)} \mathbf{i}^{\mu}(k), \ \forall k \in \mathcal{N}_0, \ 0 \leq \mu \leq \nu. \tag{3.56}$$

This is the compact *IO* system description in terms of the deviations. We will use mainly (3.56).

3.5.3 *ISO* systems

By referring to Note 3.4 (in Subsection 3.4.3) we accept that the number M of the input variables (i.e., the dimension of the input vector $\mathbf{I}$) not less than the number N of the output variables (i.e., the dimension of the output vector $\mathbf{Y}$) of the ISO system (2.14a) and (2.14b) (in Section 2.2) $N \leq M$.

Equations (2.14a) and (2.14b) (in Section 2.2), i.e., (3.57a) and (3.57b),

$$\mathbf{X}(k+1) = \boldsymbol{A}\mathbf{X}(k) + \boldsymbol{B}\mathbf{I}(k),\ \forall k \in \mathcal{N}_0, \tag{3.57a}$$

$$\mathbf{Y}(k) = \boldsymbol{C}\mathbf{X}(k) + \boldsymbol{D}\mathbf{I}(k),\ \forall k \in \mathcal{N}_0, \tag{3.57b}$$

hold for the ISO system in an arbitrary regime, i.e., in a real regime. The system description in a nominal regime relative to $\mathbf{Y}_d(\cdot)$ is determined in Theorem 3.2 by Equations (3.37a) and (3.37b) (in Subsection 3.4.3),

$$\mathbf{X}_N(k+1) = \boldsymbol{A}\mathbf{X}_N(k) + \boldsymbol{B}\mathbf{I}_N(k),\ \forall k \in \mathcal{N}_0, \tag{3.58a}$$

$$\mathbf{Y}_d(k) = \boldsymbol{C}\mathbf{X}_N(k) + \boldsymbol{D}\mathbf{I}_N(k),\ \forall k \in \mathcal{N}_0. \tag{3.58b}$$

Assumption 3.2 *The desired output response of the ISO system (2.14a) and (2.14b), i.e., (3.57a) and (3.57b), is realizable.*

We accept this assumption to hold. Its application to (3.58a) and (3.58b), which we subtract from (3.57a) and (3.57b) results in:

$$\begin{gathered}[\mathbf{X}(k+1) - \mathbf{X}_N(k+1)] = \\ = \boldsymbol{A}\left[\mathbf{X}(k) - \mathbf{X}_N(k)\right] + \boldsymbol{B}\left[\mathbf{I}(k) - \mathbf{I}_N(k)\right],\ \forall k \in \mathcal{N}_0,\end{gathered} \tag{3.59a}$$

$$\begin{gathered}\mathbf{Y}(k) - \mathbf{Y}_d(k) = \boldsymbol{C}\left[\mathbf{X}(k) - \mathbf{X}_N(k)\right] + \\ +\boldsymbol{D}\left[\mathbf{I}(k) - \mathbf{I}_N(k)\right],\ \forall k \in \mathcal{N}_0.\end{gathered} \tag{3.59b}$$

The Lyapunov coordinate transformation $\mathbf{x} = \mathbf{X} - \mathbf{X}_N$ (3.51), $\mathbf{y} = \mathbf{Y} - \mathbf{Y}_d$ (3.45) and $\mathbf{i} = \mathbf{I} - \mathbf{I}_N$ (3.49) (in Subsection 3.5.1), to (3.59a) and (3.59b) permit for the ISO system (2.14a) and (2.14b), i.e., (3.57a) and (3.57b), to determine the system mathematical model in terms of the deviations:

$$\mathbf{x}(k+1) = \boldsymbol{A}\mathbf{x}(k) + \boldsymbol{B}\mathbf{i}(k),\ \forall k \in \mathcal{N}_0, \tag{3.60a}$$

$$\mathbf{y}(k) = \boldsymbol{C}\mathbf{x}(k) + \boldsymbol{D}\mathbf{i}(k),\ \forall k \in \mathcal{N}_0. \tag{3.60b}$$

Note 3.7 *This mathematical model in terms of the deviations of all variables of the ISO system has exactly the same form, the same order, and the same matrices as the system model expressed in total values of the* *variables (2.14a) and (2.14b), i.e., (3.57a) and (3.57b). They possess the same characteristics and properties by noting once more that* $\mathbf{x} = \mathbf{0}$ *replaces* $\mathbf{X} = \mathbf{X}_N$ *and* $\mathbf{y} = \mathbf{0}$ *stands for* $\mathbf{Y} = \mathbf{Y}_d$*. For example, they have the same transfer function matrices; the stability properties of* $\mathbf{x} = \mathbf{0}$ *of (3.60a) are the same as of* $\mathbf{X} = \mathbf{X}_N$ *of (2.14a); i.e., of (3.57a), and the tracking properties of* $\mathbf{y} = \mathbf{0}$ *of (3.60a) and (3.60b) are simultaneously the same tracking properties of* $\mathbf{Y}_d(\cdot)$ *of (2.14a) and (2.14b), i.e., of (3.57a) and (3.57b). Therefore, we will continue with the ISO system description in terms of the deviations (3.60a) and (3.60b).*

3.5.4 *IIO* systems

We suppose, in view of (Note 3.5, in Subsection 3.4.4), that the *IIO* system (2.18a) and (2.18b) (in Section 2.3), is well designed. Hence, the dimension N of the output vector $\mathbf{Y}$ is not bigger than the dimension M of the input vector $\mathbf{I}$, $N \leq M$. We suppose also that Condition 2.2 holds (Section 2.3).

The mathematical model (2.18a) and (2.18b), i.e., (3.61a) and (3.61b) of the *IIO* system in terms of the total values of the variables is valid for any regime,

$$\sum_{r=0}^{r=\alpha} \boldsymbol{Q}_r E^r \mathbf{R}(k) = \sum_{r=0}^{r=\beta} \boldsymbol{P}_r E^r \mathbf{I}(k), \det \boldsymbol{Q}_\alpha \neq 0, \forall k \in \mathcal{N}_0, \tag{3.61a}$$
$$\alpha \geq 1, \alpha \geq \beta \geq 0,$$

$$\sum_{r=0}^{r=\nu} \boldsymbol{E}_r E^r \mathbf{Y}(k) = \sum_{r=0}^{r=\alpha} \boldsymbol{R}_r E^r \mathbf{R}(k) + \sum_{r=0}^{r=\mu} \boldsymbol{T}_r E^r \mathbf{I}(k), \det \boldsymbol{E}_\nu \neq 0 \tag{3.61b}$$
$$\forall k \in \mathcal{N}_0, \nu \geq 0, \nu \geq \mu \geq 0.$$

If the system is in a nominal regime then (3.61a) and (3.61b) reads [due to Theorem 3.3, Equations (3.41a) and (3.41b), Subsection 3.4.4]:

$$\sum_{r=0}^{r=\alpha} \boldsymbol{Q}_r E^r \mathbf{R}_N(k) = \sum_{r=0}^{r=\beta} \boldsymbol{P}_r E^r \mathbf{I}_N(k), \forall k \in \mathcal{N}_0, \tag{3.62a}$$

$$\sum_{r=0}^{r=\nu} \boldsymbol{E}_r E^r \mathbf{Y}_d(k) = \sum_{r=0}^{r=\alpha} \boldsymbol{R}_r E^r \mathbf{R}_N(k) + \sum_{r=0}^{r=\mu} \boldsymbol{T}_r E^r \mathbf{I}_N(k), \forall k \in \mathcal{N}_0. \tag{3.62b}$$

Assumption 3.3 *The desired output response of the IIO system (2.18a) and (2.18b), i.e., (3.61a) and (3.61b) is realizable.*

This assumption justifies the subtraction of (3.62a) and (3.62b) from (3.61a) and (3.61b),

$$\sum_{r=0}^{r=\alpha} \boldsymbol{Q}_r E^r \left[\mathbf{R}(k) - \mathbf{R}_N(k)\right] = \sum_{r=0}^{r=\beta} \boldsymbol{P}_r E^r \left[\mathbf{I}(k) - \mathbf{I}_N(k)\right],$$

$$\sum_{r=0}^{r=\nu} \boldsymbol{E}_r E^r \left[\mathbf{Y}(k) - \mathbf{Y}_d(k)\right] = \sum_{r=0}^{r=\alpha} \boldsymbol{R}_r E^r \left[\mathbf{R}(k) - \mathbf{R}_N(k)\right] +$$

$$+ \sum_{r=0}^{r=\mu} \boldsymbol{T}_r E^r \left[\mathbf{I}(k) - \mathbf{I}_N(k)\right], \forall k \in \mathcal{N}_0,$$

and to apply $\mathbf{r} = \mathbf{R} - \mathbf{R}_N$ (3.50), $\mathbf{y} = \mathbf{Y} - \mathbf{Y}_d$ (3.45), and $\mathbf{i} = \mathbf{I} - \mathbf{I}_N$ (3.49) (in Subsection 3.5.1),

$$\sum_{r=0}^{r=\alpha} \boldsymbol{Q}_r E^r \mathbf{r}(k) =$$

$$= \sum_{r=0}^{r=\beta\leq\alpha} \boldsymbol{P}_r E^r \mathbf{i}(k), \det \boldsymbol{Q}_\alpha \neq 0, \forall k \in \mathcal{N}_0, \ \alpha \geq 1, \alpha \geq \beta \geq 0, \tag{3.63a}$$

$$\sum_{r=0}^{r=\nu} \boldsymbol{E}_r E^r \mathbf{y}(k) = \sum_{r=0}^{r=\alpha} \boldsymbol{R}_r E^r \mathbf{r}(k) + \sum_{r=0}^{r=\mu} \boldsymbol{T}_r E^r \mathbf{i}(k), \tag{3.63b}$$

$$\forall k \in \mathcal{N}_0, \nu \geq 0, \mu \geq 0.$$

Note 3.8 *This is the IIO system model in terms of the deviations of all variables. It has exactly the same form, order, and matrices as the system model expressed in total values of the variables (2.18a) and (2.18b), i.e., (3.61a) and (3.61b). The systems expressed in both forms possess the same characteristics and properties by noting again that* $\mathbf{r} = \mathbf{0} \Longleftrightarrow \mathbf{R} = \mathbf{R}_N$ *and* $\mathbf{y} = \mathbf{0}$ *represents* $\mathbf{Y} = \mathbf{Y}_d$. *They have the same transfer function matrices; the stability properties of* $\mathbf{r} = \mathbf{0}$ *of (3.63a) and (3.63b) are the same as the stability properties of* $\mathbf{R} = \mathbf{R}_N$ *of (3.62a), and the tracking properties of* $\mathbf{y} = \mathbf{0}$ *of (3.63a) and (3.63b) are simultaneously the same tracking properties of* $\mathbf{Y}_d(k)$ *of (2.18a) and (2.18b), i.e., (3.61a) and (3.61b). Therefore, we will continue with the IIO system description in terms of the deviations (3.63a) and (3.63b).*

The compact form of (3.63a) and (3.63b) reads, in view of (2.20a) to

(2.20e) (in Section 2.3), as follows:

$$\boldsymbol{Q}^{(\alpha)}\mathbf{r}^{\alpha}(k) = \boldsymbol{P}^{(\beta)}\mathbf{i}^{\beta}(k), \det \boldsymbol{Q}_{\alpha} \neq 0, \forall k \in \mathcal{N}_0,\ \alpha \geq 1, \alpha \geq \beta \geq 0, \quad (3.64a)$$

$$\boldsymbol{E}^{(\nu)}\mathbf{y}^{\nu}(k) = \boldsymbol{R}^{(\alpha)}\mathbf{r}^{\alpha}(k) + \boldsymbol{T}^{(\mu)}\mathbf{i}^{\mu}(k), \det \boldsymbol{E}_{\nu} \neq 0, \quad (3.64b)$$
$$\forall k \in \mathcal{N}_0, \nu \geq 0, \mu \geq 0.$$

This follows also from (2.31a) and (2.31b) (in Section 2.3), when it is written for the nominal regime, due to $\mathbf{r} = \mathbf{R} - \mathbf{R}_N$ (3.50), $\mathbf{y} = \mathbf{Y} - \mathbf{Y}_d$ (3.45) and $\mathbf{i} = \mathbf{I} - \mathbf{I}_N$ (3.49).

Note 3.9 *The time evolution of* $\mathbf{r}^{\alpha-1}\left(k; \mathbf{r}_0^{\alpha-1}; \mathbf{i}\right)$ *reflects explicitly the state evolution of the IIO system (3.64a) and (3.64b). The time evolution of both* $\mathbf{r}^{\alpha-1}\left(k; \mathbf{r}_0^{\alpha-1}; \mathbf{i}\right)$ *and* $\mathbf{y}^{\nu-1}\left(k; \mathbf{r}_0^{\alpha-1}; \mathbf{y}_0^{\nu-1}; \mathbf{i}\right)$, *i.e., of*

$$\begin{bmatrix} \mathbf{r}^{\alpha-1}\left(k; \mathbf{r}_0^{\alpha-1}; \mathbf{i}\right) \\ \mathbf{y}^{\nu-1}\left(k; \mathbf{r}_0^{\alpha-1}; \mathbf{y}_0^{\nu-1}; \mathbf{i}\right) \end{bmatrix},$$

expresses explicitly the complete (full) internal and output dynamics, IOD (for short: the full dynamics, or shortest: dynamics, D) of the IIO system (3.64a) and (3.64b). The time evolution of $\mathbf{y}^{\nu-1}\left(k; \mathbf{r}_0^{\alpha-1}; \mathbf{y}_0^{\nu-1}; \mathbf{i}\right)$ *expresses explicitly the output dynamics (OD) of the system. The time evolution of* $\mathbf{y}\left(k; \mathbf{r}_0^{\alpha-1}; \mathbf{y}_0^{\nu-1}; \mathbf{i}\right)$ *is the output response (OR) of the system.*

Equations (2.18a) and (2.18b), or (3.61a) and (3.61b):

- together with (2.32a) and (2.32b) (in Section 2.3), result in the *PMD* system form in terms of deviations:

$$\boldsymbol{Q}^{(\alpha)}\mathbf{r}^{\alpha}(k) = \boldsymbol{P}^{(\beta)}\mathbf{i}^{\beta}(k), \forall k \in \mathcal{N}_0, \quad (3.65a)$$
$$\mathbf{y}(k) = \boldsymbol{R}^{(\alpha)}\mathbf{r}^{\alpha}(k) + \boldsymbol{T}^{(\mu)}\mathbf{i}^{\mu}(k), \forall k \in \mathcal{N}_0; \quad (3.65b)$$

- together with (2.33a) and (2.33b) (in Section 2.3), give directly the compact description in terms of the deviations of the Rosenbrock system (2.33a) and (2.33b),

$$E^1\mathbf{x}(k) = \boldsymbol{A}\mathbf{x}(k) + \boldsymbol{B}\mathbf{i}(k),\ \forall k \in \mathcal{N}_0, \quad (3.66a)$$
$$\mathbf{y}(k) = \boldsymbol{C}\mathbf{x}(k) + \boldsymbol{T}^{(\mu)}\mathbf{i}^{\mu}(k),\ \forall k \in \mathcal{N}_0; \quad (3.66b)$$

- together with (2.35a) and (2.35b) (in Section 2.3), determine the description in terms of the deviations of the $GISO$ system,

$$\sum_{r=0}^{r=\alpha} \boldsymbol{Q}_r E^r \mathbf{r}(k) = \sum_{r=0}^{r=\beta} \boldsymbol{P}_r E^r \mathbf{i}(k), \ \forall k \in \mathcal{N}_0, \tag{3.67a}$$

$$\mathbf{y}(k) = \sum_{r=0}^{r=\alpha} \boldsymbol{R}_r E^r \mathbf{r}(k) + \boldsymbol{T}\mathbf{i}(k), \ \forall k \in \mathcal{N}_0, \tag{3.67b}$$

or, simpler, in the compact form,

$$\boldsymbol{Q}^{(\alpha)} \mathbf{r}^{\alpha}(k) = \boldsymbol{P}^{(\beta)} \mathbf{i}^{\beta}(k), \ k \in \mathcal{N}_0, \tag{3.68a}$$

$$\mathbf{y}(k) = \boldsymbol{R}^{(\alpha)} \mathbf{r}^{\alpha}(k) + \boldsymbol{T}\mathbf{i}(k), \forall k \in \mathcal{N}_0. \tag{3.68b}$$

3.6 Stationary and nonstationary regimes

3.6.1 Introduction

With respect to the criterion: *the type of the system behavior* the system internal dynamical, i.e., the internal state, behavior and its output dynamical behavior can be:

- constant, unchangeable, over a time interval, which is its *steady*, or *stationary*, behavior over that time interval,

- occasionally, or permanently, variable, nonconstant, which can be repeatable after some time interval, or nonrepeatable. If it is repeatable, then it is *periodic*. If it is not repeatable, it is *aperiodic* or also called *transient*. They are *nonstationary* system behaviors.

A stationary state vector represents the system stationary behavior. Motions around the stationary state vector represent nonstationary behaviors.

A singular point characterizes a stationary regime. The singular point can be a **stationary point** in a forced regime, or an **equilibrium point** in a free regime. We will explore their existence and system behaviors around them.

We will present precise definitions of stationary and nonstationary system behaviors separately for IO, ISO and IIO systems.

3.6.2 *IO* systems

Definition 3.10 *The IO system (3.55), i.e., (3.56) (in Subsection 3.5.2),*

a) *is in* ***a stationary (steady) regime since*** $k_\sigma \in \mathcal{N}_0$ *(i.e.,* ***on*** $[k_\sigma, \infty[$*)* ***relative to*** $\mathbf{i}(\cdot)$ *if and only if its response is constant all the time on* $[k_\sigma, \infty[$*, that is that*

$$\mathbf{y}\left(k; \mathbf{y}_0^{\nu-1}; \mathbf{i}\right) = \mathbf{y}\left(k_\sigma; \mathbf{y}_0^{\nu-1}; \mathbf{i}\right) = \mathbf{y}_{k_\sigma} = \text{const}, \forall k \in [k_\sigma, \infty[\,. \quad (3.69)$$

If and only if this holds for $k_\sigma = 0$ *then the system is in* ***a stationary (steady) regime relative to*** $\mathbf{i}(\cdot)$*.*

b) *Otherwise, the system is in* ***a nonstationary regime relative to*** $\mathbf{i}(\cdot)$*, i.e.,*

$$\forall k_\sigma \in \mathcal{N}_0, \exists k_\tau \in \,]k_\sigma, \infty[\, \Longrightarrow \mathbf{y}\left(k_\tau; \mathbf{y}_{k_\sigma}^{\nu-1}; \mathbf{i}\right) \neq \mathbf{y}_{k_\sigma}. \quad (3.70)$$

In a nonstationary regime, the system is in

– ***a periodic regime relative to*** $\mathbf{i}(\cdot)$ *if and only if there is such* $\sigma \in \mathcal{N}$ *that*

$$\mathbf{y}\left(k + \sigma; \mathbf{y}_0^{\nu-1}; \mathbf{i}\right) = \mathbf{y}\left(k; \mathbf{y}_0^{\nu-1}; \mathbf{i}\right), \forall k \in \mathcal{N}_0, \quad (3.71)$$

where such minimal σ *is called* ***the period relative to*** $\mathbf{i}(\cdot)$ *of the periodic regime and it is denoted by* T_p*,*

$$T_p = \min\left\{\sigma \in \mathcal{N} : \ \mathbf{y}\left(k + \sigma; \mathbf{y}_0^{\nu-1}; \mathbf{i}\right) = \right.$$
$$\left. = \mathbf{y}\left(k; \mathbf{y}_0^{\nu-1}; \mathbf{i}\right), \forall k \in \mathcal{N}_0\right\}; \quad (3.72)$$

– ***a transient (aperiodic) regime relative to*** $\mathbf{i}(\cdot)$ *if and only if it is not in a periodic regime, i.e.,*

$$\exists k \in \mathcal{N}_0, \ \forall \sigma \in \mathcal{N}, \ \exists i \in \{1, 2, ..., n, ..\} \Longrightarrow$$
$$\mathbf{y}(k + i\sigma; \mathbf{y}_0^{\nu-1}; \mathbf{i}) \neq \mathbf{y}(k; \mathbf{y}_0^{\nu-1}; \mathbf{i}). \quad (3.73)$$

A stationary or a nonstationary regime are independent of a periodic regime, of a transient regime, of a free regime, and of a forced regime. All combinations of these regimes are possible.

Definition 3.11 ***A vector*** $\mathbf{y}^{*^{\nu-1}} \in \mathcal{R}^{\nu N}$ *is* ***a stationary vector of the*** *IO* ***system (3.55) relative to*** $\mathbf{i}(\cdot)$, $\mathbf{y}^{*^{\nu-1}} = \mathbf{y}^{*^{\nu-1}}(\mathbf{i})$, *if and only if:*

$$\mathbf{y}^{\nu-1}\left(k;\mathbf{y}^{*^{\nu-1}};\mathbf{i}\right) = \mathbf{y}^{*^{\nu-1}}(\mathbf{i}),\ \forall k \in \mathcal{N}_0. \tag{3.74}$$

It is denoted by the subscript "s" :

$$\mathbf{y}^{*^{\nu-1}}(\mathbf{i}) = \mathbf{y}_s^{\nu-1}(\mathbf{i}).$$

This definition shows that a vector $\mathbf{y}^{*^{\nu-1}}(\mathbf{i})$ can be a stationary vector of the *IO* system (3.55) relative to one input vector function $\mathbf{i}(\cdot) = \mathbf{i}_1(\cdot)$, $\mathbf{y}^{*^{\nu-1}}(\mathbf{i}_1) = \mathbf{y}_s^{\nu-1}(\mathbf{i}_1)$, but not relative to another input vector function $\mathbf{i}(\cdot) = \mathbf{i}_2(.)$, $\mathbf{y}_s^{\nu-1}(\mathbf{i}_1) \neq \mathbf{y}_s^{\nu-1}(\mathbf{i}_2)$.

Theorem 3.4 *In order for a vector* $\mathbf{y}^{*^{\nu-1}} \in \mathcal{R}^{N\nu}$ *to be a stationary vector of the IO system (3.55) relative to* $\mathbf{i}(\cdot)$, $\mathbf{y}^{*^{\nu-1}} = \mathbf{y}_s^{\nu-1}(\mathbf{i}) \in \mathcal{R}^{N\nu}$, *it is necessary and sufficient that*

$$\mathbf{y}^{\nu-1}\left(0;\mathbf{y}^{*^{\nu-1}};\mathbf{i}\right) = \mathbf{y}_0^{\nu-1} = \mathbf{y}^{*^{\nu-1}} = \\ = \begin{bmatrix} \mathbf{y}^{*^T} & \mathbf{y}^{*^T} & \cdots & \mathbf{y}^{*^T} \end{bmatrix}^T \in \mathcal{R}^{N\nu}, \tag{3.75}$$

and

$$\left(\boldsymbol{A}_0 + \boldsymbol{A}_1 + \cdots + \boldsymbol{A}_\nu\right)\mathbf{y}^* = \boldsymbol{B}^\mu \mathbf{i}^\mu(k) = \sum_{r=0}^{r=\mu} \boldsymbol{B}_r E^r \mathbf{i}(k),\ \forall k \in \mathcal{N}_0. \tag{3.76}$$

Proof. *Necessity*: Let $\mathbf{y}^{*^{\nu-1}} = \mathbf{y}_s^{\nu-1}(\mathbf{i}) \in \mathcal{R}^{N\nu}$ be a stationary vector of the *IO* system (3.55) relative to $\mathbf{i}(\cdot)$. Hence, (3.74) holds. By shifting $\mathbf{y}(k;\mathbf{y}^{*^{\nu-1}};\mathbf{i})$ from (3.74) $\nu-$times where $\mathbf{y}^*$ is constant, we verify

$$E^r\mathbf{y}(k;\mathbf{y}^{*^{\nu-1}};\mathbf{i}) = \mathbf{y}^*,\ \forall r = 1,2,...,\nu,\ \forall k \in \mathcal{N}_0 \Longrightarrow \tag{3.77}$$

$$E^r\mathbf{y}(0;\mathbf{y}^{*^{\nu-1}};\mathbf{i}) = E^r\mathbf{y}^* = \mathbf{y}^*,\ \forall r = 1,2,...,\nu. \tag{3.78}$$

The result (3.78) and (3.74) imply (3.75). Besides, (3.77) and (3.55), i.e., (3.56), imply (3.76).

Sufficiency, Let (3.75) and (3.76) be valid. We subtract (3.76) from (3.55), and use

$$\Delta E^r\mathbf{y}(k;\mathbf{y}^{*^{\nu-1}};\mathbf{0}) = E^r\mathbf{y}(k;\mathbf{y}^{*^{\nu-1}};\mathbf{i}) - E^r\mathbf{y}^*(\mathbf{i}) = \\ = E^r\mathbf{y}(k;\mathbf{y}^{*^{\nu-1}};\mathbf{i}) - \mathbf{y}^*(\mathbf{i}), \forall k \in \mathcal{N}_0, \tag{3.79}$$

to obtain

$$\sum_{r=0}^{r=\nu} \boldsymbol{A}_r \Delta E^r \mathbf{y}(k) = \mathbf{0}.$$

This is the homogenous linear discrete equation with the initial conditions equal to $\mathbf{y}^*$ due to (3.75) and (3.79). Its solution is trivial,

$$\Delta E^r \mathbf{y}(k; \mathbf{y}^{*^{\nu-1}}; \mathbf{0}) = E^r \mathbf{y}(k; \mathbf{y}^{*^{\nu-1}}; \mathbf{i}) - E^r \mathbf{y}^*(\mathbf{i}) = \mathbf{0},$$
$$\forall k \in \mathcal{N}_0, \ \forall r = 0, 1, ..., \nu - 1,$$

i.e.,

$$\mathbf{y}^{\nu-1}(k; \mathbf{y}^{*^{\nu-1}}; \mathbf{i}) = \mathbf{y}^{*^{\nu-1}}(\mathbf{i}), \ \forall k \in \mathcal{N}_0.$$

This is (3.74), which completes the proof. ■

Comment 3.3 *Notice that the conditions (3.75), (3.76) for a stationary point of the discrete-time IO system (3.55) are essentially different from the analogous conditions for a stationary vector of the related continuous-time IO system [68], [69].*

Note 3.10 *The IO system (3.55), i.e., (3.56), can be in a stationary regime in spite the input vector is time-varying [see (3.76)]. For example:*

$$\sum_{r=0}^{r=\nu} \boldsymbol{A}_r E^r \mathbf{y}(k) = \boldsymbol{B}_0 \mathbf{i}(k) + \boldsymbol{B}_1 E^1 \mathbf{i}(k), \ \boldsymbol{B}_1 = -\boldsymbol{B}_0,$$

for $\mathbf{i}(k) = k\mathbf{1}_M$, *where the vector* $\mathbf{1}_M = \begin{bmatrix} 1 & 1 & \cdots & 1 \end{bmatrix}^T \in \mathcal{R}^M$, *the preceding discrete equation reduces to*

$$\sum_{r=0}^{r=\nu} \boldsymbol{A}_r E^r \mathbf{y}(k) = k\boldsymbol{B}_0 \mathbf{1}_M - \boldsymbol{B}_0 (k+1) \mathbf{1}_M = -\boldsymbol{B}_0 \mathbf{1}_M,$$

so that the condition (3.76) takes the following form:

$$(\boldsymbol{A}_0 + \boldsymbol{A}_1 + \cdots + \boldsymbol{A}_\nu) \mathbf{y}^* (\mathbf{i}) = -\boldsymbol{B}_0 \mathbf{1}_M.$$

If $(\boldsymbol{A}_0 + \boldsymbol{A}_1 + \cdots \boldsymbol{A}_\nu)$ *is nonsingular, then the system has a unique stationary vector determined by*

$$\mathbf{y}^*(\mathbf{i}) = \mathbf{y}_s(\mathbf{i}) = -(\boldsymbol{A}_0 + \boldsymbol{A}_1 + \cdots + \boldsymbol{A}_\nu)^{-1} \boldsymbol{B}_0 \mathbf{1}_M,$$
$$\mathbf{y}^{*^{\nu-1}}(\mathbf{i}) = \mathbf{y}_s^{\nu-1}(\mathbf{i}) = \begin{bmatrix} \mathbf{y}^{*^T} & \mathbf{y}^{*^T} & \cdots & \mathbf{y}^{*^T} \end{bmatrix}^T =$$

$$= -\Bigg[\underbrace{I_N \ \ I_N \ \ \cdots \ \ I_N}_{\nu-times}\Bigg]^T (\boldsymbol{A}_0 + \boldsymbol{A}_1 + \cdots + \boldsymbol{A}_\nu)^{-1} \boldsymbol{B}_0 \mathbf{1}_M = \text{const}.$$

in view of (3.75). However, if the matrix $(\boldsymbol{A}_0 + \boldsymbol{A}_1 + \cdots \boldsymbol{A}_\nu)$ *is the zero matrix* $\boldsymbol{O}_N$ $(\boldsymbol{A}_0 + \boldsymbol{A}_1 + \cdots \boldsymbol{A}_\nu) = \boldsymbol{O}_N$*, then the system is in a stationary regime if and only if*

$$-\boldsymbol{B}_0 \mathbf{1}_M = \mathbf{0}_M.$$

Then, every vector $\mathbf{y}^{*^{\nu-1}} = \begin{bmatrix} \mathbf{y}^{*^T} & \mathbf{y}^{*^T} & \cdots & \mathbf{y}^{*^T} \end{bmatrix}^T \in \mathcal{R}^{N\nu}$ *is the stationary vector relative to the input vector* $\mathbf{i}(k)$*. The system rests in the initial output vector* $\mathbf{y}_0^{\nu-1} = \begin{bmatrix} \mathbf{y}_0^T & \mathbf{y}_0^T & \cdots & \mathbf{y}_0^T \end{bmatrix}^T \in \mathcal{R}^{N\nu}$ *all the time, i.e., the initial output vector* $\mathbf{y}_0^{\nu-1}$ *is the stationary vector relative to the input vector* $\mathbf{i}$*,*

$$(\boldsymbol{A}_0 + \boldsymbol{A}_1 + \cdots + \boldsymbol{A}_\nu) = \boldsymbol{O}_N \text{ and } -\boldsymbol{B}_0 \mathbf{1}_M = \mathbf{0}_M \Longrightarrow$$
$$\mathbf{y}_s^{\nu-1}(\mathbf{i}) = \mathbf{y}_0^{\nu-1} = \begin{bmatrix} \mathbf{y}_0^T & \mathbf{y}_0^T & \cdots & \mathbf{y}_0^T \end{bmatrix}^T \in \mathcal{R}^{N\nu}, \forall \mathbf{y}_0 \in \mathcal{R}^N.$$

To be specific let $N = 3$*,* $M = 4$ *and*

$$(\boldsymbol{A}_0 + \boldsymbol{A}_1 + \cdots + \boldsymbol{A}_\nu) = \boldsymbol{O}_3 \text{ and } \boldsymbol{B}_0 = -\boldsymbol{B}_1 =$$
$$= \begin{bmatrix} 3 & 1 & -6 & 2 \\ 5 & -7 & -2 & 4 \\ 11 & 9 & -24 & 4 \end{bmatrix} \in \mathcal{R}^{3\times 4} \Longrightarrow$$

$$\boldsymbol{B}_0 \mathbf{1}_4 = \begin{bmatrix} 3 & 1 & -6 & 2 \\ 5 & -7 & -2 & 4 \\ 11 & 9 & -24 & 4 \end{bmatrix} \begin{bmatrix} 1 \\ 1 \\ 1 \\ 1 \end{bmatrix} = \begin{bmatrix} 0 \\ 0 \\ 0 \\ 0 \end{bmatrix} = \mathbf{0}_4.$$

The stationary vector relative to the input vector $\mathbf{i}(k)$ *is the initial output vector*

$$\mathbf{y}^{\nu-1}(\mathbf{i}) = \mathbf{y}_0^{\nu-1} = \begin{bmatrix} \mathbf{y}_0^T & \mathbf{y}_0^T & \cdots & \mathbf{y}_0^T \end{bmatrix}^T, \forall \mathbf{y}_0 \in \mathcal{R}^3, \mathbf{i}(k) = k\mathbf{1}_4,$$

whatever the initial output vector

$$\mathbf{y}_0^{\nu-1} = \begin{bmatrix} \mathbf{y}_0^T & \mathbf{y}_0^T & \cdots & \mathbf{y}_0^T \end{bmatrix}^T \in \mathcal{R}^{3\nu}$$

is.

Note 3.11 *If* $M \geq N$*,* $\text{rank}\boldsymbol{B}_0 = N$ *and* $\boldsymbol{B}_r = \boldsymbol{O}$*,* $\forall r = 1, 2, ..., \mu$*, then* $\det(\boldsymbol{B}_0 \boldsymbol{B}_0^T) \neq 0$ *so that (3.76) becomes*

$$(\boldsymbol{A}_0 + \boldsymbol{A}_1 + \cdots + \boldsymbol{A}_\nu)\,\mathbf{y}_s = \boldsymbol{B}_0 \mathbf{i}(k), \ \forall k \in \mathcal{N}_0. \tag{3.80}$$

It is easy to verify that

$$\mathbf{i}(k) = \boldsymbol{B}_0^T \left(\boldsymbol{B}_0 \boldsymbol{B}_0^T\right)^{-1} (\boldsymbol{A}_0 + \boldsymbol{A}_1 + \cdots + \boldsymbol{A}_\nu) \mathbf{y}_s = \text{const}, \forall k \in \mathcal{N}_0$$

obeys Equation (3.80). In this case only a constant input vector can force the system to stay in a stationary regime.

Note 3.12 *The condition (3.76) shows that for the IO system (3.55), i.e., (3.56), to have the unique stationary vector* $\mathbf{y}_s(\mathbf{i})$ *relative to* $\mathbf{i}$*, it is necessary and sufficient that* $(\boldsymbol{A}_0 + \boldsymbol{A}_1 + \cdots \boldsymbol{A}_\nu)$ *is nonsingular and*

$$(\boldsymbol{A}_0 + \boldsymbol{A}_1 + \cdots + \boldsymbol{A}_\nu)^{-1} \boldsymbol{B}^\mu \mathbf{i}^\mu(k) =$$
$$= (\boldsymbol{A}_0 + \boldsymbol{A}_1 + \cdots + \boldsymbol{A}_\nu)^{-1} \sum_{r=0}^{r=\mu} \boldsymbol{B}_r E^r \mathbf{i}(k) = \text{const}, \ \forall k \in \mathcal{N}_0.$$

Then, the unique stationary vector relative to $\mathbf{i}$ *is given by*

$$\mathbf{y}_s(\mathbf{i}) = (\boldsymbol{A}_0 + \boldsymbol{A}_1 + \cdots + \boldsymbol{A}_\nu)^{-1} \boldsymbol{B}^\mu \mathbf{i}^\mu(k) =$$
$$= (\boldsymbol{A}_0 + \boldsymbol{A}_1 + \cdots + \boldsymbol{A}_\nu)^{-1} \sum_{r=0}^{r=\mu} \boldsymbol{B}_r E^r \mathbf{i}(k).$$

3.6.3 *ISO* systems

Definition 3.12 *The ISO system (3.60a) and (3.60b) (in Subsection 3.5.3),*

a) is in ***a stationary (steady) regime since*** $k_\sigma \in \mathcal{N}_0$ *(or,* **on** $[k_\sigma, \infty[$*)* ***relative to*** $\mathbf{i}(\cdot)$ *if and only if the following two conditions (i) and (ii) are valid:*

(i) its state vector (its motion) is constant all the time on $[k_\sigma, \infty[$*:*

$$\mathbf{x}(k) = \mathbf{x}(k; \mathbf{x}_0; \mathbf{i}) = \text{const} = \mathbf{x}(k_\sigma; \mathbf{x}_0; \mathbf{i}) = \mathbf{x}_{k_\sigma}, \quad \forall k \in [k_\sigma, \infty[, \tag{3.81}$$

(ii) its response is constant all the time on $[k_\sigma, \infty[$*:*

$$\mathbf{y}(k; \mathbf{x}_0; \mathbf{i}) = \text{const} = \mathbf{y}(k_\sigma; \mathbf{x}_0; \mathbf{i}) = \mathbf{y}_{k_\sigma}, \forall k \in [k_\sigma, \infty[. \tag{3.82}$$

If and only if this holds for $k_\sigma = 0$*, then the system is in* ***a stationary (steady) regime relative to*** $\mathbf{i}(\cdot)$***.***

b) *Otherwise, the system is in* ***a nonstationary regime relative to*** $\mathbf{i}(\cdot)$, *i.e.*,

$$\forall k_\sigma \in \mathcal{N}_0,\ \exists k_\tau \in]k_\sigma, \infty[\Longrightarrow \mathbf{x}(k_\tau; \mathbf{x}_{k_\sigma}; \mathbf{i}) \neq \mathbf{x}_{k_\sigma}, \quad \textit{and/or } \mathbf{y}(k_\tau; \mathbf{x}_{k_\sigma}; \mathbf{i}) \neq \mathbf{y}_{k_\sigma}. \tag{3.83}$$

In a nonstationary regime the system is in

- ***a periodic regime relative to*** $\mathbf{i}(\cdot)$ *if and only if there is such* $\sigma \in \mathcal{N}$ *that both (3.84) and (3.85) hold,*

$$\mathbf{x}(k; \mathbf{x}_0; \mathbf{i}) = \mathbf{x}(k + \sigma; \mathbf{x}_0; \mathbf{i}),\ \forall k \in \mathcal{N}_0, \tag{3.84}$$

$$\mathbf{y}(k; \mathbf{x}_0; \mathbf{i}) = \mathbf{y}(k + \sigma; \mathbf{x}_0; \mathbf{i}),\ \forall k \in \mathcal{N}_0, \tag{3.85}$$

  where such minimal σ *is called* ***the period relative to*** $\mathbf{i}(\cdot)$ *of the periodic regime and it is denoted by* T_p,

$$T_p = \min\{\sigma \in \mathcal{N}:\ \mathbf{x}(k; \mathbf{x}_0; \mathbf{i}) = \mathbf{x}(k + \sigma; \mathbf{x}_0; \mathbf{i}) \textit{ and } \mathbf{y}(k; \mathbf{x}_0; \mathbf{i}) = \mathbf{y}(k + \sigma; \mathbf{x}_0; \mathbf{i}),\ \forall k \in \mathcal{N}_0\}; \tag{3.86}$$

- ***a transient (aperiodic) regime relative to*** $\mathbf{i}(\cdot)$ *if and only if it is not in a periodic regime, i.e.*,

$$\exists k \in \mathcal{N}_0,\ \forall \sigma \in \mathcal{N},\ \exists i \in \{1, 2, \cdots, n, \cdots\} \Longrightarrow \mathbf{x}(k; \mathbf{x}_0; \mathbf{i}) \neq \mathbf{x}(k + i\sigma; \mathbf{x}_0; \mathbf{i}) \textit{ and/or } \mathbf{y}(k; \mathbf{x}_0; \mathbf{i}) \neq \mathbf{y}(k + i\sigma; \mathbf{x}_0; \mathbf{i}). \tag{3.87}$$

The system can be in a stationary or in a nonstationary, regime either in a free regime, or in a forced regime.

Definition 3.13 *A vector* $\mathbf{x}^* \in \mathcal{R}^n$ *is* ***a stationary vector (a stationary state) of the*** *ISO* ***system (3.60a) and (3.60b) relative to*** $\mathbf{i}(\cdot)$, $\mathbf{x}^* = \mathbf{x}^*(\mathbf{i})$, *if and only if both (i) and (ii) hold:*

(*i*)

$$\mathbf{x}(k; \mathbf{x}^*; \mathbf{i}) = \mathbf{x}(0; \mathbf{x}^*; \mathbf{i}) = \mathbf{x}_0 = \mathbf{x}^* = \text{const}, \forall k \in \mathcal{N}_0, \tag{3.88}$$

(*ii*)

$$\mathbf{y}(k; \mathbf{x}^*; \mathbf{i}) = \mathbf{y}(0; \mathbf{x}^*; \mathbf{i}) = \mathbf{y}_0(\mathbf{x}^*; \mathbf{i}) = \text{const},\ \forall k \in \mathcal{N}_0. \tag{3.89}$$

It is denoted by the subscript "s" :

$$\mathbf{x}^*(\mathbf{i}) = \mathbf{x}_s(\mathbf{i}).$$

A vector $\mathbf{x}^*(\mathbf{i})$ can be a stationary vector of the ISO system (3.60a) and (3.60b) with respect to an input vector function $\mathbf{i}(\cdot) = \mathbf{i}_1(\cdot)$, $\mathbf{x}^*(\mathbf{i}_1) = \mathbf{x}_s(\mathbf{i}_1)$, but not with respect to another input vector function $\mathbf{i}(\cdot) = \mathbf{i}_2(\cdot)$, $\mathbf{x}_s(\mathbf{i}_1) \neq \mathbf{x}_s(\mathbf{i}_2)$.

Theorem 3.5 *In order for a vector* $\mathbf{x}^* = \mathbf{x}^*(\mathbf{i}) \in \mathcal{R}^n$ *to be a stationary vector of the ISO system (3.60a) and (3.60b) relative to* $\mathbf{i}(\cdot)$ *it is necessary and sufficient that*

$$(\boldsymbol{A} - \boldsymbol{I})\mathbf{x}^* + \boldsymbol{B}\mathbf{i}(k) = \mathbf{0}_n, \ \forall k \in \mathcal{N}_0, \tag{3.90}$$

and

$$\boldsymbol{C}\mathbf{x}^* + \boldsymbol{D}\mathbf{i}(k) = \text{const}, \ \forall k \in \mathcal{N}_0. \tag{3.91}$$

Proof. *Necessity*: Let $\mathbf{x}^* = \mathbf{x}^*(\mathbf{i}) \in \mathcal{R}^n$ be a stationary vector of the ISO system (3.60a) and (3.60b) relative to $\mathbf{i}(\cdot)$. Hence, (3.81) holds. By shifting it once and applying (3.88) we verify

$$E^1\mathbf{x}(k;\mathbf{x}^*;\mathbf{i}) = E^1\mathbf{x}^* = \mathbf{x}^* = \boldsymbol{A}\mathbf{x}(k;\mathbf{x}^*;\mathbf{i}) + \boldsymbol{B}\mathbf{i}(k) = \boldsymbol{A}\mathbf{x}^* + \boldsymbol{B}\mathbf{i}(k), \forall k \in \mathcal{N}_0,$$

which implies (3.90). This, Equations (3.60b), (3.88) and (3.89) yield (3.91).

Sufficiency, Let (3.90) and (3.91) hold. We subtract (3.90) from (3.60a), and we use

$$\begin{gathered}\Delta\mathbf{x}(k;\mathbf{x}^*;\mathbf{0}_M) = \mathbf{x}(k;\mathbf{x}^*;\mathbf{i}) - \mathbf{x}^*(\mathbf{i}), \ \forall k \in \mathcal{N}_0, \Longrightarrow \\ \Delta\mathbf{x}(0;\mathbf{x}^*;\mathbf{0}) = \mathbf{x}^*(\mathbf{i}) - \mathbf{x}^*(\mathbf{i}) = \mathbf{0}_n,\end{gathered} \tag{3.92}$$

to derive

$$E^1\Delta\mathbf{x}(k;\mathbf{x}^*;\mathbf{0}) = \boldsymbol{A}\Delta\mathbf{x}(k;\mathbf{x}^*;\mathbf{0}).$$

This is the homogenous linear discrete equation with the zero initial conditions due to (3.92). Its solution is trivial,

$$\Delta\mathbf{x}(k;\mathbf{x}^*;\mathbf{0}) = \mathbf{x}(k;\mathbf{x}^*;\mathbf{i}) - \mathbf{x}^*(\mathbf{i}) = \mathbf{0}, \ \forall k \in \mathcal{N}_0,$$

i.e.,

$$\mathbf{x}(k;\mathbf{x}^*;\mathbf{i}) = \mathbf{x}^*(\mathbf{i}), \ \forall k \in \mathcal{N}_0.$$

This is (3.88), which implies (3.89) due to (3.60b), and completes the proof. ∎

Comment 3.4 *Notice that the conditions (3.90) and (3.91) for a stationary point of the discrete-time ISO system (3.60a) and (3.60b) are essentially different from the analogous conditions for a stationary vector of the related continuous-time ISO system [68], [69].*

Note 3.13 *The conditions (3.90) and (3.91) permit the existence of a stationary vector of the ISO system (3.60a) and (3.60b) even relative to a time varying input vector* $\mathbf{i}(k)$. *For example,*

$$\boldsymbol{A}=\begin{bmatrix} 2 & 2 \\ 3 & 5 \end{bmatrix},\ \boldsymbol{B}=\begin{bmatrix} 2 & -2 \\ 1 & -1 \end{bmatrix},\ \boldsymbol{C}=\begin{bmatrix} 2 & 6 \\ 7 & 4 \end{bmatrix},$$
$$\boldsymbol{D}=\begin{bmatrix} 5 & -5 \\ -1 & 1 \end{bmatrix},\ \mathbf{i}(k)=2\begin{bmatrix} ke^k+1 \\ ke^k \end{bmatrix},$$

and (3.90) imply

$$(\boldsymbol{A}-\boldsymbol{I})\,\mathbf{x}^*+\boldsymbol{B}\mathbf{i}(k)=$$
$$=\begin{bmatrix} 1 & 2 \\ 3 & 4 \end{bmatrix}\mathbf{x}^*+\begin{bmatrix} 2 & -2 \\ 1 & -1 \end{bmatrix}2\begin{bmatrix} ke^k+1 \\ ke^k \end{bmatrix}=\mathbf{0}_n,\ \forall k\in\mathcal{N}_0,$$

i.e.,

$$\mathbf{x}^*=-\frac{1}{2}\begin{bmatrix} 4 & -2 \\ -3 & 1 \end{bmatrix}\begin{bmatrix} 2 & -2 \\ 1 & -1 \end{bmatrix}2\begin{bmatrix} ke^k+1 \\ ke^k \end{bmatrix}=$$
$$=-\begin{bmatrix} 6 & -6 \\ -5 & 5 \end{bmatrix}\begin{bmatrix} ke^k+1 \\ ke^k \end{bmatrix}=\begin{bmatrix} -6 \\ 5 \end{bmatrix},\ \forall k\in\mathcal{N}_0.$$

This solution $\mathbf{x}^*$ *is constant. In order to be a stationary vector relative to* $\mathbf{i}(k)$, *it should fulfill (3.91). Let us verify,*

$$\boldsymbol{C}\mathbf{x}^*+\boldsymbol{D}\mathbf{i}(k)=\begin{bmatrix} 2 & 6 \\ 7 & 4 \end{bmatrix}\begin{bmatrix} -6 \\ 5 \end{bmatrix}+\begin{bmatrix} 5 & -5 \\ -1 & 1 \end{bmatrix}2\begin{bmatrix} ke^k+1 \\ ke^k \end{bmatrix}=$$
$$=\begin{bmatrix} 18 \\ -22 \end{bmatrix}+2\begin{bmatrix} 5 \\ -1 \end{bmatrix}=\begin{bmatrix} 28 \\ -24 \end{bmatrix}=\text{const},\forall k\in\mathcal{N}_0.$$

The constant vector $\mathbf{x}^*=\begin{bmatrix} -6 & 5 \end{bmatrix}^T$ *satisfies (3.90) and (3.91). It is the stationary vector of the system relative to time-varying input vector* $\mathbf{i}(k)=2\begin{bmatrix} ke^k+1 & ke^k \end{bmatrix}^T$, $\mathbf{x}^*=\begin{bmatrix} -6 & 5 \end{bmatrix}^T=\mathbf{x}_s(\mathbf{i})$.

3.6.4 *IIO* systems

Definition 3.14 *The IIO system (3.63a) and (3.63b), i.e., (3.64a) and (3.64b) (in Subsection 3.5.4),*

a) is in ***a stationary (steady) regime since*** $k_\sigma \in \mathcal{N}_0$ *(i.e.,* ***on*** $[k_\sigma, \infty[$*)* ***relative to*** $\mathbf{i}(\cdot)$ *if and only if the following two conditions (i) and (ii) are valid:*

(i) its internal substate vector (its internal dynamical behavior) is constant all the time on $[k_\sigma, \infty[$*:*

$$\mathbf{r}(k; \mathbf{r}_0^{\alpha-1}; \mathbf{i}) = \text{const} = \mathbf{r}(k_\sigma; \mathbf{r}_0^{\alpha-1}; \mathbf{i}) = \mathbf{r}_{k_\sigma}, \quad \forall k \in [k_\sigma, \infty[, \tag{3.93}$$

(ii) its response is constant all the time on $[k_\sigma, \infty[$*:*

$$\mathbf{y}(k; \mathbf{r}_0^{\alpha-1}; \mathbf{y}_0^{\nu-1}; \mathbf{i}) = \text{const} = \mathbf{y}(k_\sigma; \mathbf{r}_0^{\alpha-1}; \mathbf{y}_0^{\nu-1}; \mathbf{i}) = \mathbf{y}_{k_\sigma}, \quad \forall k \in [k_\sigma, \infty[. \tag{3.94}$$

If and only if this holds for $k_\sigma = 0$*, then the system is in* ***a stationary (steady) regime relative to*** $\mathbf{i}(\cdot)$*.*

b) Otherwise, the system is in ***a nonstationary regime relative to*** $\mathbf{i}(\cdot)$*, i.e.,*

$$\forall k_\sigma \in \mathcal{N}_0, \ \exists k_\tau \in]k_\sigma, \infty[\Longrightarrow \mathbf{r}(k_\tau; \mathbf{r}_0^{\alpha-1}; \mathbf{i}) \neq \mathbf{r}_{k_\sigma}, \quad \textit{and/or } \mathbf{y}(k_\tau; \mathbf{r}_\sigma^{\alpha-1}; \mathbf{y}_\sigma^{\nu-1}; \mathbf{i}) \neq \mathbf{y}_{k_\sigma}. \tag{3.95}$$

In a nonstationary regime the system is in

- ***a periodic regime relative to*** $\mathbf{i}(\cdot)$ *if and only if there is such* $\sigma \in \mathcal{N}$ *that both*

$$\mathbf{r}(k; \mathbf{r}_0^{\alpha-1}; \mathbf{i}) = \mathbf{r}(k+\sigma; \mathbf{r}_0^{\alpha-1}; \mathbf{i}), \ \forall k \in \mathcal{N}_0, \tag{3.96a}$$

and

$$\mathbf{y}(k; \mathbf{r}_0^{\alpha-1}; \mathbf{y}_0^{\nu-1}; \mathbf{i}) = \mathbf{y}(k+\sigma; \mathbf{r}_0^{\alpha-1}; \mathbf{y}_0^{\nu-1}; \mathbf{i}), \ \forall k \in \mathcal{N}_0, \tag{3.96b}$$

hold, where such minimal σ *is called* ***the period relative to*** $\mathbf{i}(\cdot)$ *of the periodic regime and it is denoted by* T_p*,*

$$T_p = \min\left\{\sigma \in \mathcal{N} : \mathbf{r}(k; \mathbf{r}_0^{\alpha-1}; \mathbf{i}) = \mathbf{r}(k+\sigma; \mathbf{r}_0^{\alpha-1}; \mathbf{i}), \ \forall k \in \mathcal{N}_0, \ \textit{and} \ \mathbf{y}(k; \mathbf{r}_0^{\alpha-1}; \mathbf{y}_0^{\nu-1}; \mathbf{i}) = \mathbf{y}(k+\sigma; \mathbf{r}_0^{\alpha-1}; \mathbf{y}_0^{\nu-1}; \mathbf{i}), \ \forall k \in \mathcal{N}_0\right\}; \tag{3.97}$$

- ***a transient (aperiodic) regime relative to*** $\mathbf{i}(\cdot)$ *if and only if it is not in a periodic regime, i.e.,*

$$\forall k \in \mathcal{N}_0,\ \forall \sigma \in \mathcal{N},\ \exists i \in \{1, 2, \cdots, n, \cdots\} \Longrightarrow$$
$$\mathbf{r}(k; \mathbf{r}_0^{\alpha-1}; \mathbf{i}) \neq \mathbf{r}(k + i\sigma; \mathbf{r}_0^{\alpha-1}; \mathbf{i}),\ \textit{and/or}$$
$$\mathbf{y}(k; \mathbf{r}_0^{\alpha-1}; \mathbf{y}_0^{\nu-1}; \mathbf{i}) \neq \mathbf{y}(k + i\sigma; \mathbf{r}_0^{\alpha-1}; \mathbf{y}_0^{\nu-1}; \mathbf{i}). \tag{3.98}$$

The stationary and nonstationary system regimes can take place either in free regimes or in forced regimes of the system.

Definition 3.15 *A vector* $\left[\left(\mathbf{r}^{*^{\alpha-1}}\right)^T \ \left(\mathbf{y}^{*^{\nu-1}}\right)^T \right]^T \in \mathcal{R}^{\alpha\rho+\nu N}$ ***is a stationary vector (a stationary point) of the** IIO* ***system (3.63a), (3.63b), i.e., (3.64a), (3.64b), relative to*** $\mathbf{i}(\cdot)$,

$$\left[\left(\mathbf{r}^{*^{\alpha-1}}\right)^T \ \left(\mathbf{y}^{*^{\nu-1}}\right)^T \right]^T = \left[\left[\mathbf{r}^{*^{\alpha-1}}(\mathbf{i})\right]^T \ \left[\mathbf{y}^{*^{\nu-1}}(\mathbf{i})\right]^T \right]^T,$$

if and only if both (i) and (ii) hold:

(i)

$$\mathbf{r}^{\alpha-1}(k; \mathbf{r}_0^{*^{\alpha-1}}; \mathbf{i}) = \text{const} = \mathbf{r}_0^{*^{\alpha-1}}, \forall k \in \mathcal{N}_0,\ \textit{and} \tag{3.99}$$

(ii)

$$\mathbf{y}^{\nu-1}(k; \mathbf{r}_0^{*^{\alpha-1}}; \mathbf{y}_0^{*^{\nu-1}}; \mathbf{i}) = \text{const} = \mathbf{y}_0^{*^{\nu-1}},\ \forall k \in \mathcal{N}_0. \tag{3.100}$$

It is denoted by the subscript "s" :

$$\begin{bmatrix} \mathbf{r}^{*^{\alpha-1}}(\mathbf{i}) \\ \mathbf{y}^{*^{\nu-1}}(\mathbf{i}) \end{bmatrix} = \begin{bmatrix} \mathbf{r}_s^{\alpha-1}(\mathbf{i}) \\ \mathbf{y}_s^{\nu-1}(\mathbf{i}) \end{bmatrix}.$$

The stationary vector $\left[\left[\mathbf{r}_s^{\alpha-1}(\mathbf{i})\right]^T \ \left[\mathbf{y}_s^{\nu-1}(\mathbf{i})\right]^T \right]^T$ is relative to the input vector function $\mathbf{i}(\cdot)$. It can be stationary relative to $\mathbf{i}_1(\cdot)$, but not relative to $\mathbf{i}_2(\cdot) \neq \mathbf{i}_1(\cdot)$.

Theorem 3.6 *In order for a vector* $\left[\left[\mathbf{r}^{*^{\alpha-1}}\right]^T \ \left[\mathbf{y}^{*^{\nu-1}}\right]^T \right]^T \in \mathcal{R}^{\alpha\rho+\nu N}$ *to be a stationary vector of the IIO system (3.63a) and (3.63b), i.e., (3.64a) and (3.64b), relative to* $\mathbf{i}(\cdot)$,

$$\begin{bmatrix} \mathbf{r}^{*^{\alpha-1}} \\ \mathbf{y}^{*^{\nu-1}} \end{bmatrix} = \begin{bmatrix} \mathbf{r}_s^{\alpha-1}(\mathbf{i}) \\ \mathbf{y}_s^{\nu-1}(\mathbf{i}) \end{bmatrix},$$

it is necessary and sufficient that

$$\mathbf{r}^{\alpha-1}(0;\mathbf{r}^{*^{\alpha-1}};\mathbf{i}) = \mathbf{r}_0^{\alpha-1} = \mathbf{r}^{*^{\alpha-1}} = \begin{bmatrix} \mathbf{r}^{*^T} & \mathbf{r}^{*^T} & \cdots & \mathbf{r}^{*^T} \end{bmatrix}^T \in \mathcal{R}^{\alpha\rho},$$
$$\mathbf{y}^{\nu-1}(0;\mathbf{r}^{*^{\alpha-1}};\mathbf{y}^{*^{\nu-1}};\mathbf{i}) = \mathbf{y}_0^{\nu-1} = \mathbf{y}^{*^{\nu-1}} =$$
$$= \begin{bmatrix} \mathbf{y}^{*^T} & \mathbf{y}^{*^T} & \cdots & \mathbf{y}^{*^T} \end{bmatrix}^T \in \mathcal{R}^{N\nu}, \tag{3.101}$$

$$(\boldsymbol{Q}_0 + \boldsymbol{Q}_1 + \cdots + \boldsymbol{Q}_\alpha)\,\mathbf{r}^* =$$
$$= \sum_{r=0}^{r=\beta\leq\alpha} \boldsymbol{P}_r E^r \mathbf{i}(k) = \boldsymbol{P}^{(\beta)}\mathbf{i}^\beta(k),\ \forall k \in \mathcal{N}_0, \tag{3.102}$$

and

$$-(\boldsymbol{R}_0 + \boldsymbol{R}_1 + \cdots + \boldsymbol{R}_\alpha)\,\mathbf{r}^* + (\boldsymbol{E}_0 + \boldsymbol{E}_1 + \cdots + \boldsymbol{E}_\alpha)\,\mathbf{y}^* =$$
$$= \sum_{r=0}^{r=\mu} \boldsymbol{T}_r E^r \mathbf{i}(k) = \boldsymbol{T}^{(\mu)}\mathbf{i}^\mu(k),\ \forall k \in \mathcal{N}_0, \tag{3.103}$$

hold.

Proof. *Necessity*: Let $\begin{bmatrix} \left[\mathbf{r}^{*^{\alpha-1}}\right]^T & \left[\mathbf{y}^{*^{\nu-1}}\right]^T \end{bmatrix}^T \in \mathcal{R}^{\alpha\rho+\nu N}$ be a stationary vector of the *IIO* system (3.63a) and (3.63b), i.e., (3.64a) and (3.64b), relative to $\mathbf{i}(\cdot)$. Hence, (3.99) and (3.100) hold. By shifting $\mathbf{r}(k;\mathbf{r}^{*^{\alpha-1}};\mathbf{i})$ and $\mathbf{y}(k;\mathbf{r}_0^{*^{\alpha-1}};\mathbf{y}_0^{*^{\nu-1}};\mathbf{i})$ from them we verify

$$E^r\mathbf{r}(k;\mathbf{r}^{*^{\alpha-1}};\mathbf{i}) = \mathbf{r}^*,\ \forall r = 1,2,...,\alpha,\ \forall k \in \mathcal{N}_0 \Longrightarrow \tag{3.104}$$
$$E^r\mathbf{r}(0;\mathbf{r}^{*^{\alpha-1}};\mathbf{i}) = E^r\mathbf{r}^* = \mathbf{r}^*,\ \forall r = 1,2,...,\alpha, \tag{3.105}$$
$$E^r\mathbf{y}(k;\mathbf{r}_0^{*^{\alpha-1}};\mathbf{y}_0^{*^{\nu-1}};\mathbf{i}) = \mathbf{y}^*,\ \forall r = 1,2,...,\nu,\ \forall k \in \mathcal{N}_0 \Longrightarrow \tag{3.106}$$
$$E^r\mathbf{y}(0;\mathbf{r}_0^{*^{\alpha-1}};\mathbf{y}_0^{*^{\nu-1}};\mathbf{i}) = E^r\mathbf{y}^* = \mathbf{y}^*,\ \forall k = 1,2,...,\nu. \tag{3.107}$$

The results (3.105) and (3.107), together with (3.99) and (3.100), imply (3.101). Besides, (3.104), (3.106), (3.63a), and (3.63b), i.e., (3.64a) and (3.64b), imply (3.102) and (3.103).

Sufficiency, Let (3.101), (3.102) and (3.103) hold. We subtract (3.102) and (3.103) from (3.63a) and (3.63b), i.e., from (3.64a) and (3.64b), and we use

$$\Delta E^r \mathbf{r}(k; \mathbf{r}^{*^{\alpha-1}}; \mathbf{0}) = E^r \mathbf{r}(k; \mathbf{r}^{*^{\alpha-1}}; \mathbf{i}) - E^r \mathbf{r}^*(\mathbf{i}) = \\ = E^r \mathbf{r}(k; \mathbf{r}^{*^{\alpha-1}}; \mathbf{i}) - \mathbf{r}^*(\mathbf{i}), \forall k \in \mathcal{N}_0, \tag{3.108}$$

$$\Delta E^r \mathbf{y}(k; \mathbf{r}^{*^{\alpha-1}}; \mathbf{y}^{*^{\nu-1}}; \mathbf{0}) = E^r \mathbf{y}(k; \mathbf{r}^{*^{\alpha-1}}; \mathbf{y}^{*^{\nu-1}}; \mathbf{i}) - E^r \mathbf{y}^*(\mathbf{i}) = \\ = E^r \mathbf{y}(k; \mathbf{r}^{*^{\alpha-1}}; \mathbf{y}^{*^{\nu-1}}; \mathbf{i}) - \mathbf{y}^*(\mathbf{i}), \forall k \in \mathcal{N}_0, \tag{3.109}$$

to derive

$$\sum_{r=0}^{r=\alpha} \boldsymbol{Q}_r \Delta E^r \mathbf{r}(k) = \mathbf{0}, \ \forall k \in \mathcal{N}_0,$$

$$-\sum_{r=0}^{r=\alpha} \boldsymbol{R}_r \Delta E^r \mathbf{r}(k) + \sum_{r=0}^{r=\nu} \boldsymbol{E}_r \Delta E^r \mathbf{y}(k) = \mathbf{0}, \ \forall k \in \mathcal{N}_0.$$

This is the system of the homogenous linear discrete equations with the zero initial conditions due to (3.108) and (3.109). Their solutions are trivial,

$$\Delta E^r \mathbf{r}(k; \mathbf{r}^{*^{\alpha-1}}; \mathbf{0}) = E^r \mathbf{r}(k; \mathbf{r}^{*^{\alpha-1}}; \mathbf{i}) - E^r \mathbf{r}^*(\mathbf{i}) = \mathbf{0}, \forall k \in \mathcal{N}_0, \\ \forall r = 0, 1, ..., \alpha - 1,$$

$$\Delta E^r \mathbf{y}(k; \mathbf{r}^{*^{\alpha-1}}; \mathbf{y}^{*^{\nu-1}}; \mathbf{0}) = E^r \mathbf{y}(k; \mathbf{r}^{*^{\alpha-1}}; \mathbf{y}^{*^{\nu-1}}; \mathbf{i}) - E^r \mathbf{y}^*(\mathbf{i}) = \mathbf{0}, \ \forall k \in \mathcal{N}_0, \\ \forall r = 0, 1, ..., \nu - 1,$$

i.e.,

$$\mathbf{r}^{\alpha-1}(k; \mathbf{r}^{*^{\alpha-1}}; \mathbf{i}) = \mathbf{r}^{*^{\alpha-1}}(\mathbf{i}), \ \forall k \in \mathcal{N}_0,$$

$$\mathbf{y}^{\nu-1}(k; \mathbf{r}^{*^{\alpha-1}}; \mathbf{y}^{*^{\nu-1}}; \mathbf{i}) = \mathbf{y}^{*^{\nu-1}}(\mathbf{i}), \ \forall k \in \mathcal{N}_0.$$

These are (3.99) and (3.100). They complete the proof. ■

Comment 3.5 *Notice that the conditions (3.101) and (3.102) for a stationary point of the discrete-time IIO system (3.63a) and (3.63b) are essentially different from the analogous conditions for a stationary vector of the related continuous-time IIO system [68], [69].*

3.7 Equilibrium regime

3.7.1 Introduction

We distinguish ***static*** **equilibrium regime** from ***dynamic*** **equilibrium regime**. The temporal characteristic of the total desired (nominal) behavior $\mathbf{Y}_d(\cdot)$ of the system determined the type of the equilibrium regime. The equilibrium regime is **static** if and only if $\mathbf{Y}_d(k)$ is constant. Then **the equilibrium point** in the deviation coordinates frame represents **a static equilibrium point**. The equilibrium regime is **dynamic** if and only if the desired (nominal) behavior $\mathbf{Y}_d(k)$ of the system is time-varying. Then the equilibrium point in the deviation coordinates frame represents a **dynamic equilibrium point**.

In the total coordinates frame an equilibrium regime concerns the system behavior under the total nominal input vector $\mathbf{I}_N(\cdot)$ and under all nominal total initial conditions.

If the system description is in terms of the deviations, then an equilibrium regime concerns the system behavior in a free regime. Since the mathematical model is time-invariant and linear (in the framework of this book), then its solution in the free regime under all zero initial conditions is trivial, i.e., it is identically equal to the zero deviation vector.

Conclusion 3.3 *The zero deviation vector represents an equilibrium vector (an equilibrium point) of every linear time-invariant mathematical model in terms of the deviations.*

Problem 3.4 *Does there exist another nontrivial equilibrium vector of a given linear time-invariant system described in terms of the deviations?*

We will prove that the reply is affirmative.

Problem 3.5 *What are conditions for the existence of two or more nontrivial equilibrium points of a given linear time-invariant system described in terms of the deviations?*

We will prove also complete solutions to this problem for *IO*, *ISO* and *IIO* systems.

3.7.2 *IO* systems

We will prove that the *IO* system (3.55), i.e., (3.56) (in Subsection 3.5.2),

$$A^{(\nu)}\mathbf{y}^{\nu}(k) = B^{(\mu)}\mathbf{i}^{\mu}(k),\ \forall k \in \mathcal{N}_0, \tag{3.110}$$

can have a single, unique equilibrium vector, but need not.

Definition 3.16 *The IO system (3.110) is in* ***an equilibrium regime*** *if and only if it is in a free regime and its response is constant all the time on* $\mathcal{N}_0$*, that is that*

$$\mathbf{y}^*(k;\mathbf{y}^{*^{\nu-1}};\mathbf{0}_M) = \mathbf{const.} = \mathbf{y}(0;\mathbf{y}^{*^{\nu-1}};\mathbf{0}_M) = \mathbf{y}^*, \forall k \in \mathcal{N}_0. \tag{3.111}$$

If and only if this holds then the vector $\mathbf{y}^{*^{\nu-1}} = \begin{bmatrix} \mathbf{y}^{*^T} & \mathbf{y}^{*^T} & \cdots & \mathbf{y}^{*^T} \end{bmatrix}^T \in \mathcal{R}^{N\nu}$ ***is an equilibrium vector (an equilibrium point) of the system,*** *which is denoted by the subscript "e,"*

$$\mathbf{y}^{*^{\nu-1}} = \begin{bmatrix} \mathbf{y}^{*^T} & \mathbf{y}^{*^T} & \cdots & \mathbf{y}^{*^T} \end{bmatrix}^T = \mathbf{y}_e^{\nu-1} = \begin{bmatrix} \mathbf{y}_e^T & \mathbf{y}_e^T & \cdots & \mathbf{y}_e^T \end{bmatrix}^T.$$

This definition and Definitions 3.10 and 3.11, (in Subsection 3.6.2), imply that an equilibrium regime is a stationary regime with respect to the zero deviation input vector $\mathbf{i}(k) \equiv \mathbf{0}_M$, and that an equilibrium vector is a stationary vector with respect to the zero deviation input vector $\mathbf{i}(k) \equiv \mathbf{0}_M$.

Notice that, by definition, the equilibrium vector of the system is independent of the real input vector. It depends only on the system itself. More precisely:

Theorem 3.7 *In order for a vector* $\mathbf{y}^{*^{\nu-1}} = \begin{bmatrix} \mathbf{y}^{*^T} & \mathbf{y}^{*^T} & \cdots & \mathbf{y}^{*^T} \end{bmatrix}^T \in \mathcal{R}^{N\nu}$ *to be an equilibrium vector of the IO system (3.110), it is necessary and sufficient that*

$$(\boldsymbol{A}_0 + \boldsymbol{A}_1 + \cdots + \boldsymbol{A}_\nu)\,\mathbf{y}^* = \mathbf{0}_N. \tag{3.112}$$

Proof. This theorem results directly from Theorem 3.4 (in Subsection 3.6.2), in view of Definitions 3.10 and 3.16 ■

Comment 3.6 *Notice that the conditions (3.112) for an equilibrium vector of the discrete-time IO system (3.55) are essentially different from the analogous conditions for an equilibrium vector of the related continuous-time IO system [68], [69].*

Theorem 3.8 *a) In order for a vector*

$$\mathbf{y}^{*^{\nu-1}} = \begin{bmatrix} \mathbf{y}^{*^T} & \mathbf{y}^{*^T} & \cdots & \mathbf{y}^{*^T} \end{bmatrix}^T \in \mathcal{R}^{N\nu}$$

to be the unique equilibrium vector of the IO system (3.110), it is necessary and sufficient that the matrix $(\boldsymbol{A}_0 + \boldsymbol{A}_1 + \cdots + \boldsymbol{A}_\nu) \in \mathcal{R}^{N \times N}$ *is nonsingular,*

$$\det(\boldsymbol{A}_0 + \boldsymbol{A}_1 + \cdots + \boldsymbol{A}_\nu) \neq 0. \tag{3.113}$$

b) *In order for the system (3.110) to have several different equilibrium vectors, it is necessary and sufficient that the matrix*

$$(\boldsymbol{A}_0 + \boldsymbol{A}_1 + \cdots + \boldsymbol{A}_\nu) \in \mathcal{R}^{N \times N}$$

is singular,

$$\det(\boldsymbol{A}_0 + \boldsymbol{A}_1 + \cdots + \boldsymbol{A}_\nu) = 0. \tag{3.114}$$

Then, and only then, the system has infinitely many different equilibrium vectors that constitute the hyperplane $S_{eIO} \subset \mathcal{R}^{\nu N}$ *of all the equilibrium vectors, which passes through the origin* $\mathbf{y}^{\nu-1} = \mathbf{0}_{\nu N}$*,*

$$S_{eIO} = \left\{ \mathbf{y}^{\nu-1} : \mathbf{y}^{\nu-1} = \begin{bmatrix} \mathbf{y}^T & \mathbf{y}^T & \cdots & \mathbf{y}^T \end{bmatrix}^T \in \mathcal{R}^{\nu N}, \right.$$
$$\left. (\boldsymbol{A}_0 + \boldsymbol{A}_1 + \cdots + \boldsymbol{A}_\nu)\,\mathbf{y} = \mathbf{0}_N \right\}. \tag{3.115}$$

Proof. The statement of the theorem follows directly from Theorem 3.7 and the well-known theorem on the number of the solutions of the homogeneous linear vector algebraic Equation (3.112). ■

Theorems 3.7 and 3.8 imply that we can easily determine the equilibrium vector(s) of the system (3.110) by solving Equation 3.112.

Note 3.14 *Equation (3.115) holds also in the case (3.113). Then, and only then, the set of all the system equilibrium states is singleton,*

$$S_{eIO} = \left\{ \mathbf{y}^{\nu-1} : \mathbf{y}^{\nu-1} = \begin{bmatrix} \mathbf{y}^T & \mathbf{y}^T & \cdots & \mathbf{y}^T \end{bmatrix}^T \in \mathcal{R}^{\nu N}, \right.$$
$$\left. (\boldsymbol{A}_0 + \boldsymbol{A}_1 + \cdots + \boldsymbol{A}_\nu)\,\mathbf{y} = \mathbf{0}_N \right\} = \left\{ \mathbf{y}^{\nu-1} : \mathbf{y}^{\nu-1} = \mathbf{0}_{\nu N} \right\} =$$
$$= \left\{ \mathbf{0}_{\nu N} \right\} \Longleftrightarrow \det(\boldsymbol{A}_0 + \boldsymbol{A}_1 + \cdots + \boldsymbol{A}_\nu) \neq 0.$$

Note 3.15 *If the matrix* $(\boldsymbol{A}_0 + \boldsymbol{A}_1 + \cdots + \boldsymbol{A}_\nu)$ *is the zero matrix* $\boldsymbol{O}_N$*,*

$$(\boldsymbol{A}_0 + \boldsymbol{A}_1 + \cdots + \boldsymbol{A}_\nu) = \boldsymbol{O}_N,$$

then every output vector $\mathbf{y}^{\nu-1} = \begin{bmatrix} \mathbf{y}^T & \mathbf{y}^T & \cdots & \mathbf{y}^T \end{bmatrix}^T \in \mathcal{R}^{\nu N}$ *is the equilibrium vector* $\mathbf{y}_e^{\nu-1} \in \mathcal{R}^{\nu N}$*. The system rests in the initial output vector*

$$\mathbf{y}_0^{\nu-1} = \begin{bmatrix} \mathbf{y}^T & \mathbf{y}^T & \cdots & \mathbf{y}^T \end{bmatrix}^T,$$

i.e.,

$$(\boldsymbol{A}_0+\boldsymbol{A}_1+\cdots+\boldsymbol{A}_\nu)=\boldsymbol{O}_N \Longrightarrow$$
$$\mathbf{y}_e^{\nu-1}=\mathbf{y}_0^{\nu-1}=\left[\begin{array}{cccc}\mathbf{y}_0^T & \mathbf{y}_0^T & \cdots & \mathbf{y}_0^T\end{array}\right]^T,\ \forall \mathbf{y}_0\in\mathcal{R}^N,$$

whatever the initial output vector $\mathbf{y}_0\in\mathcal{R}^N$ *is.*

3.7.3 *ISO* systems

The *ISO* system (3.60a) and (3.60b) (in Subsection 3.5.3),

$$\mathbf{x}(k+1)=\boldsymbol{A}\mathbf{x}(k)+\boldsymbol{B}\mathbf{i}(k),\ \forall k\in\mathcal{N}_0, \tag{3.116a}$$
$$\mathbf{y}(k)=\boldsymbol{C}\mathbf{x}(k)+\boldsymbol{D}\mathbf{i}(k),\ \forall k\in\mathcal{N}_0, \tag{3.116b}$$

can be in a single, unique equilibrium regime, but need not.

Definition 3.17 *The ISO system (3.116a) and (3.116b) is in* ***an equilibrium regime*** *if and only if its state vector (its motion) is constant all the time on* $\mathcal{N}_0$ *in a free regime, that is that*

$$\mathbf{x}(k;\mathbf{x}^*;\mathbf{0}_M)=\mathbf{const.}=\mathbf{x}(0;\mathbf{x}^*;\mathbf{0}_M)=\mathbf{x}^*,\forall k\in\mathcal{N}_0. \tag{3.117}$$

If and only if this holds then the vector $\mathbf{x}^*\in\mathcal{R}^n$ *is* ***an equilibrium vector (an equilibrium state) of the system,*** *which is denoted by the subscript "e,"*

$$\mathbf{x}^*=\mathbf{x}_e.$$

A stationary regime in a free regime and a stationary state relative to the zero input vector $\mathbf{i}(k)\equiv\mathbf{0}_M$, i.e., in a free regime, represent an equilibrium regime and an equilibrium state, respectively, of the *ISO* system (3.116a) and (3.116b) in view of the preceding definition, and Definitions 3.12 and 3.13 (in Subsection 3.6.3). This fact and Theorem 3.5 (in Subsection 3.6.3) imply directly the following:

Theorem 3.9 *In order for a vector* $\mathbf{x}^*\in\mathcal{R}^n$ *to be an equilibrium vector of the ISO system (3.116a) and (3.116b), it is necessary and sufficient that*

$$(\boldsymbol{I}-\boldsymbol{A})\,\mathbf{x}^*=\mathbf{0}_n. \tag{3.118}$$

Comment 3.7 *Notice that the conditions (3.118) for an equilibrium vector of the discrete-time ISO system (3.60a) and (3.60b) are essentially different from the analogous conditions for an equilibrium vector of the related continuous-time ISO system [68], [69].*

The theorem 3.9 and the theorem on the number of the solutions of the linear homogeneous vector algebraic equation prove:

Theorem 3.10 *a) In order for a vector* $\mathbf{x}^* \in \mathcal{R}^n$ *to be the unique equilibrium state of the ISO system (3.116a) and (3.116b), it is necessary and sufficient that the matrix* $(\boldsymbol{I} - \boldsymbol{A}) \in \mathcal{R}^{n \times n}$ *is nonsingular,*

$$\det(\boldsymbol{I} - \boldsymbol{A}) \neq 0. \tag{3.119}$$

b) In order for the ISO system (3.116a) and (3.116b) to have several different equilibrium vectors, it is necessary and sufficient that the matrix $(\boldsymbol{I} - \boldsymbol{A}) \in \mathcal{R}^{n \times n}$ *is singular,*

$$\det(\boldsymbol{I} - \boldsymbol{A}) = 0. \tag{3.120}$$

Then, and only then, the system has infinitely many different equilibrium vectors that constitute the hyperplane $S_{eISO} \subset \mathcal{R}^n$ *of all the equilibrium vectors, which passes through the origin* $\mathbf{x} = \mathbf{0}_n$*,*

$$S_{eISO} = \{\mathbf{x} : \ \mathbf{x} \in \mathcal{R}^n, \ (\boldsymbol{I} - \boldsymbol{A})\mathbf{x} = \mathbf{0}_n\}. \tag{3.121}$$

Note 3.16 *Equation (3.121) of* S_{eISO} *holds regardless of the singularity (*$\det(\boldsymbol{I} - \boldsymbol{A}) = 0$*) or the regularity (*$\det(\boldsymbol{I} - \boldsymbol{A}) \neq 0$*) of the matrix* $(\boldsymbol{I} - \boldsymbol{A})$*. If and only if the matrix* $(\boldsymbol{I} - \boldsymbol{A})$ *is nonsingular, i.e.,* $\det(\boldsymbol{I} - \boldsymbol{A}) \neq 0$*, then* S_{eISO} *is singleton,*

$$S_{eISO} = \{\mathbf{x} : \ \mathbf{x} \in \mathcal{R}^n, \ \mathbf{x} = \mathbf{0}_n\} = \{\mathbf{0}_n\} \Longleftrightarrow \det(\boldsymbol{I} - \boldsymbol{A}) \neq 0.$$

Note 3.17 *If the matrix* $(\boldsymbol{I} - \boldsymbol{A})$ *is the zero matrix,* $(\boldsymbol{I} - \boldsymbol{A}) = \boldsymbol{O}_n$*, then* S_{eISO} *is the whole state space,*

$$(\boldsymbol{I} - \boldsymbol{A}) = \boldsymbol{O}_n \Longrightarrow S_{eISO} = \mathcal{R}^n.$$

For example, the matrix $(\boldsymbol{I} - \boldsymbol{A})$ *of the system*

$$\mathbf{x}(k+1) = \mathbf{x}(k) + \boldsymbol{B}\mathbf{i}(k), \ \mathbf{y}(k) = \boldsymbol{C}\mathbf{x}(k) + \boldsymbol{D}\mathbf{i}(k)$$

is the zero matrix, $(\boldsymbol{I} - \boldsymbol{A}) = \boldsymbol{O}_n$*. The set* S_{eISO} *of all the system equilibrium states is the whole state space* $\mathcal{R}^n$*,* $S_{eISO} = \mathcal{R}^n$*. The system rests in its initial state* $\mathbf{x}_0$ *in the free regime, whatever the initial state* $\mathbf{x}_0 \in \mathcal{R}^n$ *is,*

$$\mathbf{x}_e = \mathbf{x}_0 \in \mathcal{R}^n.$$

Notice also that this system is in the stationary regime if and only if

$$\mathbf{Bi}(k) = \mathbf{0}_n, \ \forall k \in \mathcal{N}_0.$$

Then, every state can be a stationary state of this system, which implies that then the system rests in its initial state forever, whatever the initial state is. Notice that the preceding condition does not imply the free regime, i.e., it is possible to hold for $\mathbf{i}(k) \neq \mathbf{0}_n, \ \forall k \in \mathcal{N}_0$. *To be specific, let*

$$\boldsymbol{A} = \boldsymbol{I}_2, \ \boldsymbol{B} = \begin{bmatrix} -2 & 2 \\ 1 & -1 \end{bmatrix}, \ \boldsymbol{C} = \begin{bmatrix} 2 & 6 \\ 7 & 4 \end{bmatrix},$$
$$\boldsymbol{D} = \begin{bmatrix} 5 & -5 \\ -1 & 1 \end{bmatrix}, \ \mathbf{i}(k) = \begin{bmatrix} ke^k + 1 \\ ke^k + 1 \end{bmatrix} \neq \mathbf{0}_n, \ \forall k \in \mathcal{N}_0.$$

This and (3.90) (Subsection 3.6.3) imply

$$(\boldsymbol{A} - \boldsymbol{I})\,\mathbf{x}^* + \boldsymbol{B}\mathbf{i}(k) = \boldsymbol{O}_2\mathbf{x}^* + \boldsymbol{B}\mathbf{i}(k) = \boldsymbol{B}\mathbf{i}(k) = \begin{bmatrix} -2 & 2 \\ 1 & -1 \end{bmatrix}\begin{bmatrix} ke^k + 1 \\ ke^k + 1 \end{bmatrix} =$$
$$= \begin{bmatrix} -2ke^k - 2 + 2ke^k + 2 \\ ke^k + 1 - ke^k - 1 \end{bmatrix} = \begin{bmatrix} 0 \\ 0 \end{bmatrix} = \mathbf{0}_n, \ \forall k \in \mathcal{N}_0.$$

The system is in the forced regime and every state can be its stationary state $\mathbf{x}_s$. *Hence, the system rests in its initial state* $\mathbf{x}_0 \in \mathcal{R}^2$ *although it is in the forced regime under the action of the time varying input*

$$\mathbf{i}(k) = \begin{bmatrix} ke^k + 1 \\ ke^k + 1 \end{bmatrix},$$

whatever its initial state $\mathbf{x}_0 \in \mathcal{R}^2$ *is,*

$$\mathbf{x}_s(\mathbf{i}) = \mathbf{x}_0, \ \forall \mathbf{x}_0 \in \mathcal{R}^n.$$

3.7.4 *IIO* systems

For the *IIO* system (3.63a) and (3.63b), i.e., (3.64a) and (3.64b) (in Subsection 3.5.4),

$$\boldsymbol{Q}^{(\alpha)}\mathbf{r}^{\alpha}(k) = \boldsymbol{P}^{(\beta)}\mathbf{i}^{\beta}(k), \det \boldsymbol{Q}_{\alpha} \neq 0, \forall k \in \mathcal{N}_0, \tag{3.122a}$$
$$\boldsymbol{E}^{(\nu)}\mathbf{y}^{\nu}(k) = \boldsymbol{R}^{(\alpha)}\mathbf{r}^{\alpha}(k) + \boldsymbol{T}^{(\mu)}\mathbf{i}^{\mu}(k), \det \boldsymbol{E}_{\nu} \neq 0, \ \forall k \in \mathcal{N}_0. \tag{3.122b}$$

the following definition is adequate:

Definition 3.18 *The IIO system (3.122a) and (3.122b) is in* ***an equilibrium regime*** *if and only if its complete dynamics vector (its complete dynamic behavior) is constant in a free regime all the time on $\mathcal{N}_0$:*

$$\begin{bmatrix} \mathbf{r}^{\alpha-1}(k;\mathbf{r}^{*\alpha-1};\mathbf{0}_M) \\ \mathbf{y}^{\nu-1}(k;\mathbf{r}^{*\alpha-1};\mathbf{y}^{*\nu-1};\mathbf{0}_M) \end{bmatrix} = \mathbf{const.} = \begin{bmatrix} \mathbf{r}^{\alpha-1}(0;\mathbf{r}^{*\alpha-1};\mathbf{0}_M) \\ \mathbf{y}^{\nu-1}(0;\mathbf{r}^{*\alpha-1};\mathbf{y}^{*\nu-1};\mathbf{0}_M) \end{bmatrix} =$$
$$= \begin{bmatrix} \mathbf{r}^{*\alpha-1} \\ \mathbf{y}^{*\nu-1} \end{bmatrix}, \forall k \in \mathcal{N}_0. \tag{3.123}$$

If and only if this holds

– *and if $\nu \geq 1$ then the vector $\begin{bmatrix} \left(\mathbf{r}^{*\alpha-1}\right)^T & \left(\mathbf{y}^{*\nu-1}\right)^T \end{bmatrix}^T$,*

$$\begin{bmatrix} \mathbf{r}^{*\alpha-1} \\ \mathbf{y}^{*\nu-1} \end{bmatrix} = \begin{bmatrix} \begin{bmatrix} \mathbf{r}^{*T} & \mathbf{r}^{*T} & \cdots & \mathbf{r}^{*T} \end{bmatrix}^T \\ \begin{bmatrix} \mathbf{y}^{*T} & \mathbf{y}^{*T} & \cdots & \mathbf{y}^{*T} \end{bmatrix}^T \end{bmatrix} \in \mathcal{R}^{\alpha\rho+N\nu},$$

is ***an equilibrium vector (an equilibrium point) of the system*** *(3.122a) and (3.122b) which is denoted by the subscript "e,"*

$$\begin{bmatrix} \mathbf{r}^{*\alpha-1} \\ \mathbf{y}^{*\nu-1} \end{bmatrix} = \begin{bmatrix} \mathbf{r}_e^{\alpha-1} \\ \mathbf{y}_e^{\nu-1} \end{bmatrix} = \begin{bmatrix} \begin{bmatrix} \mathbf{r}_e^T & \mathbf{r}_e^T & \cdots & \mathbf{r}_e^T \end{bmatrix}^T \\ \begin{bmatrix} \mathbf{y}_e^T & \mathbf{y}_e^T & \cdots & \mathbf{y}_e^T \end{bmatrix}^T \end{bmatrix},$$

– *and if $\nu = 0$ then the vector $\mathbf{r}^{*\alpha-1}$,*

$$\mathbf{r}^{*\alpha-1} = \begin{bmatrix} \mathbf{r}^{*T} & \mathbf{r}^{*T} & \cdots & \mathbf{r}^{*T} \end{bmatrix}^T \in \mathcal{R}^{\alpha\rho},$$

is ***an equilibrium vector (an equilibrium point) of the system*** *(3.122a) and (3.122b) hence of the Rosenbrock system (3.66a) and (3.66b) and of the GISO system (3.68a) and (3.68b) (in Subsection 3.5.4), which is denoted by the subscript "e,"*

$$\mathbf{r}^{*\alpha-1} = \mathbf{r}_e^{\alpha-1} = \begin{bmatrix} \mathbf{r}_e^T & \mathbf{r}_e^T & \cdots & \mathbf{r}_e^T \end{bmatrix}^T.$$

The preceding definition and Definitions 3.14 and 3.15 (in Subsection 3.6.4) imply that a stationary regime and a stationary vector relative to the zero input $\mathbf{i}(k) \equiv \mathbf{0}_M$ represent an equilibrium regime and an equilibrium vector of the system, respectively.

Theorem 3.11 *a) If $\nu \geq 1$, then in order for a vector*

$$\left[\left(\mathbf{r}^{*\alpha-1}\right)^T \ \left(\mathbf{y}^{*\nu-1}\right)^T \right]^T,$$

$$\left[\left(\mathbf{r}^{*\alpha-1}\right)^T \ \left(\mathbf{y}^{*\nu-1}\right)^T \right]^T =$$

$$= \left[\left[\mathbf{r}^{*T} \ \mathbf{r}^{*T} \ \cdots \ \mathbf{r}^{*T} \right] \left[\mathbf{y}^{*T} \ \mathbf{y}^{*T} \ \cdots \ \mathbf{y}^{*T} \right] \right]^T \in$$
$$\in \mathcal{R}^{\alpha\rho+\nu N},$$

to be an equilibrium vector of the IIO system (3.122a) and (3.122b), it is necessary and sufficient that

$$\begin{bmatrix} \boldsymbol{Q}_0 + \boldsymbol{Q}_1 + \cdots + \boldsymbol{Q}_\alpha & \boldsymbol{O}_{\rho,N} \\ -\left(\boldsymbol{R}_0 + \boldsymbol{R}_1 + \cdots + \boldsymbol{R}_\alpha\right) & \boldsymbol{E}_0 + \boldsymbol{E}_1 + \cdots + \boldsymbol{E}_\nu \end{bmatrix} \begin{bmatrix} \mathbf{r}^* \\ \mathbf{y}^* \end{bmatrix} =$$
$$= \begin{bmatrix} \mathbf{0}_\rho \\ \mathbf{0}_N \end{bmatrix}. \tag{3.124}$$

b) If $\nu = 0$, then in order for a vector $\mathbf{r}^{*\alpha-1}$,

$$\mathbf{r}^{*\alpha-1} = \left[\mathbf{r}^{*T} \ \mathbf{r}^{*T} \ \cdots \ \mathbf{r}^{*T} \right]^T \in \mathcal{R}^{\alpha\rho},$$

to be an equilibrium vector of the IIO system (3.122a) and (3.122b) hence of the Rosenbrock system (3.66a) and (3.66b), and of the IIDO system (3.68a) and (3.68b), it is necessary and sufficient that

$$\left(\boldsymbol{Q}_0 + \boldsymbol{Q}_1 + \cdots + \boldsymbol{Q}_\alpha\right)\mathbf{r}^* = \mathbf{0}_\rho. \tag{3.125}$$

Definitions 3.14, 3.15, 3.18, and Theorem 3.6 (in Subsection 3.6.4) imply Theorem 3.11.

Comment 3.8 *Notice that the conditions (3.124) and (3.125) for an equilibrium vector of the discrete-time IIO system (3.63a) and (3.63b) are essentially different from the analogous conditions for an equilibrium vector of the related continuous-time IIO system [68], [69].*

It is necessary to solve Equation (3.124) or Equation (3.125), respectively, in order to find an equilibrium vector of the *IIO* system (3.122a),

(3.122b) or to verify whether a vector

$$\begin{bmatrix} \mathbf{r}^{*^{\alpha-1}} \\ \mathbf{y}^{*^{\nu-1}} \end{bmatrix} = \begin{bmatrix} \begin{bmatrix} \mathbf{r}^{*^T} & \mathbf{r}^{*^T} & \cdots & \mathbf{r}^{*^T} \end{bmatrix}^T \\ \begin{bmatrix} \mathbf{y}^{*^T} & \mathbf{y}^{*^T} & \cdots & \mathbf{y}^{*^T} \end{bmatrix}^T \end{bmatrix}, \ \nu \geq 1,$$

$$\mathbf{r}^{*^{\alpha-1}} = \begin{bmatrix} \mathbf{r}^{*^T} & \mathbf{r}^{*^T} & \cdots & \mathbf{r}^{*^T} \end{bmatrix}^T, \ \nu = 0,$$

is its equilibrium vector.

Theorem 3.12 *a) In order for:*

– *a vector* $\begin{bmatrix} \left(\mathbf{r}^{*^{\alpha-1}}\right)^T & \left(\mathbf{y}^{*^{\nu-1}}\right)^T \end{bmatrix}^T$ *if* $\nu \geq 1$, *where*

$$\begin{bmatrix} \left(\mathbf{r}^{*^{\alpha-1}}\right)^T & \left(\mathbf{y}^{*^{\nu-1}}\right)^T \end{bmatrix}^T =$$

$$= \begin{bmatrix} \begin{bmatrix} \mathbf{r}^{*^T} & \mathbf{r}^{*^T} & \cdots & \mathbf{r}^{*^T} \end{bmatrix} & \begin{bmatrix} \mathbf{y}^{*^T} & \mathbf{y}^{*^T} & \cdots & \mathbf{y}^{*^T} \end{bmatrix} \end{bmatrix}^T \in$$

$$\in \mathcal{R}^{\alpha\rho+\nu N},$$

to be the unique equilibrium vector of the IIO system (3.122a) and (3.122b), it is necessary and sufficient that the matrix

$$\begin{bmatrix} \boldsymbol{Q}_0 + \boldsymbol{Q}_1 + \cdots + \boldsymbol{Q}_\alpha & \boldsymbol{O}_{\rho N} \\ -(\boldsymbol{R}_0 + \boldsymbol{R}_1 + \cdots + \boldsymbol{R}_\alpha) & \boldsymbol{E}_0 + \boldsymbol{E}_1 + \cdots + \boldsymbol{E}_\nu \end{bmatrix} \in$$

$$\in \mathcal{R}^{(\rho+N)\times(\rho+N)}$$

is nonsingular,

$$\det \begin{bmatrix} \boldsymbol{Q}_0 + \boldsymbol{Q}_1 + \cdots + \boldsymbol{Q}_\alpha & \boldsymbol{O}_{\rho N} \\ -(\boldsymbol{R}_0 + \boldsymbol{R}_1 + \cdots + \boldsymbol{R}_\alpha) & \boldsymbol{E}_0 + \boldsymbol{E}_1 + \cdots + \boldsymbol{E}_\nu \end{bmatrix} \neq 0, \tag{3.126}$$

– *a vector* $\mathbf{r}^{*^{\alpha-1}}$ *if* $\nu = 0$, *where*

$$\mathbf{r}^{*^{\alpha-1}} = \begin{bmatrix} \mathbf{r}^{*^T} & \mathbf{r}^{*^T} & \cdots & \mathbf{r}^{*^T} \end{bmatrix}^T \in \mathcal{R}^{\alpha\rho},$$

to be the unique equilibrium vector of the IIO system (3.122a) and (3.122b) hence of the Rosenbrock system (3.66a) and (3.66b), and of the GISO system (3.68a) and (3.68b) it is necessary and sufficient that the matrix

$$(\boldsymbol{Q}_0 + \boldsymbol{Q}_1 + \cdots + \boldsymbol{Q}_\alpha) \in \mathcal{R}^{\rho \times \rho}$$

is nonsingular,

$$\det(\boldsymbol{Q}_0 + \boldsymbol{Q}_1 + \cdots + \boldsymbol{Q}_\alpha) \neq 0. \tag{3.127}$$

b) *In order for the system (3.122a) and (3.122b) to have several different equilibrium vectors, it is necessary and sufficient that*

- *the matrix*

$$\begin{bmatrix} \boldsymbol{Q}_0 + \boldsymbol{Q}_1 + \cdots + \boldsymbol{Q}_\alpha & \boldsymbol{O}_{\rho N} \\ -(\boldsymbol{R}_0 + \boldsymbol{R}_1 + \cdots + \boldsymbol{R}_\alpha) & \boldsymbol{E}_0 + \boldsymbol{E}_1 + \cdots + \boldsymbol{E}_\nu \end{bmatrix} \in \\ \in \mathcal{R}^{(\rho+N)\times(\rho+N)}$$

is singular if $\nu \geq 1$, i.e.,

$$\det \begin{bmatrix} \boldsymbol{Q}_0 + \boldsymbol{Q}_1 + \cdots + \boldsymbol{Q}_\alpha & \boldsymbol{O}_{\rho N} \\ -(\boldsymbol{R}_0 + \boldsymbol{R}_1 + \cdots + \boldsymbol{R}_\alpha) & \boldsymbol{E}_0 + \boldsymbol{E}_1 + \cdots + \boldsymbol{E}_\nu \end{bmatrix} = 0; \tag{3.128}$$

- *the matrix*

$$(\boldsymbol{Q}_0 + \boldsymbol{Q}_1 + \cdots + \boldsymbol{Q}_\alpha) \in \mathcal{R}^{\rho \times \rho}$$

is singular if $\nu = 0$, i.e.,

$$\det(\boldsymbol{Q}_0 + \boldsymbol{Q}_1 + \cdots + \boldsymbol{Q}_\alpha) = 0. \tag{3.129}$$

Then, and only then, the system has infinitely many different equilibrium vectors that constitute the hyperplane $S_{eIIO} \subset \mathcal{R}^{\alpha\rho+\nu N}$ of all the equilibrium vectors, which passes through:

- *the origin* $\begin{bmatrix} \mathbf{r}^{\alpha-1^T} & \mathbf{y}^{\nu-1^T} \end{bmatrix}^T = \mathbf{0}_{\alpha\rho+\nu N}$ *if $\nu \geq 1$,*

$$S_{eIIO} = \left\{ \begin{bmatrix} \mathbf{r}^{\alpha-1} \\ \mathbf{y}^{\nu-1} \end{bmatrix} : \begin{bmatrix} \mathbf{r}^{\alpha-1} \\ \mathbf{y}^{\nu-1} \end{bmatrix} = \begin{bmatrix} \begin{bmatrix} \mathbf{r}^T & \mathbf{r}^T & \cdots & \mathbf{r}^T \end{bmatrix}^T \\ \begin{bmatrix} \mathbf{y}^T & \mathbf{y}^T & \cdots & \mathbf{y}^T \end{bmatrix}^T \end{bmatrix}, \right. \\ \begin{bmatrix} \boldsymbol{Q}_0 + \boldsymbol{Q}_1 + \cdots + \boldsymbol{Q}_\alpha & \boldsymbol{O}_{\rho N} \\ -(\boldsymbol{R}_0 + \boldsymbol{R}_1 + \cdots + \boldsymbol{R}_\alpha) & \boldsymbol{E}_0 + \boldsymbol{E}_1 + \cdots + \boldsymbol{E}_\nu \end{bmatrix} \begin{bmatrix} \mathbf{r} \\ \mathbf{y} \end{bmatrix} = \\ \left. = \begin{bmatrix} \mathbf{0}_\rho \\ \mathbf{0}_N \end{bmatrix} \right\}, S_{eIIO} \subseteq \mathcal{R}^{\alpha\rho+\nu N}; \tag{3.130}$$

– *the origin* $\mathbf{r}^{\alpha-1} = \mathbf{0}_{\alpha\rho}$ *if* $\nu = 0$,

$$S_{eIIO} = \left\{ \mathbf{r}^{\alpha-1} : \mathbf{r}^{\alpha-1} = \begin{bmatrix} \mathbf{r}^T & \mathbf{r}^T & \cdots & \mathbf{r}^T \end{bmatrix}^T \in \mathcal{R}^{\alpha\rho}, \right.$$
$$\left. \left(\boldsymbol{Q}_0 + \boldsymbol{Q}_1 + \cdots + \boldsymbol{Q}_\alpha \right) \mathbf{r} = \mathbf{0}_\rho \right\}. \tag{3.131}$$

Proof. Theorem 3.11 and the theorem on the number of the solutions of the homogeneous linear vector algebraic equation imply this theorem. ■

Note 3.18 *The set* S_{eIIO} *of the equilibrium states of the IIO system (3.122a) and (3.122b) becomes the singleton if and only if*

$$\det \begin{bmatrix} \boldsymbol{Q}_0 + \boldsymbol{Q}_1 + \cdots + \boldsymbol{Q}_\alpha & \boldsymbol{O}_{\rho N} \\ -\left(\boldsymbol{R}_0 + \boldsymbol{R}_1 + \cdots + \boldsymbol{R}_\alpha \right) & \boldsymbol{E}_0 + \boldsymbol{E}_1 + \cdots + \boldsymbol{E}_\nu \end{bmatrix} \neq 0, \ \nu \geq 1,$$
$$\det \left(\boldsymbol{Q}_0 + \boldsymbol{Q}_1 + \cdots + \boldsymbol{Q}_\alpha \right) \neq 0, \ \nu = 0.$$

The IIO system (3.122a) and (3.122b) then, and only then, has the unique equilibrium vector, i.e., the set S_{eIIO} *of all system equilibrium vectors is singleton:*

$$S_{eIIO} = \left\{ \begin{bmatrix} \mathbf{r}^{\alpha-1} \\ \mathbf{y}^{\nu-1} \end{bmatrix} : \begin{bmatrix} \mathbf{r}^{\alpha-1} \\ \mathbf{y}^{\nu-1} \end{bmatrix} = \begin{bmatrix} \begin{bmatrix} \mathbf{r}^T & \mathbf{r}^T & \cdots & \mathbf{r}^T \end{bmatrix}^T \\ \begin{bmatrix} \mathbf{y}^T & \mathbf{y}^T & \cdots & \mathbf{y}^T \end{bmatrix}^T \end{bmatrix} = \right.$$
$$\left. = \begin{bmatrix} \mathbf{0}_{\alpha\rho} \\ \mathbf{0}_{\nu N} \end{bmatrix} \in \mathcal{R}^{\alpha\rho+\nu N} \right\} = \left\{ \begin{bmatrix} \mathbf{0}_{\alpha\rho} \\ \mathbf{0}_{\nu N} \end{bmatrix} \right\} = \left\{ \mathbf{0}_{\alpha\rho+\nu N} \right\} \subset \mathcal{R}^{\alpha\rho+\nu N}, \nu \geq 1,$$
$$S_{eIIO} = \left\{ \mathbf{r}^{\alpha-1} : \mathbf{r}^{\alpha-1} = \begin{bmatrix} \mathbf{r}^T & \mathbf{r}^T & \cdots & \mathbf{r}^T \end{bmatrix}^T \right\} = \left\{ \mathbf{0}_{\alpha\rho} \right\} \subset \mathcal{R}^{\alpha\rho}, \ \nu = 0.$$

Chapter 4

Transfer function matrix $G(z)$

The transfer function $G(z)$ of a $SISO$ system is defined as the ratio of the $Z-$transform of the system output to the $Z-$transform of the system input *under all zero initial conditions*. It is well known that the transfer function is the $Z-$transform of the unit impulse response of the system *under all zero initial conditions.* By following that definition, the transfer function matrix of a $MIMO$ system is defined as the matrix composed of the system transfer functions, which relates the $Z-$transform of the output vector to the $Z-$transform of the input vector *under all zero initial conditions* [2], [6], [8], [14], [18], [41], [61], [80], [81], [89], [96], [101], [108], [110], [122]. That definition expresses the physical sense of the transfer function matrix from the point of view of the transmission and the transformation of the input vector onto and into the output system response, respectively.

Another definition of the transfer function matrix, perhaps due to Desoer [22], defines it as the Laplace transform of the corresponding matrix of the system unit impulse responses *under all zero initial conditions* [5], [16], [18], [84], [125], [126], [127] in the framework of continuous-time systems. Analogous definition of the $Z-$transform function matrix, in the framework of discrete-time systems, is possible. That definition contains the physical interpretation of the transfer function matrix from the point of view of the output system response to the specific (unit impulse) input vector.

Those two definitions of $G(z)$ and of $\mathbf{G}(z)$ are equivalent. Both assume *all zero initial conditions. They cannot relate the $Z-$transform of the system output to the $Z-$transform of its input as soon as any initial condition is not equal to zero.*

The validity of the transfer function matrix $\boldsymbol{G}(z)$ exclusively for all zero initial conditions constraints sharply its applications, validity and significance. Its application to the problems induced by nonzero initial conditions gives sometimes so wrong result that is contrary to the correct result. For example, its application to the Lyapunov stability test gives the wrong result if there is the same zero and pole with their moduli greater or equal to one, which are cancelled, and the moduli of all other poles are smaller than one. Moreover, the transfer function matrix $\boldsymbol{G}(z)$ is not applicable to the determination of the complete system response due to the influence of the input vector and nonzero initial conditions. The real initial conditions are seldom equal to zero. The following part of the book will show how we can very effectively overcome this severe restriction imposed on $\boldsymbol{G}(z)$.

Part II

FULL TRANSFER FUNCTION MATRIX $F(z)$ AND SYSTEM REALIZATION

Chapter 5

Problem statement

The validity of the superposition principle induced by the linearity of the systems has permitted us to study separately systems reactions and properties under zero initial conditions in a forced regime from the study of systems reactions and properties under zero input variables, i.e., in a free regime. This has simplified the study of many problems among which are the zero-state system equivalency and the zero-input system equivalency, see [5] for continuous-time systems. However, the zero-state systems equivalency and the zero-input system equivalency do not necessarily imply the system equivalence, which was discovered and proved by P. J. Antsaklis and A. N. Michel in [5, p. 171] for continuous-time systems.

Antsaklis and Michel [5, p. 387] discovered also that different state-space realizations of the system transfer function matrix yield the same zero-state system response, but the corresponding zero-input (hence, also complete) system response can be very different.

The discoveries and observations by Antsaklis and Michel reflect the reality of the system surrounding and system past. Consequently, we should also investigate the simultaneous influence of arbitrary initial conditions and any permitted inputs in context of discrete-time systems. The significance of this conclusion is crucial not only for the system overall (i.e., complete) both motion and response, but also for the system equivalence, for the system realization and the system minimal realization, and for various system dynamical properties (i.e., $BIBO$ and L-stability if the system was not at rest at the initial moment, system tracking and system optimality). This creates the *Main problem* that is another, more specific form of the *Basic problem* 3.1 (in Subsection 3.3.4):

Problem 5.1 ***Main problem***

What is a compact mathematical description, in the complex domain $\mathcal{C}$, of how the system transfers, in the course of time, the influence of any (permitted) input vector and of any initial conditions on its state $\mathbf{s}(\cdot)$ and/or on its output response $\mathbf{y}(\cdot)$?

Can such a mathematical description be invariant relative to the system input vector and to all initial conditions, i.e., can it be fully independent of the system input vector and of all initial conditions?

The solutions to this complex problem for the IO, ISO, and IIO systems represent a fundamental contribution of this book. The basis of the problem solutions are the introduction, the definition and the derivation of ***the complete (full) matrix transfer function*** $\boldsymbol{F}(\cdot)$, and its complex matrix value: ***the complete (full) transfer function matrix*** $\boldsymbol{F}(z)$, for IO, and ISO systems in [61]. For the hystorical review of the creation of $\boldsymbol{F}(s)$ (for continuous-time systems) see [67] and [69]. The introduction, the definition and the derivation of the full transfer function matrix $\boldsymbol{F}(z)$ of discrete-time time-invariant systems was done through the course on *Control of industrial processes* (2002, 2003) [61] at the University of Technology Belfort-Montbéliard, Belfort, France. Then, all that was realized for IO and ISO, but not for IIO systems. We will show in the sequel, additionally, by referring to both [67] and [69], how the system full transfer function matrix $\boldsymbol{F}(z)$ can be applied to study various dynamical properties of the systems and to solving the corresponding problems.

The notions and properties of *degenerate* and *nondegenerate matrix functions*, and the knowledge of their differences from the well-known reducible and irreducible matrices, are indispensable for the study stability properties of the systems by using their full transfer function matrices $\boldsymbol{F}(z)$.

Chapter 6

Nondegenerate matrices

If the reader is not familiar with the definitions and properties of the greatest common (left, right) divisors of the matrix polynomials, with the unimodular matrix polynomials and with their (left, right) coprimeness, then the reader should consult the appropriate literature, for example, the books by P. J. Antsaklis and A. N. Michel [5, pp. 526-528, 535-540], C.-T. Chen [18, pp. 591-599] and/or by T. Kailath [84, pp. 373-382].

A rational matrix function

$$\boldsymbol{M}(\cdot) = \boldsymbol{M}_D^{-1}(\cdot)\,\boldsymbol{M}_N(\cdot)\ \left[\boldsymbol{M}(\cdot) = \boldsymbol{M}_N(\cdot)\,\boldsymbol{M}_D^{-1}(\cdot)\right]$$

is *irreducible* [18, p. 605], [84, p. 370] if, and only if its polynomial matrices $\boldsymbol{M}_D(\cdot)$ and $\boldsymbol{M}_N(\cdot)$ are (left and/or right) coprime. Such a definition of the irreducible matrices corresponds well to a scalar case, i.e., if $\boldsymbol{M}(\cdot)$ is a rational function; $\boldsymbol{M}_D(\cdot)$ and $\boldsymbol{M}_N(\cdot)$ are then scalar polynomials. Its generalization to rational matrix functions is not adequate because the greatest common (left $\boldsymbol{L}(\cdot)$, and right $\boldsymbol{R}(\cdot)$) divisor of $\boldsymbol{M}_D(\cdot)$ and of $\boldsymbol{M}_N(\cdot)$ cancels itself in $\boldsymbol{M}(\cdot)$, in spite $\boldsymbol{L}(\cdot)$ and $\boldsymbol{R}(\cdot)$ are unimodular polynomial matrices,

$$\boldsymbol{M}_D(z) = \boldsymbol{L}(z)\,\boldsymbol{D}(z),\ \boldsymbol{M}_N(z) = \boldsymbol{L}(z)\,\boldsymbol{N}(z) \Longrightarrow$$
$$\boldsymbol{M}(\cdot) = \boldsymbol{M}_D^{-1}(\cdot)\,\boldsymbol{M}_N(\cdot) = \boldsymbol{D}^{-1}(z)\,\boldsymbol{L}^{-1}(z)\,\boldsymbol{L}(z)\,\boldsymbol{N}(z) = \boldsymbol{D}^{-1}(z)\,\boldsymbol{N}(z),$$

$$\boldsymbol{M}_D(z) = \boldsymbol{D}(z)\,\boldsymbol{R}(z),\ \boldsymbol{M}_N(z) = \boldsymbol{N}(z)\,\boldsymbol{R}(z) \Longrightarrow$$
$$\boldsymbol{M}(\cdot) = \boldsymbol{M}_N(\cdot)\,\boldsymbol{M}_D^{-1}(\cdot) = \boldsymbol{N}(z)\,\boldsymbol{R}(z)\,\boldsymbol{R}^{-1}(z)\,\boldsymbol{D}^{-1}(z) = \boldsymbol{N}(z)\,\boldsymbol{D}^{-1}(z).$$

This opens the task to define *nondegenerate* and *degenerate* rational matrix functions. By following [34], [68], [69] we introduce

Definition 6.1 *A rational matrix function* $\mathbf{M}(\cdot) = \mathbf{M}_D^{-1}(\cdot)\mathbf{M}_N(\cdot)$ *[respectively,* $\mathbf{M}(\cdot) = \mathbf{M}_N(\cdot)\mathbf{M}_D^{-1}(\cdot)$*] is*

a) ***row nondegenerate*** *if and only if respectively:*

(i) *the greatest common left [right] divisor of* $\mathbf{M}_D(\cdot)$ *and of* $\mathbf{M}_N(\cdot)$ *is a nonzero constant matrix, and*

(ii) *the greatest common scalar factors of* $\det \mathbf{M}_D(z)$ *and of all elements of every row of* $(\operatorname{adj}\mathbf{M}_D(z))\mathbf{M}_N(z)$ *[respectively, of all elements of every row of* $\mathbf{M}_N(z)(\operatorname{adj}\mathbf{M}_D(z))$*] are nonzero constants.*

Otherwise, $\mathbf{M}(\cdot)$ *is* ***row degenerate****;*

b) ***column nondegenerate*** *if and only if respectively:*

(i) *the greatest common left [right] divisor of* $\mathbf{M}_D(\cdot)$ *and* $\mathbf{M}_N(\cdot)$ *is a nonzero constant matrix, and*

(ii) *the greatest common scalar factors of* $\det \mathbf{M}_D(z)$ *and of all elements of every column of* $(\operatorname{adj}\mathbf{M}_D(z))\mathbf{M}_N(z)$ *[respectively, of all elements of every column of* $\mathbf{M}_N(z)(\operatorname{adj}\mathbf{M}_D(z))$*] are nonzero constants.*

Otherwise, $\mathbf{M}(\cdot)$ *is* ***column degenerate****;*

c) ***nondegenerate*** *if and only if respectively:*

(i) *the greatest common left [right] divisor of* $\mathbf{M}_D(\cdot)$ *and* $\mathbf{M}_N(\cdot)$ *is a nonzero constant matrix, and*

(ii) *the greatest common scalar factor of* $\det \mathbf{M}_D(z)$ *and of all elements of* $(\operatorname{adj}\mathbf{M}_D(z))\mathbf{M}_N(z)$ *[respectively, of all elements of the product* $\mathbf{M}_N(z)(\operatorname{adj}\mathbf{M}_D(z))$*] is a nonzero constant.*

Otherwise, $\mathbf{M}(\cdot)$ *is* ***degenerate****.*

This definition implies the following:

Note 6.1 *If a rational matrix function* $\mathbf{M}(\cdot) = \mathbf{M}_D^{-1}(\cdot)\mathbf{M}_N(\cdot)$ *[respectively,* $\mathbf{M}(\cdot) = \mathbf{M}_N(\cdot)\mathbf{M}_D^{-1}(\cdot)$*] is either row nondegenerate or column nondegenerate, or both, then it is also nondegenerate.*

Example 6.1 *Let*

$$M(z) = M_D^{-1}(z)\, M_N(z) = [(z+1)(z+2)]^{-1} \begin{bmatrix} z+1 & z+1 \\ z+2 & z+2 \end{bmatrix}.$$

It is both column nondegenerate and nondegenerate despite the fact that it is row degenerate. The greatest common factor of $\det M_D(z) = (z+1)(z+2)$ *and of all elements of the first row of* $M_N(z)$,

$$M_N(z) = \begin{bmatrix} z+1 & z+1 \\ z+2 & z+2 \end{bmatrix},$$

is $z+1$. *The greatest common factor of* $\det M_D(z) = (z+1)(z+2)$ *and of all elements of the second row of* $M_N(z)$ *is* $z+2$ *that is different from* $z+1$. *The greatest common factor of* $\det M_D(z) = (z+1)(z+2)$ *and of all elements of the first column of* $M_N(z)$ *is* 1. *The same holds for the greatest common factor of* $\det M_D(z)$ *and of all elements of the second column of* $M_N(z)$, *as well as for the greatest common factor of* $\det M_D(z)$ *and of all elements of* $M_N(z)$. *The given* $M(z) = M_D^{-1}(z) M_N(z)$ *is both column nondegenerate and nondegenerate despite the fact that it is row degenerate.*

Example 6.2 *Let*

$$M(z) = \frac{1}{(z+1)(z+2)} \begin{bmatrix} z+1 & z+2 \\ z+1 & z+2 \end{bmatrix}.$$

It is both row nondegenerate and nondegenerate despite the fact that it is column degenerate.

Note 6.2 *If a rational matrix function* $M(\cdot) = M_D^{-1}(\cdot) M_N(\cdot)$ *[respectively,* $M(\cdot) = M_N(\cdot) M_D^{-1}(\cdot)$*] is degenerate, then it is both row degenerate and column degenerate.*

Example 6.3 *Let*

$$M(z) = \frac{1}{(z+1)(z+2)} \begin{bmatrix} z(z+1) & (z+3)(z+1) \\ 4(z+1) & (z+1)(z+2) \end{bmatrix}.$$

It is degenerate, and both column degenerate and row degenerate. Its nondegenerate form $M_{nd}(z)$,

$$M_{nd}(z) = \frac{1}{z+2} \begin{bmatrix} z & z+3 \\ 4 & z+2 \end{bmatrix}$$

is in this case also both column and row nondegenerate.

Note 6.3 *If a rational matrix function* $\boldsymbol{M}(\cdot) = \boldsymbol{M}_D^{-1}(\cdot)\boldsymbol{M}_N(\cdot)$ *[respectively,* $\boldsymbol{M}(\cdot) = \boldsymbol{M}_N(\cdot)\boldsymbol{M}_D^{-1}(\cdot)$*] is nondegenerate, then it is also irreducible, but the opposite does not hold in general (i.e., it can be irreducible but need not be nondegenerate).*

Example 6.4 *Let* $\boldsymbol{M}(\cdot) = \boldsymbol{M}_D^{-1}(\cdot)\boldsymbol{M}_N(\cdot)$,

$$\boldsymbol{M}_D(z) = \begin{bmatrix} z^2 + z - 8 & 2z^2 + 3z - 2 \\ 2z^2 - 2z - 12 & 4z^2 - 2z \end{bmatrix} =$$
$$= \begin{bmatrix} z+2 & 6 \\ 2z & 12 \end{bmatrix} \begin{bmatrix} z-1 & 2z-1 \\ -1 & 0 \end{bmatrix},$$

$$\boldsymbol{M}_N(s) = \begin{bmatrix} 2z^2 + 3z - 2 & 6z^2 + 9z - 6 & 4z^2 + 6z - 4 \\ 4z^2 - 2z & 12z^2 - 6z & 8z^2 - 4z \end{bmatrix} =$$
$$= \begin{bmatrix} z+2 & 6 \\ 2z & 12 \end{bmatrix} \begin{bmatrix} 2z-1 & 6z-3 & 4z-2 \\ 0 & 0 & 0 \end{bmatrix}.$$

The polynomial matrices $\boldsymbol{M}_D(\cdot)$ *and* $\boldsymbol{M}_N(\cdot)$ *are left coprime. Their greatest left common divisor* $\boldsymbol{L}(\cdot)$,

$$\boldsymbol{L}(z) = \begin{bmatrix} z+2 & 6 \\ 2z & 12 \end{bmatrix}, \ \boldsymbol{L}^{-1}(z) = \frac{1}{24}\begin{bmatrix} 12 & -6 \\ -2z & z+2 \end{bmatrix}, \ \det \boldsymbol{L}(z) = 24,$$

is unimodular and cancels itself in $\boldsymbol{M}(\cdot)$*. The rational matrix function* $\boldsymbol{M}(\cdot)$ *is irreducible in the sense of the definition in [18, p. 605], [84, p. 370]. However, it is really further reducible, i.e., it is degenerate. The reduced form of* $\boldsymbol{M}(z)$ *obtained after the cancellation of* $\boldsymbol{L}(z)$ *reads*

$$\boldsymbol{M}_{irr}(z) = \frac{1}{2z-1}\begin{bmatrix} 0 & -2z+1 \\ 1 & z-1 \end{bmatrix} \begin{bmatrix} 2z-1 & 6z-3 & 4z-2 \\ 0 & 0 & 0 \end{bmatrix} =$$
$$= \frac{1}{2z-1}\begin{bmatrix} 0 & 0 & 0 \\ 2z-1 & 6z-3 & 4z-2 \end{bmatrix} =$$
$$= \frac{1}{2z-1}\begin{bmatrix} 0 & 0 & 0 \\ 2z-1 & 3(2z-1) & 2(2z-1) \end{bmatrix}.$$

It is degenerate because the polynomial $2z-1$ *is common to* $\det \boldsymbol{M}_{irrD}(z) = 2z-1$ *and to all elements of* $(\operatorname{adj} \boldsymbol{M}_{irrD}(z))\boldsymbol{M}_{irrN}(z)$:

$$(\operatorname{adj} \boldsymbol{M}_{irrD}(z))\, \boldsymbol{M}_{irrN}(z) = \begin{bmatrix} 0 & 0 & 0 \\ 2z-1 & 3(2z-1) & 2(2z-1) \end{bmatrix}.$$

Evidently, the polynomial $2z-1$ is not constant. We can cancel it in the denominator and in all entries of the matrix $\mathbf{M}_{irr}(z)$,

$$\mathbf{M}_{irr}(z) = \frac{1}{2z-1}\begin{bmatrix} 0 & 0 & 0 \\ 2z-1 & 3(2z-1) & 2(2z-1) \end{bmatrix} =$$

$$= \frac{1}{2z-1}(2z-1)\begin{bmatrix} 0 & 0 & 0 \\ 1 & 3 & 2 \end{bmatrix}.$$

The final, completely reduced form, hence fully irreducible form, i.e., the nondegenerate form $\mathbf{M}_{nd}(z)$, of $\mathbf{M}(z)$ reads

$$\mathbf{M}_{nd}(z) = \begin{bmatrix} 0 & 0 & 0 \\ 1 & 3 & 2 \end{bmatrix}.$$

It is different from the irreducible form $\mathbf{M}_{irr}(z)$,

$$\mathbf{M}_{nd}(z) = \begin{bmatrix} 0 & 0 & 0 \\ 1 & 3 & 2 \end{bmatrix} \neq$$

$$\neq \frac{1}{2z-1}\begin{bmatrix} 0 & 0 & 0 \\ 2z-1 & 3(2z-1) & 2(2z-1) \end{bmatrix} = \mathbf{M}_{irr}(z).$$

Example 6.5 *Let us consider 1×7 row matrix $\mathbf{M}_{IIO}(z)$,*

$$\mathbf{M}_{IIO}(z) = \left[(z-1)^2(z+2)(z+5)\right]^{-1}\begin{bmatrix} (z-1)^2(z-6)(17z+10) \\ -(z-1)(z-7)(17z+10) \\ -(z-1)(13z-10) \\ 10(z-1) \\ -5(z-1)(2z+5) \\ (z-1)(z^2+6z+5) \\ (z-1)(z+5) \end{bmatrix}^T. \quad (6.1)$$

It is (row) degenerate because $(z-1)$ is common factor to the denominator polynomial $(z-1)^2(z+2)(z+5)$ and to all entries of the numerator polynomial matrix that is row vector. Its (row) nondegenerate form $\mathbf{M}_{IIOrnd}(z)$ results after the cancellation of $(z-1)$:

$$\mathbf{M}_{IIOrnd}(z) = [(z-1)(z+2)(z+5)]^{-1}\begin{bmatrix} (z-1)(z-6)(17z+10) \\ -(z-7)(17z+10) \\ -(13z-10) \\ 10 \\ -5(2z+5) \\ (z^2+6z+5) \\ (z+5) \end{bmatrix}^T. \quad (6.2)$$

It is also irreducible.

Note 6.4 *Smith-McMillan [5, pp. 298-299] form of a rational matrix is by definition of nondegenerate matrix and irreducible matrix.*

Lemma 6.1 ***Basic lemma***

Let $\boldsymbol{M}(\cdot)$ be a real rational proper matrix function of z. Let $\mathbf{Z}(\cdot)$, and $\mathbf{W}(\cdot)$ be real rational proper vector functions of z, which are interrelated via $\boldsymbol{M}(\cdot)$,

$$\mathbf{Z}(z) = \boldsymbol{M}(z)\,\mathbf{W}(z),\ \ \mathbf{Z}(z) \in \mathcal{C}^p,\ \ \boldsymbol{M}(z) \in \mathcal{C}^{p\times q},\ \ \mathbf{W}(z) \in \mathcal{C}^q. \tag{6.3}$$

1) *Any equal pole and zero common to all elements of the same row of $\boldsymbol{M}(z)$ do not influence the character of the original $\mathbf{z}(t)$ of $\mathbf{Z}(z)$ and may be cancelled.*

2) *Any equal pole and zero of any entry of $\mathbf{W}(z)$ do not influence the character of the corresponding entry of the original $\mathbf{z}(t)$ of $\mathbf{Z}(z)$ and may be cancelled.*

3) *Any equal pole and zero of any entry of $\boldsymbol{M}(z)\mathbf{W}(z)$ do not influence the character of the corresponding entry of the original $\mathbf{z}(t)$ of $\mathbf{Z}(z)$ and may be cancelled.*

4) *The poles of the row nondegenerate form $[\boldsymbol{M}(z)\mathbf{W}(z)]_{nd}$ of $\boldsymbol{M}(z)\mathbf{W}(z)$ determine the character of the original $\mathbf{z}(t)$ of $\mathbf{Z}(z)$, where $\mathbf{z}(t)$ is the inverse $Z-$transform of $\mathbf{Z}(z)$,*

$$\mathbf{z}(t) = \mathcal{Z}^{-1}\left\{\mathbf{Z}(z)\right\} = \mathcal{Z}^{-1}\left\{[\boldsymbol{M}(z)\mathbf{W}(z)]_{rnd}\right\}. \tag{6.4}$$

5) *If every zero of every element of every row of $\boldsymbol{M}(z)$ is different from every pole of the corresponding entry of $\mathbf{W}(z)$, and every pole of every element of every row of $\boldsymbol{M}(z)$ is different from every zero of the corresponding entry of $\mathbf{W}(z)$, then the row nondegenerate form $[\boldsymbol{M}(z)\mathbf{W}(z)]_{rnd}$ of $\boldsymbol{M}(z)\mathbf{W}(z)$ becomes the product of the row nondegenerate forms $\boldsymbol{M}(z)_{rnd}$ and $\mathbf{W}(z)_{rnd}$ of $\boldsymbol{M}(z)$ and $\mathbf{W}(z)$,*

$$[\boldsymbol{M}(z)\mathbf{W}(z)]_{rnd} = \boldsymbol{M}(z)_{rnd}\mathbf{W}(z)_{rnd}. \tag{6.5}$$

Then (6.4) reduces to

$$\mathbf{z}(t) = \mathcal{Z}^{-1}\left\{\mathbf{Z}(z)\right\} = \mathcal{Z}^{-1}\left\{\boldsymbol{M}(z)_{rnd}\mathbf{W}(z)_{rnd}\right\}. \tag{6.6}$$

Proof. Let the conditions of this Lemma 6.1 hold. From (6.3) and (6.4) we determine the original $\mathbf{z}(t)$. Let the ij-th element of $\boldsymbol{M}(z)$ be $m_{ij}(z)$, the i-th element of $\mathbf{z}(t)$ be $z_i(t)$, the i-th element of $\mathbf{Z}(z)$ be $\varsigma_i(z)$, and the j-th element of $\mathbf{W}(z)$ be $w_j(z)$, so that

$$z_i(t) = \mathcal{Z}^{-1}\{\varsigma_i(z)\} = \mathcal{Z}^{-1}\left\{\sum_j m_{ij}(z)w_j(z)\right\}. \tag{6.7}$$

Since $\boldsymbol{M}(\cdot)$ is a real rational proper matrix function of z, $\mathbf{Z}(\cdot)$ and $\mathbf{W}(\cdot)$ are real rational proper vector functions of z, then the same holds for their entries, which can be presented in the factorized forms,

$$m_{ij}(z) = \frac{\prod_{k=1}^{\mu_{ij}}\left(z - z_{mk}^{0ij}\right)}{\prod_{k=1}^{\nu_{ij}}\left(z - z_{mk}^{*ij}\right)},\ \mu_{ij} \leq \nu_{ij},\ w_j(z) = \frac{\prod_{k=1}^{\upsilon_j}\left(z - z_{wk}^{0j}\right)}{\prod_{k=1}^{\omega_j}\left(z - z_{wk}^{*j}\right)},\ \upsilon_j \leq \omega_j.$$

These equations set (6.7) into the following form

$$z_i(t) = \mathcal{Z}^{-1}\{\varsigma_i(z)\} = \mathcal{Z}^{-1}\left\{\sum_j \frac{\prod_{k=1}^{\mu_{ij}}\left(z - z_{mk}^{0ij}\right)\prod_{k=1}^{\upsilon_j}\left(z - z_{wk}^{0j}\right)}{\prod_{k=1}^{\nu_{ij}}\left(z - z_{mk}^{*ij}\right)\prod_{k=1}^{\omega_j}\left(z - z_{wk}^{*j}\right)}\right\}.$$

We can conclude as follows. All residues of Heaviside expansion of $\varsigma_i(z) = \mathcal{Z}\{z_i(t)\}$ are equal to zero in a pole that is equal to a zero of $\varsigma_i(z)$. They can be cancelled. If $z_{mk}^{0ij} = z_{mk}^{*ij}$, $\forall j = 1, 2, \cdots, q$, then they should be cancelled. This proves 1). If $z_{wk}^{0j} = z_{wk}^{*j}$, $\forall j = 1, 2, \cdots, q$, then they should be also cancelled. This proves 2). The equal poles and zeros of $m_{ij}(z)w_j(z)$, $\forall j = 1, 2, \cdots, q$, do not influence $z_i(t)$. They should be cancelled, too. This proves 3). The equal poles and zeros of any entry of $\boldsymbol{M}(z)\mathbf{W}(z)$ do not influence $\mathbf{z}(t)$. They should be cancelled. The result is the row nondegenerate form $[\boldsymbol{M}(z)\mathbf{W}(z)]_{rnd}$ of $\boldsymbol{M}(z)\mathbf{W}(z)$,

$$\mathbf{z}(t) = \mathcal{Z}^{-1}\{\mathbf{Z}(z)\} = \mathcal{Z}^{-1}\{[\boldsymbol{M}(z)\mathbf{W}(z)]_{rnd}\},$$

i.e., (6.4). They determine the original $\mathbf{z}(t)$. The claim under 4) is correct. Let every zero of every element of every row of $\boldsymbol{M}(z)$ be different from every pole of the corresponding entry of $\mathbf{W}(z)$, and let every pole of every element of every row of $\boldsymbol{M}(z)$ be different from every zero of the corresponding entry of $\mathbf{W}(z)$. Then, the possible zero-pole cancellation is possible only among zeros and poles of the elements of the rows of $\boldsymbol{M}(z)$, and independently of them among zeros and poles of the members of the entries of $\mathbf{W}(z)$. The cross-cancellations of the zeros/poles of the elements of the rows of $\boldsymbol{M}(z)$ with poles/zeros of the members of the entries of $\mathbf{W}(z)$ is not possible. After carrying out all possible cancellations in the elements of the rows of $\boldsymbol{M}(z)$ and in the components of the entries of $\mathbf{W}(z)$ we get the row nondegenerate form $[\boldsymbol{M}(z)\mathbf{W}(z)]_{rnd}$ of $\boldsymbol{M}(z)\mathbf{W}(z)$ as the product of the row nondegenerate forms $\boldsymbol{M}(z)_{rnd}$ and $\mathbf{W}(z)_{rnd}$ of $\boldsymbol{M}(z)$ and $\mathbf{W}(z)$,

$$[\boldsymbol{M}(z)\mathbf{W}(z)]_{rnd} = \boldsymbol{M}(z)_{rnd}\mathbf{W}(z)_{rnd},$$

which is (6.5). This and (6.4) imply (6.6). ■

Example 6.6 *The IO system*

$$\begin{bmatrix} 1 & 1 \\ 1 & 2 \end{bmatrix} E^2\mathbf{y}(k) - \begin{bmatrix} 1 & 1 \\ 1 & 2 \end{bmatrix} \mathbf{y}(k) =$$

$$= \begin{bmatrix} -1 & 2 \\ 0 & -3 \end{bmatrix} \mathbf{i}(k) + \begin{bmatrix} 1 & 1 \\ -1 & 1 \end{bmatrix} E^1\mathbf{i}(k) + \begin{bmatrix} 0 & 0 \\ 1 & 0 \end{bmatrix} E^2\mathbf{i}(k)$$

has the transfer function matrix

$$\boldsymbol{G}_{IO}(z) = \frac{z^2-1}{(z^2-1)^2} \begin{bmatrix} -(z-1)(z-2) & z+7 \\ (z-1)^2 & -5 \end{bmatrix}.$$

Its nondegenerate form $\boldsymbol{G}_{IOnd}(z)$ reads

$$\boldsymbol{G}_{IOnd}(z) = \frac{1}{z^2-1} \begin{bmatrix} -(z-1)(z-2) & z+7 \\ (z-1)^2 & -5 \end{bmatrix}.$$

It is also its row nondegenerate form $\boldsymbol{G}_{IOrnd}(z)$,

$$\boldsymbol{G}_{IOnd}(z) = \boldsymbol{G}_{IOrnd}(z).$$

However, its column nondegenerate form $\boldsymbol{G}_{IOcnd}(z)$ is different from them,

$$\boldsymbol{G}_{IOcnd}(z) = \frac{1}{z+1} \begin{bmatrix} -(z-2) & z+7 \\ z-1 & -5 \end{bmatrix}.$$

If we wish to determine the system output response under all zero initial conditions we should use the row nondegenerate form $\boldsymbol{G}_{IOrnd}(z)$ of $\boldsymbol{G}_{IO}(z)$, and we may not use its column nondegenerate form $\boldsymbol{G}_{IOcnd}(z)$ because the pole $z^* = 1$ cannot be cancelled in the rows of $\boldsymbol{G}_{IOrnd}(z)$, although it can be cancelled in its columns.

Chapter 7

Definition of $\boldsymbol{F}(z)$

7.1 Definition of $\boldsymbol{F}(z)$ in general

The general definition of the system *full transfer function matrix* $\boldsymbol{F}(z)$ reflects the substance of its meaning:

Definition 7.1 ***The full (complete) input-output (IO) transfer function matrix of the dynamical system in general***

The full (complete) input-output (IO) transfer function matrix of a linear time-invariant discrete-time dynamical system, *which is denoted by* $\boldsymbol{F}(z)$, $\boldsymbol{F}(z) \in \mathcal{C}^{N\times(M+\varsigma)}$, *is the complex matrix value of* ***the system full (complete) input-output (IO) matrix transfer function*** $\boldsymbol{F}(\cdot)$, $\boldsymbol{F}(\cdot) : \mathcal{C} \to \mathcal{C}^{N\times(M+\varsigma)}$, *which is a matrix function of the complex variable* z *such that it determines uniquely the* $Z-$*transform* $\mathbf{Y}(z)$ *of the system output* $\mathbf{y}(k)$ *as a homogenous linear function of the* $Z-$*transform* $\mathbf{I}(z)$ *of the system input vector* $\mathbf{i}(k)$ *for an arbitrary variation of* $\mathbf{i}(k)$, *for arbitrary initial vector values* $\mathbf{i}_0^{\mu-1}$, *and/or* $\mathbf{r}_0^{\alpha-1}$, *and/or* $\mathbf{x}_0$, *and/or* $\mathbf{y}_0^{\nu-1}$ *of the extended input vector* $\mathbf{i}^{\mu-1}(k)$, *the state vector* $\mathbf{r}^{\alpha-1}(k)$, *the state vector* $\mathbf{x}(k)$, *and the extended output vector* $\mathbf{y}^{\nu-1}(k)$ *at* $k = 0$, *respectively:*

$$\mathbf{Y}(z) = \boldsymbol{F}(z)\left[\begin{array}{cccc} \mathbf{I}^T(z) & \left(\mathbf{i}_0^{\mu-1}\right)^T & \mathbf{x}_0^T & \left(\mathbf{y}_0^{\nu-1}\right)^T \end{array}\right]^T, \tag{7.1}$$

see Fig. 7.1.

In order to broaden and to generalize the notations, definitions and techniques of the system transfer function matrix $\boldsymbol{G}(z)$ and of the classical

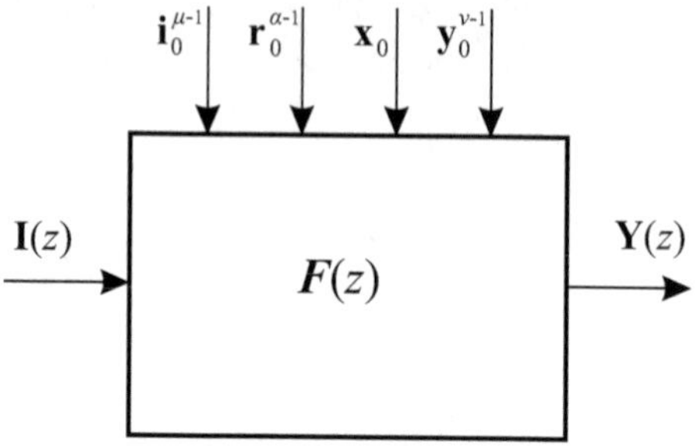

Figure 7.1: The full block of the system.

system block diagram induced by $\boldsymbol{G}(z)$ to the notations, definitions and techniques of the system *full transfer function matrix* $\boldsymbol{F}(z)$ and of the system *full block diagram* induced by $\boldsymbol{F}(z)$ we introduce the action vector function $\mathbf{v}(\cdot) : \mathcal{N}_0 \longrightarrow \mathcal{R}^{M+\varsigma}$:

$$\mathbf{v}(k) = \begin{bmatrix} \mathbf{i}(k) \\ \delta_d(k)\mathbf{C}_0 \end{bmatrix}. \tag{7.2}$$

It consists of the input vector function $\mathbf{i}(k)$ and of the vector $\mathbf{C}_0 \in \mathcal{R}^\varsigma$ of all initial conditions: of the input initial conditions ($E^0\mathbf{i}_0$, $E^1\mathbf{i}_0$, $\cdots$, $E^{\mu-1}\mathbf{i}_0$, i.e., $\mathbf{i}_0^{\mu-1}$), of the state initial conditions ($E^0\mathbf{r}_0$, $E^1\mathbf{r}_0$, $\cdots$, $E^{\alpha-1}\mathbf{r}_0$, i.e., $\mathbf{r}_0^{\alpha-1}$, or $\mathbf{x}_0$) and of the output initial conditions ($E^0\mathbf{y}_0$, $E^1\mathbf{y}_0$, $\cdots$, $E^{\nu-1}\mathbf{y}_0$, i.e., $\mathbf{y}_0^{\nu-1}$) of the system, in general,

$$\mathbf{C}_0 = \mathbf{C}_0\left(\mathbf{i}_0^{\mu-1}, \mathbf{r}_0^{\alpha-1}, \mathbf{x}_0, \mathbf{y}_0^{\nu-1}\right). \tag{7.3}$$

The vector $\mathbf{C}_0$ has the following form

- for the *IO* systems:

$$\mathbf{C}_0 = \mathbf{C}_0\left(\mathbf{i}_0^{\mu-1}, \mathbf{y}_0^{\nu-1}\right) = \begin{bmatrix} \mathbf{i}_0^{\mu-1} \\ \mathbf{y}_0^{\nu-1} \end{bmatrix} \in \mathcal{R}^\varsigma, \ \varsigma = \mu M + \nu N; \tag{7.4}$$

- for the *ISO* systems:

$$\mathbf{C}_0 = \mathbf{C}_0(\mathbf{x}_0) = \mathbf{x}_0 \in \mathcal{R}^\varsigma, \ \varsigma = n; \tag{7.5}$$

- for the *IIO* systems:

$$\mathbf{C}_0 = \mathbf{C}_0\left(\mathbf{i}_0^{\mu-1}, \mathbf{r}_0^{\alpha-1}, \mathbf{y}_0^{\nu-1}\right) = \begin{bmatrix} \mathbf{i}_0^{\mu-1} \\ \mathbf{r}_0^{\alpha-1} \\ \mathbf{y}_0^{\nu-1} \end{bmatrix} \in \mathcal{R}^\varsigma,$$

$$\varsigma = \mu M + \alpha\rho + \nu N. \tag{7.6}$$

The $Z-$transform of $\mathbf{v}(\cdot)$ is $\mathbf{V}(\cdot):\mathcal{C}\longrightarrow\mathcal{C}^{M+\varsigma}$,

$$\mathbf{V}(z)=\begin{bmatrix}\mathbf{I}(z)\\ \mathbf{C}_0\end{bmatrix}\in\mathcal{R}^{M+\varsigma}. \tag{7.7}$$

The equivalent definition of the general Definition 7.1 reads:

Definition 7.2 ***The full (complete) input-output (IO) transfer function matrix of the dynamical system in general***

The full (complete) input-output (IO) transfer function matrix of a linear time-invariant discrete-time dynamical system, *which is denoted by* $\mathbf{F}(z)$, $\mathbf{F}(z)\in\mathcal{C}^{N\times(M+\varsigma)}$, *is the complex matrix value of* ***the system full (complete) input-output (IO) matrix transfer function*** $\mathbf{F}(\cdot)$, $\mathbf{F}(\cdot):\mathcal{C}\rightarrow\mathcal{C}^{N\times(M+\varsigma)}$, *which is a matrix function of the complex variable* z *such that it determines uniquely the* $Z-$*transform* $\mathbf{Y}(z)$ *of the system output* $\mathbf{y}(k)$ *as a homogenous linear function of the* $Z-$*transform* $\mathbf{V}(z)$ *of the overall system action vector* $\mathbf{v}(k)$ *for its arbitrary value and its variation,*

$$\mathbf{Y}(z)=\mathbf{F}(z)\mathbf{V}(z), \tag{7.8}$$

see Fig. 7.2.

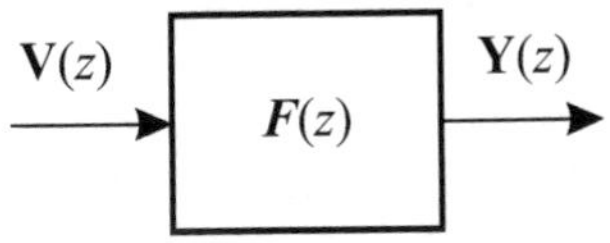

Figure 7.2: The full block of the system in the compact form.

Comment 7.1 *We present Fig. 7.1 and Fig. 7.2 as 7.3 in view of (7.7).*

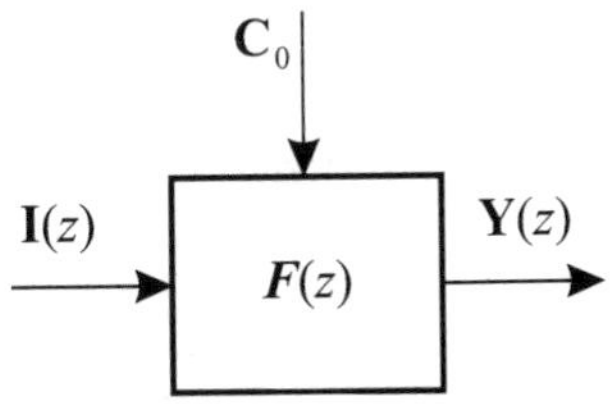

Figure 7.3: The full block of the system in the slightly extended form.

Chapter 9 introduces *the full block diagram as* the generalization of the block diagram and establishes *the algebra of the full block diagrams of the systems.*

7.2 Definition of $\boldsymbol{F}(z)$ of the IO system

The full transfer function matrix of the IO system (3.56) (in Subsection 3.5.2),

$$\boldsymbol{A}^{(\nu)}\mathbf{y}^{\nu}(k) = \boldsymbol{B}^{(\mu)}\mathbf{i}^{\mu}(k), \forall k \in \mathbb{Z}, \tag{7.9}$$

shows in the complex domain $\mathcal{C}$ how the system transfers in the course of *time* a simultaneous influence of arbitrary both input and output initial conditions $E^0\mathbf{i}_0$, $E^1\mathbf{i}_0$, $\cdots$, $E^{\mu-1}\mathbf{i}_0$, and $E^0\mathbf{y}_0$, $E^1\mathbf{y}_0$, $\cdots$, $E^{\nu-1}\mathbf{y}_0$, and of variations of the input vector function $\mathbf{i}(\cdot)$ on the system output response $\mathbf{y}(\cdot)$; see Fig. 7.4. We will use the notation $\operatorname{Int}\mathcal{N}_0$ for the *interior* of the

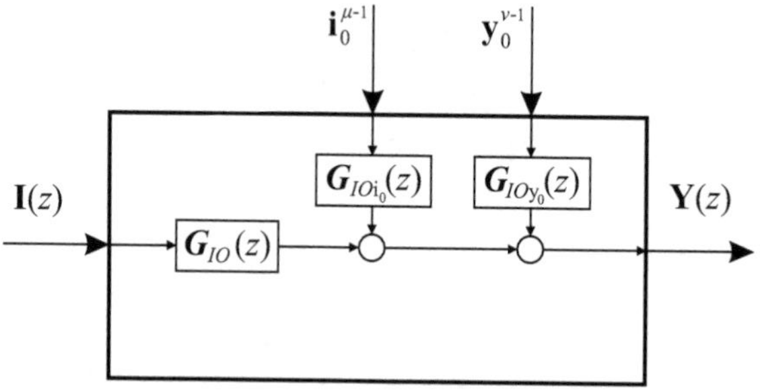

Figure 7.4: The block diagram of the IO system shows the system transfer function matrices relative to the input and initial conditions.

set $\mathcal{N}_0$, $\operatorname{Int}\mathcal{N}_0 = \mathcal{N}$, $\mathcal{N}$ is the set of natural numbers,

$$\operatorname{Int}\mathcal{N}_0 = \{k : \ k \in \mathcal{N}_0, \ k > 0\} \Longrightarrow \tag{7.10a}$$

$$\operatorname{Int}\mathcal{N}_0 = \mathcal{N}. \tag{7.10b}$$

Definition 7.3 a) ***The full (complete) input-output (IO) transfer function matrix of the IO system (7.9)****, denoted by* $\boldsymbol{F}_{IO}(z)$, $\boldsymbol{F}_{IO}(z) \in \mathcal{C}^{N\times[(\mu+1)M+\nu N]}$, *is the complex matrix value of* ***the system full (complete) IO matrix transfer function*** $\boldsymbol{F}_{IO}(\cdot)$, $\boldsymbol{F}_{IO}(\cdot) : \mathcal{C} \rightarrow \mathcal{C}^{N\times[(\mu+1)M+\nu N]}$, *which is a matrix function of the complex variable* z *such that it determines uniquely the* $Z-$*transform* $\mathbf{Y}(z)$ *of the system output* $\mathbf{y}(k)$ *as a homogenous linear function of the* $Z-$*transform* $\mathbf{I}(z)$ *of the system input* $\mathbf{i}(k)$ *for an arbitrary variation of* $\mathbf{i}(k)$, *of arbitrary initial vector values* $\mathbf{i}_0^{\mu-1}$ *and* $\mathbf{y}_0^{\nu-1}$ *of the extended input vector* $\mathbf{i}^{\mu-1}(k)$ *and of the extended output vector* $\mathbf{y}^{\nu-1}(k)$

at $k = 0$, respectively:

$$\mathbf{Y}(z) = \boldsymbol{F}_{IO}(z)\left[\begin{array}{ccc} \mathbf{I}^T(z) & \left(\mathbf{i}_0^{\mu-1}\right)^T & \left(\mathbf{y}_0^{\nu-1}\right)^T \end{array} \right]^T \Longrightarrow \tag{7.11a}$$

$$\mathbf{Y}(z) = \boldsymbol{F}_{IO}(z)\left[\begin{array}{cc} \mathbf{I}^T(z) & \mathbf{C}_{0IO}^T \end{array} \right]^T = \boldsymbol{F}_{IO}(z)\mathbf{V}_{IO}(z), \tag{7.11b}$$

$$\mathbf{C}_{0IO} = \left\{ \begin{array}{c} \left[\begin{array}{cc} \left(\mathbf{i}_0^{\mu-1}\right)^T & \left(\mathbf{y}_0^{\nu-1}\right)^T \end{array} \right]^T, \ \mu \geq 1, \\ \mathbf{y}_0^{\nu-1}, \ \mu = 0 \end{array} \right. , \tag{7.11c}$$

$$\mathbf{V}_{IO}(z) = \left[\begin{array}{cc} \mathbf{I}^T(z) & \mathbf{C}_{0IO}^T \end{array} \right]^T. \tag{7.11d}$$

b) ***The input-output (IO) transfer function matrix of the IO system (7.9)**, which is denoted by $\boldsymbol{G}_{IO}(z)$, $\boldsymbol{G}_{IO}(z) \in \mathcal{C}^{N\times M}$, is the complex matrix value of **the system IO matrix transfer function $\boldsymbol{G}_{IO}(\cdot)$, $\boldsymbol{G}_{IO}(\cdot) : \mathcal{C} \rightarrow \mathcal{C}^{N\times M}$**, which is a matrix function of the complex variable z such that it determines uniquely the $Z-$transform $\mathbf{Y}(z)$ of the system output vector $\mathbf{y}(k)$ as a homogenous linear function of the $Z-$transform $\mathbf{I}(z)$ of the system input vector $\mathbf{i}(k)$ for an arbitrary variation of $\mathbf{i}(k)$, and under all zero initial conditions, that is that the initial vector values $\mathbf{i}_0^{\mu-1} = \mathbf{0}_{\mu M}$ and $\mathbf{y}_0^{\nu-1} = \mathbf{0}_{\nu N}$ of the extended input vector $\mathbf{i}^{\mu-1}(k)$ and of the extended output vector $\mathbf{y}^{\nu-1}(k)$ at $k = 0$ are equal to zero vectors, respectively:*

$$\mathbf{Y}(z) = \boldsymbol{G}_{IO}(z)\mathbf{I}(z), \ \mathbf{i}_0^{\mu-1} = \mathbf{0}_{\mu M}, \ \mathbf{y}_0^{\nu-1} = \mathbf{0}_{\nu N}. \tag{7.12}$$

c) ***The input-output transfer function matrix (IOIC) relative to $\mathbf{i}_0^{\mu-1}$ of the IO system (7.9)**, which is denoted by $\boldsymbol{G}_{IOi_0}(z)$, $\boldsymbol{G}_{IOi_0}(z) \in \mathcal{C}^{N\times \mu M}$, is the complex matrix value of **the system IOIC matrix transfer function $\boldsymbol{G}_{IOi_0}(\cdot)$ relative to $\mathbf{i}_0^{\mu-1}$**, $\boldsymbol{G}_{IOi_0}(\cdot) : \mathcal{C} \rightarrow \mathcal{C}^{N\times \mu M}$, which is a matrix function of the complex variable z such that it determines uniquely, respectively, the $Z-$transform $\mathbf{Y}(z)$ of the system output $\mathbf{y}(k)$ as a homogenous linear function of an arbitrary initial vector $\mathbf{i}_0^{\mu-1} \neq \mathbf{0}_{\mu M}$ of the extended input vector $\mathbf{i}^{\mu-1}(k)$ at $k = 0$ in the free regime on $\mathcal{N}$ and for all zero output initial conditions, i.e., for $\mathbf{i}(k) = \mathbf{0}_M$, $\forall k \in \mathcal{N}$, and $\mathbf{y}_0^{\nu-1} \equiv \mathbf{0}_{\nu N}$:*

$$\mathbf{Y}(z) = \boldsymbol{G}_{IOi_0}(z)\mathbf{i}_0^{\mu-1}, \ \mathbf{i}(k) = \mathbf{0}_M, \ \forall k \in \mathcal{N}, \ \mathbf{y}_0^{\nu-1} \equiv \mathbf{0}_{\nu N}. \tag{7.13}$$

d) ***The input-output transfer function matrix (IOIY) relative to $\mathbf{y}_0^{\nu-1}$ of the IO system (7.9)**, which is denoted by $\boldsymbol{G}_{IOy_0}(z)$,*

$\boldsymbol{G}_{IOy_0}(z) \in \mathcal{C}^{N\times \nu N}$, *is the complex matrix value of* **the system IOIY matrix transfer function** $\boldsymbol{G}_{IOy_0}(\cdot)$ **relative to** $\mathbf{y}_0^{\nu-1}$, $\boldsymbol{G}_{IOy_0}(\cdot) : \mathcal{C} \rightarrow \mathcal{C}^{N\times \nu N}$, *which is a matrix function of the complex variable* z *such that it determines uniquely, respectively, the* $Z-$*transform* $\mathbf{Y}(z)$ *of the system output* $\mathbf{y}(k)$ *as a homogenous linear function of an arbitrary initial vector* $\mathbf{y}_0^{\nu-1}$ *of the extended output vector* $\mathbf{y}^{\nu-1}(k)$ *at* $k=0$ *for the system in a free regime and under all zero input initial conditions, i.e., for* $\mathbf{i}(k) \equiv \mathbf{0}_M$ *and* $\mathbf{i}_0^{\mu-1} = \mathbf{0}_{\mu M}$:

$$\mathbf{Y}(z) = \boldsymbol{G}_{IOy_0}(z)\mathbf{y}_0^{\nu-1},\ \mathbf{i}(k) \equiv \mathbf{0}_M,\ \mathbf{i}_0^{\mu-1} = \mathbf{0}_{\mu M}. \tag{7.14}$$

e) ***The input-output transfer function matrix relative to all initial conditions (IORAI) of the IO system (7.9)****, which is denoted by* $\boldsymbol{G}_{IO_0}(z)$, $\boldsymbol{G}_{IO_0}(z) \in \mathcal{C}^{N\times(\mu M+\nu N)}$, *is the complex matrix value of* **the system IORAI matrix transfer function** $\boldsymbol{G}_{IO_0}(\cdot)$ **relative to** $\left[\begin{array}{cc}\left(\mathbf{i}_0^{\mu-1}\right)^T & \left(\mathbf{y}_0^{\nu-1}\right)^T\end{array}\right]^T$, $\boldsymbol{G}_{IO_0}(\cdot) : \mathcal{C} \rightarrow \mathcal{C}^{N\times(\mu M+\nu N)}$, *which is a matrix function of the complex variable* z *such that it determines uniquely, respectively, the* $Z-$*transform* $\mathbf{Y}(z)$ *of the system output* $\mathbf{y}(k)$ *as a homogenous linear function of an arbitrary overall initial vector* $\mathbf{C}_{0IO} = \left[\begin{array}{cc}\left(\mathbf{i}_0^{\mu-1}\right)^T & \left(\mathbf{y}_0^{\nu-1}\right)^T\end{array}\right]^T \in \mathcal{R}^{\mu M+\nu N}$ *composed of the extended input vector* $\mathbf{i}^{\mu-1}(k)$ *and of the extended output vector* $\mathbf{y}^{\nu-1}(k)$ *at* $k=0$ *for the system in a free regime on* $\mathcal{N}$, *i.e., for* $\mathbf{i}(k) = \mathbf{0}_M$, $\forall k \in \mathcal{N}$:

$$\mathbf{Y}(z) = \boldsymbol{G}_{IO_0}(z)\left[\begin{array}{c}\mathbf{i}_0^{\mu-1}\\ \mathbf{y}_0^{\nu-1}\end{array}\right],\ \mathbf{i}(k) = \mathbf{0}_M,\ \forall k \in \mathcal{N}. \tag{7.15}$$

Note 7.1 *The matrices* $\boldsymbol{G}_{IO}(z)$, $\boldsymbol{G}_{IOi_0}(z)$, *and* $\boldsymbol{G}_{IOy_0}(z)$, *i.e.,* $\boldsymbol{G}_{IO_0}(z)$, *are submatrices of* $\boldsymbol{F}_{IO}(z)$*; see Fig. 7.5, which follows from the superposition principle:*

$$\boldsymbol{F}_{IO}(z) = \left[\begin{array}{cc}\boldsymbol{G}_{IO}(z) & \underbrace{\boldsymbol{G}_{IOi_0}(z)\ \ \boldsymbol{G}_{IOy_0}(z)}_{\boldsymbol{G}_{IO_0}(z)}\end{array}\right] = \left[\begin{array}{cc}\boldsymbol{G}_{IO}(z) & \boldsymbol{G}_{IO_0}(z)\end{array}\right],$$

$$\boldsymbol{G}_{IO_0}(z) = \left[\begin{array}{cc}\boldsymbol{G}_{IOi_0}(z) & \boldsymbol{G}_{IOy_0}(z)\end{array}\right]. \tag{7.16}$$

The matrices $\boldsymbol{G}_{IOi_0}(z)$ *and* $\boldsymbol{G}_{IOy_0}(z)$ *form the transfer function matrix* $\boldsymbol{G}_{IO_0}(z)$ *with respect to all initial conditions, The matrices* $\boldsymbol{G}_{IOi_0}(z)$ *and*

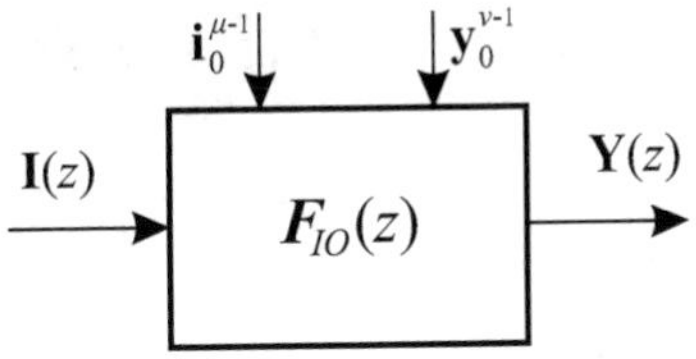

Figure 7.5: The block of the IO system shows the full transfer function matrix $\boldsymbol{F}_{IO}(z)$

$\boldsymbol{G}_{IOy_0}(z)$ form the transfer function matrix $\mathbf{G}_{IO_0}(z)$ with respect to all initial conditions,

$$\boldsymbol{G}_{IO_0}(z) = \left[\begin{array}{cc} \boldsymbol{G}_{IOi_0}(z) & \boldsymbol{G}_{IOy_0}(z) \end{array} \right]. \tag{7.17}$$

Note 7.2 *We replace the system transfer function matrix $\mathbf{G}_{IO}(z)$ by its full transfer function matrix $\boldsymbol{F}_{IO}(z)$ and we use the vector $\mathbf{V}_{IO}(z)$ instead of $\mathbf{I}(z)$ in order to generalize directly the classical block diagram technique, Fig. 7.6,*

$$\mathbf{V}_{IO}(z) = \left[\begin{array}{cc} \mathbf{I}^T(z) & \mathbf{C}_{0IO}^T \end{array} \right]^T = \left[\begin{array}{ccc} \mathbf{I}^T(z) & \left(\mathbf{i}_0^{\mu-1}\right)^T & \left(\mathbf{y}_0^{\nu-1}\right)^T \end{array} \right]^T,$$

$$\mathbf{C}_{0IO} = \left[\begin{array}{cc} \left(\mathbf{i}_0^{\mu-1}\right)^T & \left(\mathbf{y}_0^{\nu-1}\right)^T \end{array} \right]^T,$$

or in the slightly extended form, Fig. 7.7.

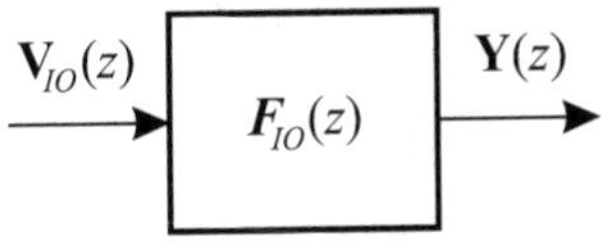

Figure 7.6: The full block of the IO system in the compact form.

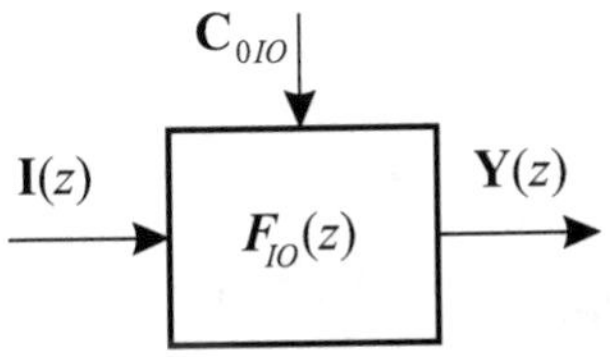

Figure 7.7: The slightly extended full block of the IO system.

7.3 Definition of $\boldsymbol{F}(z)$ of the *ISO* system

In order to ease the reading the *ISO* system (3.60a) and (3.60b) (in Subsection 3.5.3) is presented as (7.18) and (7.19),

$$\frac{d\mathbf{x}(t)}{dt} = \boldsymbol{A}\mathbf{x}(k) + \boldsymbol{B}\mathbf{i}(k), \forall k \in \mathcal{N}_0, \tag{7.18}$$

$$\mathbf{y}(k) = \boldsymbol{C}\mathbf{x}(k) + \boldsymbol{D}\mathbf{i}(k), \forall k \in \mathcal{N}_0. \tag{7.19}$$

and its full block diagram is presented in Fig. 7.8. We remind ourselves

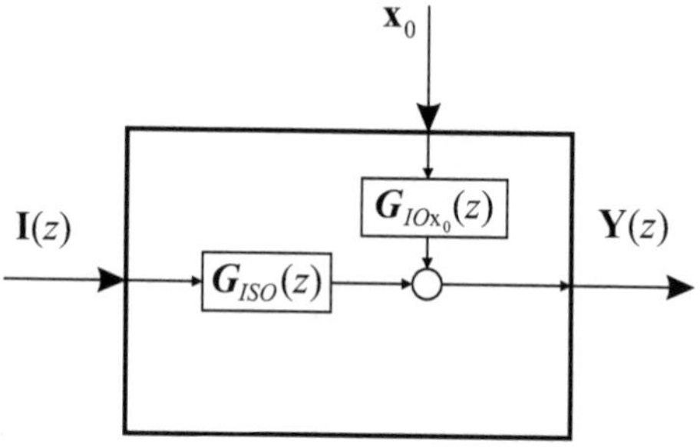

Figure 7.8: The full block diagram of the *ISO* system shows the system transfer function matrices.

that the system *full transfer function matrix* describes, in general, in the complex domain $\mathcal{C}$ how the system transmits and transfers in the course of time a simultaneous influence of an any initial state vector $\mathbf{x}_0$, and of variations of the input vector function $\mathbf{i}(\cdot)$ on the system output response $\mathbf{y}(\cdot)$, Fig. 7.8.

The input vector and the output vector of a given physical system are invariant relative to a form of the system mathematical model. The system transfer function matrix $\boldsymbol{G}(z)$ is independent of the type of the system description. Contrary to this, initial conditions have different meanings, i.e., forms, for the *IO* mathematical description (7.9) (in Section 7.2) and for the *ISO* mathematical description (7.18) and (7.19) of the same physical system. The form of mathematical model determines the form of the *ISO* system full transfer function matrix in general.

Definition 7.4 *a)* ***The full (complete) input-output (IO) transfer function matrix of the ISO system (7.18)*** *and* ***(7.19),*** *which is denoted by* $\boldsymbol{F}_{ISOIO}(z)$*,* $\boldsymbol{F}_{ISOIO}(z) \in \mathcal{C}^{N\times(M+n)}$*, for short* $\boldsymbol{F}_{ISO}(s)$*, is the complex matrix value of* ***the system full (complete)***

IO matrix transfer function $\boldsymbol{F}_{ISO}(\cdot)$, $\boldsymbol{F}_{ISO}(\cdot) : \mathcal{C} \to \mathcal{C}^{N\times(M+n)}$, *which is a matrix function of the complex variable* z *such that it determines uniquely the* $Z-$*transform* $\mathbf{Y}(z)$ *of the system output* $\mathbf{y}(k)$ *as a homogenous linear function of the* $Z-$*transform* $\mathbf{I}(z)$ *of the system input* $\mathbf{i}(k)$ *for an arbitrary variation of* $\mathbf{i}(k)$, *and of arbitrary initial vector values* $\mathbf{x}_0$ *of the state vector* $\mathbf{x}(k)$ *at* $k = 0$, *respectively:*

$$\mathbf{Y}(z) = \boldsymbol{F}_{ISO}(z) \begin{bmatrix} \mathbf{I}^T(z) & \mathbf{x}_0^T \end{bmatrix}^T \Longrightarrow \tag{7.20a}$$

$$\mathbf{Y}(z) = \boldsymbol{F}_{ISO}(z) \begin{bmatrix} \mathbf{I}^T(z) & \mathbf{C}_{0ISO}^T \end{bmatrix}^T = \boldsymbol{F}_{ISO}(z)\mathbf{V}_{ISO}(z), \tag{7.20b}$$

$$\mathbf{C}_{0ISO} = \mathbf{x}_0, \ \mathbf{V}_{ISO}(z) = \begin{bmatrix} \mathbf{I}^T(z) & \mathbf{C}_{0ISO}^T \end{bmatrix}^T. \tag{7.20c}$$

b) ***The input-output (IO) transfer function matrix of the ISO system (7.18)*** *and* ***(7.19),*** *which is denoted by* $\boldsymbol{G}_{ISO}(z)$, $\boldsymbol{G}_{ISO}(z) \in \mathcal{C}^{N\times M}$, *is the complex matrix value of* ***the system IO matrix transfer function*** $\boldsymbol{G}_{ISO}(\cdot)$, $\boldsymbol{G}_{ISO}(\cdot) : \mathcal{C} \to \mathcal{C}^{N\times M}$, *which is a matrix function of the complex variable* z *such that it determines uniquely the* $Z-$*transform* $\mathbf{Y}(z)$ *of the system output* $\mathbf{y}(k)$ *as a homogenous linear function of the* $Z-$*transform* $\mathbf{I}(z)$ *of the system input* $\mathbf{i}(k)$ *for an arbitrary variation of* $\mathbf{i}(k)$, *and for zero initial state vector* $\mathbf{x}_0$ *of the state vector* $\mathbf{x}(k)$ *at* $k = 0$, *respectively:*

$$\mathbf{Y}(z) = \boldsymbol{G}_{ISO}(z)\mathbf{I}(z), \ \mathbf{x}_0 = \mathbf{0}_n. \tag{7.21}$$

c) ***The input, initial state-output (IISO) transfer function matrix relative to*** $\mathbf{x}_0$ ***of the ISO system (7.18)*** *and* ***(7.19)****, which is denoted by* $\boldsymbol{G}_{ISOx_0}(z)$, $\boldsymbol{G}_{ISOx_0}(z) \in \mathcal{C}^{N\times n}$, *is the complex matrix value of* ***the system IISO matrix transfer function*** $\boldsymbol{G}_{ISOx_0}(\cdot)$ ***relative to*** $\mathbf{x}_0$, $\boldsymbol{G}_{ISOx_0}(\cdot) : \mathcal{C} \to \mathcal{C}^{N\times n}$, *which is a matrix function of the complex variable* z *such that it determines uniquely the* $Z-$*transform* $\mathbf{Y}(z)$ *of the system output* $\mathbf{y}(k)$ *as a homogenous linear function of an arbitrary initial vector value* $\mathbf{x}_0$ *of the state vector* $\mathbf{x}(k)$ *at* $k = 0$ *for the system in a free regime (i.e., for* $\mathbf{i}(k) \equiv \mathbf{0}_M$*):*

$$\mathbf{Y}(z) = \boldsymbol{G}_{ISOx_0}(z)\mathbf{x}_0, \ \mathbf{i}(k) \equiv \mathbf{0}_M, \ \boldsymbol{G}_{ISOx_0}(z) = \boldsymbol{G}_{ISO0}(z). \tag{7.22}$$

Note 7.3 *The superposition principle enables Equations (7.20a)-(7.20c) to imply (7.21) and (7.22), and vice versa.*

Note 7.4 *The system full transfer function matrix* $\boldsymbol{F}_{ISO}(z)$ *is composed of* $\boldsymbol{G}_{ISO}(z)$ *and* $\boldsymbol{G}_{ISO0}(z)$,

$$\boldsymbol{F}_{ISO}(z) = \begin{bmatrix} \boldsymbol{G}_{ISO}(z) & \boldsymbol{G}_{ISO0}(z) \end{bmatrix}, \ \boldsymbol{G}_{ISO0}(z) = \boldsymbol{G}_{ISOx_0}(z). \tag{7.23}$$

Fig. 7.9 and Fig. 7.10 represent the block of the ISO system (7.18) and (7.19).

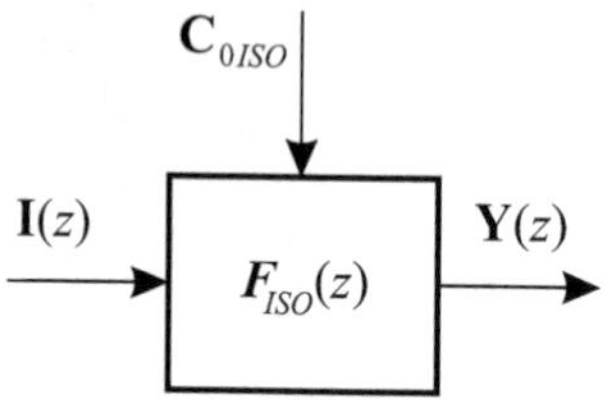

Figure 7.9: The slightly extended block of the ISO system.

Note 7.5 *When we use* $\mathbf{V}_{ISO}(z)$,

$$\mathbf{V}_{ISO}(z) = \begin{bmatrix} \mathbf{I}^T(z) & \mathbf{C}_{0ISO}^T \end{bmatrix}^T = \begin{bmatrix} \mathbf{I}^T(z) & \mathbf{x}_0^T \end{bmatrix}^T$$

instead of $\mathbf{I}(z)$, $\boldsymbol{F}_{ISO}(z)$ *instead of* $\mathbf{G}_{ISO}(z)$, *then we can apply the extended form of the full block of the ISO system (7.18) and (7.19), which is shown in Fig. 7.9 or the full block of the system in the compact form; see Fig. 7.10.*

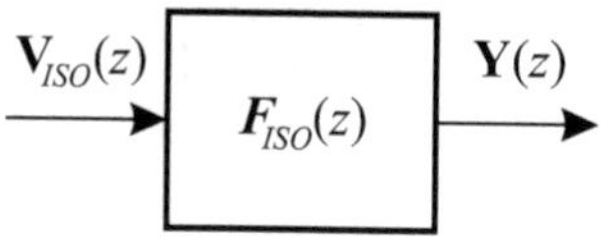

Figure 7.10: The full block of the ISO system in the compact form.

Note 7.6 *The initial state vector* $\mathbf{x}_0 = \mathbf{C}_{0ISO}$ *contains in itself both the initial input and the initial output conditions. The initial output conditions do not appear explicitly in (7.20a). The ISO system (7.18) and (7.19) does not have the transfer function matrices relative to the initial input and output conditions because they are contained in* $\mathbf{x}_0 = \mathbf{C}_{0ISO}$.

Note 7.7 *The full IO transfer function matrix* $\boldsymbol{F}_{IOISO}(z)$ *of the ISO system is the full transfer function matrix obtained from the equivalent IO system obtained from the given ISO system (in Subsection 2.4.2). The definition of* $\boldsymbol{F}_{IOISO}(z)$ *is in Section 7.2. It is determined from the full transfer function matrix of the obtained IO system equivalent to the given ISO system (in Section 8.1).*

It is necessary to distinguish the full IO transfer function matrix $\boldsymbol{F}_{IOISO}(z)$ *of the ISO system from the full ISO transfer function matrix* $\boldsymbol{F}_{ISO}(z) \equiv \boldsymbol{F}_{ISOIO}(z)$ *of the same ISO system,* $\boldsymbol{F}_{IOISO}(z) \neq \boldsymbol{F}_{ISO}(z)$. *The latter is the full transfer function matrix obtained from the given ISO model of the system.*

Note 7.8 *Let us consider the system transfer function matrices relative to the system state vector. They are defined as follows:*

Definition 7.5 *a)* ***The full (complete) input-state (IS) transfer function matrix of the ISO system (7.18)*** *and* ***(7.19),*** *which is denoted by* $\boldsymbol{F}_{ISOIS}(z)$, $\boldsymbol{F}_{ISOIS}(z) \in \mathcal{C}^{n\times(M+n)}$, *is the complex matrix value of* ***the system IS matrix transfer function*** $\boldsymbol{F}_{ISOIS}(\cdot)$, $\boldsymbol{F}_{ISOIS}(\cdot) : \mathcal{C} \to \mathcal{C}^{n\times(M+n)}$, *which is a matrix function of the complex variable* z *such that it determines uniquely the* $Z-$*transform* $\mathbf{X}(z)$ *of the system state vector* $\mathbf{x}(k)$ *as a homogenous linear function of the* $Z-$*transform* $\mathbf{I}(z)$ *of the system input* $\mathbf{i}(k)$ *for an arbitrary variation of* $\mathbf{i}(k)$, *and of arbitrary initial vector value* $\mathbf{x}_0$ *of the state vector* $\mathbf{x}(k)$ *at* $k = 0$, *respectively:*

$$\mathbf{X}(z) = \boldsymbol{F}_{ISOIS}(z) \left[\begin{array}{cc} \mathbf{I}^T(z) & \mathbf{x}_0^T \end{array} \right]^T = \boldsymbol{F}_{ISOIS}(z)\mathbf{V}_{ISO}(z). \quad (7.24)$$

b) ***The IS transfer function matrix of the ISO system (7.18)*** *and* ***(7.19),*** *which is denoted by* $\boldsymbol{G}_{ISOIS}(z)$, $\boldsymbol{G}_{ISOIS}(z) \in \mathcal{C}^{n\times M}$, *is the complex matrix value of* ***the system IS matrix transfer function*** $\boldsymbol{G}_{ISOIS}(\cdot)$, $\boldsymbol{G}_{ISOIS}(\cdot) : \mathcal{C} \to \mathcal{C}^{n\times M}$, *which is a matrix function of the complex variable* z *such that it determines uniquely the* $Z-$*transform* $\mathbf{X}(z)$ *of the system state vector* $\mathbf{x}(k)$ *as a homogenous linear function of the* $Z-$*transform* $\mathbf{I}(z)$ *of the system input* $\mathbf{i}(k)$ *for an arbitrary variation of* $\mathbf{i}(k)$, *and for zero initial state vector* $\mathbf{x}_0$ *of the state vector* $\mathbf{x}(k)$ *at* $k = 0$, *respectively:*

$$\mathbf{X}(z) = \boldsymbol{G}_{ISOIS}(z)\mathbf{I}(z),\ \mathbf{x}_0 = \mathbf{0}_n. \quad (7.25)$$

c) ***The state-state (SS) transfer function matrix of the ISO system (7.18)*** *and* ***(7.19),*** *which is denoted by* $\boldsymbol{G}_{ISOSS}(z)$, $\boldsymbol{G}_{ISOSS}(z) \in \mathcal{C}^{n\times n}$, *is the complex matrix value of* ***the system SS matrix transfer function*** $\boldsymbol{G}_{ISOSS}(\cdot)$ ***relative to the initial state*** $\mathbf{x}_0$, $\boldsymbol{G}_{ISOSS}(\cdot) : \mathcal{C} \to \mathcal{C}^{n\times n}$, *which is a matrix function of the complex variable* z *such that it determines uniquely the* $Z-$*transform* $\mathbf{X}(z)$ *of* *the system state vector* $\mathbf{x}(k)$ *as a homogenous linear function of an arbitrary initial vector value* $\mathbf{x}_0$ *of the state vector* $\mathbf{x}(k)$ *at* $k=0$ *for the system in a free regime (i.e., for* $\mathbf{i}(k)\equiv\mathbf{0}_M$, $\forall k\in\mathcal{N}$*):*

$$\mathbf{X}(z)=\boldsymbol{G}_{ISOSS}(z)\mathbf{x}_0,\ \mathbf{i}(k)\equiv\mathbf{0}_M,\forall k\in\mathcal{N}. \tag{7.26}$$

Note 7.9 *The system IS full transfer function matrix* $\boldsymbol{F}_{ISOIS}(z)$ *has two submatrices:* $\boldsymbol{G}_{ISOIS}(z)$ *and* $\boldsymbol{G}_{ISOSS}(z)$*,*

$$\boldsymbol{F}_{ISOIS}(z)=\left[\begin{array}{cc}\boldsymbol{G}_{ISOIS}(z) & \boldsymbol{G}_{ISOSS}(z)\end{array}\right]. \tag{7.27}$$

7.4 Definition of $\boldsymbol{F}(z)$ of the *IIO* system

For definition of γ see (2.36) (in Section 2.3) or (7.28):

$$\gamma=\max\{\beta,\mu\}. \tag{7.28}$$

This definition of γ simplifies the determination of *the full transfer function matrix* of the *IIO* system (3.64a) and (3.64b) (in Subsection 3.5.4) repeated as

$$\boldsymbol{Q}^{(\alpha)}\mathbf{r}^{\alpha}(k)=\boldsymbol{P}^{(\beta)}\mathbf{i}^{\beta}(k),\ \forall k\in\mathcal{N}_0, \tag{7.29a}$$

$$\boldsymbol{E}^{(\nu)}\mathbf{y}^{\nu}(k)=\boldsymbol{R}^{(\alpha)}\mathbf{r}^{\alpha}(k)+\boldsymbol{T}^{(\mu)}\mathbf{i}^{\mu}(k),\ \forall k\in\mathcal{N}_0. \tag{7.29b}$$

The system transmits and transfers differently the actions of the input vector $\mathbf{i}(k)$ and of the initial extended input vector $\mathbf{i}_0^{\gamma-1}$, of the initial internal state vector $\mathbf{r}_0^{\alpha-1}$, and of the initial output state vector $\mathbf{y}_0^{\nu-1}$. The system has several different transfer function matrices. They are related to different external actions and to the same system output; see Fig. 7.11.

Definition 7.6 *a)* ***The full (complete) input-output transfer (IO) function matrix of the IIO system (7.29a) and (7.29b)****, which is denoted by* $\boldsymbol{F}_{IIO}(z)$, $\boldsymbol{F}_{IIO}(z)\in\mathcal{C}^{N\times[(\gamma+1)M+\alpha\rho+\nu N]}$*, is the complex matrix value of* ***the system full IO matrix transfer function*** $\boldsymbol{F}_{IIO}(\cdot)$, $\boldsymbol{F}_{IIO}(\cdot):\mathcal{C}\to\mathcal{C}^{N\times[(\gamma+1)M+\alpha\rho+\nu N]}$*, which is matrix function of the complex variable* z *such that it determines uniquely the* $Z-$*transform* $\mathbf{Y}(z)$ *of the system output* $\mathbf{y}(k)$ *as a homogenous linear function of the* $Z-$*transform* $\mathbf{I}(z)$ *of the system input* $\mathbf{i}(k)$ *for an arbitrary variation of* $\mathbf{i}(k)$*, of arbitrary initial vector values* $\mathbf{i}_0^{\gamma-1}$*,*

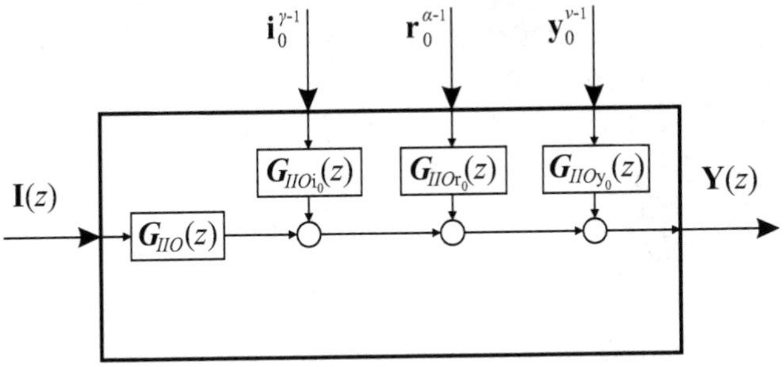

Figure 7.11: The block diagram of the IIO system shows the system transfer function matrices related to the input and to the extended initial vectors.

$\mathbf{r}_0^{\alpha-1}$, *and* $\mathbf{y}_0^{\nu-1}$ *of the extended input vector* $\mathbf{i}^{\gamma-1}(k)$, *of the state vector* $\mathbf{r}^{\alpha-1}(k)$, *and of the extended output vector* $\mathbf{y}^{\nu-1}(k)$ *at* $k = 0$, *respectively:*

$$\mathbf{Y}(z) =$$

$$= \boldsymbol{F}_{IIO}(z)\left[\begin{array}{cccc} \mathbf{I}^T(z) & \left(\mathbf{i}_0^{\gamma-1}\right)^T & \left(\mathbf{r}_0^{\alpha-1}\right)^T & \left(\mathbf{y}_0^{\nu-1}\right)^T \end{array}\right]^T \Longrightarrow \tag{7.30a}$$

$$\mathbf{Y}(z) = \boldsymbol{F}_{IIO}(z)\left[\begin{array}{cc} \mathbf{I}^T(z) & \mathbf{C}_{0IIO}^T \end{array}\right]^T = \boldsymbol{F}_{IO}(z)\mathbf{V}_{IIO}(z), \tag{7.30b}$$

$$\mathbf{C}_{0IIO} = \left[\begin{array}{ccc} \left(\mathbf{i}_0^{\gamma-1}\right)^T & \left(\mathbf{r}_0^{\alpha-1}\right)^T & \left(\mathbf{y}_0^{\nu-1}\right)^T \end{array}\right]^T, \tag{7.30c}$$

$$\mathbf{V}_{IIO}(z) = \left[\begin{array}{cc} \mathbf{I}^T(z) & \mathbf{C}_{0IIO}^T \end{array}\right]^T. \tag{7.30d}$$

b) ***The input-output (IO) transfer function matrix of the IIO system (7.29a) and (7.29b)***, *which is denoted by* $G_{IIO}(z)$, $G_{IIO}(z) \in \mathcal{C}^{N\times M}$, *is the complex matrix value of* ***the system IO matrix transfer function*** $\boldsymbol{G_{IIO}(\cdot)}$, $\boldsymbol{G_{IIO}(\cdot) : \mathcal{C} \rightarrow \mathcal{C}^{N\times M}}$, *which is a matrix function of the complex variable* z *such that it determines uniquely the* $Z-$*transform* $\mathbf{Y}(z)$ *of the system output* $\mathbf{y}(k)$ *as a homogenous linear function of the* $Z-$*transform* $\mathbf{I}(z)$ *of the system input* $\mathbf{i}(k)$ *for an arbitrary variation of* $\mathbf{i}(k)$, *and for all zero initial conditions, i.e., for* $\mathbf{i}_0^{\gamma-1} = \mathbf{0}_{\gamma M}$, $\mathbf{r}_0^{\alpha-1} = \mathbf{0}_{\alpha\rho}$, *and* $\mathbf{y}_0^{\nu-1} = \mathbf{0}_{\nu N}$ *of the extended input vector* $\mathbf{i}^{\gamma-1}(k)$, *of the state vector* $\mathbf{r}^{\alpha-1}(k)$, *and of the extended output vector* $\mathbf{y}^{\nu-1}(k)$ *at* $k = 0$, *respectively:*

$$\mathbf{Y}(z) = G_{IIO}(z)\mathbf{I}(z),\ \mathbf{i}_0^{\gamma-1} = \mathbf{0}_{\gamma M},\ \mathbf{r}_0^{\alpha-1} = \mathbf{0}_{\alpha\rho},\ \mathbf{y}_0^{\nu-1} = \mathbf{0}_{\nu N}. \tag{7.31}$$

c) ***The input-output (IICO) transfer function matrix relative to***

$\mathbf{i}_0^{\gamma-1}$ ***of the IIO system (7.29a) and (7.29b)****, which is denoted by $\mathbf{G}_{IIOi_0}(z)$, $\mathbf{G}_{IIOi_0}(z) \in \mathcal{C}^{N\times(\gamma+1)M}$, is the complex matrix value of* ***the system IICO matrix transfer function $\mathbf{G}_{IIOi_0}(\cdot)$ relative to $\mathbf{i}_0^{\gamma-1}$****, $\mathbf{G}_{IIOi_0}(\cdot): \mathcal{C} \rightarrow \mathcal{C}^{N\times(\gamma+1)M}$, which is matrix function of the complex variable z such that it determines uniquely the $Z-$transform $\mathbf{Y}(z)$ of the system output $\mathbf{y}(k)$ as a homogenous linear function of $\mathbf{i}_0^{\gamma-1}$ in the free regime on $\mathcal{N}$ (i.e., for $\mathbf{i}(k) = \mathbf{0}_M$, $\forall k \in \mathcal{N}$), and for all other zero initial conditions, i.e., for $\mathbf{r}_0^{\alpha-1} = \mathbf{0}_{\alpha\rho}$, and $\mathbf{y}_0^{\nu-1} = \mathbf{0}_{\nu N}$ of the state vector $\mathbf{r}^{\alpha-1}(k)$ and of the extended output vector $\mathbf{y}^{\nu-1}(k)$ at $k = 0$, respectively:*

$$\mathbf{Y}(z) = \mathbf{G}_{IIOi_0}(z)\mathbf{i}_0^{\gamma-1},$$
$$\mathbf{i}(k) = \mathbf{0}_M, \forall k \in \mathcal{N}, \ \mathbf{r}_0^{\alpha-1} = \mathbf{0}_{\alpha\rho}, \ \mathbf{y}_0^{\nu-1} = \mathbf{0}_{\nu N}. \tag{7.32}$$

d) ***The input-output (IIRO) transfer function matrix relative to $\mathbf{r}_0^{\alpha-1}$ of the IIO system (7.29a) and (7.29b)****, which is denoted by $\mathbf{G}_{IIOr_0}(z)$, $\mathbf{G}_{IIOr_0}(z) \in \mathcal{C}^{N\times\alpha\rho}$, is the complex matrix value of* ***the system IIRO matrix transfer function $\mathbf{G}_{IIOr_0}(\cdot)$ relative to $\mathbf{r}_0^{\alpha-1}$****, $\mathbf{G}_{IIOr_0}(\cdot): \mathcal{C} \rightarrow \mathcal{C}^{N\times\alpha\rho}$, which is a matrix function of the complex variable z such that it determines uniquely the $Z-$transform $\mathbf{Y}(z)$ of the system output $\mathbf{y}(k)$ as a homogenous linear function of $\mathbf{r}_0^{\alpha-1}$ in the free regime (i.e., for $\mathbf{i}(k) \equiv \mathbf{0}_M$), and for all other zero initial conditions, i.e., for $\mathbf{i}_0^{\gamma-1} = \mathbf{0}_{\gamma M}$, and $\mathbf{y}_0^{\nu-1} = \mathbf{0}_{\nu N}$ of the extended input vector $\mathbf{i}^{\gamma-1}(k)$ and of the extended output vector $\mathbf{y}^{\nu-1}(k)$ at $k = 0$, respectively:*

$$\mathbf{Y}(z) = \mathbf{G}_{IIOr_0}(z)\mathbf{r}_0^{\alpha-1},$$
$$\mathbf{i}(k) \equiv \mathbf{0}_M, \ \mathbf{i}_0^{\gamma-1} = \mathbf{0}_{\gamma M}, \ \mathbf{y}_0^{\nu-1} = \mathbf{0}_{\nu N}. \tag{7.33}$$

e) ***The input-output (IIYO) transfer function matrix relative to $\mathbf{y}_0^{\nu-1}$ of the IIO system (7.29a) and (7.29b)****, which is denoted by $\mathbf{G}_{IIOy_0}(z)$, $\mathbf{G}_{IIOy_0}(z) \in \mathcal{C}^{N\times\nu N}$, is the complex matrix value of* ***the system IIYO matrix transfer function $\mathbf{G}_{IIOy_0}(\cdot)$ relative to $\mathbf{y}_0^{\nu-1}$****, $\mathbf{G}_{IIOy_0}(\cdot): \mathcal{C} \rightarrow \mathcal{C}^{N\times\nu N}$, which is a matrix function of the complex variable z such that it determines uniquely the $Z-$transform $\mathbf{Y}(z)$ of the system output $\mathbf{y}(k)$ as a homogenous linear function of $\mathbf{y}_0^{\nu-1}$ in the free regime (i.e., for $\mathbf{i}(k) \equiv \mathbf{0}_M$), and for all other zero initial conditions, i.e., for $\mathbf{i}_0^{\gamma-1} = \mathbf{0}_{\gamma M}$, and $\mathbf{r}_0^{\alpha-1} = \mathbf{0}_{\alpha\rho}$ of the extended input vector $\mathbf{i}^{\gamma-1}(k)$ and of the state vector $\mathbf{r}^{\alpha-1}(k)$ at $k = 0$,* *respectively:*

$$\mathbf{Y}(z) = \boldsymbol{G}_{IIOy_0}(z)\mathbf{y}_0^{\nu-1},$$
$$\mathbf{i}(k) \equiv \mathbf{0}_M,\ \mathbf{i}_0^{\gamma-1} = \mathbf{0}_{\gamma M},\ \mathbf{r}_0^{\alpha-1} = \mathbf{0}_{\alpha\rho}. \tag{7.34}$$

Note 7.10 *This definition is general. It holds also for the RS systems (2.33a) and (2.33b), the PMD systems (2.32a) and (2.32b), and the GISO systems (2.34a) and (2.34b) (all in Section 2.3) because they are special cases of the IIO systems (Subsection 2.4.3).*

Note 7.11 *The system full transfer function matrix $\boldsymbol{F}_{IIO}(z)$ contains the transfer function matrices $\boldsymbol{G}_{IIO}(z)$, $\boldsymbol{G}_{IIOi_0}(z)$, $\boldsymbol{G}_{IIOr_0}(z)$, and $\boldsymbol{G}_{IIOy_0}(z)$:*

$$\boldsymbol{F}_{IIO}(z) = \begin{bmatrix} \boldsymbol{G}_{IIO}(z) & \boldsymbol{G}_{IIOi_0}(z) & \boldsymbol{G}_{IIOr_0}(z) & \boldsymbol{G}_{IIOy_0}(z) \end{bmatrix} \Longrightarrow \tag{7.35a}$$
$$\boldsymbol{F}_{IIO}(z) = \begin{bmatrix} \boldsymbol{G}_{IIO}(z) & \boldsymbol{G}_{IIO0}(z) \end{bmatrix}, \tag{7.35b}$$
$$\boldsymbol{G}_{IIO0}(z) = \begin{bmatrix} \boldsymbol{G}_{IIOi_0}(z) & \boldsymbol{G}_{IIOr_0}(z) & \boldsymbol{G}_{IIOy_0}(z) \end{bmatrix}. \tag{7.35c}$$

This results from the system linearity, Fig. 7.12. $\boldsymbol{G}_{IIO0}(z)$ is the system transfer function matrix relative to all initial conditions. Equations (7.31)

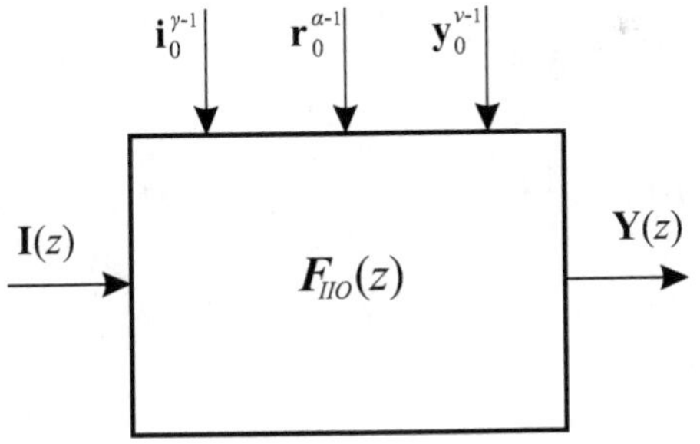

Figure 7.12: The extended full block of the *IIO* system.

through (7.34) imply Equations (7.35a) through (7.35c), and vice versa.

Note 7.12 *If the system full transfer function matrix $\boldsymbol{F}_{IIO}(z)$ replaces the system transfer function matrix $\boldsymbol{G}_{IIO}(z)$ and the vector $\mathbf{V}_{IIO}(z)$ is used instead of $\mathbf{I}(z)$ the very useful block diagram method stays effective,*

$$\mathbf{V}_{IIO}(z) = \begin{bmatrix} \mathbf{I}^T(s) & \mathbf{C}_{0IIO}^T \end{bmatrix}^T =$$
$$= \begin{bmatrix} \mathbf{I}^T(s) & \left(\mathbf{i}_0^{\gamma-1}\right)^T & \left(\mathbf{r}_0^{\alpha-1}\right)^T & \left(\mathbf{y}_0^{\nu-1}\right)^T \end{bmatrix}^T.$$

The $Z-$transform $\mathbf{V}_{IIO}(z)$ of the action vector $\mathbf{v}(k)$ plays the role of the $Z-$transform $\mathbf{I}(z)$ of the input vector $\mathbf{i}(k)$. Therefore,

$$\mathbf{Y}(z) = \begin{bmatrix} \boldsymbol{G}_{IIO}(z) & \boldsymbol{G}_{IIO0}(z) \end{bmatrix} \begin{bmatrix} \mathbf{I}^T(s) & \mathbf{C}_{0IIO}^T \end{bmatrix}^T = \boldsymbol{F}_{IIO}(z)\mathbf{V}_{IIO}(z).$$

There exist various transfer function matrices related only to the internal state of the IIO system:

Definition 7.7 *a)* ***The full (complete) input-state (IS) transfer function matrix of the IIO system (7.29a) and (7.29b)****, which is denoted by $\boldsymbol{F}_{IIOIS}(z)$, where $\boldsymbol{F}_{IIOIS}(z) \in \mathcal{C}^{\rho \times [(\beta+1)M+\alpha\rho]}$, is the complex matrix value of* ***the system full IS matrix transfer function $\boldsymbol{F}_{IIOIS}(\cdot)$****, $\boldsymbol{F}_{IIOIS}(\cdot) : \mathcal{C} \to \mathcal{C}^{\rho \times [(\beta+1)M+\alpha\rho]}$, which is a matrix function of the complex variable z such that it determines uniquely the $Z-$transform $\mathbf{R}(z)$ of the system substate vector $\mathbf{r}(k)$ as a homogenous linear function of the $Z-$transform $\mathbf{I}(z)$ of the system input $\mathbf{i}(k)$ for an arbitrary variation of $\mathbf{i}(k)$, of arbitrary initial vector values $\mathbf{i}_0^{\beta-1}$, and $\mathbf{r}_0^{\alpha-1}$ of the extended input vector $\mathbf{i}^{\beta-1}(k)$ and of the state vector $\mathbf{r}^{\alpha-1}(k)$ at $k=0$, respectively:*

$$\mathbf{R}(z) = \boldsymbol{F}_{IIOIS}(z) \left[\begin{array}{ccc} \mathbf{I}^T(z) & \left(\mathbf{i}_0^{\beta-1}\right)^T & \left(\mathbf{r}_0^{\alpha-1}\right)^T \end{array} \right]^T. \tag{7.36}$$

b) ***The IRIS transfer function matrix of the IIO system (7.29a) and (7.29b)****, which is denoted by $\boldsymbol{G}_{IIOIS}(z)$, $\boldsymbol{G}_{IIOIS}(z) \in \mathcal{C}^{\rho \times M}$, is the complex matrix value of* ***the system IRIS matrix transfer function $\boldsymbol{G}_{IIOIS}(\cdot)$****, $\boldsymbol{G}_{IIOIS}(\cdot) : \mathcal{C} \to \mathcal{C}^{\rho \times M}$, which is a matrix function of the complex variable z such that it determines uniquely the $Z-$transform $\mathbf{R}(z)$ of the system substate vector $\mathbf{r}(k)$ as a homogenous linear function of the $Z-$transform $\mathbf{I}(z)$ of the system input $\mathbf{i}(k)$ for an arbitrary variation of $\mathbf{i}(k)$, and for all zero initial conditions, i.e., for $\mathbf{i}_0^{\beta-1} = \mathbf{0}_{\beta M}$ of the extended input vector $\mathbf{i}^{\beta-1}(k)$ and of the state vector $\mathbf{r}^{\alpha-1}(k)$ at $k=0$:*

$$\mathbf{R}(z) = \boldsymbol{G}_{IIOIS}(z)\mathbf{I}(z),\ \mathbf{i}_0^{\beta-1} = \mathbf{0}_{\beta M},\ \mathbf{r}_0^{\alpha-1} = \mathbf{0}_{\alpha\rho}. \tag{7.37}$$

c) ***The IRII transfer function matrix relative to $\mathbf{i}_0^{\beta-1}$ of the IIO system (7.29a) and (7.29b)****, which is denoted by $\boldsymbol{G}_{IIOi_0IS}(z)$, $\boldsymbol{G}_{IIOi_0IS}(z) \in \mathcal{C}^{\rho \times (\beta+1)M}$, is the complex matrix value of* ***the system IRII matrix transfer function $\boldsymbol{G}_{IIOi_0IS}(\cdot)$ relative to $\mathbf{i}_0^{\beta-1}$****,*

$G_{IIOi_0IS}(\cdot): \mathcal{C} \rightarrow \mathcal{C}^{\rho\times(\beta+1)M}$, *which is a matrix function of the complex variable* z *such that it determines uniquely the* $Z-$*transform* $\mathbf{R}(z)$ *of the system substate vector* $\mathbf{r}(k)$ *as a homogenous linear function of* $\mathbf{i}_0^{\beta-1}$ *in the free regime on* $\mathcal{N}$ *(i.e., for* $\mathbf{i}(k) = \mathbf{0}_M$, $\forall k \in \mathcal{N}$*), and for all other zero initial conditions, i.e., for* $\mathbf{r}_0^{\alpha-1} = \mathbf{0}_{\alpha\rho}$, *of the state vector* $\mathbf{r}^{\alpha-1}(k)$ *at* $k = 0$, *respectively:*

$$\mathbf{R}(z) = G_{IIOi_0IS}(z)\mathbf{i}_0^{\beta-1},\ \mathbf{r}_0^{\alpha-1} = \mathbf{0}_{\alpha\rho},\ \mathbf{i}(k) = \mathbf{0}_M, \forall k \in \mathcal{N}. \quad (7.38)$$

d) ***The IRIR transfer function matrix relative to*** $\mathbf{r}_0^{\alpha-1}$ ***of the IIO system (7.29a) and (7.29b)***, *which is denoted by* $G_{IIOr_0IS}(z)$, $G_{IIOr_0IS}(z) \in \mathcal{C}^{\rho\times\alpha\rho}$, *is the complex matrix value of* ***the system IRIR matrix transfer function*** $G_{IIOr_0IS}(\cdot)$ ***relative to*** $\mathbf{r}_0^{\alpha-1}$, $G_{IIOr_0IS}(\cdot): \mathcal{C} \rightarrow \mathcal{C}^{\rho\times\alpha\rho}$, *which is a matrix function of the complex variable* z *such that it determines uniquely the* $Z-$*transform* $\mathbf{R}(z)$ *of the system substate vector* $\mathbf{r}(k)$ *as a homogenous linear function of* $\mathbf{r}_0^{\alpha-1}$ *in the free regime (i.e., for* $\mathbf{i}(k) \equiv \mathbf{0}_M$*), and for all other zero initial conditions, i.e., for* $\mathbf{i}_0^{\beta-1} = \mathbf{0}_{\beta M}$, *of the extended input vector* $\mathbf{i}^{\beta-1}(k)$ *at* $k = 0$:

$$\mathbf{R}(z) = G_{IIOr_0IS}(z)\mathbf{r}_0^{\alpha-1},\ \mathbf{i}(t) \equiv \mathbf{0}_M,\ \mathbf{i}_0^{\beta-1} = \mathbf{0}_{\beta M}. \quad (7.39)$$

Note 7.13 *The submatrices of the full IRI transfer function matrix* $\boldsymbol{F}_{IIOIS}(z)$ *are the transfer function matrices* $G_{IIOIS}(s)$, $G_{IIOi_0IS}(s)$, *and* $G_{IIOr_0IS}(s)$,

$$\boldsymbol{F}_{IIOIS}(z) = \left[\begin{array}{cc} G_{IIOIS}(z) & \underbrace{G_{IIOi_0IS}(z) \quad G_{IIOr_0IS}(z)}_{G_{IIO0IS}(z)} \end{array}\right] =$$
$$= \left[\begin{array}{cc} G_{IIOIS}(z) & G_{IIO0IS}(z) \end{array}\right],$$
$$G_{IIO0IS}(z) = \left[\begin{array}{cc} G_{IIOi_0IS}(z) & G_{IIOr_0IS}(z) \end{array}\right]. \quad (7.40)$$

Note 7.14 $\boldsymbol{F}_{IIOIS}(z) \neq \boldsymbol{F}_{IIO}(z)$.

Chapter 8

Determination of $\boldsymbol{F}(z)$

8.1 $\boldsymbol{F}(z)$ of the $\boldsymbol{IO}$ system

In order to determine the transfer function matrix $\boldsymbol{G}_{IO}(z)$ of the IO system (3.55), i.e., (3.56) (in Subsection 3.5.2):

$$\boldsymbol{A}^{(\nu)}\mathbf{y}^{\nu}(k) = \boldsymbol{B}^{(\mu)}\mathbf{i}^{\mu}(k), \ \forall k \in \mathcal{N}_0, \tag{8.1}$$

let it be set into the elegant *compact form* [36], [61]:

$$\boldsymbol{G}_{IO}(z) = \left(\boldsymbol{A}^{(\nu)}\boldsymbol{S}_N^{(\nu)}(z)\right)^{-1}\left(\boldsymbol{B}^{(\mu)}\boldsymbol{S}_M^{(\mu)}(z)\right), \tag{8.2}$$

for which we use the matrices $\boldsymbol{A}^{(\nu)}$ and $\boldsymbol{B}^{(\mu)}$ (2.2a), (2.2b) (in Section 2.1),

$$\boldsymbol{A}^{(\nu)} = \begin{bmatrix} \boldsymbol{A}_0 & \boldsymbol{A}_1 & \cdots & \boldsymbol{A}_\nu \end{bmatrix} \in \mathcal{R}^{N\times(\nu+1)N}, \tag{8.3a}$$

$$\boldsymbol{B}^{(\mu)} = \begin{bmatrix} \boldsymbol{B}_0 & \boldsymbol{B}_1 & \cdots & \boldsymbol{B}_\mu \end{bmatrix} \in \mathcal{R}^{N\times(\mu+1)M}, \tag{8.3b}$$

as well as the complex matrix function $\boldsymbol{S}_i^{(r)}(\cdot) : \mathcal{C} \longrightarrow \mathcal{C}^{i(r+1)\times i}$, (3.29) (in Subsection 3.4.2),

$$\boldsymbol{S}_i^{(r)}(z) = \begin{bmatrix} z^0\boldsymbol{I}_i & z^1\boldsymbol{I}_i & z^2\boldsymbol{I}_i & \cdots & z^r\boldsymbol{I}_i \end{bmatrix}^T \in \mathcal{C}^{i(r+1)\times i},$$
$$\boldsymbol{I}_i = \text{diag}\left\{ 1 \ \ 1 \ \ \cdots \ \ 1 \right\} \in \mathcal{R}^{i\times i}, \ (r,i) \in \{(\mu, M), \ (\nu, N)\}. \tag{8.4}$$

The application of the $Z-$transform to (8.1) and the joint application of (8.3a), (8.3b) and (8.4) yield the compact form (8.2) of $\boldsymbol{G}_{IO}(z)$:

$$\left(\sum_{r=0}^{r=\nu}\boldsymbol{A}_r z^r\right)^{-1}\left(\sum_{r=0}^{r=\mu\leq\nu}\boldsymbol{B}_r z^r\right) = \left(\boldsymbol{A}^{(\nu)}\boldsymbol{S}_N^{(\nu)}(z)\right)^{-1}\left(\boldsymbol{B}^{(\mu)}\boldsymbol{S}_M^{(\mu)}(z)\right).$$

Another breaking step to determine effectively the compact form of the system full transfer function matrix $\boldsymbol{F}_{IO}(z)$ is the application also of the matrix function $\boldsymbol{Z}_r^{(\varsigma)}(\cdot) : \mathcal{C} \rightarrow \mathcal{C}^{(\varsigma+1)r \times \varsigma r}$, (3.30) (in Subsection 3.4.2),

$$\boldsymbol{Z}_r^{(\varsigma)}(z) = \begin{cases} \begin{bmatrix} \boldsymbol{O}_r & \boldsymbol{O}_r & \boldsymbol{O}_r & \cdots & \boldsymbol{O}_r \\ z^1\boldsymbol{I}_r & \boldsymbol{O}_r & \boldsymbol{O}_r & \cdots & \boldsymbol{O}_r \\ \vdots & \vdots & \vdots & \vdots & \vdots \\ z^{\varsigma-0}\boldsymbol{I}_r & z^{\varsigma-1}\boldsymbol{I}_r & z^{\varsigma-2}\boldsymbol{I}_r & \cdots & z^1\boldsymbol{I}_r \end{bmatrix}, \varsigma \geq 1, \\ \text{not defined for } \varsigma < 1 \end{cases}$$

$$\boldsymbol{Z}_r^{(\varsigma)}(z) \in \mathcal{C}^{(\varsigma+1)r \times \varsigma r}, \quad (\varsigma, r) \in \{(\mu, M), \ (\nu, N)\}. \tag{8.5}$$

Let us recall Note 3.2 (in Subsection 3.4.2) that helps us to be clear and precise:

Note 8.1 *The matrix* $\boldsymbol{Z}_r^{(\varsigma)}(z) = \boldsymbol{Z}_r^{(0)}(z)$ *has to be completely omitted if* $\varsigma = 0$. *It may not be replaced by the zero matrix because* $\boldsymbol{Z}_r^{(\varsigma)}(z)$ *is not defined for* $\zeta \leq 0$. *It does not exist if* $\varsigma = 0$.

Theorem 8.1 *a) The IO system (8.1) full IO transfer function matrix* $\boldsymbol{F}_{IO}(z)$ *reads:*

$$\text{If } \mu \geq 1, \text{ then } \boldsymbol{F}_{IO}(z) = \boldsymbol{F}_{IOD}^{-1}(z)\boldsymbol{F}_{ION}(z) =$$
$$= \left(\boldsymbol{A}^{(\nu)}\boldsymbol{S}_N^{(\nu)}(z)\right)^{-1} \left[\begin{array}{ccc} \boldsymbol{B}^{(\mu)}\boldsymbol{S}_M^{(\mu)}(z) & -\boldsymbol{B}^{(\mu)}\boldsymbol{Z}_M^{(\mu)}(z) & \boldsymbol{A}^{(\nu)}\boldsymbol{Z}_N^{(\nu)}(z) \end{array}\right] =$$
$$= \left[\begin{array}{cc} \boldsymbol{G}_{IO}(z) & \underbrace{\boldsymbol{G}_{IOi_0}(z) \quad \boldsymbol{G}_{IOy_0}(z)}_{\boldsymbol{G}_{IO_0}(z)} \end{array}\right],$$
$$\boldsymbol{G}_{IO_0}(z) = \left[\begin{array}{cc} \boldsymbol{G}_{IOi_0}(z) & \boldsymbol{G}_{IOy_0}(z) \end{array}\right], \tag{8.6a}$$
$$\text{If } \mu = 0, \text{ then } \boldsymbol{F}_{IO}(z) = \boldsymbol{F}_{IOD}^{-1}(z)\boldsymbol{F}_{ION}(z) =$$
$$= \left(\boldsymbol{A}^{(\nu)}\boldsymbol{S}_N^{(\nu)}(z)\right)^{-1} \left[\begin{array}{cc} \boldsymbol{B}^{(\mu)}\boldsymbol{S}_M^{(\mu)}(z) & \boldsymbol{A}^{(\nu)}\boldsymbol{Z}_N^{(\nu)}(z) \end{array}\right] =$$
$$= \left(\boldsymbol{A}^{(\nu)}\boldsymbol{S}_N^{(\nu)}(z)\right)^{-1} \left[\begin{array}{cc} \boldsymbol{B}_0 & \boldsymbol{A}^{(\nu)}\boldsymbol{Z}_N^{(\nu)}(z) \end{array}\right] =$$
$$= \left[\begin{array}{cc} \boldsymbol{G}_{IO}(z) & \boldsymbol{G}_{IOy_0}(z) \end{array}\right], \ \boldsymbol{G}_{IOy_0}(z) = \boldsymbol{G}_{IO_0}(z). \tag{8.6b}$$

This implies

$$\mathbf{Y}(z) = \boldsymbol{F}_{IO}(z) \left\{ \begin{array}{c} \left[\begin{array}{ccc} \mathbf{I}^T(z) & \left(\mathbf{i}_0^{\mu-1}\right)^T & \left(\mathbf{y}_0^{\nu-1}\right)^T \end{array} \right]^T \textit{ if } \mu \geq 1, \\ \left[\begin{array}{cc} \mathbf{I}^T(z) & \left(\mathbf{y}_0^{\nu-1}\right)^T \end{array} \right]^T \textit{ if } \mu = 0 \end{array} \right. =$$

$$= \boldsymbol{F}_{IO}(z)\mathbf{V}_{IO}(z), \ \mathbf{V}_{IO}(z) = \left[\begin{array}{c} \mathbf{I}(z) \\ \mathbf{C}_{0IO} \end{array} \right], \tag{8.7a}$$

$$\mathbf{C}_{0IO} = \left\{ \begin{array}{c} \left[\begin{array}{c} \mathbf{i}_0^{\mu-1} \\ \mathbf{y}_0^{\nu-1} \end{array} \right], \textit{ if } \mu \geq 1, \\ \mathbf{y}_0^{\nu-1}, \textit{ if } \mu = 0 \end{array} \right. . \tag{8.7b}$$

b) *The system (3.55) IO transfer function matrix* $\boldsymbol{G}_{IO}(z)$ *reads:*

$$\boldsymbol{G}_{IO}(z) = \left(\boldsymbol{A}^{(\nu)} \boldsymbol{S}_N^{(\nu)}(z)\right)^{-1} \boldsymbol{B}^{(\mu)} \boldsymbol{S}_M^{(\mu)}(z). \tag{8.8}$$

c) *The system (3.55) IOIC transfer function matrix* $\boldsymbol{G}_{IOi_0}(z)$ *is given by:*

$$\boldsymbol{G}_{IOi_0}(z) = \left(\boldsymbol{A}^{(\nu)} \boldsymbol{S}_N^{(\nu)}(z)\right)^{-1} \left\{ \begin{array}{c} -\boldsymbol{B}^{(\mu)} \boldsymbol{Z}_M^{(\mu)}(z), \textit{ if } \mu \geq 1 \\ \boldsymbol{O}, \textit{ if } \mu = 0 \end{array} \right. . \tag{8.9}$$

d) *The system (3.55) IOIY transfer function matrix* $\boldsymbol{G}_{IOy_0}(z)$ *has the following form:*

$$\boldsymbol{G}_{IOy_0}(z) = \left(\boldsymbol{A}^{(\nu)} \boldsymbol{S}_N^{(\nu)}(z)\right)^{-1} \boldsymbol{A}^{(\nu)} \boldsymbol{Z}_N^{(\nu)}(z). \tag{8.10}$$

e) *The system (3.55) IORAI transfer function matrix* $\boldsymbol{G}_{IO_0}(z)$ *is found to be:*

$$\boldsymbol{G}_{IO_0}(z) = \left(\boldsymbol{A}^{(\nu)} \boldsymbol{S}_N^{(\nu)}(z)\right)^{-1} \cdot$$

$$\cdot \left\{ \begin{array}{c} \left[\begin{array}{cc} -\boldsymbol{B}^{(\mu)} \boldsymbol{Z}_M^{(\mu)}(z) & \boldsymbol{A}^{(\nu)} \boldsymbol{Z}_N^{(\nu)}(z) \end{array} \right], \textit{ if } \mu \geq 1, \\ \boldsymbol{A}^{(\nu)} \boldsymbol{Z}_N^{(\nu)}(z), \textit{ if } \mu = 0 \end{array} \right. . \tag{8.11}$$

Proof. The application of the $Z-$transform $\mathcal{Z}\{\cdot\}$ (Appendix B) of the left-hand side of (3.55) or of (3.56) (Subsection 3.5.2), i.e., of (8.1) (Section 8.1), yields the following:

$$\mathcal{Z}\left\{\sum_{r=0}^{r=\nu} \boldsymbol{A}_r E^r \mathbf{y}(k)\right\} = \mathcal{Z}\left\{\boldsymbol{A}^{(\nu)} \mathbf{y}^{\nu}(k)\right\} =$$

$$= \mathcal{Z}\left\{ \boldsymbol{A}_0 \overbrace{\mathbf{y}(k)}^{E^0\mathbf{y}(k)} + \boldsymbol{A}_1 \overbrace{\mathbf{y}(k+1)}^{E^1\mathbf{y}(k)} + \cdots + \boldsymbol{A}_\nu \overbrace{\mathbf{y}(k+\nu)}^{E^\nu\mathbf{y}(k)} \right\} =$$

$$= \overbrace{\left(\boldsymbol{A}_0 z^0 + \boldsymbol{A}_1 z^1 + ... + \boldsymbol{A}_\nu z^\nu\right)}^{\boldsymbol{A}^{(\nu)}\boldsymbol{S}_N^{(\nu)}(z)} \mathbf{y}(z) - \boldsymbol{A}_0 \mathbf{0}_N - \boldsymbol{A}_1 z^{1-0} \overbrace{E^0\mathbf{y}(0)}^{\mathbf{y}(0)} -$$

$$-\boldsymbol{A}_2 \left[z^{2-0} E^0 \mathbf{y}(0) + z^{2-1} \overbrace{E^1\mathbf{y}(0)}^{\mathbf{y}(1)} \right] - \cdots -$$

$$-\boldsymbol{A}_r \left[z^{r-0} E^0 \mathbf{y}(0) + z^{r-1} E^1 \mathbf{y}(0) + \cdots + z^{r-i} \overbrace{E^i\mathbf{y}(0)}^{\mathbf{y}(i)} + \cdots + \right.$$

$$\left. + z^{r-(r-1)} \overbrace{E^{r-1}\mathbf{y}(0)}^{\mathbf{y}(r-1)} \right] -$$

$$- \cdots -$$

$$-\boldsymbol{A}_\nu \left[z^{\nu-0} E^0 \mathbf{y}\,(0) + z^{\nu-1} E^1 \mathbf{y}\,(0) + \cdots + z^{\nu-i} E^i \mathbf{y}(0) + \cdots + \right.$$

$$\left. + z^{\nu-(\nu-1)} \overbrace{E^{\nu-1}\mathbf{y}(0)}^{\mathbf{y}(\nu-1)} \right] =$$

$$= \boldsymbol{A}^{(\nu)} \boldsymbol{S}_N^{(\nu)}(z) \mathbf{y}(z) - \overbrace{\left[\begin{array}{ccccccc} \boldsymbol{A}_0 & \boldsymbol{A}_1 & \boldsymbol{A}_2 & \cdots & \boldsymbol{A}_r & \cdots & \boldsymbol{A}_\nu \end{array}\right]}^{\boldsymbol{A}^{(\nu)},\ (2.2\text{a})} \cdot$$

$$\cdot \left[\begin{array}{c} \mathbf{0}_N \\ z^{1-0}\mathbf{y}(0) \\ z^{2-0}\mathbf{y}(0) + z^{2-1}\mathbf{y}(1) \\ \vdots \\ z^{r-0}\mathbf{y}\,(0) + z^{r-1}\mathbf{y}\,(1) + \cdots + z^{r-i}\mathbf{y}(i) + \cdots + z^{r-(r-1)}\mathbf{y}\,(r-1) \\ \vdots \\ z^{\nu-0}\mathbf{y}\,(0) + z^{\nu-1}\mathbf{y}\,(1) + \cdots + z^{\nu-i}\mathbf{y}(i) + \cdots + z^{\nu-(\nu-1)}\mathbf{y}\,(\nu-1) \end{array} \right] =$$

$$= \boldsymbol{A}^{(\nu)}\boldsymbol{S}_N^{(\nu)}(z)\mathbf{y}(z) - \overbrace{\begin{bmatrix} \boldsymbol{A}_0 & \boldsymbol{A}_1 & \boldsymbol{A}_2 & \cdots & \boldsymbol{A}_r & \cdots & \boldsymbol{A}_\nu \end{bmatrix}}^{\boldsymbol{A}^{(\nu)},\ (2.2a)} \cdot$$

$$\cdot \underbrace{\begin{bmatrix} \boldsymbol{O}_N & \boldsymbol{O}_N & \cdots & \boldsymbol{O}_N & \cdots & \boldsymbol{O}_N \\ z^{1-0}\boldsymbol{I}_N & \boldsymbol{O}_N & \cdots & \boldsymbol{O}_N & \cdots & \boldsymbol{O}_N \\ z^{2-0}\boldsymbol{I}_N & z^{2-1}\boldsymbol{I}_N & \cdots & \boldsymbol{O}_N & \cdots & \boldsymbol{O}_N \\ \vdots & \vdots & \vdots & \vdots & \vdots & \vdots \\ z^{r-0}\boldsymbol{I}_N & z^{r-1}\boldsymbol{I}_N & \cdots & z^{r-i}\boldsymbol{I}_N & \cdots & \boldsymbol{O}_N \\ \vdots & \vdots & \vdots & \vdots & \vdots & \vdots \\ z^{\nu-0}\boldsymbol{I}_N & z^{\nu-1}\boldsymbol{I}_N & \cdots & z^{r-i}\boldsymbol{I}_N & & z^{\nu-(\nu-1)}\boldsymbol{I}_N \end{bmatrix}}_{Z_N^{(\nu)}(z),\ (8.5)} \cdot$$

$$\cdot \underbrace{\begin{bmatrix} E^0\mathbf{y}(0) = \mathbf{y}(0) \\ E^1\mathbf{y}(0) = \mathbf{y}(1) \\ E^2\mathbf{y}(0) = \mathbf{y}(2) \\ \vdots \\ E^{r-1}\mathbf{y}(0) = \mathbf{y}(r-1) \\ \vdots \\ E^{\nu-1}\mathbf{y}(0) = \mathbf{y}(\nu-1) \end{bmatrix}}_{\mathbf{y}_0^{\nu-1}(0),\ (2.4b)} =$$

$$= \boldsymbol{A}^{(\nu)}\boldsymbol{S}_N^{(\nu)}(z)\mathbf{y}(z) - \boldsymbol{A}^{(\nu)}\boldsymbol{Z}_N^{(\nu)}(z)\,\mathbf{y}^{\nu-1}(0). \tag{8.12}$$

By repeating the above procedure applied to

$$\mathcal{Z}\left\{\sum_{r=0}^{r=\mu}\boldsymbol{B}_r E^r\mathbf{i}(k)\right\} = \mathcal{Z}\left\{\boldsymbol{B}^{(\mu)}\mathbf{i}^{\mu}(k)\right\},$$

and in view of (8.5)

$$\mathcal{Z}\left\{\sum_{r=0}^{r=\mu}\boldsymbol{B}_r E^r\mathbf{i}(k)\right\} =$$

$$= \begin{cases} \boldsymbol{B}^{(\mu)}\boldsymbol{S}_M^{(\mu)}(z)\,\mathbf{i}(z) - \boldsymbol{B}^{(\mu)}\boldsymbol{Z}_M^{(\mu)}(z)\,\mathbf{i}^{\mu-1}(0), & \mu \geq 1 \\ \boldsymbol{B}^{(\mu)}\boldsymbol{S}_M^{(\mu)}(z)\,\mathbf{i}(z) = \boldsymbol{B}_0\mathbf{i}(z), & \mu = 0. \end{cases} \tag{8.13}$$

These results imply the following compact form of the $Z-$transform $\mathcal{Z}\{\cdot\}$ of both (3.55) and (3.56):

$$\boldsymbol{A}^{(\nu)}\boldsymbol{S}_N^{(\nu)}(z)\mathbf{y}(z) - \boldsymbol{A}^{(\nu)}\boldsymbol{Z}_N^{(\nu)}(z)\,\mathbf{y}^{\nu-1}(0) =$$

$$= \begin{cases} \boldsymbol{B}^{(\mu)}\boldsymbol{S}_M^{(\mu)}(z)\,\mathbf{i}(z) - \boldsymbol{B}^{(\mu)}\boldsymbol{Z}_M^{(\mu)}(z)\,\mathbf{i}^{\mu-1}(0),\ \mu \geq 1 \\ \boldsymbol{B}^{(\mu)}\boldsymbol{S}_M^{(\mu)}(z)\,\mathbf{i}(z) = \boldsymbol{B}_0\mathbf{i}(z),\ \mu = 0. \end{cases} \quad (8.14)$$

This determines $\mathbf{Y}(z)$ linearly in terms of the vector function

$$\begin{bmatrix} \mathbf{i}(z) \\ \mathbf{i}^{\mu-1}(0) \\ \mathbf{y}^{\nu-1}(0) \end{bmatrix}$$

as follows:

$$\mathbf{Y}(z) = \left(\boldsymbol{A}^{(\nu)}\boldsymbol{S}_N^{(\nu)}(z)\right)^{-1}\cdot$$

$$\cdot \begin{cases} \begin{bmatrix} \boldsymbol{B}^{(\mu)}\boldsymbol{S}_M^{(\mu)}(z) & -\boldsymbol{B}^{(\mu)}\boldsymbol{Z}_M^{(\mu)}(z) & \boldsymbol{A}^{(\nu)}\boldsymbol{Z}_N^{(\nu)}(z) \end{bmatrix} \begin{bmatrix} \mathbf{i}(z) \\ \mathbf{i}^{\mu-1}(0) \\ \mathbf{y}^{\nu-1}(0) \end{bmatrix}, \mu \geq 1 \\ \begin{bmatrix} \boldsymbol{B}_0 & \boldsymbol{A}^{(\nu)}\boldsymbol{Z}_N^{(\nu)}(z) \end{bmatrix} \begin{bmatrix} \mathbf{i}(z) \\ \mathbf{y}^{\nu-1}(0) \end{bmatrix},\ \mu = 0. \end{cases}$$

(8.15)

The definition of $\boldsymbol{F}_{IO}(z)$ (Definition 7.2, Section 7.2) and this equation prove the statement under *a*) of the theorem. The statement under *b*) results directly from *a*), and the definition of $\boldsymbol{G}_{IO}(z)$ (7.12). The formulae under *c*) through *e*) result directly from (7.13) through (7.15) linked with (8.12) through (8.15) Q.E.D. ■

Note 8.2 *If* $\boldsymbol{A}_r \in \mathcal{R}^{N\times N}$, $\boldsymbol{B}_r \in \mathcal{R}^{N\times M}$, $r = 0, 1, \cdots, \nu$, $\boldsymbol{A}_\nu \neq \boldsymbol{O}_N$, *(3.55), and (8.5) then:*

$$\deg\left[\boldsymbol{A}^{(\nu)}\boldsymbol{Z}_N^{(\nu)}(z)\right] = \nu,\ \textit{and}\ \mu \geq 1 \Longrightarrow \deg\left[\boldsymbol{B}^{(\mu)}\boldsymbol{Z}_M^{(\mu)}(z)\right] = \mu. \quad (8.16)$$

Comment 8.1 *The overall action vector function* $\mathbf{v}_{IO}(\cdot)$ *and the* $Z-$*trans-*

form $\mathbf{V}_{IO}(z)$ *of the IO system result from Equation (8.7a),*

$$\mathbf{v}_{IO}(k) = \begin{cases} \left[\begin{array}{ccc} \mathbf{i}^T(k) & \delta_d(k)\left(\mathbf{i}_0^{\mu-1}\right)^T & \delta_d(k)\left(\mathbf{y}_0^{\nu-1}\right)^T \end{array} \right]^T, \ \mu \geq 1 \\ \left[\begin{array}{cc} \mathbf{i}^T(k) & \delta_d(k)\left(\mathbf{y}_0^{\nu-1}\right)^T \end{array} \right]^T, \ \mu = 0 \end{cases}, \tag{8.17a}$$

$$\mathbf{V}_{IO}(z) = \begin{cases} \left[\begin{array}{cc} \mathbf{I}^T(z) & \underbrace{\left(\mathbf{i}_0^{\mu-1}\right)^T \quad \left(\mathbf{y}_0^{\nu-1}\right)^T}_{\mathbf{C}_{0IO}^T} \end{array} \right]^T, \ \mu \geq 1 \\ \left[\begin{array}{cc} \mathbf{I}^T(z) & \left(\mathbf{y}_0^{\nu-1}\right)^T \end{array} \right]^T = \left[\begin{array}{cc} \mathbf{I}^T(z) & \mathbf{C}_{0IO}^T \end{array} \right]^T, \ \mu = 0 \end{cases} =$$

$$= \begin{cases} \mathbf{V}_{IO}(z; \mathbf{i}_0^{\mu-1}, \mathbf{y}_0^{\nu-1}), \ \mu \geq 1 \\ \mathbf{V}_{IO}(z; \mathbf{y}_0^{\nu-1}), \ \mu = 0. \end{cases} \tag{8.17b}$$

Here $\delta_d(k)$ *is the discrete unit Dirac impulse (Appendix B.2). Equation (8.7a) becomes now, for all nonzero initial conditions, the classical one for* $\mathbf{Y}(z)$ *determined by accepting all zero initial conditions,*

$$\mathbf{Y}(z) = \boldsymbol{F}_{IO}(z)\mathbf{V}_{IO}(z). \tag{8.18}$$

This yields the system full (complete) block, which holds in general, for arbitrary initial conditions, to be in the classical form well known for the zero initial conditions; see Fig. 8.1. The vector functions $\mathbf{V}_{IO}(\cdot) \equiv$

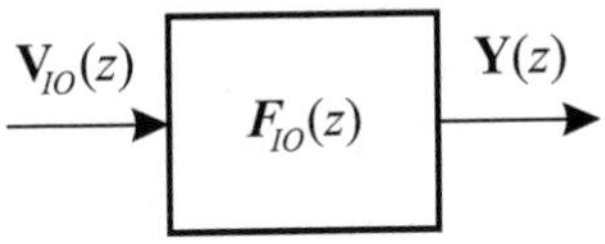

Figure 8.1: The full block of the IO system in the compact form.

$\mathbf{V}_{IO}(\cdot; \mathbf{i}_0^{\mu-1}, \mathbf{y}_0^{\nu-1})$ *and* $\mathbf{V}_{IO}(\cdot) = \mathbf{V}_{IO}(\cdot; \mathbf{y}_{0^-}^{\nu-1})$ *contain hidden the nonzero initial conditions* $\mathbf{i}_0^{\mu-1}$ *and* $\mathbf{y}_0^{\nu-1}$*:*

$$\mathbf{Y}(z) = \mathbf{Y}(z; \mathbf{i}_0^{\mu-1}, \mathbf{y}_0^{\nu-1}) = \boldsymbol{F}_{IO}(z)\mathbf{V}_{IO}(z; \mathbf{i}_0^{\mu-1}, \mathbf{y}_0^{\nu-1}), \mu \geq 1, \tag{8.19a}$$

$$\mathbf{Y}(z) = \mathbf{Y}(z; \mathbf{y}_0^{\nu-1}) = \boldsymbol{F}_{IO}(z)\mathbf{V}_{IO}(z; \mathbf{y}_0^{\nu-1}), \ \mu = 0. \tag{8.19b}$$

Comment 8.2 ***The full block diagram generalizes the block diagram technique***

The use of the generalized input vector function $\mathbf{v}_{IO}(\cdot)$*, (8.17a), enables us to apply the unchanged classical block diagram method if the full transfer* *function matrices* $\boldsymbol{F}_{IOi}(z)$ *of all system subsystems* S_i *replaces their transfer function matrices* $\boldsymbol{G}_{IOi}(z)$*, Fig. 8.1. We will develop the algebra of the full block diagrams in Chapter 9 for the characteristic system structures.*

Comment 8.3 *An important feature of the system full transfer function matrix* $\boldsymbol{F}_{IO}(z)$ *is to be the system dynamical invariant.* $\boldsymbol{F}_{IO}(z)$ *is fully determined by the system order, dimension, and parameters. Another important feature of* $\boldsymbol{F}_{IO}(z)$ *is its independence of the input vector and of all initial conditions, that is that it is independent of the generalized input vector* $\mathbf{v}_{IO}(k)$*, thus, of its the* $Z-$*transform* $\mathbf{V}_{IO}(z)$*. The same principal characterizes* $\boldsymbol{F}_{IO}(z)$*, which characterizes the system transfer function matrix* $\boldsymbol{G}_{IO}(z)$ *although it is valid only if all initial conditions are equal to zero.*

Note 8.3 *The system transfer function matrix* $\boldsymbol{G}_{IO}(z)$ *is a submatrix of the system full transfer function matrix* $\boldsymbol{F}_{IO}(z)$*:*

$$\boldsymbol{F}_{IO}(z) = \boldsymbol{F}_{IOD}^{-1}(z)\boldsymbol{F}_{ION}(z) = \underbrace{\left(\boldsymbol{A}^{(\nu)}\boldsymbol{S}_N^{(\nu)}(z)\right)^{-1}}_{\boldsymbol{F}_{IOD}(z)=\boldsymbol{G}_{IOD}(z)} \cdot$$

$$\cdot \underbrace{\left\{\begin{array}{l} \left[\begin{array}{ccc} \overbrace{\boldsymbol{B}^{(\mu)}\boldsymbol{S}_M^{(\mu)}(z)}^{\boldsymbol{G}_{ION}(z)} & -\boldsymbol{B}^{(\mu)}\boldsymbol{Z}_M^{(\mu)}(z) & \boldsymbol{A}^{(\nu)}\boldsymbol{Z}_N^{(\nu)}(z) \end{array}\right], \ \mu \geq 1 \\ \left[\begin{array}{cc} \overbrace{\boldsymbol{B}^{(\mu)}\boldsymbol{S}_M^{(\mu)}(z)}^{\boldsymbol{G}_{ION}(z)} & \boldsymbol{A}^{(\nu)}\boldsymbol{Z}_N^{(\nu)}(z) \end{array}\right], \ \mu = 0 \end{array}\right.}_{\boldsymbol{F}_{ION}(z)} =$$

$$= \left\{\begin{array}{l} \left[\begin{array}{c} \left(\boldsymbol{G}_{IO}(z)\right)^T \\ \left[\left(\boldsymbol{A}^{(\nu)}\boldsymbol{S}_N^{(\nu)}(z)\right)^{-1}\left(-\boldsymbol{B}^{(\mu)}\boldsymbol{Z}_M^{(\mu)}(z)\right)\right]^T \\ \left[\left(\boldsymbol{A}^{(\nu)}\boldsymbol{S}_N^{(\nu)}(z)\right)^{-1}\left(\boldsymbol{A}^{(\nu)}\boldsymbol{Z}_N^{(\nu)}(z)\right)\right]^T \end{array}\right]^T, \ \mu \geq 1, \\ \left[\begin{array}{ccc} \boldsymbol{G}_{IO}(z) & \left(\boldsymbol{A}^{(\nu)}\boldsymbol{S}_N^{(\nu)}(z)\right)^{-1}\left(\boldsymbol{A}^{(\nu)}\boldsymbol{Z}_N^{(\nu)}(z)\right) \end{array}\right], \ \mu = 0. \end{array}\right. \quad (8.20)$$

The denominator and the numerator polynomial matrices of the system full transfer function matrix $\boldsymbol{F}_{IO}(z)$ *are* $\boldsymbol{F}_{IOD}(z)$ *and* $\boldsymbol{F}_{ION}(z)$,

$$\boldsymbol{F}_{IOD}(z) = \boldsymbol{A}^{(\nu)} \boldsymbol{S}_N^{(\nu)}(z),$$

$$\boldsymbol{F}_{ION}(z) = \left[\begin{array}{ccc} \boldsymbol{B}^{(\mu)} \boldsymbol{S}_M^{(\mu)}(z) & -\boldsymbol{B}^{(\mu)} \boldsymbol{Z}_M^{(\mu)}(z) & \boldsymbol{A}^{(\nu)} \boldsymbol{Z}_N^{(\nu)}(z) \end{array} \right], \ \mu \geq 1$$

$$\boldsymbol{F}_{ION}(z) = \left[\begin{array}{cc} \boldsymbol{B}^{(\mu)} \boldsymbol{S}_M^{(\mu)}(z) & \boldsymbol{A}^{(\nu)} \boldsymbol{Z}_N^{(\nu)}(z) \end{array} \right], \ \mu = 0,$$

and

$$\boldsymbol{G}_{IO}(z) = \boldsymbol{G}_{IOD}^{-1}(z) \boldsymbol{G}_{ION}(z), \ \ \boldsymbol{G}_{IOD}(z) = \boldsymbol{F}_{IOD}(z).$$

Example 8.1 *The given SISO IO system is described by*

$$E^2 y(k) - E^1 y(k) - 0.75 E^0 y(k) = E^2 i(k) - 7.5 E^1 i(k) + 9 E^0 i(k).$$

Its description and characteristics in the complex domain read

$$\begin{gathered} \left(z^2 - z - 0.75\right) Y(z) - z(z-1) y(0) - z y(1) = \\ = \left(z^2 - 7.5z + 9\right) I(z) - z(z - 7.5) i(0) - z i(1) \Longrightarrow \end{gathered}$$

$$\begin{gathered} Y(z) = \left(z^2 - z - 0.75\right)^{-1} \cdot \\ \cdot \left[\begin{array}{ccccc} \left(z^2 - 7.5z + 9\right) & -z(z-7.5) & -z & z(z-1) & z \end{array} \right] \cdot \\ \cdot \left[\begin{array}{cc} I(z) & \underbrace{\underbrace{i_0 \ \ \left(E^1 i\right)_0}_{\mathbf{i}_0^{1T}} \ \ \underbrace{y_0 \ \ \left(E^1 y\right)_0}_{\mathbf{y}_0^{1T}}}_{\mathbf{C}_{0IO}^T} \end{array} \right]^T = \end{gathered}$$

$$\begin{gathered} = \boldsymbol{F}_{IO}(z) \mathbf{V}(z), \ \mathbf{V}(z) = \left[\begin{array}{cc} I(z) & \mathbf{C}_{0IO}^T \end{array} \right]^T \Longrightarrow \\ \mathbf{C}_{0IO} = \left[\begin{array}{c} \mathbf{i}_0^1 \\ \mathbf{y}_0^1 \end{array} \right], \ \boldsymbol{F}_{IO}(z) = \left(z^2 - z - 0.75\right)^{-1} \cdot \\ \cdot \left[\begin{array}{ccccc} \left(z^2 - 7.5z + 9\right) & -z(z-7.5) & -z & z(z-1) & z \end{array} \right], \end{gathered}$$

$$G_{IO}(z) = \frac{z^2 - 7.5z + 9}{z^2 - z - 0.75} = \frac{(z-1.5)(z-6)}{(z-1.5)(z+0.5)} \Longrightarrow G_{IOnd}(z) = \frac{z-6}{z+0.5}.$$

The system transfer function matrix $G_{IO}(z)$ *has the same zero* $z^0 = 1.5$ *and pole* $z^* = 1.5$. *They can be cancelled. However, the cancellation is not possible in the system full transfer function matrix* $\boldsymbol{F}_{IO}(z)$.

Example 8.2 *Let us observe the following two ratios of polynomials,*

$$\frac{z}{(z-0.3)\,(z+0.9)} \text{ and } \frac{z(z-1.5)}{(z-0.3)\,(z+0.9)(z-1.5)}, \tag{8.21}$$

which correspond to the same rational function $f(\cdot):\mathcal{N}_0 \longrightarrow \mathcal{R}$ (see $s-$complex analogy in [23, p. 58]). They have two common zeros $z_1^0=0$ and $z_2^0=\infty$, and two common poles $z_1^*=0.3$ and $z_2^*=-0.9$. The second ratio has an additional positive real identical single pole and zero, $z_3^*=1.5$, and $z_3^0=1.5$, which do not affect the form of the appropriate corresponding time function. However, if they represent the system transfer functions $G_1(\cdot)$ and $G_2(\cdot)$,

$$G_1(z)=\frac{z}{(z-0.3)\,(z+0.9)},\ G_2(z)=\frac{z(z-1.5)}{(z-0.3)\,(z+0.9)(z-1.5)}, \tag{8.22}$$

then they do not correspond to the same system. $G_1(z)$ is nondegenerate, while $G_2(z)$ is degenerate. The former is the nondegenerate form of the latter. $G_1(\cdot)$ is the transfer function of the second-order $SISO$ system described by

$$E^2y(k)+0.6E^1y(k)-0.27E^0y\,(k)=E^1i(k), \tag{8.23}$$

while $G_2(\cdot)$ is the transfer function of the third-order $SISO$ system determined by

$$\begin{aligned} E^3y(k)-0.9E^2y(k)-1.17E^1y(k)+0.405E^0y\,(k)&=\\ =-1.5E^1i(k)+E^2i(k).& \end{aligned} \tag{8.24}$$

The full transfer function matrix $\mathbf{F}_1(z)$ of the former reads

$$\mathbf{F}_1(z)=(z^2+0.6z-0.27)^{-1}\begin{bmatrix} z & -z & (z^2+0.6z) & z \end{bmatrix}, \tag{8.25}$$

while the full transfer function matrix $\mathbf{F}_2(z)$ of the latter is found as

$$\begin{aligned} \mathbf{F}_2(z)=(z^3-0.9z^2-1.17z+0.405)^{-1}\cdot\\ \cdot z\begin{bmatrix} z-1.5 & -(z-1.5) & -z & (z^2-0.9z-1.17) & (z-0.9) & 1 \end{bmatrix}. \end{aligned} \tag{8.26}$$

Both $\mathbf{F}_1(z)$ and $\mathbf{F}_2(z)$ are row nondegenerate and nondegenerate.

Example 8.3 *A $MIMO$ IO system is described by*

$$E^1\mathbf{y}(k)+\begin{bmatrix} 2 & 0\\ 0 & 1 \end{bmatrix}\mathbf{y}(k)=\begin{bmatrix} 2 & 0\\ 0 & 1.5 \end{bmatrix}\mathbf{i}(k)+E^1\mathbf{i}(k),\ \mathbf{y}=\begin{bmatrix} y_1\\ y_2 \end{bmatrix},\ \mathbf{i}=\begin{bmatrix} i_1\\ i_2 \end{bmatrix}.$$

Its transfer function matrix $\boldsymbol{G}_{IO}(z)$,

$$\boldsymbol{G}_{IO}(z) = \begin{bmatrix} 1 & 0 \\ 0 & \frac{z+1.5}{z+1} \end{bmatrix}, \ \textit{full} \ \operatorname{rank} \boldsymbol{G}_{IO}(z) = 2,$$

is rank defective for $z = -1.5$,

$$\operatorname{rank} \boldsymbol{G}_{IO}(-1.5) = 1.$$

However, the system full transfer function matrix $\boldsymbol{F}_{IO}(z)$,

$$\boldsymbol{F}_{IO}(z) = \begin{bmatrix} 1 & 0 & -\frac{z}{z+2} & 0 & \frac{z}{z+2} & 0 \\ 0 & \frac{z+1.5}{z+1} & 0 & -\frac{z}{z+1} & 0 & \frac{z}{z+1} \end{bmatrix},$$

has the full rank over the field of complex numbers z,

$$\operatorname{rank} \boldsymbol{F}_{IO}(z) \equiv \textit{full} \ \operatorname{rank} \boldsymbol{F}_{IO}(z) = 2.$$

In this example,

$$\boldsymbol{G}_{IO}(z) = \begin{bmatrix} 1 & 0 \\ 0 & \frac{z+1.5}{z+1} \end{bmatrix},$$

$$\boldsymbol{G}_{IOi_0}(z) = \begin{bmatrix} -\frac{z}{z+2} & 0 \\ 0 & -\frac{z}{z+1} \end{bmatrix},$$

$$\boldsymbol{G}_{IOy_0}(z) = \begin{bmatrix} \frac{z}{z+2} & 0 \\ 0 & \frac{z}{z+1} \end{bmatrix},$$

$$\boldsymbol{G}_{IO_0}(z) = \begin{bmatrix} -\frac{z}{z+2} & 0 & \frac{z}{z+2} & 0 \\ 0 & -\frac{z}{z+1} & 0 & \frac{z}{z+1} \end{bmatrix}.$$

Example 8.4 *A second-order* $MIMO$ IO *system is given by*

$$E^2\mathbf{y}(k) + \overbrace{\begin{bmatrix} 1 & 2 \\ 0 & 3 \end{bmatrix}}^{\boldsymbol{A}_1} E^1\mathbf{y}(k) = \underbrace{\begin{bmatrix} -1 & 2 & 2 \\ 4 & 4 & 7 \end{bmatrix}}_{\boldsymbol{B}_0} \mathbf{i}(k) + \underbrace{\begin{bmatrix} 2 & 1 & -5 \\ 3 & 1 & 6 \end{bmatrix}}_{\boldsymbol{B}_1} E^1\mathbf{i}(k).$$

It follows that $M = 3$, $N = 2$, $\nu = 2$, $\mu = 1$, $A_0 = O_2$, *and* $A_2 = I_2$. *Equations (2.2a), (2.2b), (8.4), and (8.5) become*

$$\boldsymbol{A}^{(\nu)} = \boldsymbol{A}^{(2)} = \begin{bmatrix} \boldsymbol{A}_0 & \boldsymbol{A}_1 & \boldsymbol{A}_2 \end{bmatrix} = \begin{bmatrix} 0 & 0 & 1 & 2 & 1 & 0 \\ 0 & 0 & 0 & 3 & 0 & 1 \end{bmatrix},$$

$$\boldsymbol{B}^{(\mu)} = \boldsymbol{B}^{(1)} = \begin{pmatrix} \boldsymbol{B}_0 & \boldsymbol{B}_1 \end{pmatrix} = \begin{bmatrix} -1 & 2 & 2 & 2 & 1 & -5 \\ 4 & 4 & 7 & 3 & 1 & 6 \end{bmatrix},$$

$$\boldsymbol{S}_M^{(\mu)}(z) = \boldsymbol{S}_3^{(1)}(z) = \begin{pmatrix} z^0\boldsymbol{I}_3 & z^1\boldsymbol{I}_3 \end{pmatrix}^T = \begin{bmatrix} 1 & 0 & 0 & z & 0 & 0 \\ 0 & 1 & 0 & 0 & z & 0 \\ 0 & 0 & 1 & 0 & 0 & z \end{bmatrix}^T,$$

$$\boldsymbol{S}_N^{(\nu)}(z) = \boldsymbol{S}_2^{(2)}(z) = \begin{pmatrix} z^0\boldsymbol{I}_2 & z^1\boldsymbol{I}_2 & z^2\boldsymbol{I}_2 \end{pmatrix}^T = \begin{bmatrix} 1 & 0 & z & 0 & z^2 & 0 \\ 0 & 1 & 0 & z & 0 & z^2 \end{bmatrix}^T,$$

$$\boldsymbol{A}^{(2)}\boldsymbol{S}_2^{(2)}(z) = \begin{bmatrix} 0 & 0 & 1 & 2 & 1 & 0 \\ 0 & 0 & 0 & 3 & 0 & 1 \end{bmatrix} \begin{bmatrix} 1 & 0 \\ 0 & 1 \\ z & 0 \\ 0 & z \\ z^2 & 0 \\ 0 & z^2 \end{bmatrix} = \begin{bmatrix} z^2+z & 2z \\ 0 & z^2+3z \end{bmatrix} \Rightarrow$$

$$\left(\boldsymbol{A}^{(2)}\boldsymbol{S}_2^{(2)}(z)\right)^{-1} = \frac{1}{(z^2+z)(z^2+3z)} \begin{bmatrix} z^2+3z & -2z \\ 0 & z^2+z \end{bmatrix} =$$

$$= \begin{bmatrix} \frac{z(z+3)}{z^2(z+1)(z+3)} & \frac{-2z}{z^2(z+1)(z+3)} \\ 0 & \frac{z(z+1)}{z^2(z+1)(z+3)} \end{bmatrix} = \frac{z}{z^2(z+1)(z+3)} \begin{bmatrix} z+3 & -2 \\ 0 & z+1 \end{bmatrix} =$$

$$= \begin{bmatrix} \frac{z+3}{z(z+1)(z+3)} & \frac{-2}{z(z+1)(z+3)} \\ 0 & \frac{z+1}{z(z+1)(z+3)} \end{bmatrix},$$

$$\boldsymbol{B}^{(1)}\boldsymbol{S}_3^{(1)}(z) = \begin{bmatrix} -1 & 2 & 2 & 2 & 1 & -5 \\ 4 & 4 & 7 & 3 & 1 & 6 \end{bmatrix} \begin{bmatrix} 1 & 0 & 0 \\ 0 & 1 & 0 \\ 0 & 0 & 1 \\ z & 0 & 0 \\ 0 & z & 0 \\ 0 & 0 & z \end{bmatrix} =$$

$$= \begin{bmatrix} 2z-1 & z+2 & 2-5z \\ 3z+4 & z+4 & 6z+7 \end{bmatrix},$$

$$\left(\boldsymbol{A}^{(2)}\boldsymbol{S}_2^{(2)}(z)\right)^{-1}\left(\boldsymbol{B}^{(1)}\boldsymbol{S}_3^{(1)}(z)\right) = \frac{z}{z^2(z+1)(z+3)} \begin{bmatrix} z+3 & -2 \\ 0 & z+1 \end{bmatrix} \cdot$$

$$\cdot \begin{bmatrix} 2z-1 & z+2 & 2-5z \\ 3z+4 & z+4 & 6z+7 \end{bmatrix} =$$

$$=\frac{z}{z^2(z+1)(z+3)}\begin{bmatrix} 2z^2-z-11 & z^2+3z-2 & -5z^2-25z-8 \\ (z+1)(3z+4) & (z+1)(z+4) & (z+1)(6z+7) \end{bmatrix},$$

$$Z_M^{(\mu)}(z)=Z_3^{(1)}(z)=\begin{pmatrix} 0 & 0 & 0 \\ 0 & 0 & 0 \\ 0 & 0 & 0 \\ z & 0 & 0 \\ 0 & z & 0 \\ 0 & 0 & z \end{pmatrix},\ Z_N^{(\nu)}(z)=Z_2^{(2)}(z)=\begin{pmatrix} 0 & 0 & 0 & 0 \\ 0 & 0 & 0 & 0 \\ z & 0 & 0 & 0 \\ 0 & z & 0 & 0 \\ z^2 & 0 & z & 0 \\ 0 & z^2 & 0 & z \end{pmatrix},$$

$$\boldsymbol{A}^{(2)}\boldsymbol{Z}_2^{(2)}(z)=\begin{bmatrix} 0 & 0 & 1 & 2 & 1 & 0 \\ 0 & 0 & 0 & 3 & 0 & 1 \end{bmatrix}\begin{pmatrix} 0 & 0 & 0 & 0 \\ 0 & 0 & 0 & 0 \\ z & 0 & 0 & 0 \\ 0 & z & 0 & 0 \\ z^2 & 0 & z & 0 \\ 0 & z^2 & 0 & z \end{pmatrix}=$$

$$=\begin{bmatrix} z^2+z & 2z & z & 0 \\ 0 & z^2+3z & 0 & z \end{bmatrix}\Longrightarrow$$

$$\left(\boldsymbol{A}^{(2)}\boldsymbol{S}_2^{(2)}(z)\right)^{-1}\boldsymbol{A}^{(2)}\boldsymbol{Z}_2^{(2)}(z)=$$

$$=\frac{z}{z^2(z+1)(z+3)}\begin{bmatrix} z+3 & -2 \\ 0 & z+1 \end{bmatrix}\begin{bmatrix} z^2+z & 2z & z & 0 \\ 0 & z^2+3z & 0 & z \end{bmatrix}=$$

$$=\frac{z}{z^2(z+1)(z+3)}\cdot$$

$$\cdot\begin{bmatrix} z(z+1)(z+3) & 0 & z(z+3) & -2z \\ 0 & z(z+1)(z+3) & 0 & z(z+1) \end{bmatrix},$$

$$\boldsymbol{B}^{(1)}\boldsymbol{Z}_3^{(1)}(z)=\begin{bmatrix} -1 & 2 & 2 & 2 & 1 & -5 \\ 4 & 4 & 7 & 3 & 1 & 6 \end{bmatrix}\begin{pmatrix} 0 & 0 & 0 \\ 0 & 0 & 0 \\ 0 & 0 & 0 \\ z & 0 & 0 \\ 0 & z & 0 \\ 0 & 0 & z \end{pmatrix}=$$

$$=\begin{bmatrix} 2z & z & -5z \\ 3z & z & 6z \end{bmatrix}\Longrightarrow$$

$$-\left(\boldsymbol{A}^{(2)}\boldsymbol{S}_2^{(2)}(z)\right)^{-1}\boldsymbol{B}^{(1)}\boldsymbol{Z}_3^{(1)}(z)=$$

$$=-\frac{z}{z^2(z+1)(z+3)}\begin{bmatrix} z+3 & -2 \\ 0 & z+1 \end{bmatrix}\begin{bmatrix} 2z & z & -5z \\ 3z & z & 6z \end{bmatrix}=$$

$$=-\frac{z}{z^2(z+1)(z+3)}\begin{bmatrix} 2z^2 & z(z+1) & -z(5z+27) \\ 3z(z+1) & z(z+1) & 6z(z+1) \end{bmatrix}.$$

We can now determine $\boldsymbol{G}_{IO}(z)$ in view of (8.2),

$$\boldsymbol{G}_{IO}(z)=\left(\boldsymbol{A}^{(2)}\boldsymbol{S}_2^{(2)}(z)\right)^{-1}\left(\boldsymbol{B}^{(1)}\boldsymbol{S}_3^{(1)}(z)\right)= \\ =\frac{z}{z^2(z+1)(z+3)}\cdot \\ \cdot\begin{bmatrix} 2z^2-z-11 & z^2+3z-2 & -5z^2-25z-8 \\ (z+1)(3z+4) & (z+1)(z+4) & (z+1)(6z+7) \end{bmatrix}. \tag{8.27}$$

The system full transfer function matrix $\boldsymbol{F}_{IO}(z)$ has the following form due to both (8.6a) and the above equations of this example:

$$\boldsymbol{F}_{IO}(z)=\frac{z}{z^2(z+1)(z+3)}\cdot$$

$$\cdot\begin{bmatrix} \left(\underbrace{\begin{bmatrix} 2z^2-z-11 & z^2+3z-2 & -5z^2-25z-8 \\ (z+1)(3z+4) & (z+1)(z+4) & (z+1)(6z+7) \end{bmatrix}}_{\boldsymbol{G}_{IO}(z)}\right)^T \\ \left(\underbrace{\begin{bmatrix} -2z^2 & -z(z+1) & z(5z+27) \\ -3z(z+1) & -z(z+1) & -6z(z+1) \end{bmatrix}}_{\boldsymbol{G}_{IOi_0}(z)}\right)^T \\ \left(\underbrace{\begin{bmatrix} z(z+1)(z+3) & 0 & z(z+3) & -2z \\ 0 & z(z+1)(z+3) & 0 & z(z+1) \end{bmatrix}}_{\boldsymbol{G}_{IOy_0}(z)}\right)^T \end{bmatrix}^T =$$

$$=\begin{bmatrix} \boldsymbol{G}_{IO}(z) & \boldsymbol{G}_{IOi_0}(z) & \boldsymbol{G}_{IOy_0}(z) \end{bmatrix},$$

i.e.,

$$\boldsymbol{F}_{IO}(z) = \frac{z}{z^2(z+1)(z+3)} \cdot$$

$$\cdot \begin{bmatrix} 2z^2 - z - 11 & (z+1)(3z+4) \\ z^2 + 3z - 2 & (z+1)(z+4) \\ -5z^2 - 25z - 8 & (z+1)(6z+7) \\ -2z^2 & -3z(z+1) \\ -z(z+1) & -z(z+1) \\ z(5z+27) & -6z(z+1) \\ z(z+1)(z+3) & 0 \\ 0 & z(z+1)(z+3) \\ z(z+3) & 0 \\ -2z & z(z+1) \end{bmatrix}^T \in \mathcal{C}^{2\times 10}. \tag{8.28}$$

This is the degenerate form of $\boldsymbol{F}_{IO}(z)$. The nondegenerate form $\boldsymbol{F}_{IOnd}(z)$ of $\boldsymbol{F}_{IO}(z)$ reads:

$$\boldsymbol{F}_{IO}(z) = \frac{1}{z(z+1)(z+3)} \begin{bmatrix} 2z^2 - z - 11 & (z+1)(3z+4) \\ z^2 + 3z - 2 & (z+1)(z+4) \\ -5z^2 - 25z - 8 & (z+1)(6z+7) \\ -2z^2 & -3z(z+1) \\ -z(z+1) & -z(z+1) \\ z(5z+27) & -6z(z+1) \\ z(z+1)(z+3) & 0 \\ 0 & z(z+1)(z+3) \\ z(z+3) & 0 \\ -2z & z(z+1) \end{bmatrix}^T \in \mathcal{C}^{2\times 10}.$$

Example 8.5 *The IO system of Example 2.2 (in Section 2.1)*

$$\begin{bmatrix} 3 & 0 & 0 \\ 0 & 0 & 0 \\ 0 & 0 & 0 \end{bmatrix} E^2\mathbf{y}(k) + \begin{bmatrix} 0 & 0 & 0 \\ 0 & 1 & 0 \\ 0 & 0 & 0 \end{bmatrix} E^1\mathbf{y}(k) + \begin{bmatrix} 0 & 0 & 0 \\ 0 & 0 & 0 \\ 0 & 0 & 1 \end{bmatrix} E^0\mathbf{y}(k) =$$

$$= \begin{bmatrix} 2 & 0 \\ 0 & 1 \\ 1 & 0 \end{bmatrix} E^0\mathbf{i}(k) + \begin{bmatrix} 1 & 0 \\ 0 & 1 \\ 1 & 1 \end{bmatrix} E^2\mathbf{i}(k)$$

yields

$$\nu = 2,\ \mu = 2,\ N = 3,\ M = 2,\ \det \boldsymbol{A}_\nu = \det \boldsymbol{A}_2 = \det \begin{bmatrix} 3 & 0 & 0 \\ 0 & 0 & 0 \\ 0 & 0 & 0 \end{bmatrix} = 0,$$

$$\boldsymbol{A}^{(2)} = \begin{bmatrix} 0 & 0 & 0 & 0 & 0 & 0 & 3 & 0 & 0 \\ 0 & 0 & 0 & 0 & 1 & 0 & 0 & 0 & 0 \\ 0 & 0 & 1 & 0 & 0 & 0 & 0 & 0 & 0 \end{bmatrix},$$

$$\boldsymbol{S}_3^{(2)}(z) = \begin{bmatrix} 1 & 0 & 0 \\ 0 & 1 & 0 \\ 0 & 0 & 1 \\ z & 0 & 0 \\ 0 & z & 0 \\ 0 & 0 & z \\ z^2 & 0 & 0 \\ 0 & z^2 & 0 \\ 0 & 0 & z^2 \end{bmatrix},$$

$$\boldsymbol{A}^{(2)}\boldsymbol{S}_3^{(2)}(z) = \begin{bmatrix} 3z^2 & 0 & 0 \\ 0 & z & 0 \\ 0 & 0 & 1 \end{bmatrix}, \ \deg\left[\boldsymbol{A}^{(2)}\boldsymbol{S}_3^{(2)}(z)\right] = 2,$$

$$\operatorname{adj}\left(\boldsymbol{A}^{(2)}\boldsymbol{S}_3^{(2)}(z)\right) = \begin{bmatrix} z & 0 & 0 \\ 0 & 3z^2 & 0 \\ 0 & 0 & 3z^3 \end{bmatrix}, \ \deg\left[\operatorname{adj}\left(\boldsymbol{A}^{(2)}\boldsymbol{S}_3^{(2)}(z)\right)\right] = 3,$$

$$\det\left[\boldsymbol{A}^{(2)}\boldsymbol{S}_3^{(2)}(z)\right] = 3z^3, \ \deg\left\{\det\left[\boldsymbol{A}^{(2)}\boldsymbol{S}_3^{(2)}(z)\right]\right\} = 3,$$

$$\boldsymbol{B}^{(2)} = \begin{bmatrix} 2 & 0 & 0 & 0 & 1 & 0 \\ 0 & 1 & 0 & 0 & 0 & 1 \\ 1 & 0 & 0 & 0 & 1 & 1 \end{bmatrix},$$

$$\boldsymbol{S}_2^{(2)}(z) = \begin{bmatrix} 1 & 0 \\ 0 & 1 \\ z & 0 \\ 0 & z \\ z^2 & 0 \\ 0 & z^2 \end{bmatrix},$$

$$\boldsymbol{B}^{(2)}\boldsymbol{S}_2^{(2)}(z) = \begin{bmatrix} 2+z^2 & 0 \\ 0 & 1+z^2 \\ 1+z^2 & z^2 \end{bmatrix}, \ \deg\left[\boldsymbol{B}^{(2)}\boldsymbol{S}_2^{(2)}(z)\right] = 2,$$

$$\boldsymbol{B}^{(2)}\boldsymbol{Z}_2^{(2)}(z) = \begin{bmatrix} 2 & 0 & 0 & 0 & 1 & 0 \\ 0 & 1 & 0 & 0 & 0 & 1 \\ 1 & 0 & 0 & 0 & 1 & 1 \end{bmatrix} \begin{bmatrix} 0 & 0 & 0 & 0 \\ 0 & 0 & 0 & 0 \\ z & 0 & 0 & 0 \\ 0 & z & 0 & 0 \\ z^2 & 0 & z & 0 \\ 0 & z^2 & 0 & z \end{bmatrix} =$$

$$= \begin{bmatrix} z^2 & 0 & z & 0 \\ 0 & z^2 & 0 & z \\ z^2 & z^2 & z & z \end{bmatrix}, \ \deg \boldsymbol{B}^{(2)} \boldsymbol{Z}_2^{(2)}(z) = \deg\left(z^2\right) = 2,$$

$$\boldsymbol{A}^{(2)} \boldsymbol{Z}_3^{(2)}(z) = \begin{bmatrix} 0 & 0 & 0 & 0 & 0 & 0 & 3 & 0 & 0 \\ 0 & 0 & 0 & 0 & 1 & 0 & 0 & 0 & 0 \\ 0 & 0 & 1 & 0 & 0 & 0 & 0 & 0 & 0 \end{bmatrix} \begin{bmatrix} 0 & 0 & 0 & 0 & 0 & 0 \\ 0 & 0 & 0 & 0 & 0 & 0 \\ 0 & 0 & 0 & 0 & 0 & 0 \\ z & 0 & 0 & 0 & 0 & 0 \\ 0 & z & 0 & 0 & 0 & 0 \\ 0 & 0 & z & 0 & 0 & 0 \\ z^2 & 0 & 0 & z & 0 & 0 \\ 0 & z^2 & 0 & 0 & z & 0 \\ 0 & 0 & z^2 & 0 & 0 & z \end{bmatrix} =$$

$$= \begin{bmatrix} 3z^2 & 0 & 0 & 3z & 0 & 0 \\ 0 & z & 0 & 0 & 0 & 0 \\ 0 & 0 & 0 & 0 & 0 & 0 \end{bmatrix}, \ \deg\left[\boldsymbol{A}^{(2)} \boldsymbol{Z}_3^{(2)}(z)\right] = \deg\left(3z^2\right) = 2,$$

and

$$\boldsymbol{F}_{IO}(z) = \frac{\begin{bmatrix} z & 0 & 0 \\ 0 & 3z^2 & 0 \\ 0 & 0 & 3z^3 \end{bmatrix}}{3z^3} \cdot$$

$$\cdot \begin{bmatrix} 2+z^2 & 0 & -z^2 & 0 & -z & 0 & 3z^2 & 0 & 0 & 3z & 0 & 0 \\ 0 & 1+z^2 & 0 & -z^2 & 0 & -z & 0 & z & 0 & 0 & 0 & 0 \\ 1+z^2 & z^2 & -z^2 & -z^2 & -z & -z & 0 & 0 & 0 & 0 & 0 & 0 \end{bmatrix}.$$

The full transfer function matrix $\boldsymbol{F}_{IO}(z)$ is improper because the degree of its numerator matrix (which is equal to 5) exceeds the degree of its denominator polynomial (which is equal to 3).

Example 8.6 *The IO system of Example 2.3 (in Section 2.1)*

$$\begin{bmatrix} 3 & 0 & 1 \\ 2 & 0 & 0 \\ 0 & 1 & 1 \end{bmatrix} E^2\mathbf{y}(k) + \begin{bmatrix} 0 & 0 & 0 \\ 0 & 1 & 0 \\ 0 & 0 & 0 \end{bmatrix} E^1\mathbf{y}(k) + \begin{bmatrix} 0 & 0 & 0 \\ 0 & 0 & 0 \\ 0 & 0 & 1 \end{bmatrix} E^0\mathbf{y}(k) =$$

$$= \begin{bmatrix} 2 & 0 \\ 0 & 1 \\ 1 & 0 \end{bmatrix} E^0\mathbf{i}(k) + \begin{bmatrix} 1 & 0 \\ 0 & 1 \\ 1 & 1 \end{bmatrix} E^2\mathbf{i}(k)$$

induces

$$\nu = 2,\ \mu = 2,\ N = 3,\ M = 2,\ \det A_\nu = \det A_2 = \det \begin{bmatrix} 3 & 0 & 1 \\ 2 & 0 & 0 \\ 0 & 1 & 1 \end{bmatrix} = 2 \neq 0,$$

$$\boldsymbol{A}^{(2)} \boldsymbol{S}_3^{(2)}(z) = \begin{bmatrix} 3z^2 & 0 & z^2 \\ 2z^2 & z & 0 \\ 0 & z^2 & z^2+1 \end{bmatrix},\ \deg \left[\boldsymbol{A}^{(2)} \boldsymbol{S}_3^{(2)}(z) \right] = 2,$$

$$\operatorname{adj} \left(\boldsymbol{A}^{(2)} \boldsymbol{S}_3^{(2)}(z) \right) = \begin{bmatrix} z^3+z & z^4 & -z^3 \\ -2z^4-2z^2 & 3z^4+3z^2 & 2z^4 \\ 2z^4 & -3z^4 & 3z^3 \end{bmatrix},$$

$$\deg \left[\operatorname{adj} \left(\boldsymbol{A}^{(2)} \boldsymbol{S}_3^{(2)}(z) \right) \right] = 4,$$

$$\det \left[\boldsymbol{A}^{(2)} \boldsymbol{S}_3^{(2)}(z) \right] = 2z^6 + 3z^5 + 3z^3,\ \deg \left\{ \det \left[\boldsymbol{A}^{(2)} \boldsymbol{S}_3^{(2)}(z) \right] \right\} = 6,$$

$$\boldsymbol{B}^{(2)} \boldsymbol{S}_2^{(2)}(z) = \begin{bmatrix} 2+z^2 & 0 \\ 0 & 1+z^2 \\ 1+z^2 & z^2 \end{bmatrix},\ \deg \left[\boldsymbol{B}^{(2)} \boldsymbol{S}_2^{(2)}(z) \right] = 2,$$

$$\boldsymbol{B}^{(2)} \boldsymbol{Z}_2^{(2)}(z) = \begin{bmatrix} 2 & 0 & 0 & 0 & 1 & 0 \\ 0 & 1 & 0 & 0 & 0 & 1 \\ 1 & 0 & 0 & 0 & 1 & 1 \end{bmatrix} \begin{bmatrix} 0 & 0 & 0 & 0 \\ 0 & 0 & 0 & 0 \\ z & 0 & 0 & 0 \\ 0 & z & 0 & 0 \\ z^2 & 0 & z & 0 \\ 0 & z^2 & 0 & z \end{bmatrix} =$$

$$= \begin{bmatrix} z^2 & 0 & z & 0 \\ 0 & z^2 & 0 & z \\ z^2 & z^2 & z & z \end{bmatrix},\ \deg \boldsymbol{B}^{(2)} \boldsymbol{Z}_2^{(2)}(z) = \deg \left(z^2 \right) = 2,$$

$$A^{(2)} Z_3^{(2)}(z) = \begin{bmatrix} 0 & 0 & 0 & 0 & 0 & 0 & 3 & 0 & 1 \\ 0 & 0 & 0 & 0 & 1 & 0 & 2 & 0 & 0 \\ 0 & 0 & 1 & 0 & 0 & 0 & 0 & 1 & 1 \end{bmatrix} \begin{bmatrix} 0 & 0 & 0 & 0 & 0 & 0 \\ 0 & 0 & 0 & 0 & 0 & 0 \\ 0 & 0 & 0 & 0 & 0 & 0 \\ z & 0 & 0 & 0 & 0 & 0 \\ 0 & z & 0 & 0 & 0 & 0 \\ 0 & 0 & z & 0 & 0 & 0 \\ z^2 & 0 & 0 & z & 0 & 0 \\ 0 & z^2 & 0 & 0 & z & 0 \\ 0 & 0 & z^2 & 0 & 0 & z \end{bmatrix} =$$

$$= \begin{bmatrix} 3z^2 & 0 & z^2 & 3z & 0 & z \\ 2z^2 & z & 0 & 2z & 0 & 0 \\ 0 & z^2 & z^2 & 0 & z & z \end{bmatrix}, \ \deg\left[\boldsymbol{A}^{(2)}\boldsymbol{Z}_3^{(2)}(z)\right] = \deg\left(3z^2\right) = 2,$$

and

$$\boldsymbol{F}_{IO}(z) = \frac{\begin{bmatrix} z^3+z & z^4 & -z^3 \\ -2z^4-2z^2 & 3z^4+3z^2 & 2z^4 \\ 2z^4 & -3z^4 & 3z^3 \end{bmatrix}}{2z^6+3z^5+3z^3}.$$

$$\cdot \begin{bmatrix} 2+z^2 & 0 & -z^2 & 0 & -z & 0 & 3z^2 & 0 & z^2 & 3z & 0 & z \\ 0 & 1+z^2 & 0 & -z^2 & 0 & -z & 2z^2 & z & 0 & 2z & 0 & 0 \\ 1+z^2 & -z^2 & -z^2 & z^2 & -z & -z & 0 & z^2 & z^2 & 0 & z & z \end{bmatrix}.$$

The degree of the numerator matrix polynomial is equal to six which is also the degree of the denominator polynomial. The full transfer function matrix $\boldsymbol{F}_{IO}(z)$ is proper in this case.

8.2 $\boldsymbol{F}(z)$ of the *ISO* system

In order to ease the reading we repeat the *ISO* system (3.60a) and (3.60b) (in Subsection 3.5.3) as (8.29a) and (8.29b),

$$\mathbf{x}(k+1) = \boldsymbol{A}\mathbf{x}(k) + \boldsymbol{B}\mathbf{i}(k), \ \forall k \in \mathcal{N}_0, \tag{8.29a}$$

$$\mathbf{y}(k) = \boldsymbol{C}\mathbf{x}(k) + \boldsymbol{D}\mathbf{i}(k), \ \forall k \in \mathcal{N}_0. \tag{8.29b}$$

Let at first the rank of the matrix $\boldsymbol{C}$ in (8.29b) be arbitrary.

The *IO* system transfer function $\boldsymbol{G}_{ISO}(z)$ is well-known:

$$\boldsymbol{G}_{ISO}(z) = \boldsymbol{C}(z\boldsymbol{I}_n - \boldsymbol{A})^{-1}\boldsymbol{B} + \boldsymbol{D}. \tag{8.30}$$

Theorem 8.2 *a) The ISO system (8.29a) and (8.29b) full IO transfer function matrix $\boldsymbol{F}_{ISO}(z)$ reads in general:*

$$\boldsymbol{F}_{ISO}(z) = \left[\ \boldsymbol{C}(z\boldsymbol{I}_n - \boldsymbol{A})^{-1}\boldsymbol{B} + \boldsymbol{D} \quad z\boldsymbol{C}(z\boldsymbol{I}_n - \boldsymbol{A})^{-1}\ \right] = \\ = \boldsymbol{F}_{ISOD}^{-1}(z)\boldsymbol{F}_{ISON}(z), \tag{8.31}$$

which implies

$$\mathbf{Y}(z) = \boldsymbol{F}_{ISO}(z)\left[\ \mathbf{I}^T(z) \quad \mathbf{x}_0^T\ \right]^T = \boldsymbol{F}_{ISO}(z)\mathbf{V}_{ISO}(z;\mathbf{x}_0), \tag{8.32}$$

$$\mathbf{V}_{ISO}(z;\mathbf{x}_0) = \begin{bmatrix} \mathbf{I}(z) \\ \mathbf{C}_{0ISO} \end{bmatrix}, \ \mathbf{C}_{0ISO} = \mathbf{x}_0. \tag{8.33}$$

b) *The ISO system (8.29a) and (8.29b) IISO transfer function matrix* $\boldsymbol{G}_{ISOx_0}(z)$ *with respect to* $\mathbf{x}_0$ *has the following general form:*

$$\boldsymbol{G}_{ISOx_0}(z) = z\boldsymbol{C}(z\boldsymbol{I}_n - \boldsymbol{A})^{-1}. \tag{8.34}$$

c) *The ISO system (8.29a) and (8.29b) full (complete) IS transfer function matrix* $\boldsymbol{F}_{ISOIS}(z)$ *is given by*

$$\begin{aligned}\boldsymbol{F}_{ISOIS}(z) &= \left[\ (z\boldsymbol{I}_n - \boldsymbol{A})^{-1}\boldsymbol{B} \quad z(z\boldsymbol{I}_n - \boldsymbol{A})^{-1}\ \right] = \\ &= \boldsymbol{F}_{ISOISD}^{-1}(z)\boldsymbol{F}_{ISOISN}(z).\end{aligned} \tag{8.35}$$

d) *The ISO system (8.29a) and (8.29b) IS transfer function matrix* $\boldsymbol{G}_{ISOIS}(z)$ *satisfies*

$$\boldsymbol{G}_{ISOIS}(z) = (z\boldsymbol{I}_n - \boldsymbol{A})^{-1}\boldsymbol{B}\ . \tag{8.36}$$

e) *The ISO system (8.29a) and (8.29b) SS transfer function matrix* $\boldsymbol{G}_{ISOSS}(z)$ *with respect to* x_0 *obeys*

$$\boldsymbol{G}_{ISOSS}(z) = z(z\boldsymbol{I}_n - \boldsymbol{A})^{-1}. \tag{8.37}$$

The proof of Theorem 8.2 might be evident but we present it in order to illustrate the procedure of the determination of the ISO system full transfer function matrix $\boldsymbol{F}_{ISO}(z)$. The proof shows the origin of the difference between it and the ISO system transfer function matrix $\boldsymbol{G}_{ISO}(z)$.

Proof. *a*) The $Z-$transforms of (8.29a) and (8.29b) imply:

$$\mathbf{X}(z) = (z\boldsymbol{I}_n - \boldsymbol{A})^{-1}(z\mathbf{x}_0 + \boldsymbol{B}\mathbf{I}(z)) \Longrightarrow$$

$$\mathbf{X}(z) = \left[\ (z\boldsymbol{I}_n - \boldsymbol{A})^{-1}\boldsymbol{B} \quad z(z\boldsymbol{I}_n - \boldsymbol{A})^{-1}\ \right] \underbrace{\begin{bmatrix} \mathbf{I}(z) \\ \mathbf{x}_0 \end{bmatrix}}_{\mathbf{V}_{ISO}(z;\mathbf{x}_0)}$$

$$\mathbf{Y}(z) = \boldsymbol{C}\left[(z\boldsymbol{I}_n - \boldsymbol{A})^{-1}(z\mathbf{x}_0 + \boldsymbol{B}\mathbf{I}(z))\right] + \boldsymbol{D}\mathbf{I}(z) =$$

$$= \underbrace{\left[\begin{array}{cc} \overbrace{\boldsymbol{C}(z\boldsymbol{I}_n - \boldsymbol{A})^{-1}\boldsymbol{B} + \boldsymbol{D}}^{\boldsymbol{G}_{ISO}(z)} & \overbrace{z\boldsymbol{C}(z\boldsymbol{I}_n - \boldsymbol{A})^{-1}}^{\boldsymbol{G}_{ISOx_0}(z)} \end{array}\right]}_{\boldsymbol{F}_{ISO}(z)} \underbrace{\left[\begin{array}{c} \mathbf{I}(z) \\ \mathbf{x}_0 \end{array}\right]}_{\mathbf{V}_{ISO}(z;\mathbf{x}_0)} =$$

$$= \left[\det(z\boldsymbol{I}_n - \boldsymbol{A})\right]^{-1} \cdot$$

$$\cdot \left[\begin{array}{cc} \boldsymbol{C}\operatorname{adj}(z\boldsymbol{I}_n - \boldsymbol{A})\boldsymbol{B} + \boldsymbol{D}\det(z\boldsymbol{I}_n - \boldsymbol{A}) & z\boldsymbol{C}\operatorname{adj}(z\boldsymbol{I}_n - \boldsymbol{A}) \end{array}\right] \cdot$$

$$\cdot \underbrace{\left[\begin{array}{c} \mathbf{I}(z) \\ \mathbf{C}_{0ISO} \end{array}\right]}_{\mathbf{V}_{ISO}(z;\mathbf{C}_{0ISO})} = \boldsymbol{F}_{ISOD}^{-1}(z)\boldsymbol{F}_{ISON}(z)\mathbf{V}_{ISO}(z;\mathbf{x}_0),\ \mathbf{C}_{0ISO} = \mathbf{x}_0,$$

$$\boldsymbol{F}_{ISOD}(z) = \det(z\boldsymbol{I}_n - \boldsymbol{A}),$$

$$\boldsymbol{F}_{ISON}(z) = \left[\begin{array}{cc} \boldsymbol{C}\operatorname{adj}(z\boldsymbol{I}_n - \boldsymbol{A})\boldsymbol{B} + \boldsymbol{D}\det(z\boldsymbol{I}_n - \boldsymbol{A}) & z\boldsymbol{C}\operatorname{adj}(z\boldsymbol{I}_n - \boldsymbol{A}) \end{array}\right],$$

$$\boldsymbol{F}_{ISO}(z) = \left[\begin{array}{cc} \boldsymbol{C}(z\boldsymbol{I}_n - \boldsymbol{A})^{-1}\boldsymbol{B} + \boldsymbol{D} & z\boldsymbol{C}(z\boldsymbol{I}_n - \boldsymbol{A})^{-1} \end{array}\right] =$$

$$= \left[\det(z\boldsymbol{I}_n - \boldsymbol{A})\right]^{-1} \cdot$$

$$\cdot \left[\begin{array}{cc} \boldsymbol{C}\operatorname{adj}(z\boldsymbol{I}_n - \boldsymbol{A})\boldsymbol{B} + \boldsymbol{D}\det(z\boldsymbol{I}_n - \boldsymbol{A}) & z\boldsymbol{C}\operatorname{adj}(z\boldsymbol{I}_n - \boldsymbol{A}) \end{array}\right]. \quad (8.38)$$

These results and (7.20a) (Definition 7.4, in Section 7.3), verify fully the statement under a) of the theorem, i.e., Equations (8.31) and (8.32).

b) (8.34) follows directly from a) and from the definition of $\boldsymbol{G}_{ISOx_0}(z)$ (7.22), [c) of Definition 7.4, in Section 7.3].

c) The first two equations of (8.38) and a) of Definition 7.5 (in Section 7.3) yield (8.35).

d) Equation (8.36) follows from the first equation of (8.38) and b) of Definition 7.5 (in Section 7.3).

e) The first equation of (8.38) and c) of Definition 7.5 (in Section 7.3) jointly imply (8.37). ■

Comment 8.4 *All actions on the system are the input vector function* $\mathbf{i}(\cdot)$ *and the initial state vector* $\mathbf{x}_0$*. This justifies to define the action vector function* $\mathbf{v}_{ISO}(\cdot)$ *for the ISO system by*

$$\mathbf{v}_{ISO}(k;\mathbf{x}_0) = \left[\begin{array}{c} \mathbf{i}(k) \\ \delta_d(k)\mathbf{x}_0 \end{array}\right] \in \mathcal{R}^{M+n},\ \mathbf{V}_{ISO}(z;\mathbf{x}_0) = \left[\begin{array}{c} \mathbf{I}(z) \\ \mathbf{x}_0 \end{array}\right] \in \mathcal{C}^{M+n}. \quad (8.39)$$

This allows us to put (8.32) into the compact form:

$$\mathbf{Y}(z) = \mathbf{Y}(z;\mathbf{x}_0) = \boldsymbol{F}_{ISO}(z)\mathbf{V}_{ISO}(z;\mathbf{x}_0), \quad (8.40)$$

which is the well-known form of the classical relationship between $\mathbf{Y}(z)$ *and* $\mathbf{I}(z)$ *for the zero initial state vector,* $\mathbf{x}_0 = \mathbf{0}_N$*, and which is expressed via the system transfer function matrix* $\boldsymbol{G}_{ISO}(z)$*,*

$$\mathbf{Y}(z) = \mathbf{Y}(z; \mathbf{0}_N) = \boldsymbol{G}_{ISO}(z)\mathbf{I}(z), \ \mathbf{x}_0 = \mathbf{0}_N. \tag{8.41}$$

We stress out again that (8.41) holds only for the zero initial state vector, $\mathbf{x}_0 = \mathbf{0}_N$*, while (8.32), i.e., (8.40), are valid for arbitrary initial conditions.*

Comment 8.5 *The system full transfer function matrix* $\boldsymbol{F}_{ISO}(z)$ *incorporates the system transfer function matrix* $\boldsymbol{G}_{ISO}(z)$*:*

$$\boldsymbol{F}_{ISO}(z) = \left[\begin{array}{cc} \boldsymbol{G}_{ISO}(z) & \boldsymbol{G}_{ISOx_0}(z) \end{array} \right].$$

Note 8.4 ***The full block diagram technique generalizes the block diagram technique***

The action vector $\mathbf{v}_{ISO}(k; \mathbf{x}_0)$ *allows us to use the classical block diagram technique. The action vector* $\mathbf{v}_{ISO}(k; \mathbf{x}_0)$ *should be used instead of the input vector* $\mathbf{i}(k)$*, that is that* $\mathbf{V}_{ISO}(z; \mathbf{x}_0)$ *replaces* $\mathbf{I}(z)$*, and that* $\boldsymbol{F}_{ISO}(z)$ *replaces* $\boldsymbol{G}_{ISO}(z)$*; see Fig. 8.2.*

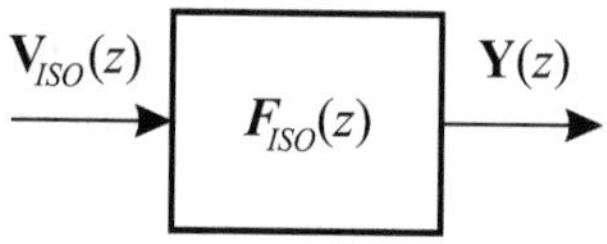

Figure 8.2: The generalized block of the ISO system.

Note 8.5 *It is well known that the system transfer function matrix* $\boldsymbol{G}_{ISO}(z)$ *is invariant with respect to a mathematical description of a given physical system. This does not hold for the system full transfer function matrix* $\boldsymbol{F}_{ISO}(z)$ *in general, because it depends on the selection of the state variables, so that it depends on the selection of the state vector* $\mathbf{x}$*. This results from Equation (8.32) that reflects the influence of the initial state vector* $\mathbf{x}_0$*, rather than the initial output vector* $\mathbf{y}_0$*, on the system output response. Hence, the transfer function matrix* $\boldsymbol{G}_{ISOx_0}(z)$ *with respect to* $\mathbf{x}_0$ *shows how* $\mathbf{x}_0$ *influences the output vector* $\mathbf{y}$*. The form of* $\boldsymbol{G}_{ISOx_0}(z)$ *is dependent of the choice of* $\mathbf{x}$*. In order to be clear let* $\boldsymbol{F}_{IOISO}(z)$ *and* $\boldsymbol{G}_{IOISO}(z)$ *be the full transfer function matrix and the transfer function matrix obtained from the IO mathematical model of the given ISO system instead of from the system original mathematical model (8.29a), (8.29b) so that*

$$\boldsymbol{F}_{IOISO}(z) \neq \boldsymbol{F}_{ISO}(z) \text{ and } \boldsymbol{G}_{IOISO}(z) = \boldsymbol{G}_{ISO}(z) \neq \boldsymbol{G}_{ISOx_0}(z) \text{ in general.}$$

When we consider a special class of the systems then we can express $\mathbf{x}_0$ in terms of $\mathbf{i}_0$ and $\mathbf{y}_0$. Let us recall $n \geq N$ so that the full rank of $\boldsymbol{C}$ is equal to N. The full rank of $\boldsymbol{C}^T\boldsymbol{C} \in \mathcal{R}^{n\times n}$ is also N that implies that $\det \boldsymbol{C}^T\boldsymbol{C} \neq 0$ is possible if and only if $N = n$. This enables us to solve uniquely (8.29b) for $\mathbf{x}$ if and only if both $N = n$ and $\det \boldsymbol{C} \neq 0$.

Theorem 8.3 *If $N = n$ and the matrix $\boldsymbol{C}$ is nonsingular, then the following statements hold:*

a) *The ISO system (8.29a) and (8.29b) full IO transfer function matrix $\boldsymbol{F}_{ISOsp}(z)$ that corresponds to the special case (the subscript "sp") has the following form:*

$$\boldsymbol{F}_{ISOsp}(z) = \begin{bmatrix} \left(\boldsymbol{C}(z\boldsymbol{I}_n - \boldsymbol{A})^{-1}\boldsymbol{B} + \boldsymbol{D}\right)^T \\ \left(-z\boldsymbol{C}(z\boldsymbol{I}_n - \boldsymbol{A})^{-1}\boldsymbol{C}^{-1}\boldsymbol{D}\right)^T \\ \left(z\boldsymbol{C}(z\boldsymbol{I}_n - \boldsymbol{A})^{-1}\boldsymbol{C}^{-1}\right)^T \end{bmatrix}^T \tag{8.42}$$

which implies

$$\mathbf{Y}(z) = \boldsymbol{F}_{ISOsp}(z) \begin{bmatrix} \mathbf{I}(z) \\ \mathbf{i}_0 \\ \mathbf{y}_0 \end{bmatrix}. \tag{8.43}$$

b) *The ISO system (8.29a) and (8.29b) transfer function matrix $\boldsymbol{G}_{ISOi_0sp}(z)$ relative to $\mathbf{i}_0$ reads:*

$$\boldsymbol{G}_{ISOi_0sp}(z) =$$
$$= -z\boldsymbol{C}(z\boldsymbol{I}_n - \boldsymbol{A})^{-1}\boldsymbol{C}^{-1}\boldsymbol{D} = -z(z\boldsymbol{I}_n - \boldsymbol{CAC}^{-1})^{-1}\boldsymbol{D}. \tag{8.44}$$

c) *The ISO system (8.29a) and (8.29b) transfer function matrix $\boldsymbol{G}_{ISOy_0}(z)$ with respect to $\mathbf{y}_0$ obeys*

$$\boldsymbol{G}_{ISOy_0sp}(z) = z\boldsymbol{C}(z\boldsymbol{I}_n - \boldsymbol{A})^{-1}\boldsymbol{C}^{-1} = z(z\boldsymbol{I}_n - \boldsymbol{CAC}^{-1})^{-1}. \tag{8.45}$$

Proof. Let us accept $N = n$ and let $\boldsymbol{C}$ be nonsingular. Now we can solve (8.29b) for $\mathbf{x}(k)$ at the initial moment $k_0 = 0$, i.e., we can solve $\mathbf{y}_0 = C\mathbf{x}_0 + D\mathbf{i}_0$ for $\mathbf{x}_0$,

$$\mathbf{x}_0 = -\boldsymbol{C}^{-1}\boldsymbol{D}\mathbf{i}_0 + \boldsymbol{C}^{-1}\mathbf{y}_0. \tag{8.46}$$

Result (8.46), (8.32) and (8.38) imply (8.42) through (8.45). ■

Comment 8.6 *We determined the action vector* $\mathbf{v}_{ISO}(k) = \mathbf{v}_{ISO}(k;\mathbf{x}_0)$ *in general and its the* $Z-$*transform* $\mathbf{V}_{ISO}(z;\mathbf{x}_0)$ *in (8.39) for the* ISO *system, which yields a special form to (8.32) for* $\det\ \boldsymbol{C}\neq 0$,

$$\mathbf{Y}(z) = \boldsymbol{F}_{ISOsp}(z)\mathbf{V}_{ISOsp}(z;\mathbf{i}_0;\mathbf{y}_0). \tag{8.47}$$

From (8.43) results that for the ISO *system in the special case the action vector* $\mathbf{v}_{ISOsp}(k) = \mathbf{v}_{ISOsp}(k;\mathbf{i}_0;\mathbf{y}_0)$ *and its the* $Z-$*transform* $\mathbf{V}_{ISOsp}(z;\mathbf{i}_0;\mathbf{y}_0)$ *obey*

$$\begin{aligned} \mathbf{v}_{ISOsp}(k;\mathbf{i}_0;\mathbf{y}_0) &= \begin{bmatrix} \mathbf{i}^T(k) & \delta_d(k)\mathbf{i}_0^T & \delta_d(k)\mathbf{y}_0^T \end{bmatrix}^T, \\ \mathbf{V}_{ISOsp}(z;\mathbf{i}_0;\mathbf{y}_0) &= \begin{bmatrix} \mathbf{I}^T(z) & \mathbf{i}_0^T & \mathbf{y}_0^T \end{bmatrix}^T. \end{aligned} \tag{8.48}$$

This permitted us to set (8.43) into the classical form (8.41),

$$\mathbf{Y}(z) = \boldsymbol{F}_{ISOsp}(z)\mathbf{V}_{ISOsp}(z;\mathbf{i}_0;\mathbf{y}_0). \tag{8.49}$$

Example 8.7 *Let us observe* $\boldsymbol{G}_{ISO}(z) = \left(z^2-1\right)^{-1}(z-1) = (z+1)^{-1}$, *and four different (state space, i.e.,* ISO*) realizations* $(\boldsymbol{A},\boldsymbol{B},\boldsymbol{C},\boldsymbol{D})$ *of* $\boldsymbol{G}_{ISO}(z)$ *by analogy to the* $s-$*complex case given in [5, p. 395]. We show at first how to determine the full transfer function matrix and other transfer function matrices for each of the four realizations.*

$$\textit{1)}\ \left\{\boldsymbol{A}_1 = \begin{bmatrix} 0 & 1 \\ 1 & 0 \end{bmatrix}, \boldsymbol{B}_1 = \begin{bmatrix} 0 \\ 1 \end{bmatrix}, \boldsymbol{C}_1 = \begin{bmatrix} -1 & 1 \end{bmatrix}, \boldsymbol{D}_1 = 0\right\} \Longrightarrow$$

$$\begin{aligned} Y_1(z) &= \begin{bmatrix} -1 & 1 \end{bmatrix} \begin{bmatrix} z & -1 \\ -1 & z \end{bmatrix}^{-1} \left\{ \begin{bmatrix} 0 \\ 1 \end{bmatrix} I(z) + z\mathbf{x}_0 \right\} = \\ &= \left(z^2-1\right)^{-1} \begin{bmatrix} z-1 & z(1-z) & z(z-1) \end{bmatrix} \begin{bmatrix} I(z) \\ x_{10} \\ x_{20} \end{bmatrix}, \end{aligned}$$

$$Y_1(z) = \boldsymbol{F}_{ISO1}(z)\mathbf{V}_{ISO1}(z;\mathbf{x}_0),\ \mathbf{V}_{ISO1}(z;\mathbf{x}_0) = \begin{bmatrix} I(z) \\ \mathbf{x}_0 \end{bmatrix} \Longrightarrow$$

$$\begin{aligned} \boldsymbol{F}_{ISO1}(z) &= \left(z^2-1\right)^{-1} \begin{bmatrix} z-1 & z(1-z) & z(z-1) \end{bmatrix} = \\ &= \begin{bmatrix} \boldsymbol{C}_1(z\boldsymbol{I}_2-\boldsymbol{A}_1)^{-1}\boldsymbol{B}_1+\boldsymbol{D}_1 & z\boldsymbol{C}_1(z\boldsymbol{I}_2-\boldsymbol{A}_1)^{-1} \end{bmatrix}, \end{aligned}$$

$$\Longrightarrow$$

$$\boldsymbol{F}_{ISO1nd}(z) = \begin{bmatrix} \frac{1}{z+1} & -\frac{z}{z+1} & \frac{z}{z+1} \end{bmatrix},$$

$$\boldsymbol{G}_{ISO1}(z) = \left[\boldsymbol{C}_1(z\boldsymbol{I}_2 - \boldsymbol{A}_1)^{-1}\boldsymbol{B}_1 + \boldsymbol{D}_1\right] = \frac{z-1}{z^2-1} \Longrightarrow$$

$$\boldsymbol{G}_{ISO1nd}(z) = \frac{1}{z+1},$$

$$\boldsymbol{G}_{ISOx_01}(z) = z\boldsymbol{C}_1(z\boldsymbol{I}_2 - \boldsymbol{A}_1)^{-1} = \frac{\left[\begin{array}{cc} z(1-z) & z(z-1) \end{array}\right]}{z^2-1},$$

$$F_{ISOIS1}(z) = \left[\begin{array}{cc} (z\boldsymbol{I}_2 - \boldsymbol{A}_1)^{-1}\boldsymbol{B}_1 & z(z\boldsymbol{I}_2 - \boldsymbol{A}_1)^{-1} \end{array}\right] = \frac{\left[\begin{array}{ccc} 1 & z^2 & z \\ z & z & z^2 \end{array}\right]}{z^2-1},$$

$$\boldsymbol{G}_{ISOIS1}(z) = (z\boldsymbol{I}_2 - \boldsymbol{A}_1)^{-1}\boldsymbol{B}_1 = \frac{\left[\begin{array}{c} 1 \\ z \end{array}\right]}{z^2-1},$$

$$\boldsymbol{G}_{ISOSS1}(z) = z(z\boldsymbol{I}_2 - \boldsymbol{A}_1)^{-1} = \frac{\left[\begin{array}{cc} z^2 & z \\ z & z^2 \end{array}\right]}{z^2-1}.$$

We cannot show the influence of the initial output value y_0 *on the system response because we cannot express the state variables* x_1 *and* x_2 *in terms of the output* y_1 *due to* $y_1 = -x_1 + x_2$*, i.e., due to* $\operatorname{rank} C_1 = 1 = M_1 < 2 = n_1$.

Notice that the given ISO system description, i.e.,

$$\left[\begin{array}{c} Ex_1(k) \\ Ex_2(k) \end{array}\right] = \left[\begin{array}{cc} 0 & 1 \\ 1 & 0 \end{array}\right]\left[\begin{array}{c} x_1(k) \\ x_2(k) \end{array}\right] + \left[\begin{array}{c} 0 \\ 1 \end{array}\right] i(k),$$

$$y_1(k) = \left[\begin{array}{cc} -1 & 1 \end{array}\right]\left[\begin{array}{c} x_1(k) \\ x_2(k) \end{array}\right]$$

allows

$$Ex_1(k) = x_2(k),\ Ex_2(k) = x_1(k) + i(k),$$
$$y_1(k) = -x_1(k) + x_2(k) \Longrightarrow$$
$$Ey_1(k) = -Ex_1(k) + Ex_2(k) = -x_2(k) + x_1(k) + i(k) =$$
$$= -y_1(k) + i(k) \Longrightarrow$$
$$Ey_1(k) + y_1(k) = i(k).$$

This yields

$$zY_1(z) - zy_{10} + Y_1(z) = I(z) \Longrightarrow Y_1(z) = \frac{1}{z+1}\left[\begin{array}{cc} 1 & z \end{array}\right]\left[\begin{array}{c} I(z) \\ y_{10} \end{array}\right] =$$
$$= \boldsymbol{F}_{IOISO1}(z)\mathbf{V}_{IOISO1}(z; y_{10}) \Longrightarrow$$

$$\boldsymbol{F}_{IOISO1}(z) = \left[\begin{array}{cc} \frac{1}{z+1} & \frac{z}{z+1} \end{array}\right] \neq \boldsymbol{F}_{ISO1}(z),$$

$$\boldsymbol{G}_{IOISO1}(z) = \frac{1}{z+1} = \boldsymbol{G}_{ISO1nd}(z),$$

$$\mathbf{V}_{IOISO1}(z; y_{10}) = \left[\begin{array}{c} I(z) \\ y_{10} \end{array}\right] \neq \left[\begin{array}{c} I(z) \\ \mathbf{x}_0 \end{array}\right] = \mathbf{V}_{ISO1}(z; \mathbf{x}_0).$$

$\boldsymbol{G}_{IOISO1}(z)$ is the nondegenerate form $\boldsymbol{G}_{ISO1nd}(z)$ of $\boldsymbol{G}_{ISO1}(z)$. $\boldsymbol{F}_{IOISO1}(z)$ is different from $\boldsymbol{F}_{ISO1}(z)$ and from the nondegenerate form $\boldsymbol{F}_{ISO1nd}(z)$ of $\boldsymbol{F}_{ISO1}(z)$, that is,

$$\boldsymbol{F}_{ISO1nd}(z) = \left[\begin{array}{ccc} \frac{1}{z+1} & -\frac{z}{z+1} & \frac{z}{z+1} \end{array}\right].$$

Notice that

$$Y_1(z) = F_{ISO1}(z) V_{ISO1}(z; x_0) =$$
$$= \left[\begin{array}{ccc} \frac{z-1}{z^2-1} & -\frac{z(z-1)}{z^2-1} & \frac{z(z-1)}{z^2-1} \end{array}\right] \left[\begin{array}{c} I(z) \\ \mathbf{x}_0 \end{array}\right] =$$
$$= \left[\begin{array}{ccc} \frac{1}{z+1} & -\frac{z}{z+1} & \frac{z}{z+1} \end{array}\right] \left[\begin{array}{c} I(z) \\ x_{10} \\ x_{20} \end{array}\right] = \boldsymbol{F}_{ISO1nd}(z) \left[\begin{array}{c} I(z) \\ \mathbf{x}_0 \end{array}\right],$$

and

$$y_1 = -x_1 + x_2 \Longrightarrow y_{10} = -x_{10} + x_{20}$$

imply

$$Y_1(z) = \left[\begin{array}{cc} \frac{1}{z+1} & \frac{z}{z+1} \end{array}\right] \left[\begin{array}{c} I(z) \\ y_{10} \end{array}\right] = \boldsymbol{F}_{IOISO1}(z) \mathbf{V}_{IOISO1}(z; y_{10}).$$

This shows the equivalence between $\boldsymbol{F}_{ISO1}(z)$, $\boldsymbol{F}_{ISO1nd}(z)$ and $\boldsymbol{F}_{IOISO1}(z)$ in this example,

$$\boldsymbol{F}_{ISO1}(z)\mathbf{V}_{ISO1}(z; \mathbf{x}_0) = \boldsymbol{F}_{IOISO1}(z)\mathbf{V}_{IOISO1}(z; y_{10}) = Y_1(z).$$

2) $$\left\{\boldsymbol{A}_2 = \left[\begin{array}{cc} 0 & 1 \\ 1 & 0 \end{array}\right], \ \boldsymbol{B}_2 = \left[\begin{array}{c} -1 \\ 1 \end{array}\right], \ \boldsymbol{C}_2 = \left[\begin{array}{cc} 0 & 1 \end{array}\right], \ \boldsymbol{D}_2 = 0\right\} \Longrightarrow$$

$$Y_2(z) = \left[\begin{array}{cc} 0 & 1 \end{array}\right] \left[\begin{array}{cc} z & -1 \\ -1 & z \end{array}\right]^{-1} \left\{\left[\begin{array}{c} -1 \\ 1 \end{array}\right] I(z) + z\mathbf{x}_0\right\} =$$
$$= \underbrace{\left(z^2-1\right)^{-1} \left[\begin{array}{ccc} z-1 & z & z^2 \end{array}\right]}_{\boldsymbol{F}_{ISO2}(z)} \left[\begin{array}{c} I(z) \\ x_{10} \\ x_{20} \end{array}\right],$$

$$Y_2(z) = \boldsymbol{F}_{ISO2}(z)\mathbf{V}_{ISO2}(z;\mathbf{x}_0),\ \mathbf{V}_{ISO2}(z;\mathbf{x}_0) = \begin{bmatrix} I(z) \\ \mathbf{x}_0 \end{bmatrix} \Longrightarrow$$

$$\boldsymbol{F}_{ISO2}(z) = \left(z^2-1\right)^{-1}\begin{bmatrix} z-1 & z & z^2 \end{bmatrix} = \boldsymbol{F}_{ISO2nd}(z),$$

$$\boldsymbol{G}_{ISO2}(z) = \frac{z-1}{z^2-1},\ \boldsymbol{G}_{ISO2nd}(z) = \frac{1}{z+1},$$

$$\boldsymbol{G}_{ISOx02}(z) = \frac{\begin{bmatrix} z & z^2 \end{bmatrix}}{z^2-1},\ \boldsymbol{F}_{ISOIS2}(z) = \frac{\begin{bmatrix} 1-z & z^2 & z \\ z-1 & z & z^2 \end{bmatrix}}{z^2-1},$$

$$\boldsymbol{G}_{ISOIS2}(z) = \frac{\begin{bmatrix} 1-z \\ z-1 \end{bmatrix}}{z^2-1},\ \boldsymbol{G}_{ISOSS2}(z) = \frac{\begin{bmatrix} z^2 & z \\ z & z^2 \end{bmatrix}}{z^2-1}.$$

We find the IO system model as follows:

$$Ex_1(k) = x_2(k) - i(k),\ Ex_2(k) = x_1(k) + i(k),\ y_2(k) = x_2(k) \Rightarrow$$
$$Ey_2(k) = x_1(k) + i(k) \Longrightarrow E^2y_2(k) = x_2(k) - i(k) + Ei(k) =$$
$$= y_2(k) - i(k) + Ei(k) \Longrightarrow$$
$$E^2y_2(k) - y_2(k) = Ei(k) - i(k).$$

This IO system model implies

$$z^2Y_2(z) - z^2y_{20} - z\left(E^1y_2\right)_0 - Y_2(z) = zI(z) - zi_0 - I(z) \Longrightarrow$$
$$Y_2(z) = \frac{1}{z^2-1}\begin{bmatrix} z-1 & -z & z^2 & z \end{bmatrix}\begin{bmatrix} I(z) \\ i_0 \\ \mathbf{y}_{20}^1 \end{bmatrix} =$$
$$= \boldsymbol{F}_{IOISO2}(z)\mathbf{V}_{IOISO2}(z;i_0;\mathbf{y}_{20}^1) \Longrightarrow$$

$$\boldsymbol{F}_{IOISO2}(z) = \begin{bmatrix} \frac{z-1}{z^2-1} & -\frac{z}{z^2-1} & \frac{z^2}{z^2-1} & \frac{z}{z^2-1} \end{bmatrix} \neq \boldsymbol{F}_{ISO2}(z),$$

$$\mathbf{V}_{IOISO2}(z;i_0;\mathbf{y}_{20}^1) = \begin{bmatrix} I(z) \\ i_0 \\ \mathbf{y}_{20}^1 \end{bmatrix} \neq \begin{bmatrix} I(z) \\ \mathbf{x}_0 \end{bmatrix} = \mathbf{V}_{ISO2}(z;\mathbf{x}_0),$$

$$\boldsymbol{G}_{IOISO2}(z) = \frac{z-1}{z^2-1} = \boldsymbol{G}_{ISO2}(z).$$

Besides,

$$Y_2(z)=\frac{1}{z^2-1}\left[\begin{array}{cccc} z-1 & -z & z^2 & z \end{array}\right]\left[\begin{array}{c} I(z) \\ i_0 \\ \mathbf{y}_{20}^1 \end{array}\right]=$$

$$=\boldsymbol{F}_{IOISO2}(z)\mathbf{V}_{IOISO2}(z;i_0;\mathbf{y}_{20}^1)=$$

$$=\frac{1}{z^2-1}\left[(z-1)\,I(z)-zi_0+z^2x_{20}+zx_{10}+zi_0\right]=$$

$$=\frac{1}{z^2-1}\left[(z-1)\,I\,(z)+zx_{10}+z^2x_{20}\right]=$$

$$=\left[\begin{array}{ccc} \frac{z-1}{z^2-1} & \frac{z}{z^2-1} & \frac{z^2}{z^2-1} \end{array}\right]\left[\begin{array}{c} I(z) \\ x_{10} \\ x_{20} \end{array}\right]=\boldsymbol{F}_{ISO2}(z)\mathbf{V}_{ISO2}(z;\mathbf{x}_0).$$

This shows the equivalence between $\boldsymbol{F}_{ISO2}(z)$ *and* $\boldsymbol{F}_{IOISO2}(z)$ *in this case.*

3) $\left\{\boldsymbol{A}_3=\left[\begin{array}{cc} 1 & 0 \\ 0 & -1 \end{array}\right],\ \boldsymbol{B}_3=\left[\begin{array}{c} 0 \\ 1 \end{array}\right],\ \boldsymbol{C}_3=\left[\begin{array}{cc} 0 & 1 \end{array}\right],\ \boldsymbol{D}_3=0\right\}\Longrightarrow$

$$Y_3(z)=\left[\begin{array}{cc} 0 & 1 \end{array}\right]\left[\begin{array}{cc} z-1 & 0 \\ 0 & z+1 \end{array}\right]^{-1}\left\{\left[\begin{array}{c} 0 \\ 1 \end{array}\right]I(z)+z\mathbf{x}_0\right\}=$$

$$=\left(z^2-1\right)^{-1}\left[\begin{array}{ccc} z-1 & 0 & z\,(z-1) \end{array}\right]\left[\begin{array}{c} I(z) \\ x_{10} \\ x_{20} \end{array}\right]_{x_{20}=y_{30}}$$

$$=\underbrace{\left(z^2-1\right)^{-1}\left[\begin{array}{ccc} z-1 & 0 & z\,(z-1) \end{array}\right]}_{\boldsymbol{F}_{ISO3}(z)}\left[\begin{array}{c} I(z) \\ x_{10} \\ y_{30} \end{array}\right],$$

$$\boldsymbol{F}_{ISO3}(z)=\left(z^2-1\right)^{-1}\left[\begin{array}{ccc} z-1 & 0 & z\,(z-1) \end{array}\right]\Longrightarrow$$
$$\boldsymbol{F}_{ISO3nd}(z)=\left[\begin{array}{ccc} \frac{1}{z+1} & 0 & \frac{z}{z+1} \end{array}\right].$$

$$Y_3(z)=\boldsymbol{F}_{ISO3}(z)\mathbf{V}_{ISO3}(z;\mathbf{x}_0),\ \mathbf{V}_{ISO3}(z;\mathbf{x}_0)=\left[\begin{array}{c} I(z) \\ \mathbf{x}_0 \end{array}\right]\Longrightarrow$$

$$\boldsymbol{F}_{ISO3}(z)=\left(z^2-1\right)^{-1}\left[\begin{array}{ccc} z-1 & 0 & z\,(z-1) \end{array}\right]\Longrightarrow$$

$$\boldsymbol{G}_{ISO3}(z)=\frac{z-1}{z^2-1},\ \boldsymbol{G}_{ISOx_03}(z)=\frac{\left[\begin{array}{cc} 0 & z\,(z-1) \end{array}\right]}{z^2-1},$$

$$\boldsymbol{F}_{ISOIS3}(z)=\frac{\begin{bmatrix} 0 & z(z+1) & 0 \\ z-1 & 0 & z(z-1)\end{bmatrix}}{z^2-1}\Longrightarrow$$

$$\boldsymbol{G}_{ISOIS3}(z)=\frac{\begin{bmatrix} 0 \\ z-1\end{bmatrix}}{z^2-1},\ \boldsymbol{G}_{ISOSS3}(z)=\frac{\begin{bmatrix} z(z+1) & 0 \\ 0 & z(z-1)\end{bmatrix}}{z^2-1}.$$

The nondegenerate form $\boldsymbol{F}_{ISO3nd}(z)$ *of* $\boldsymbol{F}_{ISO3}(z)$ *reads*

$$\boldsymbol{F}_{ISO3nd}(z)=\frac{1}{z+1}\begin{bmatrix} 1 & 0 & z\end{bmatrix}.$$

We determine now the IO *model of the system,*

$$Ex_1=x_1,\ Ex_2=-x_2+i,\ y_3=x_2\Longrightarrow$$
$$Ey_3+y_3=i\Longrightarrow$$

$$Y_3(z)=\frac{1}{z+1}\begin{bmatrix} 1 & z\end{bmatrix}\begin{bmatrix} I(z) \\ y_{30}\end{bmatrix}=\boldsymbol{F}_{IOISO3}(z)\mathbf{V}_{IOISO3}(z;y_{30})\Longrightarrow$$

$$\boldsymbol{F}_{IOISO3}(z)=\begin{bmatrix} \frac{1}{z+1} & \frac{z}{z+1}\end{bmatrix},\ \mathbf{V}_{IOISO3}(z;y_{30})=\begin{bmatrix} I(z) \\ y_{30}\end{bmatrix}.$$

Notice that we can write

$$Y_3(z)=\underbrace{\frac{1}{z+1}\begin{bmatrix} 1 & z\end{bmatrix}}_{\boldsymbol{F}_{IOISO3}(z)}\underbrace{\begin{bmatrix} I(z) \\ y_{30}\end{bmatrix}}_{\mathbf{V}_{IOISO3}(z;y_{30})}=\underbrace{\frac{1}{z+1}\begin{bmatrix} 1 & 0 & z\end{bmatrix}}_{\boldsymbol{F}_{ISO3ird}(z)}\begin{bmatrix} I(z) \\ x_{10} \\ y_{30}\end{bmatrix}=$$
$$=\boldsymbol{F}_{ISO3ird}(z)\begin{bmatrix} I(z) \\ x_{10} \\ x_{20}\end{bmatrix}.$$

We may conclude the equivalence between $\boldsymbol{F}_{IOISO3}(z)$ *and the nondegenerate form* $\boldsymbol{F}_{ISO3nd}(z)$ *of* $\boldsymbol{F}_{ISO3}(z)$ *in this example.*

4) $\{\boldsymbol{A}_4=[-1],\ \boldsymbol{B}_4=[1],\ \boldsymbol{C}_4=[1],\ \boldsymbol{D}_4=0\}\Longrightarrow$

$$Y_4(z)=[1]\,[z+1]^{-1}\,\{[1]\,I(z)+z\mathbf{x}_0\}=$$
$$=(z+1)^{-1}\begin{bmatrix} 1 & z\end{bmatrix}\begin{bmatrix} I(z) \\ x_{10}\end{bmatrix}_{x_{10}=y_{40}}=$$
$$=(z+1)^{-1}\begin{bmatrix} 1 & z\end{bmatrix}\begin{bmatrix} I(z) \\ y_{40}\end{bmatrix}=$$
$$=\boldsymbol{F}_{ISO4}(z)\mathbf{V}_{ISO4}(z;\mathbf{x}_0),\ \mathbf{V}_{ISO4}(z;\mathbf{x}_0)=\begin{bmatrix} I(z) \\ x_{10}\end{bmatrix}\Longrightarrow$$

$$\boldsymbol{F}_{ISO4}(z) = (z+1)^{-1} \begin{bmatrix} 1 & z \end{bmatrix} = \boldsymbol{F}_{IO4}(z),$$
$$\boldsymbol{G}_{ISO4}(z) = \frac{1}{z+1},$$

$$\boldsymbol{G}_{ISOx_04}(z) = \frac{z}{z+1}, \ \boldsymbol{F}_{ISOIS4}(z) = \frac{\begin{bmatrix} 1 & z \end{bmatrix}}{z+1},$$
$$\boldsymbol{G}_{ISOIS4}(z) = \frac{1}{z+1}, \ \boldsymbol{G}_{ISOSS4}(z) = \frac{z}{z+1}.$$

When we replace $\mathbf{x}_0 = (x_0) = (x_{10})$ *by* y_{40} *due to* $x_{10} = y_{40}$ *then, formally, there is not an explicit influence of the initial state variable on the system output response.*

The state space model under (4) corresponds to the following first-order IO discrete equation and the full transfer function matrix $\boldsymbol{F}_{IO4}(z)$*:*

$$y_4(k+1) + y_4(k) = i(k) \Longrightarrow$$
$$(z+1)\, Y_4(z) - z y_{40} = I(z) \Longrightarrow$$

$$Y_4(z) = \underbrace{(z+1)^{-1} \begin{bmatrix} 1 & z \end{bmatrix}}_{\boldsymbol{F}_{IOISO4}(z)} \begin{bmatrix} I(z) \\ y_{40} \end{bmatrix}, \ \boldsymbol{F}_{IOISO4}(z) = \frac{1}{z+1} \begin{bmatrix} 1 & z \end{bmatrix},$$

$$Y_4(z) = \boldsymbol{F}_{IOISO4}(z) \mathbf{V}_{IOISO4}(z; y_{40}), \ \mathbf{V}_{IOISO4}(z; y_{40}) = \begin{bmatrix} I(z) \\ y_{40} \end{bmatrix} \Longrightarrow$$
$$\boldsymbol{F}_{IOISO4}(z) = (z+1)^{-1} \begin{bmatrix} 1 & z \end{bmatrix} = \boldsymbol{F}_{ISO4}(z),$$
$$\boldsymbol{G}_{IOISO4}(z) = \frac{1}{z+1} = \boldsymbol{G}_{ISO4}(z).$$

Example 8.8 *Let us observe discrete-time analogy of a time-varying continuous-time LC network considered by Kalman in [85, Example 1, pp. 163-165], which is neither completely controllable nor observable. Without losing these properties, we adopt* $C(k) \equiv C$ *and* $L(k) \equiv L$*. The system description then reads*

$$E^1 x_1 = -\frac{1}{L} x_1 + u \Longrightarrow X_1(z) = (z + \frac{1}{L})^{-1} \left[z x_{10} + U(z) \right],$$
$$E^1 x_2 = -\frac{1}{L} x_2 \Longrightarrow X_2(z) = z(z + \frac{1}{L})^{-1} x_{20},$$

$$y = \frac{2}{L} x_2 + u \Longrightarrow Y(z) = \frac{2}{L} z(z + \frac{1}{L})^{-1} x_{20} + U(z) \Longrightarrow$$
$$Y(z) = \begin{bmatrix} 1 & \frac{2}{L} z(z + \frac{1}{L})^{-1} \end{bmatrix} \begin{bmatrix} U(z) \\ x_{20} \end{bmatrix} \Longrightarrow$$

$$G_{ISO}(z) \equiv 1,\ \boldsymbol{F}_{ISO}(z) = \left[\begin{array}{cc} 1 & \frac{2}{L}z(z+\frac{1}{L})^{-1} \end{array}\right].$$

The transfer function $G_{ISO}(z)$ leads to the conclusion that the system is static. However, the full transfer function matrix $\boldsymbol{F}_{ISO}(z)$ shows that the system is dynamic. If we write $\boldsymbol{F}_{ISO}(z)$ in the form

$$\boldsymbol{F}_{ISO}(z) = (z+\frac{1}{L})^{-1}\left[\begin{array}{cc} z+\frac{1}{L} & \frac{2}{L}z \end{array}\right]$$

then

$$G_{ISO}(s) = \frac{z+\frac{1}{L}}{z+\frac{1}{L}} \equiv 1 = G_{ISOnd}(z)$$

shows also that the system is dynamic, and that it is not completely controllable and observable. $\boldsymbol{F}_{ISO}(z)$ is not either degenerate or reducible, while $G_{ISO}(z)$ is both degenerate and reducible.

Example 8.9 *Let us observe the following discrete-time analogy of the continuous-time ISO system presented also by Kalman in [85, Example 8, pp. 188, 189]:*

$$E^1\mathbf{x} = \left[\begin{array}{ccc} 0 & 1 & 0 \\ 5 & 0 & 2 \\ -2 & 0 & -2 \end{array}\right]\mathbf{x} + \left[\begin{array}{c} 0 \\ 0 \\ 0.5 \end{array}\right] i,$$
$$y = \left[\begin{array}{ccc} -2 & 1 & 0 \end{array}\right]\mathbf{x}.$$

We apply Equation (8.38)

$$\boldsymbol{F}_{ISO}(z) = [\det(z\boldsymbol{I}_n - \boldsymbol{A})]^{-1}\cdot$$
$$\cdot\left[\begin{array}{cc} \boldsymbol{C}\operatorname{adj}(z\boldsymbol{I}_n - \boldsymbol{A})\boldsymbol{B} + \boldsymbol{D}\det(z\boldsymbol{I}_n - \boldsymbol{A}) & z\boldsymbol{C}\operatorname{adj}(z\boldsymbol{I}_n - \boldsymbol{A}) \end{array}\right] =$$

$$= \left(z^3 + 2z^2 - 5z - 6\right)^{-1}\cdot$$
$$\cdot\left[\begin{array}{cccc} z-2 & z\left(-2z^2+z+6\right) & z\left(z^2-4\right) & z\left(2z-4\right) \end{array}\right] =$$
$$= \left[(z+1)(z-2)(z+3)\right]^{-1}\cdot$$
$$\cdot\left[\begin{array}{cccc} z-2 & z(z-2)(-2z-3) & z(z-2)(z+2) & 2z(z-2) \end{array}\right] =$$
$$= \frac{z-2}{z-2}\left[(z+1)(z+3)\right]^{-1}\left[\begin{array}{cccc} 1 & z(-2z-3) & z(z+2) & 2z \end{array}\right] \Longrightarrow$$

$$G_{ISO}(z) = [\det(z\boldsymbol{I}_n - \boldsymbol{A})]^{-1}\left[\boldsymbol{C}\operatorname{adj}(z\boldsymbol{I}_n - \boldsymbol{A})\boldsymbol{B} + \boldsymbol{D}\det(z\boldsymbol{I}_n - \boldsymbol{A})\right] =$$
$$= \left(z^3 + 2z^2 - 5z - 6\right)^{-1}(z-2) = \frac{z-2}{z-2}\left[(z+1)(z+3)\right]^{-1}.$$

Since $G_{ISO}(z)$ is reducible, i.e., since it degenerates to

$$G_{ISOird}(z) = \frac{1}{(z+1)(z+3)} = G_{ISOnd}(z),$$

then it follows that the system is not completely controllable and observable. In this example, $\boldsymbol{F}_{ISO}(z)$ is also both reducible and degenerate. After cancelling the same zero and pole at $z = 2$, we determine the nondegenerate form $\boldsymbol{F}_{ISOnd}(z)$ of $\boldsymbol{F}_{ISO}(z)$,

$$\boldsymbol{F}_{ISOnd}(s) = \frac{\begin{bmatrix} 1 & z(-2z-3) & z(z+2) & 2z \end{bmatrix}}{(z+1)(z+3)}.$$

It is also the irreducible form $\boldsymbol{F}_{ISOird}(z)$ of $\boldsymbol{F}_{ISO}(z)$,

$$\boldsymbol{F}_{ISOird}(z) = \boldsymbol{F}_{ISOnd}(z).$$

8.3 $\boldsymbol{F}(z)$ of the *IIO* system

In order to help the reader to easily follow the determination of the full transfer function matrices of the *IIO* system (3.63a), (3.63b) i.e., (3.64a) and (3.64b) (in Subsection 3.5.4) we repeat it as

$$\boldsymbol{Q}^{(\alpha)}\mathbf{r}^{\alpha}(k) = \boldsymbol{P}^{(\beta)}\mathbf{i}^{\beta}(k), \forall k \in \mathcal{N}_0, \tag{8.50a}$$

$$\boldsymbol{E}^{(\nu)}\mathbf{y}^{\nu}(k) = \boldsymbol{R}^{(\alpha)}\mathbf{r}^{\alpha}(k) + \boldsymbol{T}^{(\mu)}\mathbf{i}^{\mu}(k), \forall k \in \mathcal{N}_0. \tag{8.50b}$$

The system has the following two transfer function matrices:

- The full *IS* transfer function matrix $\boldsymbol{F}_{IIOIS}(z)$ that relates $\mathbf{R}(z)$ to $\mathbf{I}(z)$, $\mathbf{i}_0^{\beta-1}$ and $\mathbf{r}_0^{\alpha-1}$ (Definition 7.7, in Section 7.4),

$$\mathbf{R}(z) = \boldsymbol{F}_{IIOIS}(z)\begin{bmatrix} \mathbf{I}^T(z) & \left(\mathbf{i}_0^{\beta-1}\right)^T & \left(\mathbf{r}_0^{\alpha-1}\right)^T \end{bmatrix}^T, \tag{8.51}$$

$$\mathbf{R}(z) = \boldsymbol{F}_{IIOIS}(z)\mathbf{V}_{IIOIS}(z), \tag{8.52}$$

$$\mathbf{V}_{IIOIS}(z;\mathbf{i}_0^{\beta-1};\mathbf{r}_0^{\alpha-1}) = \begin{bmatrix} \mathbf{I}^T(z) & \underbrace{\left(\mathbf{i}_0^{\beta-1}\right)^T \quad \left(\mathbf{r}_0^{\alpha-1}\right)^T}_{\mathbf{C}_{0IIOIS}^T} \end{bmatrix}^T, \tag{8.53}$$

$$\mathbf{C}_{0IIOIS} = \begin{bmatrix} \mathbf{i}_0^{\beta-1} \\ \mathbf{r}_0^{\alpha-1} \end{bmatrix}, \ \mathbf{V}_{IIOIS}(z;\mathbf{C}_{0IIOIS}) = \begin{bmatrix} \mathbf{I}(z) \\ \mathbf{C}_{0IIOIS} \end{bmatrix}. \tag{8.54}$$

where $\mathbf{V}_{IIOIS}(z)$ is the $Z-$transform of the vector function $\mathbf{v}_{IIOIS}(\cdot)$ of all the action on the system,

$$\mathbf{v}_{IIOIS}(k;\mathbf{i}_0^{\beta-1};\mathbf{r}_0^{\alpha-1}) =$$

$$= \begin{bmatrix} \mathbf{i}^T(k) & \delta_d(k)\left(\mathbf{i}_0^{\beta-1}\right)^T & \delta_d(k)\left(\mathbf{r}_0^{\alpha-1}\right)^T \end{bmatrix}^T. \tag{8.55}$$

- The full IO transfer function matrix $\boldsymbol{F}_{IIO}(z)$ that relates $\mathbf{Y}(z)$ to $\mathbf{I}(z)$, $\mathbf{i}_0^{\gamma-1}$, $\mathbf{r}_0^{\alpha-1}$, and $\mathbf{y}_0^{\nu-1}$ (Definition 7.6, in Section 7.4),

$$\mathbf{Y}(z) = \boldsymbol{F}_{IIO}(z)\begin{bmatrix} \mathbf{I}^T(z) & \left(\mathbf{i}_0^{\gamma-1}\right)^T & \left(\mathbf{r}_0^{\alpha-1}\right)^T & \left(\mathbf{y}_0^{\nu-1}\right)^T \end{bmatrix}^T,$$

$$\mathbf{Y}(z) = \boldsymbol{F}_{IIO}(z)\mathbf{V}_{IIO}(z), \tag{8.56}$$

$$\mathbf{V}_{IIO}(z;\mathbf{i}_0^{\gamma-1};\mathbf{r}_0^{\alpha-1};\mathbf{y}_0^{\nu-1}) =$$

$$= \begin{bmatrix} \mathbf{I}^T(z) & \underbrace{\left(\mathbf{i}_0^{\gamma-1}\right)^T \quad \left(\mathbf{r}_0^{\alpha-1}\right)^T \quad \left(\mathbf{y}_0^{\nu-1}\right)^T}_{\mathbf{C}_{0IIO}^T} \end{bmatrix}^T, \tag{8.57}$$

$$\mathbf{C}_{0IIO} = \begin{bmatrix} \mathbf{i}_0^{\gamma-1} \\ \mathbf{r}_0^{\alpha-1} \\ \mathbf{y}_0^{\nu-1} \end{bmatrix} = \begin{bmatrix} \mathbf{C}_{0IIOIS} \\ \mathbf{y}_0^{\nu-1} \end{bmatrix},$$

$$\mathbf{V}_{IIO}(z;\mathbf{C}_{0IIO}) = \begin{bmatrix} \mathbf{I}(z) \\ \mathbf{C}_{0IIO} \end{bmatrix} = \begin{bmatrix} \mathbf{V}_{IIOIS}(z;\mathbf{C}_{0IIOIS}) \\ \mathbf{y}_0^{\nu-1} \end{bmatrix} \tag{8.58}$$

where $\mathbf{V}_{IIO}(z)$ is the $Z-$transform of the vector function $\mathbf{v}_{IIO}(\cdot)$ of all the actions on the system,

$$\mathbf{v}_{IIO}(k;\mathbf{i}_0^{\gamma-1};\mathbf{r}_0^{\alpha-1};\mathbf{y}_0^{\nu-1}) =$$

$$= \begin{bmatrix} \mathbf{i}^T(k) & \delta_d(k)\left(\mathbf{i}_0^{\gamma-1}\right)^T & \delta_d(k)\left(\mathbf{r}_0^{\alpha-1}\right)^T & \delta_d(k)\left(\mathbf{y}_0^{\nu-1}\right)^T \end{bmatrix}^T. \tag{8.59}$$

Theorem 8.4 *The full IS transfer function matrix $\boldsymbol{F}_{IIOIS}(z)$ of the system (3.63a) and (3.63b), i.e., of (8.50a) and (8.50b), is determined by (8.60) and (8.61),*

$$\boldsymbol{F}_{IIOIS}(z) =$$

$$= \left(\boldsymbol{Q}^{(\alpha)}\boldsymbol{S}_\rho^{(\alpha)}(z)\right)^{-1}\begin{bmatrix} \boldsymbol{P}^{(\beta)}\boldsymbol{S}_M^{(\beta)}(z) & -\boldsymbol{P}^{(\beta)}\boldsymbol{Z}_M^{(\beta)}(z) & \boldsymbol{Q}^{(\alpha)}\boldsymbol{Z}_\rho^{(\alpha)}(z) \end{bmatrix} \tag{8.60}$$

$$\boldsymbol{F}_{IIOIS}(z) = \begin{bmatrix} \boldsymbol{G}_{IIOIS}(z) & \boldsymbol{G}_{IIO0IS}(z) \end{bmatrix} = \boldsymbol{F}_{IIOISD}^{-1}(z)\boldsymbol{F}_{IIOISN}(z),$$

$$\boldsymbol{G}_{IIO0IS}(z) = \begin{bmatrix} \boldsymbol{G}_{IIOi_0IS}(z) & \boldsymbol{G}_{IIOr_0IS}(z) \end{bmatrix}, \tag{8.61}$$

together with (8.62) through (8.64):

- *IRIS transfer function matrix* $\boldsymbol{G}_{IIOIS}(z)$,

$$\boldsymbol{G}_{IIOIS}(z) = \left(\boldsymbol{Q}^{(\alpha)}\boldsymbol{S}_{\rho}^{(\alpha)}(z)\right)^{-1}\boldsymbol{P}^{(\beta)}\boldsymbol{S}_{M}^{(\beta)}(z) = \\ = \boldsymbol{G}_{IIOISD}^{-1}(z)\boldsymbol{G}_{IIOISN}(z), \tag{8.62}$$

- *IRII transfer function matrix* $\boldsymbol{G}_{IIOi_0IS}(z)$,

$$\boldsymbol{G}_{IIOi_0IS}(z) = -\left(\boldsymbol{Q}^{(\alpha)}\boldsymbol{S}_{\rho}^{(\alpha)}(z)\right)^{-1}\boldsymbol{P}^{(\beta)}\boldsymbol{Z}_{M}^{(\beta)}(z) = \\ = \boldsymbol{G}_{IIOi_0ISD}^{-1}(z)\boldsymbol{G}_{IIOi_0ISN}(z), \tag{8.63}$$

- *IRIR transfer function matrix* $\boldsymbol{G}_{IIOr_0IS}(z)$,

$$\boldsymbol{G}_{IIOr_0IS}(z) = \left(\boldsymbol{Q}^{(\alpha)}\boldsymbol{S}_{\rho}^{(\alpha)}(z)\right)^{-1}\boldsymbol{Q}^{(\alpha)}\boldsymbol{Z}_{\rho}^{(\alpha)}(z) = \\ = \boldsymbol{G}_{IIOr_0ISD}^{-1}(z)\boldsymbol{G}_{IIOr_0ISN}(z), \tag{8.64}$$

so that the denominator matrix polynomial $\boldsymbol{F}_{IIOISD}(z)$ *and the numerator matrix polynomial* $\boldsymbol{F}_{IIOISN}(z)$ *of* $\boldsymbol{F}_{IIOIS}(z)$, $\boldsymbol{F}_{IIOIS}(z) = \boldsymbol{F}_{IIOISD}^{-1}(z)\boldsymbol{F}_{IIOISN}(z)$, *read:*

$$\boldsymbol{F}_{IIOISD}(z) = \boldsymbol{Q}^{(\alpha)}\boldsymbol{S}_{\rho}^{(\alpha)}(z),$$

$$\boldsymbol{F}_{IIOISN}(z) = \\ = \left[\begin{array}{ccc} \boldsymbol{P}^{(\beta)}\boldsymbol{S}_{M}^{(\beta)}(z) & -\boldsymbol{P}^{(\beta)}\boldsymbol{Z}_{M}^{(\beta)}(z) & \boldsymbol{Q}^{(\alpha)}\boldsymbol{Z}_{\rho}^{(\alpha)}(z) \end{array}\right]. \tag{8.65}$$

Proof. The $Z-$transform of the compact form (3.64a) of (3.63a) (in Subsection 3.5.4) reads

$$\left(\boldsymbol{Q}^{(\alpha)}\boldsymbol{S}_{\rho}^{(\alpha)}(z)\right)\mathbf{R}(z) = \\ = \left[\begin{array}{ccc} \boldsymbol{P}^{(\beta)}\boldsymbol{S}_{M}^{(\beta)}(z) & -\boldsymbol{P}^{(\beta)}\boldsymbol{Z}_{M}^{(\beta)}(z) & \boldsymbol{Q}^{(\alpha)}\boldsymbol{Z}_{\rho}^{(\alpha)}(z) \end{array}\right]\left[\begin{array}{c} \mathbf{I}(z) \\ \mathbf{i}_0^{\beta-1} \\ \mathbf{r}_0^{\alpha-1} \end{array}\right] \Longrightarrow$$

$$\mathbf{R}(z) = \left(\boldsymbol{Q}^{(\alpha)}\boldsymbol{S}_{\rho}^{(\alpha)}(z)\right)^{-1} \cdot$$

$$\cdot \left[\begin{array}{ccc} \boldsymbol{P}^{(\beta)}\boldsymbol{S}_{M}^{(\beta)}(z) & -\boldsymbol{P}^{(\beta)}\boldsymbol{Z}_{M}^{(\beta)}(z) & \boldsymbol{Q}^{(\alpha)}\boldsymbol{Z}_{\rho}^{(\alpha)}(z) \end{array} \right] \left[\begin{array}{c} \mathbf{I}(z) \\ \mathbf{i}_0^{\beta-1} \\ \mathbf{r}_0^{\alpha-1} \end{array} \right], \tag{8.66}$$

i.e.,

$$\mathbf{R}(z) = \boldsymbol{F}_{IIOIS}(z) \left[\begin{array}{c} \mathbf{I}(z) \\ \mathbf{i}_0^{\beta-1} \\ \mathbf{r}_0^{\alpha-1} \end{array} \right] = \boldsymbol{F}_{IIOIS}(z)\mathbf{V}_{IIOIS}(z;\mathbf{i}_0^{\beta-1};\mathbf{r}_0^{\alpha-1}), \tag{8.67}$$

$$\mathbf{V}_{IIOIS}(z;\mathbf{i}_0^{\beta-1};\mathbf{r}_0^{\alpha-1}) = \left[\begin{array}{c} \mathbf{I}(z) \\ \mathbf{i}_0^{\beta-1} \\ \mathbf{r}_0^{\alpha-1} \end{array} \right], \tag{8.68}$$

where in view of Definition 7.7, in Section 7.4:

$$\boldsymbol{F}_{IIOIS}(z) = \left[\begin{array}{ccc} \boldsymbol{G}_{IIOIS}(z) & \boldsymbol{G}_{IIOi_0IS}(z) & \boldsymbol{G}_{IIOr_0IS}(z) \end{array} \right] =$$

$$= \left(\boldsymbol{Q}^{(\alpha)}\boldsymbol{S}_{\rho}^{(\alpha)}(z)\right)^{-1} \cdot$$

$$\cdot \left[\begin{array}{ccc} \boldsymbol{P}^{(\beta)}\boldsymbol{S}_{M}^{(\beta)}(z) & -\boldsymbol{P}^{(\beta)}\boldsymbol{Z}_{M}^{(\beta)}(z) & \boldsymbol{Q}^{(\alpha)}\boldsymbol{Z}_{\rho}^{(\alpha)}(z) \end{array} \right], \tag{8.69}$$

and

−*IRIS* transfer function matrix $\boldsymbol{G}_{IIOIS}(z)$,

$$\boldsymbol{G}_{IIOIS}(z) = \left(\boldsymbol{Q}^{(\alpha)}\boldsymbol{S}_{\rho}^{(\alpha)}(z)\right)^{-1}\boldsymbol{P}^{(\beta)}\boldsymbol{S}_{M}^{(\beta)}(z) =$$

$$= \boldsymbol{G}_{IIOISD}^{-1}(z)\boldsymbol{G}_{IIOISN}(z), \tag{8.70}$$

−*IRII* transfer function matrix $\boldsymbol{G}_{IIOi_0IS}(z)$,

$$\boldsymbol{G}_{IIOi_0IS}(z) = -\left(\boldsymbol{Q}^{(\alpha)}\boldsymbol{S}_{\rho}^{(\alpha)}(z)\right)^{-1}\boldsymbol{P}^{(\beta)}\boldsymbol{Z}_{M}^{(\beta)}(z) =$$

$$= \boldsymbol{G}_{IIOi_0ISD}^{-1}(z)\boldsymbol{G}_{IIOi_0ISN}(z), \tag{8.71}$$

−*IRIR* transfer function matrix $\boldsymbol{G}_{IIOr_0IS}(z)$,

$$\boldsymbol{G}_{IIOr_0IS}(z) = \left(\boldsymbol{Q}^{(\alpha)}\boldsymbol{S}_{\rho}^{(\alpha)}(z)\right)^{-1}\boldsymbol{Q}^{(\alpha)}\boldsymbol{Z}_{\rho}^{(\alpha)}(z) =$$

$$= \boldsymbol{G}_{IIOr_0ISD}^{-1}(z)\boldsymbol{G}_{IIOr_0ISN}(z), \tag{8.72}$$

so that the denominator matrix polynomial $\boldsymbol{F}_{IIOISD}(z)$ and the numerator matrix polynomial $\boldsymbol{F}_{IIOISN}(z)$,

$$\boldsymbol{F}_{IIOIS}(z)=\boldsymbol{F}_{IIOISD}^{-1}(z)\boldsymbol{F}_{IIOISN}(z),$$

read:

$$\boldsymbol{F}_{IIOISD}(z)=\left(\boldsymbol{Q}^{(\alpha)}\boldsymbol{S}_{\rho}^{(\alpha)}(z)\right),$$

$$\boldsymbol{F}_{IIOISN}(z)=$$
$$=\left[\begin{array}{ccc}\boldsymbol{P}^{(\beta)}\boldsymbol{S}_{M}^{(\beta)}(z) & -\boldsymbol{P}^{(\beta)}\boldsymbol{Z}_{M}^{(\beta)}(z) & \boldsymbol{Q}^{(\alpha)}\boldsymbol{Z}_{\rho}^{(\alpha)}(z)\end{array}\right]. \tag{8.73}$$

Equations (8.69) - (8.73) prove Equations (8.60) - (8.65). ■

Theorem 8.5 *Equations (8.74) determine the full IO transfer function matrix* $\boldsymbol{F}_{IIO}(z)$ *of the system (3.63a) and (3.63b), i.e., (8.50a) and (8.50b):*

$$\begin{gathered}\boldsymbol{F}_{IIO}(z)=\left[\begin{array}{cc}\boldsymbol{G}_{IIO}(z) & \boldsymbol{G}_{IIO0}(z)\end{array}\right]=\boldsymbol{F}_{IIOD}^{-1}(z)\boldsymbol{F}_{IION}(z),\\ \boldsymbol{G}_{IIO0}(z)=\left[\begin{array}{ccc}\boldsymbol{G}_{IIOi_0}(z) & \boldsymbol{G}_{IIOr_0}(z) & \boldsymbol{G}_{IIOy_0}(z)\end{array}\right],\end{gathered} \tag{8.74}$$

together with (8.75) through (8.79):

– *the system IO transfer function matrix* $\boldsymbol{G}_{IIO}(z)$,

$$\begin{gathered}\boldsymbol{G}_{IIO}(z)=\left(\boldsymbol{E}^{(\nu)}\boldsymbol{S}_{N}^{(\nu)}(z)\right)^{-1}\cdot\\ \cdot\left[\boldsymbol{R}^{(\alpha)}\boldsymbol{S}_{\rho}^{(\alpha)}(z)\left(\boldsymbol{Q}^{(\alpha)}\boldsymbol{S}_{\rho}^{(\alpha)}(z)\right)^{-1}\boldsymbol{P}^{(\beta)}\boldsymbol{S}_{M}^{(\beta)}(z)+\boldsymbol{T}^{(\mu)}\boldsymbol{S}_{M}^{(\mu)}(z)\right]=\\ =\boldsymbol{G}_{IIOD}^{-1}(z)\boldsymbol{G}_{IION}(z),\end{gathered} \tag{8.75}$$

– *the system IICO transfer function matrix* $\boldsymbol{G}_{IIOi_0}(z)$,

$$\begin{gathered}\boldsymbol{G}_{IIOi_0}(z)=-\left(\boldsymbol{E}^{(\nu)}\boldsymbol{S}_{N}^{(\nu)}(z)\right)^{-1}\cdot\\ \cdot\left\{\left[\begin{array}{cc}\boldsymbol{R}^{(\alpha)}\boldsymbol{S}_{\rho}^{(\alpha)}(z)\left(\boldsymbol{Q}^{(\alpha)}\boldsymbol{S}_{\rho}^{(\alpha)}(z)\right)^{-1}\boldsymbol{P}^{(\beta)}\boldsymbol{Z}_{M}^{(\beta)}(z) & \boldsymbol{O}_{N,(\gamma-\beta)M}\end{array}\right]+\right.\\ \left.+\left[\begin{array}{cc}\boldsymbol{T}^{(\mu)}\boldsymbol{Z}_{M}^{(\mu)}(z) & \boldsymbol{O}_{N,(\gamma-\mu)M}\end{array}\right]\right\}=\\ =\boldsymbol{G}_{IIOi_0D}^{-1}(z)\boldsymbol{G}_{IIOi_0N}(z),\end{gathered} \tag{8.76}$$

– *the system IIRO transfer function matrix* $\boldsymbol{G}_{IIOr_0}(z)$,

$$\boldsymbol{G}_{IIOr_0}(z) = \left(\boldsymbol{E}^{(\nu)} \boldsymbol{S}_N^{(\nu)}(z)\right)^{-1} \cdot$$
$$\cdot \left[\boldsymbol{R}^{(\alpha)} \boldsymbol{S}_\rho^{(\alpha)}(z) \left(\boldsymbol{Q}^{(\alpha)} \boldsymbol{S}_\rho^{(\alpha)}(z)\right)^{-1} \boldsymbol{Q}^{(\alpha)} \boldsymbol{Z}_\rho^{(\alpha)}(z) - \boldsymbol{R}^{(\alpha)} \boldsymbol{Z}_\rho^{(\alpha)}(z)\right] =$$
$$= \boldsymbol{G}_{IIOr_0D}^{-1}(z) \boldsymbol{G}_{IIOr_0N}(z), \tag{8.77}$$

– *the system IIYO transfer function matrix* $\boldsymbol{G}_{IIOy_0}(z)$,

$$\boldsymbol{G}_{IIOy_0}(z) = \left(\boldsymbol{E}^{(\nu)} \boldsymbol{S}_N^{(\nu)}(z)\right)^{-1} \boldsymbol{E}^{(\nu)} \boldsymbol{Z}_N^{(\nu)}(z) =$$
$$= \boldsymbol{G}_{IIOy_0D}^{-1}(z) \boldsymbol{G}_{IIOy_0N}(z). \tag{8.78}$$

The denominator matrix polynomial $\boldsymbol{F}_{IIOD}(z)$ *and the numerator matrix polynomial* $\boldsymbol{F}_{IION}(z)$ *of* $\boldsymbol{F}_{IIO}(z)$, $\boldsymbol{F}_{IIO}(z) = \boldsymbol{F}_{IIOD}^{-1}(z) \cdot \boldsymbol{F}_{IION}(z)$ *result from the preceding equations in the following forms:*

$$\boldsymbol{F}_{IIOD}(z) = \boldsymbol{E}^{(\nu)} \boldsymbol{S}_N^{(\nu)}(z),$$

$$\boldsymbol{F}_{IION}(z) =$$
$$= \left[\left[\boldsymbol{R}^{(\alpha)} \boldsymbol{S}_\rho^{(\alpha)}(z) \left(\boldsymbol{Q}^{(\alpha)} \boldsymbol{S}_\rho^{(\alpha)}(z)\right)^{-1} \boldsymbol{P}^{(\beta)} \boldsymbol{S}_M^{(\beta)}(z) + \boldsymbol{T}^{(\mu)} \boldsymbol{S}_M^{(\mu)}(z)\right] \ \vdots \right.$$
$$\vdots \left\{\left[-\boldsymbol{R}^{(\alpha)} \boldsymbol{S}_\rho^{(\alpha)}(z) \left(\boldsymbol{Q}^{(\alpha)} \boldsymbol{S}_\rho^{(\alpha)}(z)\right)^{-1} \boldsymbol{P}^{(\beta)} \boldsymbol{Z}_M^{(\beta)}(z) \ \vdots \ \boldsymbol{O}_{N,(\gamma-\beta)M} \right] + \right.$$
$$\left. + \left[-\boldsymbol{T}^{(\mu)} \boldsymbol{Z}_M^{(\mu)}(z) \quad \boldsymbol{O}_{N,(\gamma-\mu)M} \right]\right\} \ \vdots$$
$$\vdots \ \left[\boldsymbol{R}^{(\alpha)} \boldsymbol{S}_\rho^{(\alpha)}(z) \left(\boldsymbol{Q}^{(\alpha)} \boldsymbol{S}_\rho^{(\alpha)}(z)\right)^{-1} \boldsymbol{Q}^{(\alpha)} \boldsymbol{Z}_\rho^{(\alpha)}(z) - \boldsymbol{R}^{(\alpha)} \boldsymbol{Z}_\rho^{(\alpha)}(z)\right] \ \vdots$$
$$\left. \vdots \ \left(\boldsymbol{E}^{(\nu)} \boldsymbol{Z}_N^{(\nu)}(z)\right) \right]. \tag{8.79}$$

Proof. The $Z-$transform of the compact form (3.64b) of (3.63b) (in Subsection 3.5.4) with use of (8.66) reads

$$\left(\boldsymbol{E}^{(\nu)}\boldsymbol{S}_N^{(\nu)}(z)\right)\mathbf{Y}(z) = \begin{bmatrix} \left(\boldsymbol{R}^{(\alpha)}\boldsymbol{S}_\rho^{(\alpha)}(z)\right)^T \\ \left(\boldsymbol{T}^{(\mu)}\boldsymbol{S}_M^{(\mu)}(z)\right)^T \\ \left(-\boldsymbol{R}^{(\alpha)}\boldsymbol{Z}_\rho^{(\alpha)}(z)\right)^T \\ \left(-\boldsymbol{T}^{(\mu)}\boldsymbol{Z}_M^{(\mu)}(z)\right)^T \\ \left(\boldsymbol{E}^{(\nu)}\boldsymbol{Z}_N^{(\nu)}(z)\right)^T \end{bmatrix}^T \begin{bmatrix} \mathbf{R}(z) \\ \mathbf{I}(z) \\ \mathbf{r}_0^{\alpha-1} \\ \mathbf{i}_0^{\mu-1} \\ \mathbf{y}_0^{\nu-1} \end{bmatrix} \Longrightarrow$$

$$\left(\boldsymbol{E}^{(\nu)}\boldsymbol{S}_N^{(\nu)}(z)\right)\mathbf{Y}(z) = \begin{bmatrix} \left(\boldsymbol{R}^{(\alpha)}\boldsymbol{S}_\rho^{(\alpha)}(z)\right)^T \\ \left(\boldsymbol{T}^{(\mu)}\boldsymbol{S}_M^{(\mu)}(z)\right)^T \\ \left(-\boldsymbol{R}^{(\alpha)}\boldsymbol{Z}_\rho^{(\alpha)}(z)\right)^T \\ \left(-\boldsymbol{T}^{(\mu)}\boldsymbol{Z}_M^{(\mu)}(z)\right)^T \\ \left(\boldsymbol{E}^{(\nu)}\boldsymbol{Z}_N^{(\nu)}(z)\right)^T \end{bmatrix}^T \cdot$$

$$\cdot \begin{bmatrix} \left(\boldsymbol{Q}^{(\alpha)}\boldsymbol{S}_\rho^{(\alpha)}(z)\right)^{-1}\left[\begin{array}{ccc} \boldsymbol{P}^{(\beta)}\boldsymbol{S}_M^{(\beta)}(z) & -\boldsymbol{P}^{(\beta)}\boldsymbol{Z}_M^{(\beta)}(z) & \boldsymbol{Q}^{(\alpha)}\boldsymbol{Z}_\rho^{(\alpha)}(z) \end{array}\right] \cdot \\ \cdot \begin{bmatrix} \mathbf{I}(z) \\ \mathbf{i}_0^{\beta-1} \\ \mathbf{r}_0^{\alpha-1} \end{bmatrix} \\ \mathbf{I}(z) \\ \mathbf{r}_0^{\alpha-1} \\ \mathbf{i}_0^{\mu-1} \\ \mathbf{y}_0^{\nu-1} \end{bmatrix} \Rightarrow$$

$$\left(\boldsymbol{E}^{(\nu)}\boldsymbol{S}_N^{(\nu)}(z)\right)\mathbf{Y}(z) =$$

$$= \left[\begin{array}{c} \boldsymbol{R}^{(\alpha)}\boldsymbol{S}_\rho^{(\alpha)}(z)\left(\boldsymbol{Q}^{(\alpha)}\boldsymbol{S}_\rho^{(\alpha)}(z)\right)^{-1} \cdot \\ \cdot\left[\begin{array}{ccc} \boldsymbol{P}^{(\beta)}\boldsymbol{S}_M^{(\beta)}(z) & -\boldsymbol{P}^{(\beta)}\boldsymbol{Z}_M^{(\beta)}(z) & \boldsymbol{Q}^{(\alpha)}\boldsymbol{Z}_\rho^{(\alpha)}(z) \end{array}\right] \cdot \begin{pmatrix} \mathbf{I}(z) \\ \mathbf{i}_0^{\beta-1} \\ \mathbf{r}_0^{\alpha-1} \end{pmatrix} + \end{array} \right.$$

$$+\boldsymbol{T}^{(\mu)}\boldsymbol{S}_M^{(\mu)}(z)\mathbf{I}(z) + \left(-\boldsymbol{R}^{(\alpha)}\boldsymbol{Z}_\rho^{(\alpha)}(z)\right)\mathbf{r}_0^{\alpha-1} +$$

$$\left. + \left(-\boldsymbol{T}^{(\mu)}\boldsymbol{Z}_M^{(\mu)}(z)\right)\mathbf{i}_0^{\mu-1} + \boldsymbol{E}^{(\nu)}\boldsymbol{Z}_N^{(\nu)}(z)\mathbf{y}_0^{\nu-1} \right] \Longrightarrow$$

$$\left(\boldsymbol{E}^{(\nu)}\boldsymbol{S}_N^{(\nu)}(z)\right)\mathbf{Y}(z) =$$
$$= \left\{ \boldsymbol{R}^{(\alpha)}\boldsymbol{S}_\rho^{(\alpha)}(z)\left(\boldsymbol{Q}^{(\alpha)}\boldsymbol{S}_\rho^{(\alpha)}(z)\right)^{-1} \cdot \right.$$
$$\cdot \left[\boldsymbol{P}^{(\beta)}\boldsymbol{S}_M^{(\beta)}(z)\mathbf{I}(z) - \boldsymbol{P}^{(\beta)}\boldsymbol{Z}_M^{(\beta)}(z)\mathbf{i}_0^{\beta-1} + \boldsymbol{Q}^{(\alpha)}\boldsymbol{Z}_\rho^{(\alpha)}(z)\mathbf{r}_0^{\alpha-1}\right] +$$
$$+ \boldsymbol{T}^{(\mu)}\boldsymbol{S}_M^{(\mu)}(z)\mathbf{I}(z) + \left(-\boldsymbol{R}^{(\alpha)}\boldsymbol{Z}_\rho^{(\alpha)}(z)\right)\mathbf{r}_0^{\alpha-1} +$$
$$\left. + \left(-\boldsymbol{T}^{(\mu)}\boldsymbol{Z}_M^{(\mu)}(z)\right)\mathbf{i}_0^{\mu-1} + \boldsymbol{E}^{(\nu)}\boldsymbol{Z}_N^{(\nu)}(z)\mathbf{y}_0^{\nu-1}\right\} \Longrightarrow$$

$$\mathbf{Y}(z) = \left(\boldsymbol{E}^{(\nu)}\boldsymbol{S}_N^{(\nu)}(z)\right)^{-1} \cdot$$
$$\cdot \left\{ \left[\boldsymbol{R}^{(\alpha)}\boldsymbol{S}_\rho^{(\alpha)}(z)\left(\boldsymbol{Q}^{(\alpha)}\boldsymbol{S}_\rho^{(\alpha)}(z)\right)^{-1}\boldsymbol{P}^{(\beta)}\boldsymbol{S}_M^{(\beta)}(z) + \boldsymbol{T}^{(\mu)}\boldsymbol{S}_M^{(\mu)}(z)\right]\mathbf{I}(z) + \right.$$
$$+ \left\{\left[\begin{array}{cc} -\boldsymbol{R}^{(\alpha)}\boldsymbol{S}_\rho^{(\alpha)}(z)\left(\boldsymbol{Q}^{(\alpha)}\boldsymbol{S}_\rho^{(\alpha)}(z)\right)^{-1}\boldsymbol{P}^{(\beta)}\boldsymbol{Z}_M^{(\beta)}(z) & \boldsymbol{O}_{N,(\gamma-\beta)M} \end{array}\right] + \right.$$
$$\left. + \left[\begin{array}{cc} \left(-\boldsymbol{T}^{(\mu)}\boldsymbol{Z}_M^{(\mu)}(z)\right) & \boldsymbol{O}_{N,(\gamma-\mu)M} \end{array}\right]\right\}\mathbf{i}_0^{\gamma-1} +$$
$$+ \left[\boldsymbol{R}^{(\alpha)}\boldsymbol{S}_\rho^{(\alpha)}(z)\left(\boldsymbol{Q}^{(\alpha)}\boldsymbol{S}_\rho^{(\alpha)}(z)\right)^{-1}\boldsymbol{Q}^{(\alpha)}\boldsymbol{Z}_\rho^{(\alpha)}(z) - \boldsymbol{R}^{(\alpha)}\boldsymbol{Z}_\rho^{(\alpha)}(z)\right]\mathbf{r}_0^{\alpha-1} +$$
$$\left. + \boldsymbol{E}^{(\nu)}\boldsymbol{Z}_N^{(\nu)}(z)\mathbf{y}_0^{\nu-1} \right\} \Longrightarrow$$

These equations and (7.30a) (Definition 7.6, Section 7.4) yield

$$\boldsymbol{F}_{IIO}(z) = \left[\begin{array}{cccc} \boldsymbol{G}_{IIO}(z) & \boldsymbol{G}_{IIOi_0}(z) & \boldsymbol{G}_{IIOr_0}(z) & \boldsymbol{G}_{IIOy_0}(z) \end{array}\right],$$

$$\boldsymbol{G}_{IIO}(z) = \left(\boldsymbol{E}^{(\nu)}\boldsymbol{S}_N^{(\nu)}(z)\right)^{-1} \cdot$$
$$\cdot \left(\boldsymbol{R}^{(\alpha)}\boldsymbol{S}_\rho^{(\alpha)}(z)\left(\boldsymbol{Q}^{(\alpha)}\boldsymbol{S}_\rho^{(\alpha)}(z)\right)^{-1}\boldsymbol{P}^{(\beta)}\boldsymbol{S}_M^{(\beta)}(z) + \boldsymbol{T}^{(\mu)}\boldsymbol{S}_M^{(\mu)}(z)\right) =$$
$$= \boldsymbol{G}_{IIOD}^{-1}(z)\boldsymbol{G}_{IION}(z),$$

$$\boldsymbol{G}_{IIOi_0}(z) = -\left(\boldsymbol{E}^{(\nu)}\boldsymbol{S}_N^{(\nu)}(z)\right)^{-1} \cdot$$
$$\cdot\left\{\left[\begin{array}{cc} \boldsymbol{R}^{(\alpha)}\boldsymbol{S}_\rho^{(\alpha)}(z)\left(\boldsymbol{Q}^{(\alpha)}\boldsymbol{S}_\rho^{(\alpha)}(z)\right)^{-1}\boldsymbol{P}^{(\beta)}\boldsymbol{Z}_M^{(\beta)}(z) & \boldsymbol{O}_{N,(\gamma-\beta)M} \end{array}\right]+\right.$$
$$\left.+\left[\begin{array}{cc} \boldsymbol{T}^{(\mu)}\boldsymbol{Z}_M^{(\mu)}(z) & \boldsymbol{O}_{N,(\gamma-\mu)M} \end{array}\right]\right\} =$$
$$= \boldsymbol{G}_{IIOi_0D}^{-1}(z)\boldsymbol{G}_{IIOi_0N}(z),$$

$$\boldsymbol{G}_{IIOr_0}(z) = \left(\boldsymbol{E}^{(\nu)}\boldsymbol{S}_N^{(\nu)}(z)\right)^{-1} \cdot$$
$$\cdot\left[\boldsymbol{R}^{(\alpha)}\boldsymbol{S}_\rho^{(\alpha)}(z)\left(\boldsymbol{Q}^{(\alpha)}\boldsymbol{S}_\rho^{(\alpha)}(z)\right)^{-1}\boldsymbol{Q}^{(\alpha)}\boldsymbol{Z}_\rho^{(\alpha)}(z) - \boldsymbol{R}^{(\alpha)}\boldsymbol{Z}_\rho^{(\alpha)}(z)\right] =$$
$$= \boldsymbol{G}_{IIOr_0D}^{-1}(z)\boldsymbol{G}_{IIOr_0N}(z),$$

$$\boldsymbol{G}_{IIOy_0}(z) = \left(\boldsymbol{E}^{(\nu)}\boldsymbol{S}_N^{(\nu)}(z)\right)^{-1}\boldsymbol{E}^{(\nu)}\boldsymbol{Z}_N^{(\nu)}(z) =$$
$$= \boldsymbol{G}_{IIOy_0D}^{-1}(z)\boldsymbol{G}_{IIOy_0N}(z). \tag{8.80}$$

Let

$$\mathbf{V}_{IIO}(z;\mathbf{i}_0^{\gamma-1};\ \mathbf{r}_0^{\alpha-1};\ \mathbf{y}_0^{\nu-1}) = \left[\begin{array}{cccc} \mathbf{I}^T(z) & \left(\mathbf{i}_0^{\gamma-1}\right)^T & \left(\mathbf{r}_0^{\alpha-1}\right)^T & \left(\mathbf{y}_0^{\nu-1}\right)^T \end{array}\right]^T. \tag{8.81}$$

Equations (8.80), (8.81), and (7.30a) (in Section 7.4) enable the following:

$$\mathbf{Y}(z) = \boldsymbol{F}_{IIO}(z)\mathbf{V}_{IIO}(z;\mathbf{i}_0^{\gamma-1};\mathbf{r}_0^{\alpha-1};\ \mathbf{y}_0^{\nu-1}), \tag{8.82}$$

$$\boldsymbol{F}_{IIO}(z) = \boldsymbol{F}_{IIOD}^{-1}(z)\boldsymbol{F}_{IION}(z),$$
$$\boldsymbol{F}_{IIOD}(z) = \boldsymbol{E}^{(\nu)}\boldsymbol{S}_N^{(\nu)}(z),$$
$$\boldsymbol{F}_{IION}(z) =$$
$$= \left[\begin{array}{c} \left[\boldsymbol{R}^{(\alpha)}\boldsymbol{S}_\rho^{(\alpha)}(z)\left(\boldsymbol{Q}^{(\alpha)}\boldsymbol{S}_\rho^{(\alpha)}(z)\right)^{-1}\boldsymbol{P}^{(\beta)}\boldsymbol{S}_M^{(\beta)}(z) + \boldsymbol{T}^{(\mu)}\boldsymbol{S}_M^{(\mu)}(z)\right]^T \\ \left\{\left[\begin{array}{cc} -\boldsymbol{R}^{(\alpha)}\boldsymbol{S}_\rho^{(\alpha)}(z)\left(\boldsymbol{Q}^{(\alpha)}\boldsymbol{S}_\rho^{(\alpha)}(z)\right)^{-1}\boldsymbol{P}^{(\beta)}\boldsymbol{Z}_M^{(\beta)}(z) & \boldsymbol{O}_{N,(\gamma-\beta)M}\end{array}\right]+\right. \\ \left.+\left[\begin{array}{cc} -\boldsymbol{T}^{(\mu)}\boldsymbol{Z}_M^{(\mu)}(z) & \boldsymbol{O}_{N,(\gamma-\mu)M}\end{array}\right]\right\}^T \\ \left[\boldsymbol{R}^{(\alpha)}\boldsymbol{S}_\rho^{(\alpha)}(z)\left(\boldsymbol{Q}^{(\alpha)}\boldsymbol{S}_\rho^{(\alpha)}(z)\right)^{-1}\boldsymbol{Q}^{(\alpha)}\boldsymbol{Z}_\rho^{(\alpha)}(z) - \boldsymbol{R}^{(\alpha)}\boldsymbol{Z}_\rho^{(\alpha)}(z)\right]^T \\ \left(\boldsymbol{E}^{(\nu)}\boldsymbol{Z}_N^{(\nu)}(z)\right)^T \end{array}\right]^T, \tag{8.83}$$

which prove (8.74) through (8.79). ■

Note 8.6 *This theorem is valid also for the RS (2.33a) and (2.33b) and the PMD systems (2.32a) and (2.32b) (in Section 2.3), because they are the special cases (in Subsection 2.4.3), of the IIO systems (2.31a) and (2.31b) (in Section 2.3), i.e., (3.64a) and (3.64b) (in Subsection 3.5.4),*

$$\boldsymbol{Q}^{(\alpha)}\mathbf{r}^{\alpha}(k) = \boldsymbol{P}^{(\beta)}\mathbf{i}^{\beta}(k), \tag{8.84a}$$

$$\boldsymbol{E}^{(\nu)}\mathbf{y}^{\nu}(k) = \boldsymbol{R}^{(\alpha)}\mathbf{r}^{\alpha}(k) + \boldsymbol{T}^{(\mu)}\mathbf{i}^{\mu}(k). \tag{8.84b}$$

Note 8.7 ***The IIO full block diagram technique generalizes the classical block diagram technique***

Equations (8.56)-(8.59) show that the general relationship between the $Z-$transform of the output vector function $\mathbf{y}(\cdot)$ and the $Z-$transform $\mathbf{V}_{IIO}(z)$ of the action vector function $\mathbf{v}(\cdot)$ of the system under ***nonzero initial conditions*** *has the same form as the relationship between the $Z-$transform $\mathbf{Y}(z)$ of the output vector function $\mathbf{y}(\cdot)$ and the $Z-$transform $\mathbf{I}(z)$ of the input vector function $\mathbf{i}(\cdot)$ under* ***all zero initial conditions****. In the former case, the full transfer function matrix $\boldsymbol{F}_{IIO}(z)$ replaces the transfer function matrix $\boldsymbol{G}_{IIO}(z)$ from the latter case; see Fig. 8.3. The usage of $\boldsymbol{F}_{IIO}(z)$ and $\mathbf{V}_{IIO}(z) = \mathbf{V}_{IIO}(z;\mathbf{i}_0^{\gamma-1};\mathbf{r}_0^{\alpha-1};\mathbf{y}_0^{\nu-1})$ enables the generalization of the block diagram technique to arbitrary initial conditions. The system full block, Fig. 8.3, holds for arbitrary initial conditions. The*

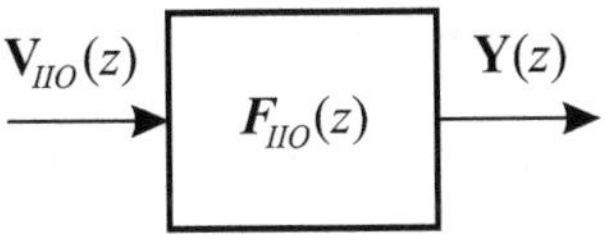

Figure 8.3: The generalized block of the IIO system.

system block is valid only under all zero initial conditions.

Example 8.10 *Let the IIO system be described by*

$$E^2r(k) - 1.5E^1r(k) - r(k) = 3E^2i(k) - 24E^1i(k) + 36i(k),$$
$$E^2y(k) - 1.8E^1y(k) - 0.4y\,(k) =$$
$$= 5E^2r(k) - 10E^1r(k) + 2E^2i(k) - 16E^1i(k) + 24i(k). \tag{8.85}$$

The application of the $Z-$transform to the system (8.85) leads to the system description and characteristics in the complex domain:

$$\left(z^2 - 1.5z - 1\right)R\,(z) - \left(z^2 - 1.5z\right)r_0 - z\,(Er)_0 =$$
$$= \left(3z^2 - 24z + 36\right)I\,(z) - \left(3z^2 - 24z\right)i_0 - 3z\,(Ei)_0\,,$$

$$\begin{gathered}\left(z^2-1.8z-0.4\right)Y(z)-\left(z^2-1.8z\right)y_0-z\,(Ey)_0=\\=\left(5z^2-10z\right)R(z)-\left(5z^2-10z\right)r_0-5z\,(Er)_0+\\+\left(2z^2-16z+24\right)I(z)-\left(2z^2-16z\right)i_0-2z\,(Ei)_0\Longrightarrow\end{gathered}$$

$$\begin{gathered}\left(z^2-1.8z-0.4\right)Y(z)-\left(z^2-1.8z\right)y_0-z\,(Ey)_0=\\=\left(5z^2-10z\right)\left(z^2-1.5z-1\right)^{-1}\cdot\\\cdot\left[\left(3z^2-24z+36\right)I(z)-\left(3z^2-24z\right)i_0-3z\,(Ei)_0+\left(z^2-1.5z\right)r_0+\right.\\\left.+z\,(Er)_0\right]-\left(5z^2-10z\right)r_0-5z\,(Er)_0+\\+\left(2z^2-16z+24\right)I(z)-\left(2z^2-16z\right)i_0-2z\,(Ei)_0\Longrightarrow\end{gathered}$$

$$\begin{gathered}Y(z)=\left(z^2-1.8z-0.4\right)^{-1}\cdot\\\cdot\left\{\left(5z^2-10z\right)\left(z^2-1.5z-1\right)^{-1}\cdot\right.\\\cdot\left[\left(3z^2-24z+36\right)I(z)-\left(3z^2-24z\right)i_0-3z\,(Ei)_0+\left(z^2-1.5z\right)r_0+\right.\\\left.+z\,(Er)_0\right]-\left(5z^2-10z\right)r_0-5z\,(Er)_0+\left(2z^2-16z+24\right)I(z)-\\\left.-\left(2z^2-16z\right)i_0-2z\,(Ei)_0+\left(z^2-1.8z\right)y_0+z\,(Ey)_0\right\}\Longrightarrow\end{gathered}$$

$$\begin{gathered}Y(z)=\left(z^2-1.8z-0.4\right)^{-1}\cdot\\\cdot\left\{\left(5z^2-10z\right)\left(z^2-1.5z-1\right)^{-1}\left(3z^2-24z+36\right)I(z)-\right.\\-\left(5z^2-10z\right)\left(z^2-1.5z-1\right)^{-1}\left(3z^2-24z\right)i_0-\\-\left(5z^2-10z\right)\left(z^2-1.5z-1\right)^{-1}3z\,(Ei)_0+\\+\left(5z^2-10z\right)\left(z^2-1.5z-1\right)^{-1}\left(z^2-1.5z\right)r_0+\\+\left(5z^2-10z\right)\left(z^2-1.5z-1\right)^{-1}z\,(Er)_0-\\-\left(5z^2-10z\right)r_0-5z\,(Er)_0+\left(2z^2-16z+24\right)I(z)-\\\left.-\left(2z^2-16z\right)i_0-2z\,(Ei)_0+\left(z^2-1.8z\right)y_0+z\,(Ey)_0\right\}\Longrightarrow\end{gathered}$$

$$
\begin{gathered}
Y(z) = \left(z^2 - 1.8z - 0.4\right)^{-1} \cdot \\
\cdot \left\{ \left[\left(5z^2 - 10z\right)\left(z^2 - 1.5z - 1\right)^{-1}\left(3z^2 - 24z + 36\right) + \left(2z^2 - 16z + 24\right) \right] \cdot \right. \\
\cdot I(z) - \left[\left(5z^2 - 10z\right)\left(z^2 - 1.5z - 1\right)^{-1}\left(3z^2 - 24z\right) + \left(2z^2 - 16z\right) \right] i_0 - \\
- \left[\left(5z^2 - 10z\right)\left(z^2 - 1.5z - 1\right)^{-1} 3z + 2z \right] (Ei)_0 + \\
+ \left[\left(5z^2 - 10z\right)\left(z^2 - 1.5z - 1\right)^{-1}\left(z^2 - 1.5z\right) - \left(5z^2 - 10z\right) \right] r_0 + \\
+ \left[\left(5z^2 - 10z\right)\left(z^2 - 1.5z - 1\right)^{-1} z - 5z \right] (Er)_0 + \left(z^2 - 1.8z\right) y_0 + \\
\left. + z(Ey)_0 \right\} \Longrightarrow
\end{gathered}
$$

$$
\begin{gathered}
Y(z) = \left(z^2 - 1.8z - 0.4\right)^{-1} \cdot \\
\cdot \begin{bmatrix}
\left(5z^2 - 10z\right)\left(z^2 - 1.5z - 1\right)^{-1}\left(3z^2 - 24z + 36\right) + \left(2z^2 - 16z + 24\right) \\
- \left[\left(5z^2 - 10z\right)\left(z^2 - 1.5z - 1\right)^{-1}\left(3z^2 - 24z\right) + \left(2z^2 - 16z\right) \right] \\
- \left[\left(5z^2 - 10z\right)\left(z^2 - 1.5z - 1\right)^{-1} 3z + 2z \right] \\
\left(5z^2 - 10z\right)\left(z^2 - 1.5z - 1\right)^{-1}\left(z^2 - 1.5z\right) - \left(5z^2 - 10z\right) \\
\left(5z^2 - 10z\right)\left(z^2 - 1.5z - 1\right)^{-1} z - 5z \\
\left(z^2 - 1.8z\right) \\
z
\end{bmatrix}^T \cdot \\
\cdot \begin{bmatrix} I(z) & i_0 & (Ei)_0 & r_0 & (Er)_0 & y_0 & (Ey)_0 \end{bmatrix}^T
\end{gathered} \tag{8.86}
$$

$$\Longrightarrow$$

$$
\begin{gathered}
\boldsymbol{F}_{IIO}(z) = \left(z^2 - 1.8z - 0.4\right)^{-1}\left(z^2 - 1.5z - 1\right)^{-1} \cdot \\
\cdot \begin{bmatrix}
\left(5z^2 - 10z\right)\left(3z^2 - 24z + 36\right) + \left(z^2 - 1.5z - 1\right)\left(2z^2 - 16z + 24\right) \\
- \left[\left(5z^2 - 10z\right)\left(3z^2 - 24z\right) + \left(z^2 - 1.5z - 1\right)\left(2z^2 - 16z\right) \right] \\
- \left[\left(5z^2 - 10z\right) 3z + \left(z^2 - 1.5z - 1\right) 2z \right] \\
\left(5z^2 - 10z\right)\left(z^2 - 1.5z\right) - \left(z^2 - 1.5z - 1\right)\left(5z^2 - 10z\right) \\
\left(5z^2 - 10z\right) z - \left(z^2 - 1.5z - 1\right) 5z \\
\left(z^2 - 1.5z - 1\right)\left(z^2 - 1.8z\right) \\
\left(z^2 - 1.5z - 1\right) z
\end{bmatrix}^T
\end{gathered}
$$

$$\Longrightarrow$$

$$\boldsymbol{F}_{IIO}(z) = \left[(z-2)^2 (z+0.2)(z+0.5) \right]^{-1} \cdot$$
$$\cdot \begin{bmatrix} (z-2)^2 (z-6)(17z+1) \\ -z(z-2)(z-8)(17z+1) \\ -z(z-2)(17z+1) \\ 5z(z-2) \\ -2.5z(z-2) \\ z(z-2)(z+0.5)(z-1.8) \\ z(z-2)(z+0.5) \end{bmatrix}^T . \tag{8.87}$$

We can cancel $(z-2)$ because it is common factor to the denominator polynomial and to all entries of the numerator polynomial matrix,

$$\boldsymbol{F}_{IIOnd}(z) = \left[(z-2)(z+0.2)(z+0.5) \right]^{-1} \cdot$$
$$\cdot \begin{bmatrix} (z-2)(z-6)(17z+1) \\ -z(z-8)(17z+1) \\ -z(17z+1) \\ 5z \\ -2.5z \\ z(z+0.5)(z-1.8) \\ z(z+0.5) \end{bmatrix}^T . \tag{8.88}$$

This is the nondegenerate form $\boldsymbol{F}_{IIOnd}(z)$ of $\boldsymbol{F}_{IIO}(z)$. The system transfer function $\boldsymbol{G}_{IIO}(\cdot)$ follows now from (8.87),

$$\boldsymbol{G}_{IIO}(z) = \left[(z-2)^2 (z+0.2)(z+0.5) \right]^{-1} \cdot$$
$$\cdot (z-2)^2 (z-6)(17z+1) . \tag{8.89}$$

We cancel now $(z-2)^2$ in $\boldsymbol{G}_{IIO}(z)$, which is not possible either in $\boldsymbol{F}_{IIO}(z)$ or in its nondegenerate form $\boldsymbol{F}_{IIOnd}(z)$,

$$\boldsymbol{G}_{IIO}(z) = \left[(z+0.2)(z+0.5) \right]^{-1} (z-6)(17z+1) . \tag{8.90}$$

The $Z-$transform $\mathbf{V}_{IIO}(z)$ of the action vector function $\mathbf{v}_{IIO}(\cdot)$ and the action vector function itself result from (8.86),

$$\mathbf{V}_{IIO}(z; \mathbf{i}_0^1; \mathbf{r}_0^1; \mathbf{y}_0^1) = \left[\begin{array}{cccc} I(z) & \underbrace{i_0 \quad (Ei)_0}_{\mathbf{i}_0^{1T}} & \underbrace{r_0 \quad (Er)_0}_{\mathbf{r}_0^{1T}} & \underbrace{y_0 \quad (Ey)_0}_{\mathbf{y}_0^{1T}} \end{array} \right]^T ,$$

$$\mathbf{v}_{IIO}(k; \mathbf{i}_0^1; \mathbf{r}_0^1; \mathbf{y}_0^1) = \begin{bmatrix} i(k) \\ \delta_d(k)\mathbf{i}_0^1 \\ \delta_d(k)\mathbf{r}_0^1 \\ \delta_d(k)\mathbf{y}_0^1 \end{bmatrix} . \tag{8.91}$$

Equations (8.86), (8.87), and (8.91) yield relation between $\mathbf{Y}(z)$ *and* $\mathbf{V}_{IIO}(z;\mathbf{i}_0^1;\mathbf{r}_0^1;\mathbf{y}_0^1)$ *via* $F_{IIO}(s)$,

$$\mathbf{Y}(z) = \boldsymbol{F}_{IIO}(z)\mathbf{V}_{IIO}(z;\mathbf{i}_0^1;\mathbf{r}_0^1;\mathbf{y}_0^1). \tag{8.92}$$

8.4 Conclusion: Common general form of $\boldsymbol{F}(z)$

The preceding results, (8.6a), (8.6b), (8.7a), (8.18), (8.31), (8.32), (8.40), (8.56), and (8.74), imply that for all the treated systems, the following relationship holds between $\mathbf{Y}(z)$ and $\mathbf{V}(z)$,

$$\mathbf{Y}(z) = \boldsymbol{F}(z)\mathbf{V}(z),\ \boldsymbol{F}(z) = \boldsymbol{F}_D^{-1}(z)\boldsymbol{F}_N(z),$$
$$\boldsymbol{F}(z) \in \mathcal{C}^{N\times W},\ \mathbf{V}(z) \in \mathcal{C}^W,\ \mathbf{V}(z) = \begin{bmatrix} \mathbf{I}(z) \\ \mathbf{C}_0 \end{bmatrix},\ W = (L+1)\,U, \tag{8.93}$$

where $\mathbf{C}_0$ is the vector of all initial conditions, $\boldsymbol{F}_D(z)$ and $\boldsymbol{F}_N(z)$ are, respectively, the denominator polynomial matrix and the numerator polynomial matrix of $\boldsymbol{F}(z)$. Their general forms are:

$$\boldsymbol{F}_D(z) = \boldsymbol{A}_D^{(J)}\boldsymbol{S}_N^{(J)}(z) = \sum_{r=0}^{r=J} \boldsymbol{A}_{Dr}z^r \in \mathcal{C}^{N\times N},$$
$$\boldsymbol{A}_{Dr} \in \mathcal{R}^{N\times N},\ \boldsymbol{A}_D^{(J)} \in \mathcal{R}^{N\times(J+1)N},$$
$$\boldsymbol{F}_N(z) = \boldsymbol{B}_N^{(L)}\boldsymbol{S}_W^{(L)}(z) = \sum_{r=0}^{r=L} \boldsymbol{B}_{Nr}z^r \in \mathcal{C}^{N\times W},$$
$$\boldsymbol{B}_{Nr} \in \mathcal{R}^{N\times U},\ \boldsymbol{B}_N^{(L)} \in \mathcal{R}^{N\times(L+1)U},\ (L+1)\,U = W. \tag{8.94}$$

This shows that

$$\Delta(z) = \det \boldsymbol{F}_D(z) = \det\left[\boldsymbol{A}_D^{(J)}\boldsymbol{S}_N^{(J)}(z)\right] = \det\left[\sum_{r=0}^{r=J} \boldsymbol{A}_{Dr}z^r\right] \tag{8.95}$$

is the characteristic polynomial of the system in general.

Conclusion 8.1 ***The full block diagram technique***

The system full transfer function matrix $\boldsymbol{F}(z)$ *and the* $Z-$*transform* $\mathbf{V}(z)$ *of the generalized action vector* $\mathbf{v}(k)$ *(which contains the input vector and the vectors of all initial conditions) enabled us to generalize the classic block diagram technique so to incorporate all initial conditions. Simply, the system full transfer function matrix* $\mathbf{F}(z)$ *and the* $Z-$*transform* $\mathbf{V}(z)$ *of the action vector* $\mathbf{v}(k)$ *replace the system transfer function matrix* $\mathbf{G}(z)$ *and the* $Z-$*transform of the input vector* $\mathbf{i}(k)$, *respectively, in the system block that then becomes the system full block, Fig. 8.3.*

Chapter 9

Full block diagram algebra

9.1 Introduction

We use (8.93) (in Section 8.4) in the following form

$$\mathbf{Y}(z) = \boldsymbol{F}(z)\mathbf{V}(z), \tag{9.1}$$

where in general

$$\boldsymbol{F}(z) = \begin{bmatrix} \boldsymbol{G}(z) & \boldsymbol{G}_0(z) \end{bmatrix}, \tag{9.2}$$

and

$$\mathbf{V}(z) = \begin{bmatrix} \mathbf{I}(z) \\ \mathbf{C}_0 \end{bmatrix}. \tag{9.3}$$

The matrix $\boldsymbol{G}(z)$ is the system transfer function matrix with respect to the input vector,

$$\boldsymbol{G}(z) = \begin{cases} \boldsymbol{G}_{IO}(z) \text{ for the } IO \text{ system}, \\ \boldsymbol{G}_{ISO}(z) \text{ for the } ISO \text{ system}, \\ \boldsymbol{G}_{IIO}(z) \text{ for the } IIO \text{ system}. \end{cases}$$

The matrix $\boldsymbol{G}_0(z)$ is the system transfer function matrix with respect to all initial conditions at $k = 0$ acting on the system.

For:

- the IO system $\boldsymbol{G}_0(z)$ is $\boldsymbol{G}_{IO_0}(z)$, (7.17) (in Section 7.2),

$$\boldsymbol{G}_0(z) = \boldsymbol{G}_{IO_0}(z) = \begin{bmatrix} \boldsymbol{G}_{IOi_0}(z) & \boldsymbol{G}_{IOy_0}(z) \end{bmatrix}; \tag{9.4}$$

- the ISO system $\boldsymbol{G}_0(z)$ is $\boldsymbol{G}_{ISO_0}(z)$, (7.22) (in Section 7.3),

$$\boldsymbol{G}_0(z) = \boldsymbol{G}_{ISO_0}(z) = \boldsymbol{G}_{ISOx_0}(z); \tag{9.5}$$

- the *IIO* system $\boldsymbol{G}_0(z)$ is $\boldsymbol{G}_{IIO_0}(z)$, (7.35c) (in Section 7.4),

$$\boldsymbol{G}_0(z) = \boldsymbol{G}_{IIO_0}(z) = \begin{bmatrix} \boldsymbol{G}_{IIOi_0}(z) & \boldsymbol{G}_{IIOr_0}(z) & \boldsymbol{G}_{IIOy_0}(z) \end{bmatrix}. \quad (9.6)$$

The vector $\mathbf{C}_0$ composed of all initial conditions at $k = 0$ acting on the system has the special forms,

- for the *IO* system (3.56) (in Subsection 3.5.2) $\mathbf{C}_0$ is $\mathbf{C}_{0IO}$, (7.11c) (in Section 7.2),

$$\mathbf{C}_0 = \mathbf{C}_{0IO} = \left\{ \begin{array}{c} \begin{bmatrix} \mathbf{i}_0^{\mu-1} \\ \mathbf{y}_0^{\nu-1} \end{bmatrix}, \ if \ \mu \geq 1, \\ \mathbf{y}_0^{\nu-1}, \ if \ \mu = 0 \end{array} \right\}, \quad (9.7)$$

- for the *ISO* system (3.60a) and (3.60b) (in Subsection 3.5.3) $\mathbf{C}_0$ is $\mathbf{C}_{0ISO}$, (7.20c) (in Section 7.3),

$$\mathbf{C}_0 = \mathbf{C}_{0ISO} = \mathbf{x}_0, \quad (9.8)$$

- for the *IIO* system (3.64a) and (3.64b) (in Subsection 3.5.4) $\mathbf{C}_0$ is $\mathbf{C}_{0IIO}$, (7.30c) (in Section 7.4),

$$\mathbf{C}_0 = \mathbf{C}_{0IIO} = \begin{bmatrix} \mathbf{i}_0^{\gamma-1} \\ \mathbf{r}_0^{\alpha-1} \\ \mathbf{y}_0^{\nu-1} \end{bmatrix}. \quad (9.9)$$

Claim 9.1 *Equations (9.1) through (9.9) enable us to present* ***the full (complete) block diagram of the system****, see Fig. 9.1, and* ***the full***

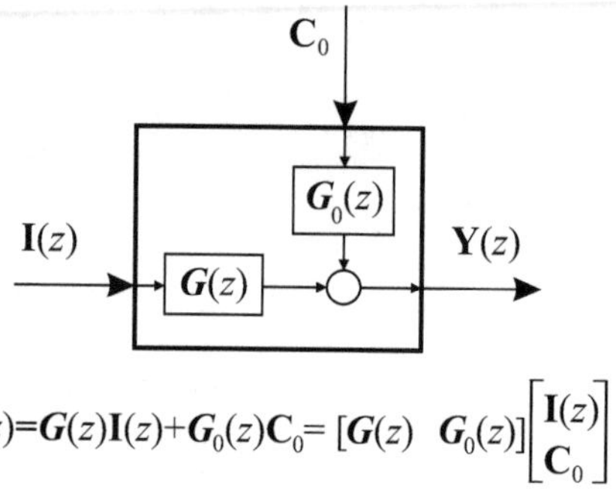

Figure 9.1: The full (complete) block diagram of the system.

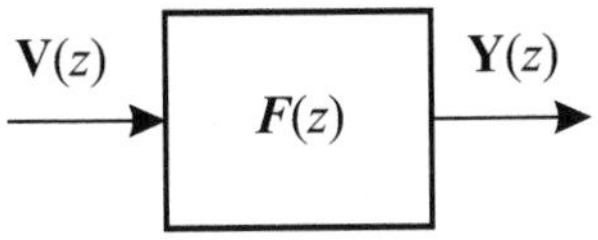

$$F(z)=[G(z)\ \ G_0(z)],\ V(z)=[I^T(z)\ \ C_0^T]^T,\ Y(z)=F(z)V(z)$$

Figure 9.2: The full (complete) block of the system.

(complete) block of the system*; see Fig. 9.2. They are expressed in terms of the system full transfer function matrix* $\boldsymbol{F}(z) = \left[\begin{array}{cc} \boldsymbol{G}(z) & \boldsymbol{G}_0(z) \end{array}\right]$ *and the vectors* $\mathbf{I}(z)$ *and* $\mathbf{C}_0$*, or in terms of the system full transfer function matrix* $\boldsymbol{F}(z)$ *and the vector* $\mathbf{V}(z)$*, respectively. They are valid for arbitrary initial conditions, hence for both zero and nonzero initial conditions. They generalize the classical block diagram and the block of the system expressed in terms of* $\boldsymbol{G}(z)$ *and* $\mathbf{I}(z)$*, which are valid exclusively for all zero initial conditions, i.e., for*

$$\mathbf{C}_0 = \mathbf{0},$$

hence only for

$$\mathbf{V}(z) = \left[\begin{array}{c} \mathbf{I}(z) \\ \mathbf{0} \end{array}\right].$$

We will establish general rules of **the algebra of the full block diagrams of the systems** in the next sections. The preceding consideration is the basis of the new technique – **the technique of the full block diagrams of the systems.**

9.2 Parallel connection

Fig. 9.3 presents the full block diagram of the parallel connection of r systems. We will discover in what follows the rule how to simplify the full block diagram of the parallel connection of r systems by determining the full transfer function matrix $\boldsymbol{F}(z)$ of the whole system, the vector $\mathbf{C}_0$ of all initial conditions and the $Z-$transform $\mathbf{V}(z)$ of the overall action vector $\mathbf{v}(k)$.

Theorem 9.1 *The full transfer function matrix* $\boldsymbol{F}(z)$ *of the parallel connection of r systems, see Fig. 9.3, the resulting IO transfer function matrix* $\boldsymbol{G}(z)$ *relative to the input vector* $\mathbf{i}$*, the resulting transfer function matrix*

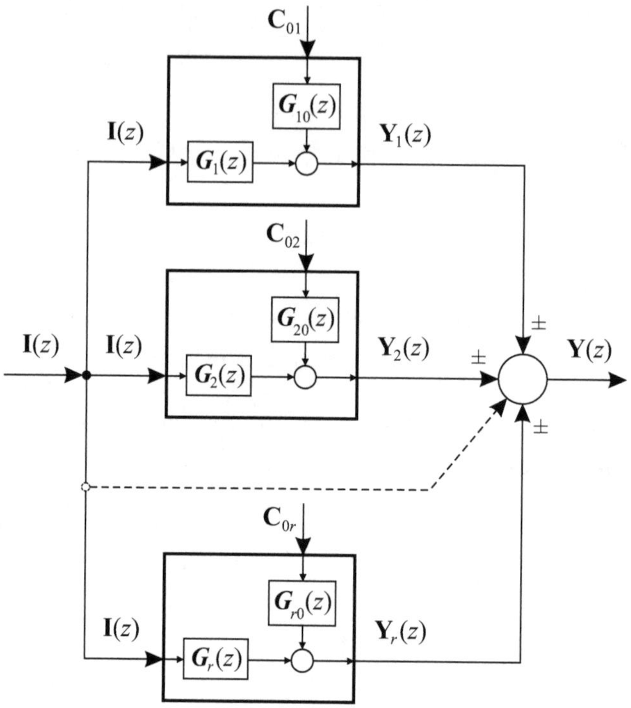

Figure 9.3: The full block diagram of the parallel connection of r systems.

$\boldsymbol{G}_0(z)$ relative to all initial conditions, the vector $\mathbf{C}_0$ of all initial conditions, and the vector $\mathbf{V}(z)$ representing the $Z-$transform of the overall action vector $\mathbf{v}(k)$, obey the following; see Fig. 9.4:

$$\boldsymbol{F}(z) = \begin{bmatrix} \boldsymbol{G}(z) & \boldsymbol{G}_0(z) \end{bmatrix}, \tag{9.10}$$

$$\boldsymbol{G}(z) = \sum_{i=1}^{i=r} (\pm \boldsymbol{G}_i(z)), \tag{9.11}$$

$$\boldsymbol{G}_0(z) = \begin{bmatrix} \pm \boldsymbol{G}_{10}(z) & \pm \boldsymbol{G}_{20}(z) & \cdots & \pm \boldsymbol{G}_{r0}(z) \end{bmatrix}, \tag{9.12}$$

$$\mathbf{C}_0 = \begin{bmatrix} \mathbf{C}_{01} \\ \mathbf{C}_{02} \\ \vdots \\ \mathbf{C}_{0r} \end{bmatrix}, \ \mathbf{V}(z) = \begin{bmatrix} \mathbf{I}(z) \\ \mathbf{C}_0 \end{bmatrix}, \tag{9.13}$$

so that

$$\mathbf{Y}(z) = \begin{bmatrix} \boldsymbol{G}(z) & \boldsymbol{G}_0(z) \end{bmatrix} \begin{bmatrix} \mathbf{I}(z) \\ \mathbf{C}_0 \end{bmatrix} = \boldsymbol{F}(z)\mathbf{V}(z). \tag{9.14}$$

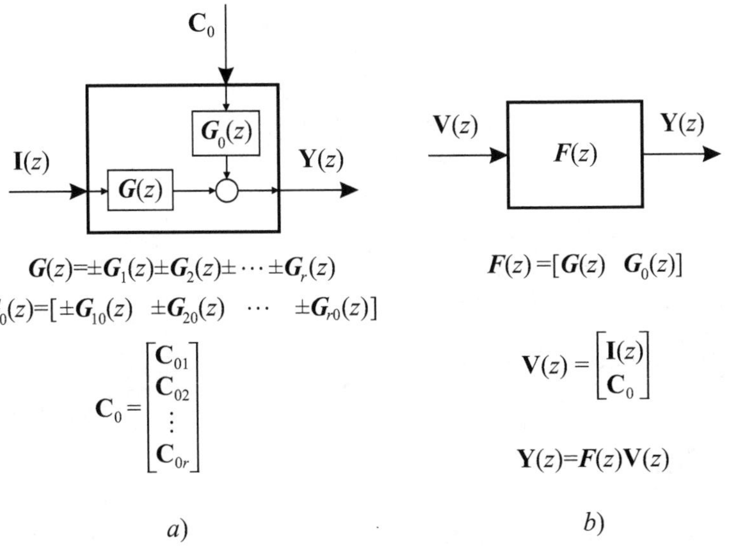

Figure 9.4: The equivalent full block diagram on the left under *a*) and the full block on the right under *b*) of the parallel connection of r systems.

Proof. We refer to the full block diagram of the parallel connection of r systems, Fig. 9.3 and to (9.1) through (9.9) (in Section 9.1),

$$\mathbf{Y}_i(z) = \begin{bmatrix} \boldsymbol{G}_i(z) & \boldsymbol{G}_{i0}(z) \end{bmatrix} \begin{bmatrix} \mathbf{I}(z) \\ \mathbf{C}_{0i} \end{bmatrix} = \boldsymbol{F}_i(z)\mathbf{V}_i(z),\ \mathbf{V}_i(z) = \begin{bmatrix} \mathbf{I}(z) \\ \mathbf{C}_{0i} \end{bmatrix},$$

$$\mathbf{Y}(z) = \sum_{i=1}^{i=r} (\pm \mathbf{Y}_i(z)),$$

$$\Longrightarrow$$

$$\mathbf{Y}(z) = \sum_{i=1}^{i=r} \begin{bmatrix} \pm \boldsymbol{G}_i(z) & \pm \boldsymbol{G}_{i0}(z) \end{bmatrix} \begin{bmatrix} \mathbf{I}(z) \\ \mathbf{C}_{0i} \end{bmatrix} = \sum_{i=1}^{i=r} (\pm \boldsymbol{F}_i(z))\, \mathbf{V}_i(z) \Longrightarrow$$

$$\mathbf{Y}(z) = \underbrace{\left(\sum_{i=1}^{i=r} \pm \boldsymbol{G}_i(z) \right)}_{\boldsymbol{G}(z)} \mathbf{I}(z) + \underbrace{\sum_{i=1}^{i=r} (\pm \boldsymbol{G}_{i0}(z))\, \mathbf{C}_{0i}}_{\boldsymbol{G}_0(z)\mathbf{C}_0}.$$

From these equations, we find the following formulae:

$$\boldsymbol{G}(z) = \sum_{i=1}^{i=r} \pm \boldsymbol{G}_i(z),$$

$$\boldsymbol{G}_0(z) = \left[\begin{array}{cccc} \pm \boldsymbol{G}_{10}(z) & \pm \boldsymbol{G}_{20}(z) & \cdots & \pm \boldsymbol{G}_{r0}(z) \end{array} \right],$$

$$\mathbf{C}_0 = \begin{bmatrix} \mathbf{C}_{01} \\ \mathbf{C}_{02} \\ \vdots \\ \mathbf{C}_{0r} \end{bmatrix}, \quad \mathbf{V}(z) = \begin{bmatrix} \mathbf{I}(z) \\ \mathbf{C}_0 \end{bmatrix},$$

which prove (9.10)-(9.14). ■

9.3 Connection in series

Fig. 9.5 shows the full block diagram of the connection of r systems in series. It will permit us to simplify the block diagram of the whole connection of the systems to be replaced by the equivalent full block represented by the resulting full transfer function matrix $\boldsymbol{F}(z)$ of the whole connection and by the $Z-$transform $\mathbf{V}(z)$ of the overall action vector $\mathbf{v}(k)$, which replaces the $Z-$transform $\mathbf{I}(z)$ of the input vector $\mathbf{i}(k)$. Besides, we will determine the vector $\mathbf{C}_0$ of all initial conditions together with the corresponding transfer function matrices $\boldsymbol{G}(z)$ and $\boldsymbol{G}_0(z)$ relative to $\mathbf{I}(z)$ and to $\mathbf{C}_0$, respectively.

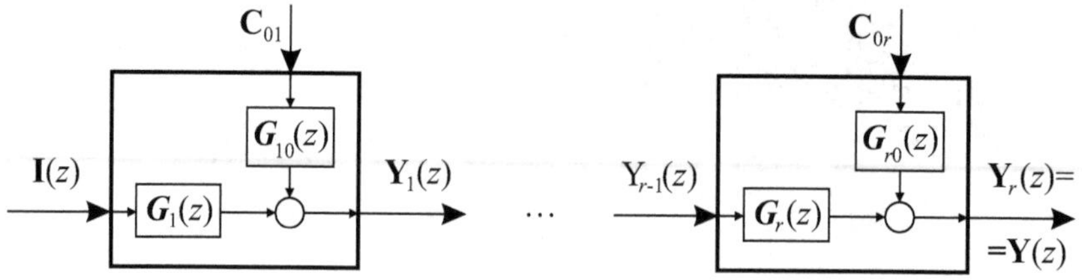

Figure 9.5: The full block diagram of the connection of r systems in series.

Theorem 9.2 *The full transfer function matrix* $\boldsymbol{F}(z)$ *of the connection of r systems in series, see Fig. 9.5, the resulting IO transfer function matrix* $\boldsymbol{G}(z)$ *relative to the input vector* $\mathbf{i}$, *the resulting transfer function matrix* $\boldsymbol{G}_0(z)$ *relative to all initial conditions, the vector* $\mathbf{C}_0$ *of all initial* *conditions, and the vector* $\mathbf{V}(z)$ *representing the* $Z-$*transform of the overall action vector* $\mathbf{v}(k)$*, obey the following; see Fig. 9.6:*

$$\boldsymbol{F}(z) = \begin{bmatrix} \boldsymbol{G}(z) & \boldsymbol{G}_0(z) \end{bmatrix}, \tag{9.15}$$

$$\boldsymbol{G}(z) = \prod_{i=r}^{i=1} \boldsymbol{G}_i(z), \tag{9.16}$$

$$\boldsymbol{G}_0(z) =$$

$$= \Bigg[\underbrace{\left(\prod_{i=r}^{i=2} \boldsymbol{G}_i(z) \right) \boldsymbol{G}_{10}(z)}_{\boldsymbol{G}_{01}(z)} \quad \underbrace{\left(\prod_{i=r}^{i=3} \boldsymbol{G}_i(z) \right) \boldsymbol{G}_{20}(z)}_{\boldsymbol{G}_{02}(z)} \quad \cdots$$

$$\cdots \quad \underbrace{\boldsymbol{G}_r(z)\, \boldsymbol{G}_{r-1,0}(z)}_{\boldsymbol{G}_{0,r-1}(z)} \quad \underbrace{\boldsymbol{G}_{r0}(z)}_{\boldsymbol{G}_{0r}(z)} \Bigg] =$$

$$= \begin{bmatrix} \boldsymbol{G}_{01}(z) & \boldsymbol{G}_{02}(z) & \cdots & \boldsymbol{G}_{0,r-1}(z) & \boldsymbol{G}_{0r}(z) \end{bmatrix}, \tag{9.17}$$

$$\boldsymbol{G}_{0j}(z) = \prod_{i=r}^{i=j+1} \boldsymbol{G}_i(z) \boldsymbol{G}_{j0}(z), \; j = 1, 2, \cdots, r-1,$$

$$\boldsymbol{G}_{0r}(z) = \boldsymbol{G}_{r0}(z), \tag{9.18}$$

$$\mathbf{C}_0 = \begin{bmatrix} \mathbf{C}_{01} \\ \mathbf{C}_{02} \\ \vdots \\ \mathbf{C}_{0k} \end{bmatrix}, \; \mathbf{V}(z) = \begin{bmatrix} \mathbf{I}(z) \\ \mathbf{C}_0 \end{bmatrix}. \tag{9.19}$$

so that

$$\mathbf{Y}(z) = \begin{bmatrix} \boldsymbol{G}(z) & \boldsymbol{G}_0(z) \end{bmatrix} \begin{bmatrix} \mathbf{I}(z) \\ \mathbf{C}_0 \end{bmatrix} = \boldsymbol{F}(z)\mathbf{V}(z). \tag{9.20}$$

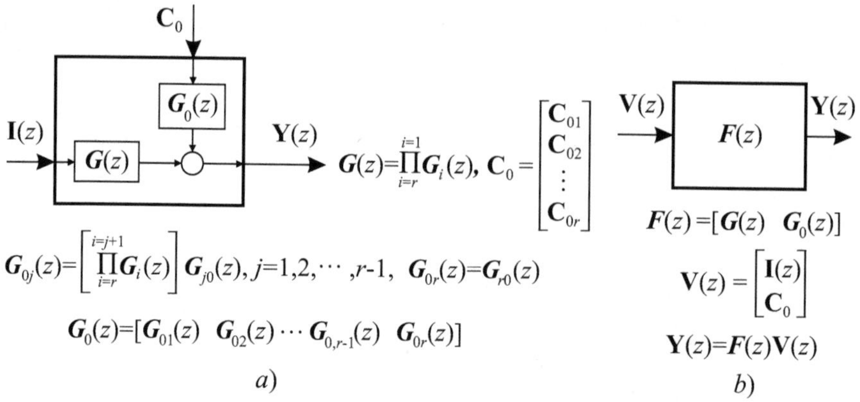

Figure 9.6: The equivalent full block diagram on the left under a) and the full block on the right under b) of the connection of r systems in series.

Proof. We refer to the full block diagram of the connection of r systems in series, Fig. 9.6 and to (9.1) through (9.9) (in Section 9.1),

$$\mathbf{Y}_1(z) = \begin{bmatrix} \boldsymbol{G}_1(z) & \boldsymbol{G}_{10}(z) \end{bmatrix} \begin{bmatrix} \mathbf{I}(z) \\ \mathbf{C}_{01} \end{bmatrix} = \boldsymbol{F}_1(z)\mathbf{V}_1(z),$$

$$\mathbf{V}_1(z) = \begin{bmatrix} \mathbf{I}(z) \\ \mathbf{C}_{01} \end{bmatrix}, \ \boldsymbol{F}_1(z) = \begin{bmatrix} \boldsymbol{G}_1(z) & \boldsymbol{G}_{10}(z) \end{bmatrix}$$

$$\mathbf{Y}_i(z) = \begin{bmatrix} \boldsymbol{G}_i(z) & \boldsymbol{G}_{i0}(z) \end{bmatrix} \begin{bmatrix} \mathbf{Y}_{i-1}(z) \\ \mathbf{C}_{0i} \end{bmatrix} = \boldsymbol{F}_i(z)\mathbf{V}_i(z),$$

$$\mathbf{V}_i(z) = \begin{bmatrix} \mathbf{Y}_{i-1}(z) \\ \mathbf{C}_{0i} \end{bmatrix}, \ \boldsymbol{F}_i(z) = \begin{bmatrix} \boldsymbol{G}_i(z) & \boldsymbol{G}_{i0}(z) \end{bmatrix},$$

$$i = 2, 3, ..., r-1,$$

$$\mathbf{Y}(z) = \mathbf{Y}_r(z) = \begin{bmatrix} \boldsymbol{G}_r(z) & \boldsymbol{G}_{r0}(z) \end{bmatrix} \begin{bmatrix} \mathbf{Y}_{r-1}(z) \\ \mathbf{C}_{0r} \end{bmatrix} = \boldsymbol{F}_r(z)\mathbf{V}_r(z),$$

$$\boldsymbol{F}_r(z) = \begin{bmatrix} \boldsymbol{G}_r(z) & \boldsymbol{G}_{r0}(z) \end{bmatrix}, \ \mathbf{V}_r(z) = \begin{bmatrix} \mathbf{Y}_{r-1}(z) \\ \mathbf{C}_{0r} \end{bmatrix} \Longrightarrow$$

$$\mathbf{Y}(z) = \boldsymbol{G}_r(z)\mathbf{Y}_{r-1}(z) + \boldsymbol{G}_{r0}(z)\mathbf{C}_{0r} \Longrightarrow$$
$$\mathbf{Y}(z) = \boldsymbol{G}_r(z)\left[\boldsymbol{G}_{r-1}(z)\mathbf{Y}_{r-2}(z) + \boldsymbol{G}_{r-1,0}(z)\mathbf{C}_{0,r-1}\right] + \boldsymbol{G}_{r0}(z)\mathbf{C}_{0r}$$
$$\Longrightarrow$$

$$\mathbf{Y}(z) = \boldsymbol{G}_r(z)\boldsymbol{G}_{r-1}(z)\mathbf{Y}_{r-2}(z) + \boldsymbol{G}_r(z)\boldsymbol{G}_{r-1,0}(z)\mathbf{C}_{0,r-1} + \boldsymbol{G}_{r0}(z)\mathbf{C}_{0r}$$
$$\Longrightarrow$$

$$\mathbf{Y}(z) = \boldsymbol{G}_r(z)\boldsymbol{G}_{r-1}(z)\left[\boldsymbol{G}_{r-2}(z)\mathbf{Y}_{r-3}(z) + \boldsymbol{G}_{r-2,0}(z)\mathbf{C}_{0,r-2}\right] + \\ +\boldsymbol{G}_r(z)\boldsymbol{G}_{r-1,0}(z)\mathbf{C}_{0,r-1} + \boldsymbol{G}_{r0}(z)\mathbf{C}_{0r} \Longrightarrow$$

$$\mathbf{Y}(z) = \boldsymbol{G}_r(z)\boldsymbol{G}_{r-1}(z)\boldsymbol{G}_{r-2}(z)\mathbf{Y}_{r-3}(z) + \boldsymbol{G}_r(z)\boldsymbol{G}_{r-1}(z)\boldsymbol{G}_{r-2,0}(z)\mathbf{C}_{0,r-2} + \\ +\boldsymbol{G}_r(z)\boldsymbol{G}_{r-1,0}(z)\mathbf{C}_{0,r-1} + \boldsymbol{G}_{r0}(z)\mathbf{C}_{0r}.$$

By continuing this calculation, we arrive at

$$\mathbf{Y}(z) = \boldsymbol{G}_r(z)\boldsymbol{G}_{r-1}(z)\boldsymbol{G}_{r-2}(z)\cdots\boldsymbol{G}_2(z)\mathbf{Y}_1(z) + \\ +\boldsymbol{G}_r(z)\boldsymbol{G}_{r-1}(z)\boldsymbol{G}_{r-2}(z)\ldots\boldsymbol{G}_3(z)\,\boldsymbol{G}_{20}(z)\mathbf{C}_{02} + \cdots + \\ +\boldsymbol{G}_r(z)\boldsymbol{G}_{r-1}(z)\boldsymbol{G}_{r-2,0}(z)\mathbf{C}_{0,r-2} + \boldsymbol{G}_r(z)\boldsymbol{G}_{r-1,0}(z)\mathbf{C}_{0,r-1} + \boldsymbol{G}_{r0}(z)\mathbf{C}_{0r}.$$

This, and

$$\mathbf{Y}_1(z) = \boldsymbol{G}_1(z)\mathbf{I}(z) + \boldsymbol{G}_{10}(z)\mathbf{C}_{01}$$

furnish

$$\mathbf{Y}(z) = \boldsymbol{G}_r(z)\boldsymbol{G}_{r-1}(z)\boldsymbol{G}_{r-2}(z)\cdots\boldsymbol{G}_2(z)\boldsymbol{G}_1(z)\mathbf{I}(z) + \\ +\boldsymbol{G}_r(z)\boldsymbol{G}_{r-1}(z)\boldsymbol{G}_{r-2}(z)\cdots\boldsymbol{G}_2(z)\boldsymbol{G}_{10}(z)\mathbf{C}_{01} + \\ +\boldsymbol{G}_r(z)\boldsymbol{G}_{r-1}(z)\boldsymbol{G}_{r-2}(z)\cdots\boldsymbol{G}_3(z)\boldsymbol{G}_{20}(z)\mathbf{C}_{02} + \cdots + \\ +\boldsymbol{G}_r(z)\boldsymbol{G}_{r-1}(z)\boldsymbol{G}_{r-2,0}(z)\mathbf{C}_{0,r-2} + \boldsymbol{G}_r(z)\boldsymbol{G}_{r-1,0}(z)\mathbf{C}_{0,r-1} + \boldsymbol{G}_{r0}(z)\mathbf{C}_{0r}.$$

We can put this in the matrix-vector form,

$$\mathbf{Y}(z) = \underbrace{\left[\boldsymbol{G}_r(z)\boldsymbol{G}_{r-1}(z)\boldsymbol{G}_{r-2}(z)\cdots\boldsymbol{G}_2(z)\boldsymbol{G}_1(z)\right]}_{\boldsymbol{G}(z)}\mathbf{I}(z) +$$

$$+\underbrace{\begin{bmatrix} \left(\boldsymbol{G}_r(z)\boldsymbol{G}_{r-1}(z)\boldsymbol{G}_{r-2}(z)\cdots\boldsymbol{G}_2(z)\boldsymbol{G}_{10}(z)\right)^T \\ \left(\boldsymbol{G}_r(z)\boldsymbol{G}_{r-1}(z)\boldsymbol{G}_{r-2}(z)\ldots\boldsymbol{G}_3(z)\boldsymbol{G}_{20}(z)\right)^T \\ \vdots \\ \left(\boldsymbol{G}_r(z)\boldsymbol{G}_{r-1}(z)\boldsymbol{G}_{r-2,0}(z)\right)^T \\ \left(\boldsymbol{G}_r(z)\boldsymbol{G}_{r-1,0}(z)\right)^T \\ \boldsymbol{G}_{r0}^T(z) \end{bmatrix}^T}_{\boldsymbol{G}_0^T(z)} \underbrace{\begin{bmatrix} \mathbf{C}_{01} \\ \mathbf{C}_{02} \\ \cdots \\ \mathbf{C}_{0,r-2} \\ \mathbf{C}_{0,r-1} \\ \mathbf{C}_{0r} \end{bmatrix}}_{\mathbf{C}_0} \Longrightarrow$$

$$\mathbf{Y}(z) = \begin{bmatrix} \boldsymbol{G}(z) & \boldsymbol{G}_0(z) \end{bmatrix} \begin{bmatrix} \mathbf{I}(z) \\ \mathbf{C}_0 \end{bmatrix} = \boldsymbol{F}(z)\mathbf{V}(z),$$

which yield

$$\boldsymbol{G}(z) = \prod_{i=r}^{i=1} \boldsymbol{G}_i(z),$$

$$\boldsymbol{G}_0(z) =$$

$$= \Bigg[\underbrace{\left(\prod_{i=r}^{i=2} \boldsymbol{G}_i(z)\right) \boldsymbol{G}_{10}(z)}_{\boldsymbol{G}_{01}(z)} \quad \underbrace{\left(\prod_{i=r}^{i=3} \boldsymbol{G}_i(z)\right) \boldsymbol{G}_{20}(z)}_{\boldsymbol{G}_{02}(z)} \quad \cdots$$

$$\cdots \quad \underbrace{\boldsymbol{G}_r(z)\boldsymbol{G}_{r-1,0}(z)}_{\boldsymbol{G}_{0,r-1}(z)} \quad \underbrace{\boldsymbol{G}_{r0}(z)}_{\boldsymbol{G}_{0r}(z)} \Bigg],$$

$$\boldsymbol{F}(z) = \begin{bmatrix} \boldsymbol{G}(z) & \boldsymbol{G}_0(z) \end{bmatrix}, \ \mathbf{C}_0 = \begin{bmatrix} \mathbf{C}_{01} \\ \mathbf{C}_{02} \\ \cdots \\ \mathbf{C}_{0,r-2} \\ \mathbf{C}_{0,r-1} \\ \mathbf{C}_{0r} \end{bmatrix}, \ \mathbf{V}(z) = \begin{bmatrix} \mathbf{I}(z) \\ \mathbf{C}_0 \end{bmatrix}.$$

These equations prove (9.16)-(9.20). ■

9.4 Feedback connection

Fig. 9.7 represents the full block diagram of the feedback connection of *two* systems. The overall system has two input vectors $\mathbf{i}_1(k)$ and $\mathbf{i}_2(k)$, and two output vectors $\mathbf{y}_1(k)$ and $\mathbf{y}_2(t)$. Their $Z-$transforms are $\mathbf{I}_1(z)$, $\mathbf{I}_2(z)$, $\mathbf{Y}_1(z)$ and $\mathbf{Y}_2(z)$, respectively. We wish to find the full transfer function matrix $\boldsymbol{F}(z)$ of the whole connection, its main transfer function submatrices $\boldsymbol{G}_{R1}(z)$, $\boldsymbol{G}_{R10}(z)$, $\boldsymbol{G}_{R2}(z)$, and $\boldsymbol{G}_{R20}(z)$, i.e., $\boldsymbol{G}(z)$, and $\boldsymbol{G}_0(z)$, as well as the corresponding vectors $\mathbf{V}_1(z)$, $\mathbf{V}_2(z)$, $\mathbf{C}_{01}$ and $\mathbf{C}_{02}$.

Theorem 9.3 *The full transfer function matrix $\boldsymbol{F}(z)$ of the feedback connection of two systems, see Fig. 9.7, the resulting IO transfer function matrix $\mathbf{G}(z)$ relative to the overall input vector $\mathbf{i}$, the resulting transfer function matrix $\mathbf{G}_0(z)$ relative to all initial conditions, the vector $\mathbf{C}_0$ of all*

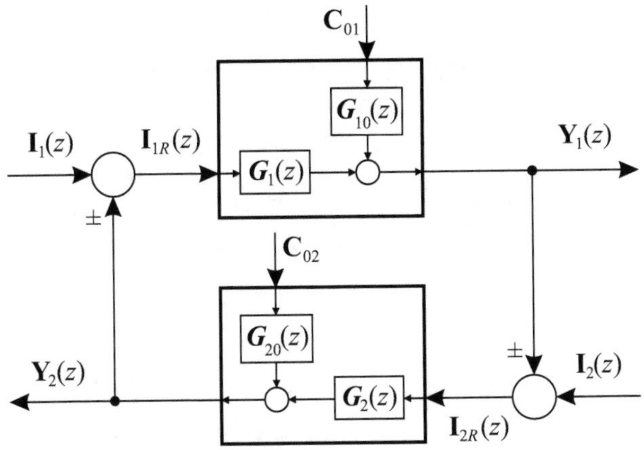

Figure 9.7: The full block diagram of the feedback connection of *two* systems.

initial conditions, and the vector $\mathbf{V}(z)$ *representing the* $Z-$*transform of the overall action vector* $\mathbf{v}(k)$*, obey the following; see Fig. 9.8, Fig. 9.9:*

$$\boldsymbol{F}(z) = \begin{bmatrix} \boldsymbol{F}_1(z) \\ \boldsymbol{F}_2(z) \end{bmatrix} = \begin{bmatrix} \boldsymbol{G}_{R1}(z) & \boldsymbol{G}_{R10}(z) \\ \boldsymbol{G}_{R2}(z) & \boldsymbol{G}_{R20}(z) \end{bmatrix} = \begin{bmatrix} \boldsymbol{G}(z) & \boldsymbol{G}_0(z) \end{bmatrix}, \quad (9.21)$$

$$\boldsymbol{G}_{R1}(z) = (\boldsymbol{I}_{N_1} \mp \boldsymbol{G}_1(z)\boldsymbol{G}_2(z))^{-1} \begin{bmatrix} \boldsymbol{G}_1(z) & \pm \boldsymbol{G}_1(z)\boldsymbol{G}_2(z) \end{bmatrix}, \quad (9.22)$$

$$\boldsymbol{G}_{R10}(z) = (\boldsymbol{I}_{N_1} \mp \boldsymbol{G}_1(z)\boldsymbol{G}_2(z))^{-1} \begin{bmatrix} \boldsymbol{G}_{10}(z) & \pm \boldsymbol{G}_1(z)\boldsymbol{G}_{20}(z) \end{bmatrix}, \quad (9.23)$$

$$\boldsymbol{G}_{R2}(z) = (\boldsymbol{I}_{N_2} \mp \boldsymbol{G}_2(z)\boldsymbol{G}_1(z))^{-1} \begin{bmatrix} \pm \boldsymbol{G}_2(z)\boldsymbol{G}_1(z) & \boldsymbol{G}_2(z) \end{bmatrix}, \quad (9.24)$$

$$\boldsymbol{G}_{R20}(z) = (\boldsymbol{I}_{N_2} \mp \boldsymbol{G}_2(z)\boldsymbol{G}_1(z))^{-1} \begin{bmatrix} \pm \boldsymbol{G}_2(z)\boldsymbol{G}_{10}(z) & \boldsymbol{G}_{20}(z) \end{bmatrix}, \quad (9.25)$$

$$\boldsymbol{G}(z) = \begin{bmatrix} \boldsymbol{G}_{R1}(z) \\ \boldsymbol{G}_{R2}(z) \end{bmatrix}, \quad \boldsymbol{G}_0(z) = \begin{bmatrix} \boldsymbol{G}_{R10}(z) \\ \boldsymbol{G}_{R20}(z) \end{bmatrix}, \quad (9.26)$$

$$\mathbf{C}_0 = \begin{bmatrix} \mathbf{C}_{01} \\ \mathbf{C}_{02} \end{bmatrix}, \quad \mathbf{V}(z) = \begin{bmatrix} \mathbf{I}(z) \\ \mathbf{C}_0 \end{bmatrix}. \quad (9.27)$$

so that

$$\mathbf{Y}(z) = \begin{bmatrix} \boldsymbol{G}(z) & \boldsymbol{G}_0(z) \end{bmatrix} \begin{bmatrix} \mathbf{I}(z) \\ \mathbf{C}_0 \end{bmatrix} = \boldsymbol{F}(z)\mathbf{V}(z). \quad (9.28)$$

Fig. 9.8, Fig. 9.9, Fig. 9.10.

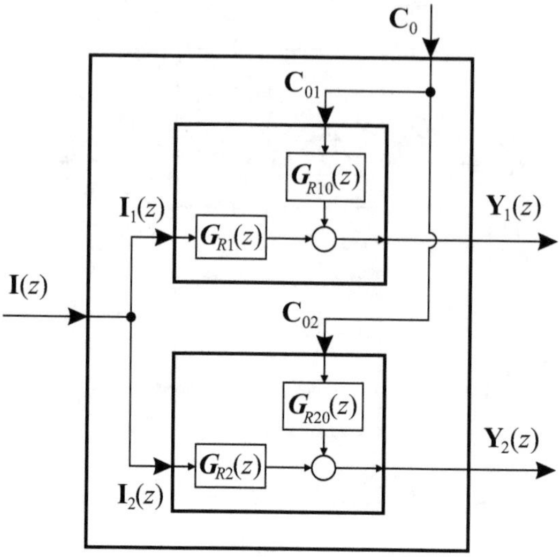

Figure 9.8: The equivalent full block diagram of the feedback connection of *two* systems.

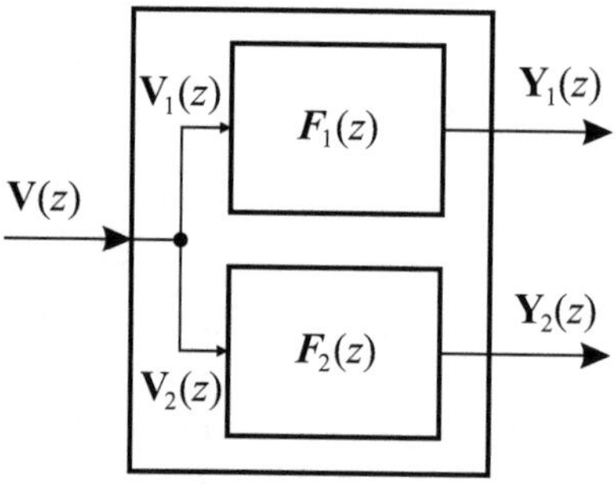

Figure 9.9: The equivalent full block diagram of the feedback connection of *two* systems.

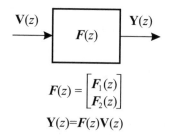

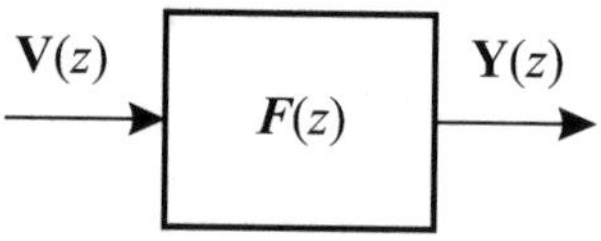

$$\boldsymbol{F}(z) = \begin{bmatrix} \boldsymbol{F}_1(z) \\ \boldsymbol{F}_2(z) \end{bmatrix}$$

$$\mathbf{Y}(z) = \boldsymbol{F}(z)\mathbf{V}(z)$$

Figure 9.10: The full block of the feedback connection of *two* systems.

Proof. We refer to the full block diagram of the feedback connection of *two* systems, Fig. 9.7 and to (9.1) through (9.9) (in Section 9.1),

$$\mathbf{Y}_i(z) = \begin{bmatrix} \boldsymbol{G}_i(z) & \boldsymbol{G}_{i0}(z) \end{bmatrix} \begin{bmatrix} \mathbf{I}_{iR}(z) \\ \mathbf{C}_{0i} \end{bmatrix} = \boldsymbol{F}_i(z)\mathbf{V}_i(z),\ i = 1,2,$$

$$\mathbf{I}_{1R}(z) = \mathbf{I}_1(z) \pm \mathbf{Y}_2(z),\ \mathbf{I}_{2R}(z) = \mathbf{I}_2(z) \pm \mathbf{Y}_1(z)$$

$$\mathbf{V}_i(z) = \begin{bmatrix} \mathbf{I}_{iR}(z) \\ \mathbf{C}_{0i} \end{bmatrix},\ \boldsymbol{F}_i(z) = \begin{bmatrix} \boldsymbol{G}_i(z) & \boldsymbol{G}_{i0}(z) \end{bmatrix},\ i = 1,2,$$

$$\mathbf{Y}_1(z) = \begin{bmatrix} \boldsymbol{G}_1(z) & \boldsymbol{G}_{10}(z) \end{bmatrix} \begin{bmatrix} \mathbf{I}_1(z) \pm \mathbf{Y}_2(z) \\ \mathbf{C}_{01} \end{bmatrix},$$

$$\mathbf{Y}_2(z) = \begin{bmatrix} \boldsymbol{G}_2(z) & \boldsymbol{G}_{20}(z) \end{bmatrix} \begin{bmatrix} \mathbf{I}_2(z) \pm \mathbf{Y}_1(z) \\ \mathbf{C}_{02} \end{bmatrix},$$

$$\Longrightarrow$$

$$\begin{aligned} \mathbf{Y}_1(z) &= \boldsymbol{G}_1(z)\left[\mathbf{I}_1(z) \pm \mathbf{Y}_2(z)\right] + \boldsymbol{G}_{10}(z)\mathbf{C}_{01} = \\ &= \boldsymbol{G}_1(z)\mathbf{I}_1(z) \pm \boldsymbol{G}_1(z)\mathbf{Y}_2(z) + \boldsymbol{G}_{10}(z)\mathbf{C}_{01}, \end{aligned} \tag{9.29}$$

$$\begin{aligned} \mathbf{Y}_2(z) &= \boldsymbol{G}_2(z)\left[\mathbf{I}_2(z) \pm \mathbf{Y}_1(z)\right] + \boldsymbol{G}_{20}(z)\mathbf{C}_{02} = \\ &= \boldsymbol{G}_2(z)\mathbf{I}_2(z) \pm \boldsymbol{G}_2(z)\mathbf{Y}_1(z) + \boldsymbol{G}_{20}(z)\mathbf{C}_{02} \end{aligned} \tag{9.30}$$

$$\Longrightarrow$$

$$\mathbf{Y}_1(z) = \boldsymbol{G}_1(z)\mathbf{I}_1(z) \pm$$
$$\pm \boldsymbol{G}_1(z)\left[\boldsymbol{G}_2(z)\mathbf{I}_2(z) \pm \boldsymbol{G}_2(z)\mathbf{Y}_1(z) + \boldsymbol{G}_{20}(z)\mathbf{C}_{02}\right] + \boldsymbol{G}_{10}(z)\mathbf{C}_{01} \Longrightarrow$$

$$\mathbf{Y}_1(z) = \boldsymbol{G}_1(z)\mathbf{I}_1(z) \pm \boldsymbol{G}_1(z)\boldsymbol{G}_2(z)\mathbf{I}_2(z) \pm \boldsymbol{G}_1(z)\boldsymbol{G}_2(z)\mathbf{Y}_1(z) \pm$$
$$\pm \boldsymbol{G}_1(z)\boldsymbol{G}_{20}(z)\mathbf{C}_{02} + \boldsymbol{G}_{10}(z)\mathbf{C}_{01}$$

$$\Longrightarrow$$

$$\left(\boldsymbol{I}_{N_1} \mp \boldsymbol{G}_1(z)\boldsymbol{G}_2(z)\right)\mathbf{Y}_1(z) =$$
$$= \boldsymbol{G}_1(z)\mathbf{I}_1(z) \pm \boldsymbol{G}_1(z)\boldsymbol{G}_2(z)\mathbf{I}_2(z) \pm \boldsymbol{G}_1(z)\boldsymbol{G}_{20}(z)\mathbf{C}_{02} + \boldsymbol{G}_{10}(z)\mathbf{C}_{01}$$

$$\Longrightarrow$$

$$\mathbf{Y}_1(z) = \left(\boldsymbol{I}_{N_1} \mp \boldsymbol{G}_1(z)\boldsymbol{G}_2(z)\right)^{-1}\left[\boldsymbol{G}_1(z)\mathbf{I}_1(z) \pm \boldsymbol{G}_1(z)\boldsymbol{G}_2(z)\mathbf{I}_2(z) \pm\right.$$
$$\left.\pm \boldsymbol{G}_1(z)\boldsymbol{G}_{20}(z)\mathbf{C}_{02} + \boldsymbol{G}_{10}(z)\mathbf{C}_{01}\right]$$

$$\Longrightarrow$$

$$\mathbf{Y}_1(z) = \underbrace{(\boldsymbol{I}_{N_1} \mp \boldsymbol{G}_1(z)\boldsymbol{G}_2(z))^{-1} \left[\begin{array}{cc} \boldsymbol{G}_1(z) & \pm \boldsymbol{G}_1(z)\boldsymbol{G}_2(z) \end{array} \right]}_{\boldsymbol{G}_{R1}(z)} \underbrace{\left[\begin{array}{c} \mathbf{I}_1(z) \\ \mathbf{I}_2(z) \end{array} \right]}_{\mathbf{I}(z)} +$$

$$+ \underbrace{(\boldsymbol{I}_{N_1} \mp \boldsymbol{G}_1(z)\boldsymbol{G}_2(z))^{-1} \left[\begin{array}{cc} \boldsymbol{G}_{10}(z) & \pm \boldsymbol{G}_1(z)\boldsymbol{G}_{20}(z) \end{array} \right]}_{\boldsymbol{G}_{R10}(z)} \underbrace{\left[\begin{array}{c} \mathbf{C}_{01} \\ \mathbf{C}_{02} \end{array} \right]}_{\mathbf{C}_0} \Longrightarrow$$

$$\boldsymbol{G}_{R1}(z) = (\boldsymbol{I}_{N_1} \mp \boldsymbol{G}_1(z)\boldsymbol{G}_2(z))^{-1} \left[\begin{array}{cc} \boldsymbol{G}_1(z) & \pm \boldsymbol{G}_1(z)\boldsymbol{G}_2(z) \end{array} \right],$$

$$\boldsymbol{G}_{R10}(z) = (\boldsymbol{I}_{N_1} \mp \boldsymbol{G}_1(z)\boldsymbol{G}_2(z))^{-1} \left[\begin{array}{cc} \boldsymbol{G}_{10}(z) & \pm \boldsymbol{G}_1(z)\boldsymbol{G}_{20}(z) \end{array} \right],$$

$$\mathbf{I}(z) = \left[\begin{array}{c} \mathbf{I}_1(z) \\ \mathbf{I}_2(z) \end{array} \right], \ \mathbf{C}_0 = \left[\begin{array}{c} \mathbf{C}_{01} \\ \mathbf{C}_{02} \end{array} \right],$$

and

$$\mathbf{Y}_1(z) = \underbrace{\left[\begin{array}{cc} \boldsymbol{G}_{R1}(z) & \boldsymbol{G}_{R10}(z) \end{array} \right]}_{\boldsymbol{F}_1(z)} \underbrace{\left[\begin{array}{c} \mathbf{I}(z) \\ \mathbf{C}_0 \end{array} \right]}_{\mathbf{V}(z)} \Longrightarrow$$

$$\boldsymbol{F}_1(z) = \left[\begin{array}{cc} \boldsymbol{G}_{R1}(z) & \boldsymbol{G}_{R10}(z) \end{array} \right], \ \mathbf{V}(z) = \left[\begin{array}{c} \mathbf{I}(z) \\ \mathbf{C}_0 \end{array} \right],$$

so that

$$\mathbf{Y}_1(z) = \boldsymbol{F}_1(z)\mathbf{V}(z).$$

Analogously, (9.29) and (9.30) yield

$$\mathbf{Y}_2(z) = \boldsymbol{G}_2(z)\mathbf{I}_2(z) \pm \boldsymbol{G}_2(z)\left[\boldsymbol{G}_1(z)\mathbf{I}_1(z) \pm \boldsymbol{G}_1(z)\mathbf{Y}_2(z) + \boldsymbol{G}_{10}(z)\mathbf{C}_{01}\right] +$$
$$+ \boldsymbol{G}_{20}(z)\mathbf{C}_{02}$$
$$\Longrightarrow$$

$$\mathbf{Y}_2(z) = \boldsymbol{G}_2(z)\mathbf{I}_2(z) \pm \boldsymbol{G}_2(z)\boldsymbol{G}_1(z)\mathbf{I}_1(z) \pm \boldsymbol{G}_2(z)\boldsymbol{G}_1(z)\mathbf{Y}_2(z) \pm$$
$$\pm \boldsymbol{G}_2(z)\boldsymbol{G}_{10}(z)\mathbf{C}_{01} + \boldsymbol{G}_{20}(z)\mathbf{C}_{02}$$
$$\Longrightarrow$$

$$(\boldsymbol{I}_{N_2} \mp \boldsymbol{G}_2(z)\boldsymbol{G}_1(z))\,\mathbf{Y}_2(z) =$$
$$= \pm \boldsymbol{G}_2(z)\boldsymbol{G}_1(z)\mathbf{I}_1(z) + \boldsymbol{G}_2(z)\mathbf{I}_2(z) \pm \boldsymbol{G}_2(z)\boldsymbol{G}_{10}(z)\mathbf{C}_{01} + \boldsymbol{G}_{20}(z)\mathbf{C}_{02}$$
$$\Longrightarrow$$

$$\mathbf{Y}_2(z) = (\boldsymbol{I}_{N_2} \mp \boldsymbol{G}_2(z)\boldsymbol{G}_1(z))^{-1} [\pm \boldsymbol{G}_2(z)\boldsymbol{G}_1(z)\mathbf{I}_1(z) + \boldsymbol{G}_2(z)\mathbf{I}_2(z) \pm \\ \pm \boldsymbol{G}_2(z)\boldsymbol{G}_{10}(z)\mathbf{C}_{01} + \boldsymbol{G}_{20}(z)\mathbf{C}_{02}] \\ \Longrightarrow$$

$$\mathbf{Y}_2(z) = \underbrace{(\boldsymbol{I}_{N_2} \mp \boldsymbol{G}_2(z)\boldsymbol{G}_1(z))^{-1} \left[\pm \boldsymbol{G}_2(z)\boldsymbol{G}_1(z) \quad \boldsymbol{G}_2(z) \right]}_{\boldsymbol{G}_{R2}(z)} \underbrace{\begin{bmatrix} \mathbf{I}_1(z) \\ \mathbf{I}_2(z) \end{bmatrix}}_{\mathbf{I}(z)} +$$

$$+ \underbrace{(\boldsymbol{I}_{N_2} \mp \boldsymbol{G}_2(z)\boldsymbol{G}_1(z))^{-1} \left[\pm \boldsymbol{G}_2(z)\boldsymbol{G}_{10}(z) \quad \boldsymbol{G}_{20}(z) \right]}_{\boldsymbol{G}_{R20}(z)} \underbrace{\begin{bmatrix} \mathbf{C}_{01} \\ \mathbf{C}_{02} \end{bmatrix}}_{\mathbf{C}_0} \Longrightarrow$$

$$\boldsymbol{G}_{R2}(z) = (\boldsymbol{I}_{N_2} \mp \boldsymbol{G}_2(z)\boldsymbol{G}_1(z))^{-1} \left[\pm \boldsymbol{G}_2(z)\boldsymbol{G}_1(z) \quad \boldsymbol{G}_2(z) \right],$$

$$\boldsymbol{G}_{R20}(z) = (\boldsymbol{I}_{N_2} \mp \boldsymbol{G}_2(z)\boldsymbol{G}_1(z))^{-1} \left[\pm \boldsymbol{G}_2(z)\boldsymbol{G}_{10}(z) \quad \boldsymbol{G}_{20}(z) \right],$$

and

$$\mathbf{Y}_2(z) = \underbrace{\left[\boldsymbol{G}_{R2}(z) \quad \boldsymbol{G}_{R20}(z) \right]}_{\boldsymbol{F}_2(z)} \underbrace{\begin{bmatrix} \mathbf{I}(z) \\ \mathbf{C}_0 \end{bmatrix}}_{\mathbf{V}(z)} \Longrightarrow$$

$$\mathbf{Y}_2(z) = \boldsymbol{F}_2(z)\mathbf{V}(z).$$

By continuing this calculation, we arrive at

$$\mathbf{Y}(z) = \begin{bmatrix} \mathbf{Y}_1(z) \\ \mathbf{Y}_2(z) \end{bmatrix} = \underbrace{\begin{bmatrix} \boldsymbol{F}_1(z) \\ \boldsymbol{F}_2(z) \end{bmatrix}}_{\boldsymbol{F}(z)} \mathbf{V}(z) = \boldsymbol{F}(z)\mathbf{V}(z) \Longrightarrow$$

$$\boldsymbol{F}(z) = \begin{bmatrix} \boldsymbol{F}_1(z) \\ \boldsymbol{F}_2(z) \end{bmatrix} = \begin{bmatrix} \boldsymbol{G}_{R1}(z) & \boldsymbol{G}_{R10}(z) \\ \boldsymbol{G}_{R2}(z) & \boldsymbol{G}_{R20}(z) \end{bmatrix}.$$

These equations prove (9.21)-(9.28). ∎

Chapter 10

Physical meaning of $\boldsymbol{F}(z)$

10.1 The *IO* system

Let us explain the physical sense of the system full transfer function matrix $\boldsymbol{F}(z)$ additionally to that expressed in its definition. The vector $\mathbf{1} = \begin{bmatrix} 1 & 1 & \cdots & 1 \end{bmatrix}^T$ is the unit vector of the appropriate dimension.

Definition 10.1 ***The full fundamental matrix function*** *of the IO system (3.56) (in Subsection 3.5.2), is a matrix function* $\boldsymbol{\Psi}_{IO}(\cdot) : \mathbb{Z} \longrightarrow \mathcal{R}^{N\times[(\mu+1)M+\nu N]}$ *if and only if it obeys both conditions (i) and (ii) for any input vector function* $\mathbf{i}(\cdot)$*, and for any initial conditions* $\mathbf{i}_0^{\mu-1}$ *and* $\mathbf{y}_0^{\nu-1}$*,*

(i)

$$\mathbf{y}(k; \mathbf{y}_0^{\nu-1}; \mathbf{i}) = \sum_{j=0}^{j=k} \boldsymbol{\Psi}_{IO}(j) \begin{bmatrix} \mathbf{i}(k-j) \\ \delta_d(k-j)\mathbf{i}_0^{\mu-1} \\ \delta_d(k-j)\mathbf{y}_0^{\nu-1} \end{bmatrix} =$$

$$= \sum_{j=0}^{j=k} \boldsymbol{\Gamma}_{IO}(j)\mathbf{i}(k-j) + \boldsymbol{\Gamma}_{IOi_0}(k)\mathbf{i}_0^{\mu-1} + \boldsymbol{\Gamma}_{IOy_0}(k)\mathbf{y}_0^{\nu-1}, \qquad (10.1)$$

$$\boldsymbol{\Psi}_{IO}(k) = \begin{bmatrix} \boldsymbol{\Gamma}_{IO}(k) & \boldsymbol{\Gamma}_{IOi_0}(k) & \boldsymbol{\Gamma}_{IOy_0}(k) \end{bmatrix},$$
$$\boldsymbol{\Gamma}_{IO}(k) \in \mathcal{R}^{N\times M},\ \boldsymbol{\Gamma}_{IOi_0}(k) \in \mathcal{R}^{N\times \mu M},\ \boldsymbol{\Gamma}_{IOy_0}(k) \in \mathcal{R}^{N\times \nu N}, \qquad (10.2)$$

(ii)

$$\boldsymbol{\Gamma}_{IOi_0}(0) = \begin{bmatrix} \boldsymbol{\Gamma}_{IOi_01}(0) & \boldsymbol{O}_{N,(\mu-1)M} \end{bmatrix} \text{ where}$$
$$\boldsymbol{\Gamma}_{IOi_01}(0) = -\boldsymbol{\Gamma}_{IO}(0)$$
$$\boldsymbol{\Gamma}_{IOy_0}(0) \equiv \begin{bmatrix} \boldsymbol{I}_N & \boldsymbol{O}_{N,(\nu-1)N} \end{bmatrix}. \qquad (10.3)$$

Note 10.1 *Equations (10.1) and (10.2) under (i) of the preceding Definition 10.1 together with the properties of $\delta_d(\cdot)$ (in Section B.2) yield*

$$\mathbf{y}(k;\mathbf{y}_0^{\nu-1};\mathbf{i}) =$$

$$= \left(\sum_{j=0}^{j=k} \boldsymbol{\Gamma}_{IO}(j)\mathbf{i}(k-j)\right) + \boldsymbol{\Gamma}_{IOi_0}(k)\mathbf{i}_0^{\mu-1} + \boldsymbol{\Gamma}_{IOy_0}(k)\mathbf{y}_0^{\nu-1},\ k \in \mathcal{N}_0. \quad (10.4)$$

Theorem 10.1 *(i) The inverse of the Z−transform of the IO system (3.56) full transfer function matrix $\boldsymbol{F}_{IO}(z)$ is its full fundamental matrix function $\boldsymbol{\Psi}_{IO}(\cdot)$,*

$$\mathcal{Z}^{-1}\{\boldsymbol{F}_{IO}(z)\} = \boldsymbol{\Psi}_{IO}(k). \quad (10.5)$$

(ii) The Z−transform of the IO system (3.56) full fundamental matrix $\boldsymbol{\Psi}_{IO}(k)$ is the full transfer function matrix $\boldsymbol{F}_{IO}(z)$

$$\mathcal{Z}\{\boldsymbol{\Psi}_{IO}(k)\} = \boldsymbol{F}_{IO}(z). \quad (10.6)$$

(iii) The inverse Z−transforms of $\boldsymbol{G}_{IO}(z)$, $\boldsymbol{G}_{IOi_0}(z)$ and $\boldsymbol{G}_{IOy_0}(z)$, are the submatrices $\boldsymbol{\Gamma}_{IO}(k)$, $\boldsymbol{\Gamma}_{IOi_0}(k)$ and $\boldsymbol{\Gamma}_{IOy_0}(k)$, respectively,

$$\boldsymbol{\Gamma}_{IO}(k) = \mathcal{Z}^{-1}\{\boldsymbol{G}_{IO}(z)\} = \mathcal{Z}^{-1}\left\{\boldsymbol{\Phi}_{IO}(z)\left[\boldsymbol{B}^{(\mu)}\boldsymbol{S}_M^{(\mu)}(z)\right]\right\}, \quad (10.7a)$$

$$\boldsymbol{\Gamma}_{IOi_0}(k) = \mathcal{Z}^{-1}\{\boldsymbol{G}_{IOi_0}(z)\} =$$
$$= -\mathcal{Z}^{-1}\left\{\boldsymbol{\Phi}_{IO}(z)\left[\boldsymbol{B}^{(\mu)}\boldsymbol{Z}_M^{(\mu)}(z)\right]\right\}, \quad (10.7b)$$

$$\boldsymbol{\Gamma}_{IOy_0}(k) = \mathcal{Z}^{-1}\{\boldsymbol{G}_{IOy_0}(z)\} =$$
$$\mathcal{Z}^{-1}\left\{\boldsymbol{\Phi}_{IO}(z)\left[\boldsymbol{A}^{(\nu)}\boldsymbol{Z}_N^{(\nu)}(z)\right]\right\}, \quad (10.7c)$$

with $\boldsymbol{\Phi}_{IO}(z)$ being the Z−transform of the IO ***system fundamental matrix*** *$\boldsymbol{\Phi}_{IO}(k)$,*

$$\boldsymbol{\Phi}_{IO}(z) = \mathcal{Z}\{\boldsymbol{\Phi}_{IO}(k)\},\ \boldsymbol{\Phi}_{IO}(k) = \mathcal{Z}^{-1}\{\boldsymbol{\Phi}_{IO}(z)\}, \quad (10.8)$$

and

$$\boldsymbol{\Phi}_{IO}(z) = \left(\boldsymbol{A}^{(\nu)}\boldsymbol{S}_N^{(\nu)}(z)\right)^{-1},$$
$$\boldsymbol{\Phi}_{IO}(k) = \mathcal{Z}^{-1}\left\{\left(\boldsymbol{A}^{(\nu)}\boldsymbol{S}_N^{(\nu)}(z)\right)^{-1}\right\}. \quad (10.9)$$

(*iv*) *The link between the IO system (3.56) full fundamental matrix* $\mathbf{\Psi}_{IO}(k)$ *and its fundamental matrix* $\mathbf{\Phi}_{IO}(k)$ *reads:*

$$\mathbf{\Psi}_{IO}(k) = $$
$$= \mathcal{Z}^{-1}\left\{\mathbf{\Phi}_{IO}(z)\left[\begin{array}{ccc} \boldsymbol{B}^{(\mu)}\boldsymbol{S}_M^{(\mu)}(z) & -\boldsymbol{B}^{(\mu)}\boldsymbol{Z}_M^{(\mu)}(z) & \boldsymbol{A}^{(\nu)}\boldsymbol{Z}_N^{(\nu)}(z) \end{array}\right]\right\}, \tag{10.10a}$$

$$\mathcal{Z}\left\{\mathbf{\Psi}_{IO}(k)\right\} = $$
$$= \mathbf{\Phi}_{IO}(z)\left[\begin{array}{ccc} \boldsymbol{B}^{(\mu)}\boldsymbol{S}_M^{(\mu)}(z) & -\boldsymbol{B}^{(\mu)}\boldsymbol{Z}_M^{(\mu)}(z) & \boldsymbol{A}^{(\nu)}\boldsymbol{Z}_N^{(\nu)}(z) \end{array}\right]. \tag{10.10b}$$

Proof. (*i*) The application of the inverse $Z-$transform to $\mathbf{Y}(z)$ enables us to find the inverse $Z-$transform of (7.11a) (in Section 7.2):

$$\mathbf{y}(k;\mathbf{y}_0^{\nu-1};\mathbf{i}) = \mathcal{Z}^{-1}\left\{\mathbf{Y}(z)\right\} = \mathcal{Z}^{-1}\left\{\boldsymbol{F}_{IO}(z)\left[\begin{array}{c} \mathbf{I}(z) \\ \mathbf{i}_0^{\mu-1} \\ \mathbf{y}_0^{\nu-1} \end{array}\right]\right\}. \tag{10.11}$$

We introduce now matrix functions defined by

$$\Xi(k) = \left[\begin{array}{ccc} \Xi_I(k) & \Xi_{i_0}(k) & \Xi_{y_0}(k) \end{array}\right] = \mathcal{Z}^{-1}\left\{\boldsymbol{F}_{IO}(z)\right\}, \tag{10.12}$$

$$\left[\begin{array}{ccc} \Xi_I(k) & \Xi_{i_0}(k) & \Xi_{y_0}(k) \end{array}\right] = $$
$$= \mathcal{Z}^{-1}\left\{\left[\begin{array}{ccc} \boldsymbol{G}_{IO}(z) & \boldsymbol{G}_{IOi_0}(z) & \boldsymbol{G}_{IOy_0}(z) \end{array}\right]\right\}, \tag{10.13a}$$
$$\mathcal{Z}\left\{\left[\begin{array}{ccc} \Xi_I(k) & \Xi_{i_0}(k) & \Xi_{y_0}(k) \end{array}\right]\right\} = $$
$$= \left[\begin{array}{ccc} \boldsymbol{G}_{IO}(z) & \boldsymbol{G}_{IOi_0}(z) & \boldsymbol{G}_{IOy_0}(z) \end{array}\right]. \tag{10.13b}$$

They transform the preceding result (10.11) as follows (see Appendix B.2):

$$\mathbf{y}(k;\mathbf{y}_0^{\nu-1};\mathbf{i}) = \mathcal{Z}^{-1}\left\{\left[\begin{array}{ccc} \boldsymbol{G}_{IO}(z) & \boldsymbol{G}_{IOi_0}(z) & \boldsymbol{G}_{IOy_0}(z) \end{array}\right]\left[\begin{array}{c} \mathbf{I}(z) \\ \mathbf{i}_0^{\mu-1} \\ \mathbf{y}_0^{\nu-1} \end{array}\right]\right\} = $$

$$= \mathcal{Z}^{-1}\left\{\boldsymbol{G}_{IO}(z)\mathbf{I}(z) + \boldsymbol{G}_{IOi_0}(z)\mathbf{i}_0^{\mu-1} + \boldsymbol{G}_{IOy_0}(z)\mathbf{y}_0^{\nu-1}\right\} = $$

$$= \mathcal{Z}^{-1}\left\{\boldsymbol{G}_{IO}(z)\mathbf{I}(z)\right\} + \mathcal{Z}^{-1}\left\{\boldsymbol{G}_{IOi_0}(z)\mathbf{i}_0^{\mu-1}\right\} + \mathcal{Z}^{-1}\left\{\boldsymbol{G}_{IOy_0}(z)\mathbf{y}_0^{\nu-1}\right\} = $$
$$= \mathcal{Z}^{-1}\left\{\boldsymbol{G}_{IO}(z)\mathbf{I}(z)\right\} + \mathcal{Z}^{-1}\left\{\boldsymbol{G}_{IOi_0}(z)\left[1 = \mathcal{Z}\left\{\delta_d(k)\right\}\right]\mathbf{i}_0^{\mu-1}\right\} +$$
$$+\mathcal{Z}^{-1}\left\{\boldsymbol{G}_{IOy_0}(z)\left[1 = \mathcal{Z}\left\{\delta_d(k)\right\}\right]\mathbf{y}_0^{\nu-1}\right\} \Longrightarrow$$

$$\mathbf{y}(k;\mathbf{y}_0^{\nu-1};\mathbf{i}) =$$
$$=\sum_{j=0}^{j=k}\Xi_I(j)i(k-j)+\sum_{j=0}^{j=k}\Xi_{i_0}(j)\delta_d(k-j)\,i_0^{\mu-1}+\sum_{j=0}^{j=k}\Xi_{y_0}(j)\delta_d(k-j)\,y_0^{\nu-1}=$$
$$=\left[\sum_{j=0}^{j=k}\boldsymbol{\Xi}_I(j)\mathbf{i}(k-j)\right]+\boldsymbol{\Xi}_{i_0}(k)\mathbf{i}_0^{\mu-1}+\boldsymbol{\Xi}_{y_0}(k)\mathbf{y}_0^{\nu-1}.$$

This result and (10.4) imply

$$\boldsymbol{\Xi}(k)=\left[\begin{array}{ccc}\boldsymbol{\Xi}_I(k) & \boldsymbol{\Xi}_{i_0}(k) & \boldsymbol{\Xi}_{y_0}(k)\end{array}\right]=$$
$$=\boldsymbol{\Psi}_{IO}(k)=\left[\begin{array}{ccc}\boldsymbol{\Gamma}_{IO}(k) & \boldsymbol{\Gamma}_{IOi_0}(k) & \boldsymbol{\Gamma}_{IOy_0}(k)\end{array}\right],$$
$$\boldsymbol{\Psi}_{IO}(k)=\mathcal{Z}^{-1}\left\{\boldsymbol{F}_{IO}(z)\right\}=$$
$$=\left[\begin{array}{ccc}\boldsymbol{\Gamma}_{IO}(k) & \boldsymbol{\Gamma}_{IOi_0}(k) & \boldsymbol{\Gamma}_{IOy_0}(k)\end{array}\right].$$

This proves (10.5).

(ii) The $Z-$transform of the preceding equations verifies (10.6).

(iii), (iv) Equations (8.6a)-(8.11), (10.1)-(10.6) prove Equations (10.7a)-(10.10b) and complete the proof. ■

The preceding theorem reflects a physical meaning of the full transfer function matrix $\boldsymbol{F}_{IO}(z)$ of the IO system (3.56), that is, that $\boldsymbol{F}_{IO}(z)$ is the $Z-$transform of the IO system full fundamental matrix $\boldsymbol{\Psi}_{IO}(k)$.

Example 10.1 *Let us continue to consider Example 8.1 (in Section 8.1). It starts with the second-order $SISO$ IO system described by*

$$E^2y(k)-E^1y(k)-0.75E^0y(k)=E^2i(k)-7.5E^1i(k)+9E^0i(k).$$

The system full transfer function matrix was determined to be

$$F_{IO}(z)=$$
$$=\left[\underbrace{\frac{z^2-7.5z+9}{z^2-z-0.75}}_{G_{IO}(z)}\quad\underbrace{\frac{-z(z-7.5)}{z^2-z-0.75}\quad\frac{-z}{z^2-z-0.75}}_{G_{IOi_0}(z)}\quad\underbrace{\frac{z(z-1)}{z^2-z-0.75}\quad\frac{z}{z^2-z-0.75}}_{G_{IOy_0}(z)}\right],$$

i.e.,

$$\boldsymbol{F}_{IO}(z) = \left[\begin{array}{ccc} G_{IO}(z) & \boldsymbol{G}_{IOi_0}(z) & \boldsymbol{G}_{IOy_0}(z) \end{array}\right] =$$
$$= \left[\begin{array}{c} \underbrace{\frac{(z-1.5)(z-6)}{(z-1.5)(z+0.5)}}_{G_{IO}(z)} \\ \underbrace{\left[\begin{array}{c} \frac{-z(z-7.5)}{(z-1.5)(z+0.5)} \\ \frac{-z}{(z-1.5)(z+0.5)} \end{array}\right]}_{\boldsymbol{G}^T_{IOi_0}(z)} \\ \underbrace{\left[\begin{array}{c} \frac{z(z-1)}{(z-1.5)(z+0.5)} \\ \frac{z}{(z-1.5)(z+0.5)} \end{array}\right]}_{\boldsymbol{G}^T_{IOy_0}(z)} \end{array}\right]^T.$$

This shows that $\boldsymbol{F}_{IO}(\cdot)$ is only proper, not strictly proper, real rational matrix function. Its inverse $Z-$transform $\mathcal{Z}^{-1}\{\boldsymbol{F}_{IO}(z)\}$ reads:

$$\mathcal{Z}^{-1}\{\boldsymbol{F}_{IO}(z)\} = \mathcal{Z}^{-1}\left\{\left[\begin{array}{ccc} G_{IO}(z) & \boldsymbol{G}_{IOi_0}(z) & \boldsymbol{G}_{IOy_0}(z) \end{array}\right]\right\} =$$
$$= \left[\begin{array}{c} \underbrace{\mathcal{Z}^{-1}\left\{\frac{(z-1.5)(z-6)}{(z-1.5)(z+0.5)}\right\}}_{\mathcal{Z}^{-1}\{G_{IO}(z)\}} \\ \underbrace{\left[\begin{array}{c} \mathcal{Z}^{-1}\left\{\frac{-z(z-7.5)}{(z-1.5)(z+0.5)}\right\} \\ \mathcal{Z}^{-1}\left\{\frac{-z}{(z-1.5)(z+0.5)}\right\} \end{array}\right]}_{\mathcal{Z}^{-1}\left\{\boldsymbol{G}^T_{IOi_0}(z)\right\}} \\ \underbrace{\left[\begin{array}{c} \mathcal{Z}^{-1}\left\{\frac{z(z-1)}{(z-1.5)(z+0.5)}\right\} \\ \mathcal{Z}^{-1}\left\{\frac{z}{(z-1.5)(z+0.5)}\right\} \end{array}\right]}_{\mathcal{Z}^{-1}\left\{\boldsymbol{G}^T_{IOy_0}(z)\right\}} \end{array}\right]^T.$$

This determines the full fundamental matrix function $\mathbf{\Psi}_{IO}(\cdot)$ of the system by referring to (10.4):

$$\mathbf{\Psi}_{IO}(k) = \begin{bmatrix} \Gamma_{IO}(k) & \mathbf{\Gamma}_{IOi_0}(k) & \mathbf{\Gamma}_{IOy_0}(k) \end{bmatrix} =$$

$$= \begin{bmatrix} \underbrace{\delta_d(k) - 6.5(-0.5)^{k-1} h_d(k-1)}_{\Gamma_{IO}(k)} \\ \underbrace{\begin{bmatrix} -\delta_d(k) + \left[4.5(1.5)^{k-1} + 2(-0.5)^{k-1}\right] h_d(k-1) \\ \left[-0.5(1.5)^k + 0.5(-0.5)^k\right] h_d(k) \end{bmatrix}}_{\mathbf{\Gamma}^T_{IOi_0}(k)} \\ \underbrace{\begin{bmatrix} \delta_d(k) + \left[0.375(1.5)^{k-1} - 0.375(-0.5)^{k-1}\right] h_d(k-1) \\ \left[0.5(1.5)^k - 0.5(-0.5)^k\right] h_d(k) \end{bmatrix}}_{\mathbf{\Gamma}^T_{IOy_0}(k)} \end{bmatrix}^T.$$

We use

$$\Gamma_{IO}(0) = \left[\delta_d(k) - 6.5(-0.5)^{k-1} h_d(k-1)\right]_{k=0} = 1,$$

$$\mathbf{\Gamma}_{IOi_0}(0) = \begin{bmatrix} -\delta_d(k) + \left[4.5(1.5)^{k-1} + 2(-0.5)^{k-1}\right] h_d(k-1) \\ \left[-0.5(1.5)^k + 0.5(-0.5)^k\right] h_d(k) \end{bmatrix}^T_{k=0} =$$

$$= \begin{bmatrix} -1 & 0 \end{bmatrix},$$

$$\mathbf{\Gamma}_{IOy_0}(0) =$$

$$= \begin{bmatrix} \delta_d(k) + \left[0.375(1.5)^{k-1} - 0.375(-0.5)^{k-1}\right] h_d(k-1) \\ \left[0.5(1.5)^k - 0.5(-0.5)^k\right] h_d(k) \end{bmatrix}^T_{k=0} =$$

$$= \begin{bmatrix} 1 & 0 \end{bmatrix} \Longrightarrow$$

$$y(0;\mathbf{y}_0^1;i) = \left(\sum_{j=0}^{j=k=0} \Gamma_{IO}(j)i(k-j)\right) + \mathbf{\Gamma}_{IOi_0}(0)\mathbf{i}_0^1 + \mathbf{\Gamma}_{IOy_0}(0)\mathbf{y}_0^1 =$$

$$= \Gamma_{IO}(0)i_0 + \mathbf{\Gamma}_{IOi_0}(0)\mathbf{i}_0^1 + \mathbf{\Gamma}_{IOy_0}(0)\mathbf{y}_0^1 =$$

$$= i_0 + \begin{bmatrix} -1 & 0 \end{bmatrix} \mathbf{i}_0^1 + \begin{bmatrix} 1 & 0 \end{bmatrix} \mathbf{y}_0^1 = i_0 - i_0 + y_0 = y_0.$$

$$\Gamma_{IO}(k) = \delta_d(k) - 6.5(-0.5)^{k-1} h_d(k-1),$$

$$\Gamma_{IOi_01}(k) = -\delta_d(k) + \left[4.5(1.5)^{k-1} + 2(-0.5)^{k-1}\right] h_d(k-1), \text{ and}$$

$$\Gamma_{IOy_{0}1}(k)=\delta_{d}\left(k\right)+\left[0.375\left(1.5\right)^{k-1}-0.375\left(-0.5\right)^{k-1}\right]h_{d}\left(k-1\right),$$

contain discrete unit impulses $\delta_d(k)$ because they are all the inverse $Z-$transform of the proper ratios

$$\frac{(z-1.5)(z-6)}{(z-1.5)(z+0.5)}=\frac{z-6}{z+0.5}=\frac{z+0.5-6.5}{z+0.5}=1-\frac{6.5}{z+0.5},$$

$$\frac{-\left(z^{2}-7.5z\right)}{z^{2}-z-0.75}=\frac{-\left(z^{2}-z-0.75+0.75-6.5z\right)}{z^{2}-z-0.75}=$$
$$=-1+\frac{6.5z-0.75}{(z-1.5)(z+0.5)}=-1+\frac{4.5}{z-1.5}+\frac{2}{z+0.5},$$

$$\frac{z^{2}-z}{z^{2}-z-0.75}=\frac{z^{2}-z-0.75+0.75}{z^{2}-z-0.75}=$$
$$=1+\frac{0.75}{(z-1.5)(z+0.5)}=1+\frac{0.375}{z-1.5}-\frac{0.375}{z+0.5},$$

respectively. This shows that $\boldsymbol{F}_{IO}(\cdot)$ is only proper, not strictly proper, real rational matrix function, but $\Gamma_{IOi_{0}2}(k)$ and $\Gamma_{IOy_{0}2}(k)$ do not contain discrete unit impulses. They are exponential functions. Their $Z-$transforms are strictly proper rational functions $G_{IOio2}(z)$ and $G_{IOyo2}(z)$, respectively.

10.2 The *ISO* system

Definition 10.2 ***The** IS **full fundamental matrix function*** *of the ISO system (3.60a) and (3.60b) (in Subsection 3.5.3), is a matrix function $\mathbf{\Psi}_{ISOIS}(\cdot):\mathbb{Z}\longrightarrow\mathcal{R}^{n\times(M+n)}$ if and only if it obeys both conditions (i) and (ii) for any input vector function $\mathbf{i}(\cdot)$, and for any initial state vector $\mathbf{x}_0$:*

(*i*)

$$\mathbf{x}(k;\mathbf{x}_{0};\mathbf{i})=\sum_{j=0}^{j=k}\mathbf{\Psi}_{ISOIS}(j)\left[\begin{array}{c}\mathbf{i}(k-j)\\ \delta_{d}(k-j)\mathbf{x}_{0}\end{array}\right]=$$
$$=\left(\sum_{j=0}^{j=k}\mathbf{\Gamma}_{ISOIS}(j)\mathbf{i}(k-j)\right)+\mathbf{\Gamma}_{ISOSS}(k)\mathbf{x}_{0},$$
$$\mathbf{\Gamma}_{ISOIS}(k)\in\mathcal{R}^{n\times M},\ \mathbf{\Gamma}_{ISOSS}(k)\in\mathcal{R}^{n\times n}, \tag{10.14}$$

(*ii*)

$$\boldsymbol{\Gamma}_{ISOIS}(0) = \boldsymbol{O}_{n,M}, \boldsymbol{\Gamma}_{ISOSS}(0) = \boldsymbol{I}_n. \tag{10.15}$$

The references [2], [6], [14], [41], [81], [89], [122], [108], [101] show that

$$\boldsymbol{\Gamma}_{ISOIS}(k) = \boldsymbol{A}^{k-1}\boldsymbol{B},\ \boldsymbol{\Gamma}_{ISOSS}(k) = \boldsymbol{A}^k = \boldsymbol{\Phi}_{ISOSS}(k), \tag{10.16}$$

with $\boldsymbol{\Phi}_{ISOSS}(k) = \boldsymbol{A}^k$ being **the** *ISO* system **state fundamental matrix**, for short its **fundamental matrix.** We can compute it from **the matrix** $z(z\boldsymbol{I}_n - \boldsymbol{A})^{-1}$, e.g., [108], of the *ISO* system (3.60a) and (3.60b),

$$\boldsymbol{\Gamma}_{ISOSS}(k) = \boldsymbol{A}^k = \boldsymbol{\Phi}_{ISOSS}(k) = \mathcal{Z}^{-1}\left\{z(z\boldsymbol{I}_n - \boldsymbol{A})^{-1}\right\}. \tag{10.17}$$

The vice versa also holds,

$$\boldsymbol{\Phi}_{ISOSS}(z) = z(z\boldsymbol{I}_n - \boldsymbol{A})^{-1} =$$
$$= \mathcal{Z}\left\{\boldsymbol{\Gamma}_{ISOSS}(k)\right\} = \mathcal{Z}\left\{\boldsymbol{A}^k\right\} = \mathcal{Z}\left\{\boldsymbol{\Phi}_{ISOSS}(k)\right\}. \tag{10.18}$$

The transfer function matrix $\boldsymbol{\Phi}_{ISOSS}(z) = \boldsymbol{G}_{ISOSS}(z)$ is simultaneously the matrix $z(z\boldsymbol{I}_n - \boldsymbol{A})^{-1}$. It is **the** *ISO* **system state fundamental transfer function matrix**, for short its **fundamental transfer function matrix.**

Theorem 10.2 (*i*) *The ISO system (3.60a) and (3.60b) IS full fundamental matrix function* $\boldsymbol{\Psi}_{ISOIS}(\cdot)$ *is the inverse Z−transform of the system full IS transfer function matrix* $\boldsymbol{F}_{ISOIS}(z)$,

$$\boldsymbol{\Psi}_{ISOIS}(k) = \mathcal{Z}^{-1}\left\{\boldsymbol{F}_{ISOIS}(z)\right\}. \tag{10.19}$$

(*ii*) *The ISO system (3.60a) and (3.60b) full transfer function matrix* $\boldsymbol{F}_{ISOIS}(z)$ *is the Z−transform of the system full fundamental matrix* $\boldsymbol{\Psi}_{ISOIS}(k)$,

$$\boldsymbol{F}_{ISOIS}(z) = \mathcal{Z}\left\{\boldsymbol{\Psi}_{ISOIS}(k)\right\}. \tag{10.20}$$

Proof. (*i*) The inverse Z−transform of (7.24) (in Section 7.3) reads:

$$\mathbf{x}(k) = \mathcal{Z}^{-1}\left\{\mathbf{X}(z)\right\} = \mathcal{Z}^{-1}\left\{\boldsymbol{F}_{ISOIS}(z)\left[\begin{array}{cc}\mathbf{I}^T(z) & \mathbf{x}_0^T\end{array}\right]^T\right\} =$$

$$= \frac{1}{2\pi j}\oint_G \left\{\boldsymbol{F}_{ISOIS}(z)\left[\begin{array}{c}\mathbf{I}(z)\\ \mathbf{x}_0\end{array}\right]\right\} z^{k-1}dz =$$

$$= \frac{1}{2\pi j}\oint_G \left\{\sum_{j=0}^{j=\infty} \mathcal{Z}^{-1}\left\{\boldsymbol{F}_{ISOIS}(z)\right\} z^{-j}\right\}\left[\begin{array}{c}\mathbf{I}(z)\\ \mathbf{x}_0\end{array}\right] z^{k-1}dz =$$

$$= \sum_{j=0}^{j=\infty} \mathcal{Z}^{-1}\left\{\boldsymbol{F}_{ISOIS}(z)\right\}\left[\frac{1}{2\pi j}\oint_G \left[\begin{array}{c}\mathbf{I}(z)\\ \left[1 = \mathcal{Z}\left\{\delta_d\left(t\right)\right\}\right]\mathbf{x}_0\end{array}\right] z^{(k-j)-1}dz\right] =$$

$$= \sum_{j=0}^{j=\infty} \mathcal{Z}^{-1}\left\{\boldsymbol{F}_{ISOIS}(z)\right\}\left[\begin{array}{c}\mathbf{i}(k-j)\\ \delta_d\left(k-j\right)\mathbf{x}_0\end{array}\right] =$$

$$= \sum_{j=0}^{j=k} \mathcal{Z}^{-1}\left\{\boldsymbol{F}_{ISOIS}(z)\right\}\left[\begin{array}{c}\mathbf{i}(k-j)\\ \delta_d\left(k-j\right)\mathbf{x}_0\end{array}\right].$$

This and (i) of Definition 10.2 imply $\boldsymbol{\Psi}_{ISOIS}(k) = \mathcal{Z}^{-1}\left\{\boldsymbol{F}_{ISOIS}(z)\right\}$, which is Equation (10.19).

(ii) Equation (10.20) is the $Z-$transform of (10.19). ■

Definition 10.3 ***The ISO system (3.60a) and (3.60b) IO full fundamental matrix function*** *is a matrix function* $\boldsymbol{\Psi}_{ISO}(\cdot) : \mathbb{Z} \longrightarrow \mathcal{R}^{N\times(M+n)}$ *if and only if it obeys both conditions (i) and (ii) for any input vector function* $\mathbf{i}(\cdot)$*, and for any initial state vector* $\mathbf{x}_0$*,*

(i)

$$\mathbf{y}(k;\mathbf{x}_0;\mathbf{i}) = \sum_{j=0}^{j=k} \boldsymbol{\Psi}_{ISO}(j)\left[\begin{array}{c}\mathbf{i}(k-j)\\ \delta_d(k-j)\mathbf{x}_0\end{array}\right] =$$

$$= \left(\sum_{j=0}^{j=k} \boldsymbol{\Gamma}_{ISO}(j)\mathbf{i}(k-j)\right) + \boldsymbol{\Gamma}_{ISOx_0}(k)\mathbf{x}_0, \tag{10.21}$$

(ii)

$$\boldsymbol{\Psi}_{ISO}(k) = \left[\begin{array}{cc}\boldsymbol{\Gamma}_{ISO}(k) & \boldsymbol{\Gamma}_{ISOx_0}(k)\end{array}\right],$$
$$\boldsymbol{\Gamma}_{ISO}(k) \in \mathcal{R}^{N\times M}, \boldsymbol{\Gamma}_{ISOx_0}(k) \in \mathcal{R}^{N\times n}. \tag{10.22}$$

Note 10.2 *The equations of Definition 10.3 and the features of* $\delta_d(\cdot)$ *(in Section B.2) lead to*

$$\mathbf{y}(k;\mathbf{x}_0;\mathbf{i}) = \left(\sum_{j=0}^{j=k} \boldsymbol{\Gamma}_{ISO}(j)\mathbf{i}(k-j)\right) + \boldsymbol{\Gamma}_{ISOx_0}(k)\mathbf{x}_0,\ k \in \mathcal{N}_0. \tag{10.23}$$

Note 10.3 *The following references [2], [6], [14], [41], [81], [89], [122], [108], [101] and (10.23) permit to write:*

$$\boldsymbol{\Gamma}_{ISO}(k) = \boldsymbol{C}\boldsymbol{A}^{k-1}\boldsymbol{B} + \delta_d(k)\boldsymbol{D},\ \boldsymbol{\Gamma}_{ISOx_0}(k) = \boldsymbol{C}\boldsymbol{A}^k. \tag{10.24}$$

Theorem 10.3 *(i) The ISO system (3.60a) and (3.60b) IO full fundamental matrix function $\boldsymbol{\Psi}_{ISO}(\cdot)$ is the inverse of the Z−transform of the system full transfer function matrix $\boldsymbol{F}_{ISO}(z)$,*

$$\boldsymbol{\Psi}_{ISO}(k) = \mathcal{Z}^{-1}\left\{\boldsymbol{F}_{ISO}(z)\right\}. \tag{10.25}$$

(ii) The ISO system (3.60a) and (3.60b) full transfer function matrix $\boldsymbol{F}_{ISO}(z)$ is the Z−transform of the system full fundamental matrix $\boldsymbol{\Psi}_{ISO}(k)$,

$$\boldsymbol{F}_{ISO}(z) = \mathcal{Z}\left\{\boldsymbol{\Psi}_{ISO}(k)\right\}. \tag{10.26}$$

(iii) The ISO system (3.60a) and (3.60b) transfer function matrices $\boldsymbol{G}_{ISO}(z)$ and $\boldsymbol{G}_{ISOx_0}(z)$ are the Z−transforms of the system full fundamental matrix $\boldsymbol{\Psi}_{ISO}(k)$ submatrices $\boldsymbol{\Gamma}_{ISO}(k)$ and $\boldsymbol{\Gamma}_{ISOx_0}(k)$, respectively,

$$\begin{gathered}
\boldsymbol{G}_{ISO}(z) = \mathcal{Z}\left\{\boldsymbol{\Gamma}_{ISO}(k)\right\} = \\
= \boldsymbol{C}(z\boldsymbol{I}_n - \boldsymbol{A})^{-1}\boldsymbol{B} + \boldsymbol{D} = z^{-1}\boldsymbol{C}\boldsymbol{\Phi}_{ISOSS}(z)\boldsymbol{B} + \boldsymbol{D}, \\
\boldsymbol{G}_{ISOx_0}(z) = \mathcal{Z}\left\{\boldsymbol{\Gamma}_{ISOx_0}(k)\right\} = \\
= z\boldsymbol{C}(z\boldsymbol{I}_n - \boldsymbol{A})^{-1} = \boldsymbol{C}\boldsymbol{\Phi}_{ISOSS}(z),
\end{gathered} \tag{10.27}$$

and vice versa,

$$\begin{gathered}
\boldsymbol{\Gamma}_{ISO}(k) = \mathcal{Z}^{-1}\left\{\boldsymbol{G}_{ISO}(z)\right\} = \\
= \boldsymbol{C}\boldsymbol{A}^{k-1}\boldsymbol{B} + \delta_d(k)\boldsymbol{D} = \boldsymbol{C}\boldsymbol{\Phi}_{ISOSS}(k-1)\boldsymbol{B} + \delta_d(k)\boldsymbol{D}, \\
\boldsymbol{\Gamma}_{ISOx_0}(k) = \mathcal{Z}^{-1}\left\{\boldsymbol{G}_{ISOx_0}(z)\right\} = \boldsymbol{C}\boldsymbol{A}^k = \boldsymbol{C}\boldsymbol{\Phi}_{ISOSS}(k).
\end{gathered} \tag{10.28}$$

Proof. *(i)* The inverse Z−transform of (8.32) (in Section 8.2) results into

$$\mathbf{y}(k) = \mathcal{Z}^{-1}\left\{\mathbf{Y}(z)\right\} = \mathcal{Z}^{-1}\left\{\boldsymbol{F}_{ISO}(z)\begin{bmatrix} \mathbf{I}^T(z) & \mathbf{x}_0^T \end{bmatrix}^T\right\} =$$

$$= \frac{1}{2\pi j}\oint_G \mathbf{Y}(z)z^{k-1}dz =$$

$$= \frac{1}{2\pi j}\oint_G \left\{\boldsymbol{F}_{ISO}(z)\begin{bmatrix} \mathbf{I}(z) \\ \mathbf{x}_0 \end{bmatrix}\right\} z^{k-1}dz =$$

$$= \frac{1}{2\pi j} \oint_G \left\{ \sum_{j=0}^{j=\infty} \mathcal{Z}^{-1} \left\{ \boldsymbol{F}_{ISO}(z) \right\} z^{-j} \right\} \begin{bmatrix} \mathbf{I}(z) \\ \mathbf{x}_0 \end{bmatrix} z^{k-1} dz =$$

$$= \sum_{j=0}^{j=\infty} \mathcal{Z}^{-1} \left\{ \boldsymbol{F}_{ISO}(z) \right\} \left[\frac{1}{2\pi j} \oint_G \begin{bmatrix} \mathbf{I}(z) \\ \left[1 = \mathcal{Z} \left\{ \delta_d (t) \right\} \right] \mathbf{x}_0 \end{bmatrix} z^{(k-j)-1} dz \right] =$$

$$= \sum_{j=0}^{j=\infty} \mathcal{Z}^{-1} \left\{ \boldsymbol{F}_{ISO}(z) \right\} \begin{bmatrix} \mathbf{i}(k-j) \\ \delta_d (k-j) \mathbf{x}_0 \end{bmatrix} =$$

$$= \sum_{j=0}^{j=k} \mathcal{Z}^{-1} \left\{ \boldsymbol{F}_{ISO}(z) \right\} \begin{bmatrix} \mathbf{i}(k-j) \\ \delta_d (k-j) \mathbf{x}_0 \end{bmatrix}.$$

This and (i) of Definition 10.3 prove $\boldsymbol{\Psi}_{ISO}(k) = \mathcal{Z}^{-1} \left\{ \boldsymbol{F}_{ISO}(z) \right\}$, which is Equation (10.25).

(ii) Equation (10.26) is the $Z-$transform of (10.25).

(iii) Equations (10.27) and (10.28) result from (10.16)-(10.18), (10.22) and (10.24) ■

Theorem 10.3 expresses a physical meaning of the *IO* full transfer function matrix of the *ISO* system (3.60a) and (3.60b).

Note 10.4 *Equations (3.60b), (10.23) and (10.24), written for* $k = 0$*, show the relationships among* $\mathbf{y}_0 = \mathbf{y}(0)$, $\mathbf{i}_0 = \mathbf{i}(0)$ *and* $\mathbf{x}_0 = \mathbf{x}(0)$,

$$\mathbf{y}_0 = \left(\sum_{j=0}^{j=k=0} \boldsymbol{\Gamma}_{ISO}(j) \mathbf{i}(k-j) \right) + \boldsymbol{\Gamma}_{ISOx_0}(0) \mathbf{x}_0 = \boldsymbol{C} \mathbf{x}_0 + \boldsymbol{D} \mathbf{i}_0 \Longrightarrow$$

$$\boldsymbol{\Gamma}_{ISO}(0) = \boldsymbol{D},\ \boldsymbol{\Gamma}_{ISOx_0}(0) = \boldsymbol{C}. \tag{10.29}$$

Note 10.5 *If we consider a special case, in which* $N = n$ *and* $\det \Gamma_{ISOx_0}(0) = \det \boldsymbol{C} \neq 0$*, then we can to solve (10.29) for* $\mathbf{x}_0$ *in terms of* $\mathbf{y}_0$,

$$\mathbf{x}_0 = \boldsymbol{C}^{-1} \left(\mathbf{y}_0 - \boldsymbol{D} \mathbf{i}_0 \right). \tag{10.30}$$

This equation transforms (8.31) and (8.32) (in Section 8.2), into

$$\mathbf{Y}(z) = \boldsymbol{F}_{ISOsp}(z) \begin{bmatrix} \mathbf{I}^T(z) & \mathbf{i}_0^T & \mathbf{y}_0^T \end{bmatrix}^T,$$

$$\boldsymbol{F}_{ISOsp}(z) =$$

$$= \begin{bmatrix} \boldsymbol{C}(z\boldsymbol{I}_n - \boldsymbol{A})^{-1}\boldsymbol{B} + \boldsymbol{D} & -z(z\boldsymbol{I}_n - \boldsymbol{CAC}^{-1})^{-1}\boldsymbol{D} & z(z\boldsymbol{I}_n - \boldsymbol{CAC}^{-1})^{-1} \end{bmatrix}, \tag{10.31}$$

and

$$\mathbf{Y}(z) = \boldsymbol{F}_{ISOsp}(z)\mathbf{V}_{ISOsp}(z),\ \mathbf{V}_{ISOsp}(z) = \left[\begin{array}{ccc} \mathbf{I}^T(z) & \mathbf{i}_0^T & \mathbf{y}_0^T \end{array}\right]^T. \quad (10.32)$$

Equation (10.31) verifies Equation (8.42) (Theorem 8.3, in Section 8.2).

10.3 The *IIO* system

Definition 10.4 *A matrix function* $\boldsymbol{\Psi}_{IIOIS}(\cdot) : \mathbb{Z} \longrightarrow \mathcal{R}^{\rho\times[(\beta+1)M+\alpha\rho]}$, $\boldsymbol{\Psi}_{IIOIS}(k) = \left[\begin{array}{ccc} \boldsymbol{\Gamma}_{IIOIS}(k) & \boldsymbol{\Gamma}_{IIOi_0IS}(k) & \boldsymbol{\Gamma}_{IIOr_oIS}(k) \end{array}\right]$*, is* ***the full*** IS ***fundamental matrix function of the*** IIO ***system (3.64a) and (3.64b)*** *(in Subsection 3.5.4) if and only if it obeys both (i) and (ii) for an arbitrary input vector function* $\mathbf{i}(\cdot)$*, and for arbitrary initial conditions* $\mathbf{i}_0^{\mu-1}$ *and* $\mathbf{r}_0^{\alpha-1}$*:*

(i)

$$\mathbf{r}(k;\mathbf{r}_0^{\alpha-1};\mathbf{i}) = \sum_{j=0}^{j=k}\boldsymbol{\Psi}_{IIOIS}(j)\left[\begin{array}{c} \mathbf{i}(k-j) \\ \delta_d(k-j)\mathbf{i}_0^{\beta-1} \\ \delta_d(k-j)\mathbf{r}_0^{\alpha-1} \end{array}\right] =$$
$$= \left(\sum_{j=0}^{j=k}\boldsymbol{\Gamma}_{IIOIS}(j)\mathbf{i}(k-j)\right) +$$
$$+\boldsymbol{\Gamma}_{IIOi_0IS}(k)\mathbf{i}_0^{\beta-1} + \boldsymbol{\Gamma}_{IIOr_0IS}(k)\mathbf{r}_0^{\alpha-1}, \quad (10.33)$$

(ii)

$$\boldsymbol{\Gamma}_{IIOi_0IS}(0) = \left[\begin{array}{cc} \boldsymbol{\Gamma}_{IIOi_0IS1}(0) & \boldsymbol{O}_{N,(\mu-1)M} \end{array}\right] \text{ where}$$
$$\boldsymbol{\Gamma}_{IIOi_0IS1}(0) = -\boldsymbol{\Gamma}_{IIOIS}(0),$$
$$\boldsymbol{\Gamma}_{IIOr_0IS}(0) \equiv \left[\begin{array}{cc} \boldsymbol{I}_\rho & \boldsymbol{O}_{\rho,(\alpha-1)\rho} \end{array}\right]. \quad (10.34)$$

Theorem 10.4 *(i) The full* IS *fundamental matrix function* $\boldsymbol{\Psi}_{IIOIS}(\cdot)$ *of the* IIO *system (3.64a) and (3.64b) is the inverse of the* $Z-$*transform of the system full* IS *transfer function matrix* $\boldsymbol{F}_{IIOIS}(z)$*,*

$$\Psi_{IIOIS}(k) = \mathcal{Z}^{-1}\left\{\boldsymbol{F}_{IIOIS}(z)\right\}. \quad (10.35)$$

(ii) The full IS *transfer function matrix* $\boldsymbol{F}_{IIOIS}(z)$ *of the* IIO *system (3.64a) and (3.64b) is the* $Z-$*transform of the system full* IS *fundamental matrix* $\boldsymbol{\Psi}_{IIOIS}(k)$*,*

$$\boldsymbol{F}_{IIOIS}(z) = \mathcal{Z}\left\{\boldsymbol{\Psi}_{IIOIS}(k)\right\}. \quad (10.36)$$

(iii) *The full IS transfer function matrix $\boldsymbol{F}_{IIOIS}(z)$ of the system (3.63a) and (3.63b) i.e., (8.50a) and (8.50b), is determined by (8.60) and (8.61),*

$$\boldsymbol{F}_{IIOIS}(z) = \left(\boldsymbol{\Phi}_{IIOIS}(z)\right)^{-1} \cdot \cdot \left[\begin{array}{ccc} \boldsymbol{P}^{(\beta)} \boldsymbol{S}_M^{(\beta)}(z) & -\boldsymbol{P}^{(\beta)} \boldsymbol{Z}_M^{(\beta)}(z) & \boldsymbol{Q}^{(\alpha)} \boldsymbol{Z}_\rho^{(\alpha)}(z) \end{array} \right] \quad (10.37)$$

$$\boldsymbol{F}_{IIOIS}(z) = \left[\begin{array}{cc} \boldsymbol{G}_{IIOIS}(z) & \boldsymbol{G}_{IIO0IS}(z) \end{array} \right] = = \boldsymbol{F}_{IIOISD}^{-1}(z)\boldsymbol{F}_{IIOISN}(z),$$

$$\boldsymbol{G}_{IIO0IS}(z) = \left[\begin{array}{cc} \boldsymbol{G}_{IIOi_0IS}(z) & \boldsymbol{G}_{IIOr_0IS}(z) \end{array} \right], \quad (10.38)$$

together with (10.39) through (10.42):

– *IRIS transfer function matrix $\boldsymbol{G}_{IIOIS}(z)$,*

$$\boldsymbol{G}_{IIOIS}(z) = \left(\boldsymbol{\Phi}_{IIOIS}(z)\right)^{-1} \boldsymbol{P}^{(\beta)} \boldsymbol{S}_M^{(\beta)}(z) = = \boldsymbol{G}_{IIOISD}^{-1}(z) \boldsymbol{G}_{IIOISN}(z), \quad (10.39)$$

– *IRII transfer function matrix $\boldsymbol{G}_{IIOi_0IS}(z)$,*

$$\boldsymbol{G}_{IIOi_0IS}(z) = -\left(\boldsymbol{\Phi}_{IIOIS}(z)\right)^{-1} \boldsymbol{P}^{(\beta)} \boldsymbol{Z}_M^{(\beta)}(z) = = \boldsymbol{G}_{IIOi_0ISD}^{-1}(z) \boldsymbol{G}_{IIOi_0ISN}(z), \quad (10.40)$$

– *IRIR transfer function matrix $\boldsymbol{G}_{IIOr_0IS}(z)$,*

$$\boldsymbol{G}_{IIOr_0IS}(z) = \left(\boldsymbol{\Phi}_{IIOIS}(z)\right)^{-1} \boldsymbol{Q}^{(\alpha)} \boldsymbol{Z}_\rho^{(\alpha)}(z) = = \boldsymbol{G}_{IIOr_0ISD}^{-1}(z) \boldsymbol{G}_{IIOr_0ISN}(z), \quad (10.41)$$

so that the denominator matrix polynomial $\boldsymbol{F}_{IIOISD}(z)$ and the numerator matrix polynomial $\boldsymbol{F}_{IIOISN}(z)$ of $\boldsymbol{F}_{IIOIS}(z)$,

$$\boldsymbol{F}_{IIOIS}(z) = \boldsymbol{F}_{IIOISD}^{-1}(z)\boldsymbol{F}_{IIOISN}(z),$$

read:

$$\boldsymbol{F}_{IIOISD}(z) = \boldsymbol{\Phi}_{IIOIS}(z), \quad \boldsymbol{F}_{IIOISN}(z) = = \left[\begin{array}{ccc} \boldsymbol{P}^{(\beta)} \boldsymbol{S}_M^{(\beta)}(z) & -\boldsymbol{P}^{(\beta)} \boldsymbol{Z}_M^{(\beta)}(z) & \boldsymbol{Q}^{(\alpha)} \boldsymbol{Z}_\rho^{(\alpha)}(z) \end{array} \right]. \quad (10.42)$$

where $\mathbf{\Phi}_{IIOIS}(z)$ *is the* $Z-$*transform of the IIO* ***system fundamental IS matrix*** $\mathbf{\Phi}_{IIOIS}(k)$,

$$\mathbf{\Phi}_{IIOIS}(z) = \mathcal{Z}\left\{\mathbf{\Phi}_{IIOIS}(k)\right\},$$
$$\mathbf{\Phi}_{IIOIS}(k) = \mathcal{Z}^{-1}\left\{\mathbf{\Phi}_{IIOIS}(z)\right\}, \tag{10.43}$$

and

$$\mathbf{\Phi}_{IIOIS}(z) = \left(\boldsymbol{Q}^{(\alpha)}\boldsymbol{S}_{\rho}^{(\alpha)}(z)\right)^{-1},$$
$$\mathbf{\Phi}_{IIOIS}(k) = \mathcal{Z}^{-1}\left\{\left(\boldsymbol{Q}^{(\alpha)}\boldsymbol{S}_{\rho}^{(\alpha)}(z)\right)^{-1}\right\}. \tag{10.44}$$

The proof of this theorem is fully analogous to the proofs of Theorem 10.1 (in Section 10.1) and Theorem 10.2 (in Section 10.2).

Definition 10.5 ***The IIO system (3.64a) and (3.64b) IO full fundamental matrix function*** *is a matrix function* $\mathbf{\Psi}_{IIO}(\cdot) : \mathbb{Z} \longrightarrow \mathcal{R}^{N\times[(\gamma+1)M+\alpha\rho+\nu N]}$ *if and only if it obeys both conditions (i) and (ii) for any input vector function* $\mathbf{i}(\cdot)$, *and for any initial conditions* $\mathbf{i}_0^{\gamma-1}$, $\mathbf{r}_0^{\alpha-1}$, *and* $\mathbf{y}_0^{\nu-1}$,

(i)

$$\mathbf{y}(k;\mathbf{r}_0^{\alpha-1};\mathbf{y}_0^{\nu-1};\mathbf{i}) =$$
$$= \sum_{j=0}^{j=k}\mathbf{\Psi}_{IIO}(j)\begin{bmatrix} \mathbf{i}(k-j) \\ \delta_d(k-j)\mathbf{i}_0^{\gamma-1} \\ \delta_d(k-j)\mathbf{r}_0^{\alpha-1} \\ \left\{\begin{array}{c} \delta_d(k-j)\mathbf{y}_0^{\nu-1},\ \nu\geq 1, \\ \mathbf{0}_{\nu N}\ ,\ \nu=0 \end{array}\right. \end{bmatrix},$$
$$\mathbf{\Psi}_{IIO}(k) = \left[\begin{array}{cccc} \mathbf{\Gamma}_{IIO}(k) & \mathbf{\Gamma}_{IIOi_0}(k) & \mathbf{\Gamma}_{IIOr_0}(k) & \mathbf{\Gamma}_{IIOy_0}(k) \end{array}\right],$$
$$\mathbf{\Gamma}_{IIO}(k)\in\mathcal{R}^{N\times M}, \mathbf{\Gamma}_{IIOi_0}(k)\in\mathcal{R}^{N\times\gamma M}, \mathbf{\Gamma}_{IIOr_0}(k)\in\mathcal{R}^{N\times\alpha\rho},$$
$$\mathbf{\Gamma}_{IIOy_0}(k)\in\mathcal{R}^{N\times\nu N}, \tag{10.45}$$

(ii)

$$\mathbf{\Gamma}_{IIOi_0}(0) = \left[\begin{array}{cc} \mathbf{\Gamma}_{IIOi_01}(0) & \boldsymbol{O}_{N,(\gamma-1)M} \end{array}\right] \text{ where}$$
$$\mathbf{\Gamma}_{IIOi_01}(0) = -\mathbf{\Gamma}_{IIO}(0),$$
$$\mathbf{\Gamma}_{IIOr_0}(0) = \boldsymbol{O}_{N,\alpha\rho},$$
$$\mathbf{\Gamma}_{IIOy_0}(0) = \left\{\begin{array}{c} \left[\begin{array}{cc} \boldsymbol{I}_N & \boldsymbol{O}_{N,(\nu-1)N} \end{array}\right],\ \nu\geq 1, \\ \boldsymbol{O}_N\ ,\ \nu=0 \end{array}\right. . \tag{10.46}$$

Note 10.6 *The equations under (i) of this definition and the properties of $\delta_d(\cdot)$ (in Section B.2) furnish*

$$\mathbf{y}(k;\mathbf{r}_0^{\alpha-1};\mathbf{y}_0^{\nu-1};\mathbf{i}) = \left(\sum_{j=0}^{j=k}\mathbf{\Gamma}_{IIO}(j)\mathbf{i}(k-j)\right) + \mathbf{\Gamma}_{IIOi_0}(k)\mathbf{i}_0^{\gamma-1} +$$

$$+\mathbf{\Gamma}_{IIOr_0}(k)\mathbf{r}_0^{\alpha-1} + \begin{cases} \mathbf{\Gamma}_{IIOy_0}(k)\mathbf{y}_0^{\nu-1},\ \nu \geq 1, \\ \mathbf{0}_N\ ,\ \nu = 0 \end{cases},\ \forall k \in \mathcal{N}_0. \tag{10.47}$$

Theorem 10.5 *(i) The IIO system (3.64a) and (3.64b) IO full fundamental matrix function $\mathbf{\Psi}_{IIO}(\cdot)$ is the inverse of the Z−transform of the system IO full transfer function matrix $\boldsymbol{F}_{IIO}(z)$,*

$$\mathbf{\Psi}_{IIO}(k) = \mathcal{Z}^{-1}\left\{\boldsymbol{F}_{IIO}(z)\right\}. \tag{10.48}$$

(ii) The IIO system (3.64a) and (3.64b) full transfer function matrix $\boldsymbol{F}_{IIO}(z)$ is the Z−transform of the system IIO full fundamental matrix $\mathbf{\Psi}_{IIO}(k)$,

$$\boldsymbol{F}_{IIO}(z) = \mathcal{Z}\left\{\mathbf{\Psi}_{IIO}(k)\right\}. \tag{10.49}$$

(iii) The IO transfer matrix functions $\mathbf{\Gamma}_{IIO}(z) = \mathcal{Z}\{\mathbf{\Gamma}_{IIO}(k)\}$, $\mathbf{\Gamma}_{IIOi_0}(z) = \mathcal{Z}\{\mathbf{\Gamma}_{IIOi_0}(k)\}$, $\mathbf{\Gamma}_{IIOr_0}(z) = \mathcal{Z}\{\mathbf{\Gamma}_{IIOr_0}(k)\}$ and $\mathbf{\Gamma}_{IIOy_0}(z) = \mathcal{Z}\{\mathbf{\Gamma}_{IIOy_0}(k)\}$ of the IIO system (3.64a) and (3.64b) are linked with the system transfer function matrices as follows:

$$\begin{aligned} \mathbf{\Gamma}_{IIO}(z) &= \boldsymbol{G}_{IIO}(z),\ \mathbf{\Gamma}_{IIOi_0}(z) = \boldsymbol{G}_{IIOi_0}(z), \\ \mathbf{\Gamma}_{IIOr_0}(z) &= \boldsymbol{G}_{IIOr_0}(z),\ \mathbf{\Gamma}_{IIOy_0}(z) = \boldsymbol{G}_{IIOy_0}(z), \end{aligned} \tag{10.50}$$

and

$$\mathbf{\Gamma}_{IIO}(k) = \mathcal{Z}^{-1}\left\{\boldsymbol{G}_{IIO}(z)\right\} = \mathcal{Z}^{-1}\left\{\mathbf{\Phi}_{IIO}(z)\cdot\right.$$

$$\left.\cdot\left[\boldsymbol{R}^{(\alpha)}\boldsymbol{S}_\rho^{(\alpha)}(z)\left(\boldsymbol{Q}^{(\alpha)}\boldsymbol{S}_\rho^{(\alpha)}(z)\right)^{-1}\boldsymbol{P}^{(\beta)}\boldsymbol{S}_M^{(\beta)}(z) + \boldsymbol{T}^{(\mu)}\boldsymbol{S}_M^{(\mu)}(z)\right]\right\} =$$

$$= \mathcal{Z}^{-1}\left\{\mathbf{\Phi}_{IO}(z)\cdot\right.$$

$$\left.\cdot\left[\boldsymbol{R}^{(\alpha)}\boldsymbol{S}_\rho^{(\alpha)}(z)\left(\boldsymbol{Q}^{(\alpha)}\boldsymbol{S}_\rho^{(\alpha)}(z)\right)^{-1}\boldsymbol{P}^{(\beta)}\boldsymbol{S}_M^{(\beta)}(z) + \boldsymbol{T}^{(\mu)}\boldsymbol{S}_M^{(\mu)}(z)\right]\right\}, \tag{10.51}$$

$$\boldsymbol{\Gamma}_{IIOi_0}(k) = \mathcal{Z}^{-1}\left\{\boldsymbol{G}_{IIOi_0}(z)\right\} = -\mathcal{Z}^{-1}\left\{\boldsymbol{\Phi}_{IIO}(z)\cdot\right.$$
$$\cdot\left\{\left[\begin{array}{cc}\boldsymbol{R}^{(\alpha)}\boldsymbol{S}_{\rho}^{(\alpha)}(z)\left(\boldsymbol{Q}^{(\alpha)}\boldsymbol{S}_{\rho}^{(\alpha)}(z)\right)^{-1}\boldsymbol{P}^{(\beta)}\boldsymbol{Z}_{M}^{(\beta)}(z) & \boldsymbol{O}_{N,(\gamma-\beta)M}\end{array}\right]+\right.$$
$$\left.\left.+\left[\begin{array}{cc}\boldsymbol{T}^{(\mu)}\boldsymbol{Z}_{M}^{(\mu)}(z) & \boldsymbol{O}_{N,(\gamma-\mu)M}\end{array}\right]\right\}\right\}, \tag{10.52}$$

$$\boldsymbol{\Gamma}_{IIOr_0}(k) = \mathcal{Z}^{-1}\left\{\boldsymbol{G}_{IIOr_0}(z)\right\} = \mathcal{Z}^{-1}\left\{\boldsymbol{\Phi}_{IIO}(z)\cdot\right.$$
$$\left.\cdot\left[\boldsymbol{R}^{(\alpha)}\boldsymbol{S}_{\rho}^{(\alpha)}(z)\left(\boldsymbol{Q}^{(\alpha)}\boldsymbol{S}_{\rho}^{(\alpha)}(z)\right)^{-1}\boldsymbol{Q}^{(\alpha)}\boldsymbol{Z}_{\rho}^{(\alpha)}(z) - \boldsymbol{R}^{(\alpha)}\boldsymbol{Z}_{\rho}^{(\alpha)}(z)\right]\right\}, \tag{10.53}$$

$$\boldsymbol{\Gamma}_{IIOy_0}(k) = \mathcal{Z}^{-1}\left\{\boldsymbol{\Phi}_{IIO}(z)\boldsymbol{E}^{(\nu)}\boldsymbol{Z}_{N}^{(\nu)}(z)\right\} \tag{10.54}$$

*where $\boldsymbol{\Phi}_{IIO}(z)$ is the $Z-$transform of the IIO **system fundamental matrix** $\boldsymbol{\Phi}_{IO}(k)$,*

$$\boldsymbol{\Phi}_{IIO}(z) = \mathcal{Z}\left\{\boldsymbol{\Phi}_{IIO}(k)\right\},\ \boldsymbol{\Phi}_{IIO}(k) = \mathcal{Z}^{-1}\left\{\boldsymbol{\Phi}_{IIO}(z)\right\}, \tag{10.55}$$

and

$$\boldsymbol{\Phi}_{IIO}(z) = \left(\boldsymbol{E}^{(\nu)}\boldsymbol{S}_{N}^{(\nu)}(z)\right)^{-1},$$
$$\boldsymbol{\Phi}_{IIO}(k) = \mathcal{Z}^{-1}\left\{\left(\boldsymbol{E}^{(\nu)}\boldsymbol{S}_{N}^{(\nu)}(z)\right)^{-1}\right\}. \tag{10.56}$$

The proof of this theorem is essentially the same as the proof of Theorem 10.1 (in Section 10.1). It shows a physical meaning of the full transfer function matrix of the IIO system (3.64a) and (3.64b). Besides, it is a guideline how to determine it effectively. Equations (8.75)-(8.78) determine the system transfer functions $\boldsymbol{G}_{IIO}(z)$, $\boldsymbol{G}_{IIOi_0}(z)$, $\boldsymbol{G}_{IIOr_0}(z)$ and $\boldsymbol{G}_{IIOy_0}(z)$.

Obviously, in view of Equations (10.50),

$$\boldsymbol{\Gamma}_{IIO}(k) = \mathcal{Z}^{-1}\left\{\boldsymbol{G}_{IIO}(z)\right\},\ \boldsymbol{\Gamma}_{IIOi_0}(k) = \mathcal{Z}^{-1}\left\{\boldsymbol{G}_{IIOi_0}(z)\right\},$$
$$\boldsymbol{\Gamma}_{IIOr_0}(k) = \mathcal{Z}^{-1}\left\{\boldsymbol{G}_{IIOr_0}(z)\right\},\ \boldsymbol{\Gamma}_{IIOy_0}(k) = \mathcal{Z}^{-1}\left\{\boldsymbol{G}_{IIOy_0}(z)\right\}. \tag{10.57}$$

Chapter 11

System matrix and equivalence

11.1 System matrix of the *IO* system

In order to find the system matrix we will present the *IO* system (3.56) (in Subsection 3.5.2) into the following form:

$$A^{(\nu)}\mathbf{r}^{\nu}(k) = B^{(\mu)}\mathbf{i}^{\mu}(k),\ \mathbf{y}(k) = \mathbf{r}(k),\ \forall k \in \mathcal{N}_0. \tag{11.1}$$

The $Z-$transforms of the preceding equations read:

$$A^{(\nu)} S_N^{(\nu)}(z)\,\mathbf{R}(z) =$$
$$= B^{(\mu)} S_M^{(\mu)}(z)\,\mathbf{I}(z) - B^{(\mu)} Z_M^{(\mu)}(z)\,\mathbf{i}_0^{\mu-1} + A^{(\nu)} Z_N^{(\nu)}(z)\,\mathbf{r}_0^{\nu-1},\ \mathbf{r}_0^{\nu-1} = \mathbf{y}_0^{\nu-1},$$
$$\mathbf{Y}(z) = \mathbf{R}(z). \tag{11.2}$$

At first we accept that *all initial conditions are equal to zero.* This condition greatly simplifies Equations (11.2) to

$$A^{(\nu)} S_N^{(\nu)}(z)\,\mathbf{R}(z) - B^{(\mu)} S_M^{(\mu)}(z)\,\mathbf{I}(z) = \mathbf{0}_N,$$
$$-\mathbf{Y}(z) = -\mathbf{R}(z)$$

or equivalently, to (11.3):

$$\underbrace{\begin{bmatrix} A^{(\nu)} S_N^{(\nu)}(z) & B^{(\mu)} S_M^{(\mu)}(z) \\ -I_N & O_{N\times M} \end{bmatrix}}_{P_{IO}(z;\mathbf{0}_{\mu M};\mathbf{0}_{\nu N})} \begin{bmatrix} \mathbf{R}(z) \\ -\mathbf{I}(z) \end{bmatrix} = \begin{bmatrix} \mathbf{0}_N \\ -\mathbf{Y}(z) \end{bmatrix}. \tag{11.3}$$

The system matrix function under all zero initial conditions of the IO system (11.1), for short: ***the system matrix function of the IO system (11.1)*** is the matrix function $\boldsymbol{P}_{IO}(\cdot;\mathbf{0}_{\mu M};\mathbf{0}_{\nu N}) : \mathcal{C} \longrightarrow \mathcal{C}^{2N\times(M+N)}$ that has the matrix value $\boldsymbol{P}_{IO}(z;\mathbf{0}_{\mu M};\mathbf{0}_{\nu N})$,

$$\boldsymbol{P}_{IO}(z;\mathbf{0}_{\mu M};\mathbf{0}_{\nu N}) \in \mathcal{C}^{2N\times(M+N)},$$

which is ***the system matrix under all zero initial conditions of the IO system (11.1)***, for short, ***the system matrix (11.4) of the IO system (11.1):***

$$\boldsymbol{P}_{IO}(z;\mathbf{0}_{\mu M};\mathbf{0}_{\nu N}) = \begin{bmatrix} \boldsymbol{A}^{(\nu)}\boldsymbol{S}_N^{(\nu)}(z) & \boldsymbol{B}^{(\mu)}\boldsymbol{S}_M^{(\mu)}(z) \\ -\boldsymbol{I}_N & \boldsymbol{O}_{N\times M} \end{bmatrix}. \tag{11.4}$$

There is the following relationship between it and the system transfer function matrix $\boldsymbol{G}_{IO}(z)$,

$$\boldsymbol{G}_{IO}(z) = \underbrace{\left[\boldsymbol{A}^{(\nu)}\boldsymbol{S}_N^{(\nu)}(z)\right]}_{\boldsymbol{G}_{IOD}(z)}{}^{-1}\underbrace{\boldsymbol{B}^{(\mu)}\boldsymbol{S}_M^{(\mu)}(z)}_{\boldsymbol{G}_{ION}(z)} = \boldsymbol{G}_{IOD}^{-1}(z)\boldsymbol{G}_{ION}(z) \Longrightarrow$$

$$\boldsymbol{P}_{IO}(z;\mathbf{0}_{\mu M};\mathbf{0}_{\nu N}) = \begin{bmatrix} \boldsymbol{G}_{IOD}(z) & \boldsymbol{G}_{ION}(z) \\ -\boldsymbol{I}_N & \boldsymbol{O}_{N\times M} \end{bmatrix}. \tag{11.5}$$

In the reality *initial conditions are arbitrary.* When we accept this then we can set Equations (11.2) into the following form:

$$\underbrace{\begin{bmatrix} \boldsymbol{A}^{(\nu)}\boldsymbol{S}_N^{(\nu)}(z) & \boldsymbol{B}^{(\mu)}\boldsymbol{S}_M^{(\mu)}(z) & -\boldsymbol{B}^{(\mu)}\boldsymbol{Z}_M^{(\mu)}(z) & \boldsymbol{A}^{(\nu)}\boldsymbol{Z}_N^{(\nu)}(z) \\ -\boldsymbol{I}_N & \boldsymbol{O}_{N\times M} & \boldsymbol{O}_{N\times \mu M} & \boldsymbol{O}_{N\times \nu N} \end{bmatrix}}_{\boldsymbol{P}_{IO}(z;\mathbf{i}_0^{\mu-1};\mathbf{y}_0^{\nu-1})}\cdot$$

$$\cdot\begin{bmatrix} \mathbf{R}(z) \\ -\mathbf{I}(z) \\ -\mathbf{i}_0^{\mu-1} \\ -\mathbf{y}_0^{\nu-1} \end{bmatrix} = \begin{bmatrix} \mathbf{0}_N \\ -\mathbf{Y}(z) \end{bmatrix}. \tag{11.6}$$

This justifies to define ***the full system matrix function***

$$\boldsymbol{P}_{IO}(\cdot;\mathbf{i}_0^{\mu-1};\mathbf{y}_0^{\nu-1}) : \mathcal{C} \longrightarrow \mathcal{C}^{2N\times[(\mu+1)M+(\nu+1)N]}$$

under arbitrary initial conditions of the IO system (11.1), for short: ***the full system matrix function $\boldsymbol{P}_{IO}(\cdot)$ of the IO system***

(11.1), as follows:

$$\boldsymbol{P}_{IO}(z) \equiv \boldsymbol{P}_{IO}(z;\mathbf{i}_0^{\mu-1};\mathbf{y}_0^{\nu-1}) =$$

$$= \left[\begin{array}{cccc} \boldsymbol{A}^{(\nu)}\boldsymbol{S}_N^{(\nu)}(z) & \boldsymbol{B}^{(\mu)}\boldsymbol{S}_M^{(\mu)}(z) & -\boldsymbol{B}^{(\mu)}\boldsymbol{Z}_M^{(\mu)}(z) & \boldsymbol{A}^{(\nu)}\boldsymbol{Z}_N^{(\nu)}(z) \\ -\boldsymbol{I}_N & \boldsymbol{O}_{N\times M} & \boldsymbol{O}_{N\times \mu M} & \boldsymbol{O}_{N\times \nu N} \end{array}\right] \Longleftrightarrow$$

$$\Longleftrightarrow \boldsymbol{P}_{IO}(z)\left[\begin{array}{c} \mathbf{R}(z) \\ -\mathbf{V}_{IO}(z) \end{array}\right] = \left[\begin{array}{c} \mathbf{0}_N \\ -\mathbf{Y}(z) \end{array}\right] \text{ and}$$

$$\mathbf{V}_{IO}(z) = \left[\begin{array}{c} \mathbf{I}(z) \\ \mathbf{i}_0^{\mu-1} \\ \mathbf{y}_0^{\nu-1} \end{array}\right] \in \mathcal{C}^{2N\times[(\mu+1)M+\nu N]}. \tag{11.7}$$

The matrix $\boldsymbol{P}_{IO}(z;\mathbf{i}_0^{\mu-1};\mathbf{y}_0^{\nu-1})$,

$$\boldsymbol{P}_{IO}(z;\mathbf{i}_0^{\mu-1};\mathbf{y}_0^{\nu-1}) \in \mathcal{C}^{2N\times[(\mu+1)M+(\nu+1)N]},$$

for short $\boldsymbol{P}_{IO}(z)$,

$$\boldsymbol{P}_{IO}(z) \equiv \boldsymbol{P}_{IO}(z;\mathbf{i}_0^{\mu-1};\mathbf{y}_0^{\nu-1}),$$

is the matrix value of $\boldsymbol{P}_{IO}(\cdot;\mathbf{i}_0^{\mu-1};\mathbf{y}_0^{\nu-1})$. It is ***the full system matrix of the IO system (11.1) under arbitrary initial conditions***, for short: ***the full system matrix*** $\boldsymbol{P}_{IO}(z)$ ***of the IO system (11.1).***

Note 11.1 *The full system matrix* $\boldsymbol{P}_{IO}(z)$ *and the full system matrix function* $\boldsymbol{P}_{IO}(\cdot)$ *should be distinguished from the system full transfer function matrix* $\boldsymbol{F}_{IO}(z)$ *and the system full matrix transfer function* $\boldsymbol{F}_{IO}(\cdot)$, *respectively. They are linked, due to (8.20) (Section 8.1), and (11.7), as follows:*

$$\boldsymbol{F}_{IO}(z) = \boldsymbol{F}_{IOD}^{-1}(z)\boldsymbol{F}_{ION}(z) =$$

$$= \underbrace{\left(\boldsymbol{A}^{(\nu)}\boldsymbol{S}_N^{(\nu)}(z)\right)}_{\boldsymbol{F}_{IOD}(z)}{}^{-1}\underbrace{\left[\begin{array}{ccc} \boldsymbol{B}^{(\mu)}\boldsymbol{S}_M^{(\mu)}(z) & -\boldsymbol{B}^{(\mu)}\boldsymbol{Z}_M^{(\mu)}(z) & \boldsymbol{A}^{(\nu)}\boldsymbol{Z}_N^{(\nu)}(z) \end{array}\right]}_{\boldsymbol{F}_{ION}(z)} \tag{11.8}$$

$$\Longrightarrow$$

$$\boldsymbol{P}_{IO}(z) = \left[\begin{array}{cc} \boldsymbol{F}_{IOD}(z) & \boldsymbol{F}_{ION}(z) \\ -\boldsymbol{I}_N & \boldsymbol{O}_{N\times[(\mu+1)M+\nu N]} \end{array}\right]. \tag{11.9}$$

Note 11.2 *The structure of the system matrix* $\boldsymbol{P}_{IO}(z;\mathbf{0}_{\mu M};\mathbf{0}_{\nu N})$ *(11.4) and of the full system matrix* $\boldsymbol{P}_{IO}(z)$ *(11.7) is the same. However, their submatrices and their dimensions are different.*

The submatrices of $\boldsymbol{P}_{IO}(z)$ are polynomial matrices in the complex variable z. Their forms are the following:

$$\boldsymbol{T}_{IO}(z) = \boldsymbol{A}^{(\nu)}\boldsymbol{S}_N^{(\nu)}(z) = \boldsymbol{F}_{IOD}(z) \in \mathcal{C}^{N\times N},$$
$$\boldsymbol{U}_{IO}(z) = \left[\begin{array}{ccc} \boldsymbol{B}^{(\mu)}\boldsymbol{S}_M^{(\mu)}(z) & -\boldsymbol{B}^{(\mu)}\boldsymbol{Z}_M^{(\mu)}(z) & \boldsymbol{A}^{(\nu)}\boldsymbol{Z}_N^{(\nu)}(z) \end{array}\right] =$$
$$= \boldsymbol{F}_{ION}(z) \in \mathcal{C}^{N\times[(\mu+1)M+\nu N]},$$
$$\boldsymbol{V}_{IO}(z) = \boldsymbol{I}_N,\ \boldsymbol{W}_{IO}(z) = \boldsymbol{O}_{N\times[(\mu+1)M+\nu N]}. \tag{11.10}$$

These submatrices are *the full Rosenbrock submatrices of* $\boldsymbol{P}_{IO}(z)$ by referring to Rosenbrock [111, p. 43] (see Section 11.2 in the sequel). They permit to set $\boldsymbol{P}_{IO}(z)$, (11.7), of the IO system (11.1), into the known *Rosenbrock form* $\boldsymbol{P}_{IOR}(z)$:

$$\boldsymbol{P}_{IOR}(z) = \left[\begin{array}{cc} \boldsymbol{T}_{IO}(z) & \boldsymbol{U}_{IO}(z) \\ -\boldsymbol{V}_{IO}(z) & \boldsymbol{W}_{IO}(z) \end{array}\right] \Longrightarrow$$
$$\boldsymbol{P}_{IOR}(z) = \left[\begin{array}{cc} \boldsymbol{F}_{IOD}(z) & \boldsymbol{F}_{ION}(z) \\ -\boldsymbol{I}_N & \boldsymbol{O}_{N\times[(\mu+1)M+\nu N]} \end{array}\right]. \tag{11.11}$$

Here $\boldsymbol{F}_{IOD}(z)$ and $\boldsymbol{F}_{ION}(z)$ are the denominator and the numerator polynomial matrices of the system full transfer function matrix $\boldsymbol{F}_{IO}(z)$, Note 8.3 (in Section 8.1). This and (11.8) lead to

$$\boldsymbol{F}_{IO}(z) = \boldsymbol{F}_{IOD}^{-1}(z)\boldsymbol{F}_{ION}(z) = \boldsymbol{V}_{IO}(z)\boldsymbol{T}_{IO}^{-1}(z)\boldsymbol{U}_{IO}(z) + \boldsymbol{W}_{IO}(z). \tag{11.12}$$

The second Equation (11.11) is really the Equation (11.9):

$$\boldsymbol{P}_{IOR}(z) = \boldsymbol{P}_{IO}(z). \tag{11.13}$$

By referring to Rosenbrock (for the definition of strictly system equivalent matrices and for Theorem 3.1, see [111, p. 52]) we introduce.

Definition 11.1 ***Strictly equivalent full system matrices***
Let $2N \times [(\mu+1)M + (\nu+1)N]$ *full system matrix* $\mathbf{P}_{IO}(z)$ *be given in its Rosenbrock form* $\mathbf{P}_{IOR}(z)$ *(11.11). Let* $\mathbf{M}(z)$ *and* $\mathbf{N}(z)$ *be* $N\times N$ *unimodular polynomial matrices. Let* $\mathbf{X}(z)$ *and* $\mathbf{Y}(z)$ *be also polynomial matrices, respectively* $N\times N$ *and* $N\times[(\mu+1)M+\nu N]$*. If* $\mathbf{P}_{IO}(z)$ *and*

$$\boldsymbol{P}_1(z) = \left[\begin{array}{cc} \boldsymbol{T}_1(z) & \boldsymbol{U}_1(z) \\ -\boldsymbol{V}_1(z) & \boldsymbol{W}_1(z) \end{array}\right] \tag{11.14}$$

are related by the transformation

$$\begin{bmatrix} \boldsymbol{M}(z) & \boldsymbol{O}_N \\ \boldsymbol{X}(z) & \boldsymbol{I}_N \end{bmatrix} \begin{bmatrix} \boldsymbol{T}_{IO}(z) & \boldsymbol{U}_{IO}(z) \\ -\boldsymbol{V}_{IO}(z) & \boldsymbol{W}_{IO}(z) \end{bmatrix} \cdot$$

$$\cdot \begin{bmatrix} \boldsymbol{N}(z) & \boldsymbol{Y}(z) \\ \boldsymbol{O}_{[(\mu+1)M+\nu N]\times N} & \boldsymbol{I}_{[(\mu+1)M+\nu N]} \end{bmatrix} = \begin{bmatrix} \boldsymbol{T}_1(z) & \boldsymbol{U}_1(z) \\ -\boldsymbol{V}_1(z) & \boldsymbol{W}_1(z) \end{bmatrix} \tag{11.15}$$

then $\boldsymbol{P}_{IO}(z)$ and $\boldsymbol{P}_1(z)$ are ***strictly system equivalent.***

Theorem 11.1 *Two full system matrices which are strictly system equivalent correspond to the same system full transfer function matrix $\boldsymbol{F}_{IO}(z)$.*

Proof. We use Definition 11.1. Let $2N\times\ [(\mu+1)M+(\nu+1)N]$ full system matrix $\boldsymbol{P}_{IO}(z)$ be given in its Rosenbrock form $\boldsymbol{P}_{IOR}(z)$ (11.11). We can write Equation (11.15) as

$$\begin{bmatrix} \boldsymbol{M}(z) & \boldsymbol{O}_N \\ \boldsymbol{X}(z) & \boldsymbol{I}_N \end{bmatrix} \begin{bmatrix} \boldsymbol{T}_{IO}(z) & \boldsymbol{U}_{IO}(z) \\ -\boldsymbol{V}_{IO}(z) & \boldsymbol{W}_{IO}(z) \end{bmatrix} \cdot$$

$$\cdot \begin{bmatrix} \boldsymbol{N}(z) & \boldsymbol{Y}(z) \\ \boldsymbol{O}_{[(\mu+1)M+\nu N]\times N} & \boldsymbol{I}_{[(\mu+1)M+\nu N]} \end{bmatrix} = \begin{bmatrix} \boldsymbol{T}_1(z) & \boldsymbol{U}_1(z) \\ -\boldsymbol{V}_1(z) & \boldsymbol{W}_1(z) \end{bmatrix} =$$

$$= \begin{bmatrix} \boldsymbol{M}(z)\boldsymbol{T}_{IO}(z) & \boldsymbol{M}(z)\boldsymbol{U}_{IO}(z) \\ \boldsymbol{X}(z)\boldsymbol{T}_{IO}(z) - \boldsymbol{V}_{IO}(z) & \boldsymbol{X}(z)\boldsymbol{U}_{IO}(z) + \boldsymbol{W}_{IO}(z) \end{bmatrix} \cdot$$

$$\cdot \begin{bmatrix} \boldsymbol{N}(z) & \boldsymbol{Y}(z) \\ \boldsymbol{O}_{[(\mu+1)M+\nu N]\times N} & \boldsymbol{I}_{[(\mu+1)M+\nu N]} \end{bmatrix} =$$

$$= \begin{bmatrix} \boldsymbol{M}\boldsymbol{T}_{IO}\boldsymbol{N} & \boldsymbol{M}\boldsymbol{T}_{IO}\boldsymbol{Y} + \boldsymbol{M}\boldsymbol{U}_{IO} \\ (\boldsymbol{X}\boldsymbol{T}_{IO} - \boldsymbol{V}_{IO})\boldsymbol{N} & (\boldsymbol{X}\boldsymbol{T}_{IO} - \boldsymbol{V}_{IO})\boldsymbol{Y} + \boldsymbol{X}\boldsymbol{U}_{IO} + \boldsymbol{W}_{IO} \end{bmatrix}.$$

This result, Equations (11.10) and (11.12) that request

$$\boldsymbol{F}_1(z) = \boldsymbol{V}_1(z)\boldsymbol{T}_1^{-1}(z)\boldsymbol{U}_1(z) + \boldsymbol{W}_1(z),$$

and (11.15) imply the following due to the nonsingularity of the unimodular matrices:

$$\begin{aligned} \boldsymbol{F}_1(z) &= -(\boldsymbol{X}\boldsymbol{T}_{IO} - \boldsymbol{V}_{IO})\boldsymbol{N}(\boldsymbol{M}\boldsymbol{T}_{IO}\boldsymbol{N})^{-1}(\boldsymbol{M}\boldsymbol{T}_{IO}\boldsymbol{Y} + \boldsymbol{M}\boldsymbol{U}_{IO}) + \\ &\quad + (\boldsymbol{X}\boldsymbol{T}_{IO} - \boldsymbol{V}_{IO})\boldsymbol{Y} + \boldsymbol{X}\boldsymbol{U}_{IO} + \boldsymbol{W}_{IO} = \\ &= -(\boldsymbol{X}\boldsymbol{T}_{IO} - \boldsymbol{V}_{IO})\boldsymbol{T}_{IO}^{-1}\boldsymbol{M}^{-1}\boldsymbol{M}(\boldsymbol{T}_{IO}\boldsymbol{Y} + \boldsymbol{U}_{IO}) + \\ &\quad + (\boldsymbol{X}\boldsymbol{T}_{IO} - \boldsymbol{V}_{IO})\boldsymbol{Y} + \boldsymbol{X}\boldsymbol{U}_{IO} + \boldsymbol{W}_{IO} = \end{aligned}$$

$$
\begin{gathered}
= -\left(\boldsymbol{X}\boldsymbol{T}_{IO} - \boldsymbol{V}_{IO}\right)\boldsymbol{T}_{IO}^{-1}\left(\boldsymbol{T}_{IO}\boldsymbol{Y} + \boldsymbol{U}_{IO}\right) + \\
+\boldsymbol{X}\boldsymbol{T}_{IO}\boldsymbol{Y} - \boldsymbol{V}_{IO}\boldsymbol{Y} + \boldsymbol{X}\boldsymbol{U}_{IO} + \boldsymbol{W}_{IO} = \\
= -\boldsymbol{X}\boldsymbol{T}_{IO}\boldsymbol{Y} - \boldsymbol{X}\boldsymbol{U}_{IO} + \boldsymbol{V}_{IO}\boldsymbol{Y} + \boldsymbol{V}_{IO}\boldsymbol{T}_{IO}^{-1}\boldsymbol{U}_{IO} + \\
+\boldsymbol{X}\boldsymbol{T}_{IO}\boldsymbol{Y} - \boldsymbol{V}_{IO}\boldsymbol{Y} + \boldsymbol{X}\boldsymbol{U}_{IO} + \boldsymbol{W}_{IO} = \\
= \boldsymbol{V}_{IO}\boldsymbol{T}_{IO}^{-1}\boldsymbol{U}_{IO} + \boldsymbol{W}_{IO} = \boldsymbol{F}_{IOD}^{-1}(z)\boldsymbol{F}_{ION}(z) = \boldsymbol{F}_{IO}(z).
\end{gathered}
$$

This ends the proof. ■

Note 11.3 *This is a double generalization of Rosenbrock's Theorem 3.1 [111, p. 52]: to the IO systems and for arbitrary initial conditions.*

Example 11.1 *From Example 8.1 we use the IO system description:*

$$
\begin{gathered}
\boldsymbol{F}_{IO}(z) = \left(z^2 - z - 0.75\right)^{-1} \cdot \\
\cdot \left[\begin{array}{ccccc} \left(z^2 - 7.5z + 9\right) & -z\left(z - 7.5\right) & -z & z\left(z-1\right) & z \end{array}\right].
\end{gathered}
$$

The full system transfer matrix $\boldsymbol{F}_{IO}(z)$ induces, due to (11.8) and (11.9), the full system matrix $\boldsymbol{P}_{IO}(z)$:

$$
\boldsymbol{P}_{IO}(z) =
$$

$$
= \left[\begin{array}{cccccc} z^2 - z - 0.75 & z^2 - 7.5z + 9 & -z\left(z - 7.5\right) & -z & z\left(z-1\right) & z \\ -1 & 0 & 0 & 0 & 0 & 0 \end{array}\right].
$$

Note 11.4 *The following two ratios of polynomials,*

$$
\frac{z}{\left(z - 0.3\right)\left(z + 0.9\right)} \text{ and } \frac{z(z - 1.5)}{\left(z - 0.3\right)\left(z + 0.9\right)(z - 1.5)}, \tag{11.16}
$$

correspond to the same rational function $f(\cdot)$ (see $s-$complex analogy in [23, p. 58]). They have two common zeros $z_1^0 = 0$ and $z_2^0 = \infty$, and two common poles $z_1^* = 0.3$ and $z_2^* = -0.9$. The second ratio has an additional positive real pole $z_3^* = 1.5$, of modulus grater of one, $|z_3^*| > 1$. However, if they represent the system transfer functions $G_1(\cdot)$ and $G_2(\cdot)$,

$$
\begin{gathered}
G_1(z) = \frac{z}{\left(z - 0.3\right)\left(z + 0.9\right)} \text{ and } \\
G_2(z) = \frac{z(z - 1.5)}{\left(z - 0.3\right)\left(z + 0.9\right)(z - 1.5)},
\end{gathered} \tag{11.17}
$$

then they do not correspond to the same system. $G_1(z)$ is the transfer function of the second-order $SISO$ system described by

$$
E^2y(k) + 0.6E^1y(k) - 0.27E^0y\left(k\right) = E^1i(k), \tag{11.18}
$$

while $G_2(\cdot)$ *is the transfer function of the third-order* $SISO$ *system determined by*

$$E^3y(k) - 0.9E^2y(k) - 1.17E^1y(k) + 0.405E^0y(k) = \\ = -1.5E^1i(k) + E^2i(k). \tag{11.19}$$

The full transfer function matrix $\boldsymbol{F}_1(z)$ *of the former reads*

$$\boldsymbol{F}_1(z) = (z^2 + 0.6z - 0.27)^{-1} \begin{bmatrix} z & -z & (z^2 + 0.6z) & z \end{bmatrix}, \tag{11.20}$$

while the full transfer function matrix $\boldsymbol{F}_2(z)$ *of the latter is found as*

$$\boldsymbol{F}_2(z) = (z^3 - 0.9z^2 - 1.17z + 0.405)^{-1} z \cdot \\ \cdot \begin{bmatrix} z - 1.5 & -(z-1.5) & -z & (z^2 - 0.9z - 1.17) & (z-0.9) & 1 \end{bmatrix}. \tag{11.21}$$

They imply the following full system matrices, respectively:

$$\boldsymbol{P}_{IO1}(z) = \begin{bmatrix} z^2 + 0.6z - 0.27 & z & -z & z^2 + 0.6z & z \\ -1 & 0 & 0 & 0 & 0 \end{bmatrix},$$

$$\boldsymbol{P}_{IO2}(z) = \begin{bmatrix} z^3 - 0.9z^2 - 1.17z + 0.405 & -1 \\ z^2 - 1.5z & 0 \\ -z^2 + 1.5z & 0 \\ -z^2 & 0 \\ z^3 - 0.9z^2 - 1.17z & 0 \\ z^2 - 0.9z & 0 \\ z & 0 \end{bmatrix}^T \neq \boldsymbol{P}_{IO1}(z).$$

Example 11.2 *For the* $MIMO$ IO *system described by*

$$E^1\mathbf{y}(k) + \begin{bmatrix} 1.5 & 0 \\ 0 & -1 \end{bmatrix} \mathbf{y}(k) = \begin{bmatrix} -1 & 0 \\ 0 & 1.5 \end{bmatrix} \mathbf{i}(k) + E^1\mathbf{i}(k),$$
$$\mathbf{y} = \begin{bmatrix} y_1 \\ y_2 \end{bmatrix}, \ \mathbf{i} = \begin{bmatrix} i_1 \\ i_2 \end{bmatrix}.$$

we find the transfer function matrix $\boldsymbol{G}_{IO}(z)$,

$$\boldsymbol{G}_{IO}(z) = \begin{bmatrix} \frac{z-1}{z+1.5} & 0 \\ 0 & \frac{z+1.5}{z-1} \end{bmatrix}, \ \textit{full rank}\ \boldsymbol{G}_{IO}(z) = 2,$$

and the full transfer function matrix $\boldsymbol{F}_{IO}(z)$,

$$\boldsymbol{F}_{IO}(z) = \begin{bmatrix} \frac{z-1}{z+1.5} & 0 & -\frac{z}{z+1.5} & 0 & \frac{z}{z+1.5} & 0 \\ 0 & \frac{z+1.5}{z-1} & 0 & -\frac{z}{z-1} & 0 & \frac{z}{z-1} \end{bmatrix} =$$

$$= \begin{bmatrix} \frac{1}{z+1.5} & 0 \\ 0 & \frac{1}{z-1} \end{bmatrix} \begin{bmatrix} z-1 & 0 & -z & 0 & z & 0 \\ 0 & z+1.5 & 0 & -z & 0 & z \end{bmatrix} =$$

$$= \underbrace{\begin{bmatrix} z+1.5 & 0 \\ 0 & z-1 \end{bmatrix}^{-1}}_{F_{IOD}(z)} \underbrace{\begin{bmatrix} z-1 & 0 & -z & 0 & z & 0 \\ 0 & z+1.5 & 0 & -z & 0 & z \end{bmatrix}}_{F_{ION}(z)}.$$

We determine now the full system matrix $\boldsymbol{P}_{IO}(z)$,

$$\boldsymbol{P}_{IO}(z) = \begin{bmatrix} z+1.5 & 0 & z-1 & 0 & -z & 0 & z & 0 \\ 0 & z-1 & 0 & z+1.5 & 0 & -z & 0 & z \\ -1 & 0 & 0 & 0 & 0 & 0 & 0 & 0 \\ 0 & -1 & 0 & 0 & 0 & 0 & 0 & 0 \end{bmatrix}.$$

11.2 System matrix of the *ISO* System

We emphasize that Rosenbrock defined in [111, p. 43] *the system matrix function* $\boldsymbol{P}_R(\cdot)$ *under all zero initial conditions* $\mathbf{x}_0 = \mathbf{0}_n$ for Rosenbrock systems *RS* (2.33a) and (2.33b) (in Section 2.3). We will denote $\boldsymbol{P}_R(\cdot)$ also by $\boldsymbol{P}_R(\cdot;\mathbf{0}_n)$ in order to underline that $\boldsymbol{P}_R(\cdot)$ is defined and valid only if all initial conditions are equal to zero,

$$\boldsymbol{P}_R(\cdot) = \boldsymbol{P}_R(\cdot;\mathbf{0}_n).$$

The *ISO* systems form a subclass of Rosenbrock systems *RS* (2.33a) and (2.33b). The Rosenbrock definition of the system matrix function $\boldsymbol{P}_{ISOR}(\cdot;\mathbf{0}_n)$ of (2.33a) and (2.33b) is applicable to the *ISO* system (3.60a) and (3.60b) (in Subsection 3.5.3) under all zero initial conditions,

$$\boldsymbol{P}_{ISOR}(z;\mathbf{0}_n) = \begin{bmatrix} z\boldsymbol{I}_n - \boldsymbol{A} & \boldsymbol{B} \\ -\boldsymbol{C} & \boldsymbol{D} \end{bmatrix} =$$

$$= \begin{bmatrix} \boldsymbol{T}_{ISOR}(z;\mathbf{0}_n) & \boldsymbol{U}_{ISOR}(z;\mathbf{0}_n) \\ -\boldsymbol{V}_{ISOR}(z;\mathbf{0}_n) & \boldsymbol{W}_{ISOR}(z;\mathbf{0}_n) \end{bmatrix} \Longleftrightarrow$$

$$\Longleftrightarrow \boldsymbol{P}_{ISOR}(z;\mathbf{0}_n) \begin{bmatrix} \mathbf{X}(z) \\ -\mathbf{I}(z) \end{bmatrix} = \begin{bmatrix} \mathbf{0}_n \\ -\mathbf{Y}(z) \end{bmatrix}. \quad (11.22)$$

The matrices $\boldsymbol{T}_{ISOR}(z;\mathbf{0}_n)$, $\boldsymbol{U}_{ISOR}(z;\mathbf{0}_n)$, $\boldsymbol{V}_{ISOR}(z;\mathbf{0}_n)$, and $\boldsymbol{W}_{ISOR}(z;\mathbf{0}_n)$ are *Rosenbrock polynomial submatrices of* $\boldsymbol{P}_{ISOR}(z;\mathbf{0}_n)$, [111, p. 52], of the *ISO* system (3.60a) and (3.60b),

$$\boldsymbol{T}_{ISOR}(z;\mathbf{0}_n) = z\boldsymbol{I}_n - \boldsymbol{A},\ \boldsymbol{U}_{ISOR}(z;\mathbf{0}_n) = \boldsymbol{B},$$
$$\boldsymbol{V}_{ISOR}(z;\mathbf{0}_n) = \boldsymbol{C},\ \boldsymbol{W}_{ISOR}(z;\mathbf{0}_n) = \boldsymbol{D}. \quad (11.23)$$

In the case of arbitrary initial conditions the Rosenbrock system matrix $\boldsymbol{P}_{ISOR}(z;\mathbf{0}_n)$ (11.22) should be replaced by the ***full system matrix*** $\boldsymbol{P}_{ISOR}(z;\mathbf{x}_0)$, $\boldsymbol{P}_{ISOR}(z;\mathbf{x}_0) \in \mathcal{C}^{(N+n)\times(M+2n)}$, ***in the general Rosenbrock form*** *that reads*

$$\boldsymbol{P}_{ISOR}(z;\mathbf{x}_0) = \begin{bmatrix} \boldsymbol{T}_{ISOR}(z;\mathbf{x}_0) & \boldsymbol{U}_{ISOR}(z;\mathbf{x}_0) \\ -\boldsymbol{V}_{ISOR}(z;\mathbf{x}_0) & \boldsymbol{W}_{ISOR}(z;\mathbf{x}_0) \end{bmatrix}, \tag{11.24}$$

$$\begin{gathered}\boldsymbol{P}_{ISOR}(z;\mathbf{x}_0) = \begin{bmatrix} z\boldsymbol{I}_n - \boldsymbol{A} & \boldsymbol{B} & z\boldsymbol{I}_n \\ -\boldsymbol{C} & \boldsymbol{D} & \boldsymbol{O}_{N\times n} \end{bmatrix} = \\ = \begin{bmatrix} \boldsymbol{P}_{ISOR}(z;\mathbf{0}_n) & \begin{matrix} z\boldsymbol{I}_n \\ \boldsymbol{O}_{N\times n} \end{matrix} \end{bmatrix} = \\ = \begin{bmatrix} \boldsymbol{T}_{ISOR}(z;\mathbf{x}_0) & \boldsymbol{U}_{ISOR}(z;\mathbf{x}_0) \\ -\boldsymbol{V}_{ISOR}(z;\mathbf{x}_0) & \boldsymbol{W}_{ISOR}(z;\mathbf{x}_0) \end{bmatrix} \Longleftrightarrow \\ \Longleftrightarrow \boldsymbol{P}_{ISOR}(z;\mathbf{x}_0) \begin{bmatrix} \mathbf{X}(z) \\ -\mathbf{V}_{ISOR}(z;\mathbf{x}_0) \end{bmatrix} = \begin{bmatrix} \mathbf{0}_n \\ -\mathbf{Y}(z) \end{bmatrix}, \\ \text{where } \mathbf{V}_{ISOR}(z;\mathbf{x}_0) = \begin{bmatrix} \mathbf{I}(z) \\ \mathbf{x}_0 \end{bmatrix} \in \mathcal{C}^{M+n}.\end{gathered} \tag{11.25}$$

The preceding equations show that the Rosenbrock submatrices changed their forms from those in (11.23) to the following *full Rosenbrock submatrices* of $\boldsymbol{P}_{ISOR}(z;\mathbf{x}_0)$,

$$\begin{gathered}\boldsymbol{T}_{ISOR}(z;\mathbf{x}_0) = (z\boldsymbol{I}_n - \boldsymbol{A}) \in \mathcal{C}^{n\times n}, \\ \boldsymbol{U}_{ISOR}(z;\mathbf{x}_0) = \begin{bmatrix} \boldsymbol{B} & z\boldsymbol{I}_n \end{bmatrix} \in \mathcal{C}^{n\times(M+n)}, \\ \boldsymbol{V}_{ISOR}(z;\mathbf{x}_0) = \boldsymbol{C} \in \mathcal{C}^{N\times n}, \\ \boldsymbol{W}_{ISOR}(z;\mathbf{x}_0) = \begin{bmatrix} \boldsymbol{D} & \boldsymbol{O}_{N\times n} \end{bmatrix} \in \mathcal{C}^{N\times(M+n)}.\end{gathered} \tag{11.26}$$

A consequence is that the equations and (8.31) (in Section 8.2) imply:

$$\begin{gathered}\boldsymbol{F}_{ISO}(z) = \boldsymbol{V}_{ISOR}(z;\mathbf{x}_0)\boldsymbol{T}_{ISOR}^{-1}(z;\mathbf{x}_0)\boldsymbol{U}_{ISOR}(z;\mathbf{x}_0) + \boldsymbol{W}_{ISOR}(z;\mathbf{x}_0) = \\ = \boldsymbol{F}_{ISOD}^{-1}(z)\boldsymbol{F}_{ISON}(z).\end{gathered} \tag{11.27}$$

We generalize Rosenbrock's definition of strictly system equivalent matrices and Theorem 3.1 of [111, p. 52] in the sequel.

Example 11.3 *Given* $\boldsymbol{G}_{ISO}(z) = \left(z^2 - 1\right)^{-1}(z-1) = (z+1)^{-1}$. *Four different (state space, i.e., ISO) realizations* $(\boldsymbol{A},\boldsymbol{B},\boldsymbol{C},\boldsymbol{D})$ *of* $\boldsymbol{G}_{ISO}(z)$ *are observed* *by analogy to the s−complex case determined in [5, p. 395], Example 8.7 (in Section 8.2). We show at first how to determine the full system matrix for each in the Rosenbrock form (11.25).*

1) $\boldsymbol{A}_1 = \begin{bmatrix} 0 & 1 \\ 1 & 0 \end{bmatrix}$, $\boldsymbol{B}_1 = \begin{bmatrix} 0 \\ 1 \end{bmatrix}$, $\boldsymbol{C}_1 = \begin{bmatrix} -1 & 1 \end{bmatrix}$, $\boldsymbol{D}_1 = 0$, $\mathbf{x} = \begin{bmatrix} x_1 \\ x_2 \end{bmatrix}$,
$\mathbf{y} = [y]$

$\Longrightarrow$

$$\boldsymbol{P}_{ISOR1}(z;\mathbf{x}_0) = \begin{bmatrix} \boldsymbol{T}_{ISOR1}(z;\mathbf{x}_0) & \boldsymbol{U}_{ISOR1}(z;\mathbf{x}_0) \\ -\boldsymbol{V}_{ISOR1}(z;\mathbf{x}_0) & \boldsymbol{W}_{ISOR1}(z;\mathbf{x}_0) \end{bmatrix} =$$

$$= \begin{bmatrix} z & -1 & 0 & z & 0 \\ -1 & z & 1 & 0 & z \\ 1 & -1 & 0 & 0 & 0 \end{bmatrix}, \ \mathbf{V}_{ISOR1}(z;\mathbf{x}_0) = \begin{bmatrix} I(z) \\ \mathbf{x}_0 \end{bmatrix}.$$

2) $\boldsymbol{A}_2 = \begin{bmatrix} 0 & 1 \\ 1 & 0 \end{bmatrix}$, $\boldsymbol{B}_2 = \begin{bmatrix} -1 \\ 1 \end{bmatrix}$, $\boldsymbol{C}_2 = \begin{bmatrix} 0 & 1 \end{bmatrix}$, $\boldsymbol{D}_2 = 0$, $\mathbf{x} = \begin{bmatrix} x_1 \\ x_2 \end{bmatrix}$,
$\mathbf{y} = [y]$

$\Longrightarrow$

$$\boldsymbol{P}_{ISOR2}(z;\mathbf{x}_0) = \begin{bmatrix} \boldsymbol{T}_{ISOR2}(z;\mathbf{x}_0) & \boldsymbol{U}_{ISOR2}(z;\mathbf{x}_0) \\ -\boldsymbol{V}_{ISOR2}(z;\mathbf{x}_0) & \boldsymbol{W}_{ISOR2}(z;\mathbf{x}_0) \end{bmatrix} =$$

$$= \begin{bmatrix} z & -1 & -1 & z & 0 \\ -1 & z & 1 & 0 & z \\ 0 & -1 & 0 & 0 & 0 \end{bmatrix}, \ \mathbf{V}_{ISOR2}(z;\mathbf{x}_0) = \begin{bmatrix} I(z) \\ \mathbf{x}_0 \end{bmatrix}.$$

3) $\boldsymbol{A}_3 = \begin{bmatrix} 1 & 0 \\ 0 & -1 \end{bmatrix}$, $\boldsymbol{B}_3 = \begin{bmatrix} 0 \\ 1 \end{bmatrix}$, $\boldsymbol{C}_3 = \begin{bmatrix} 0 & 1 \end{bmatrix}$, $\boldsymbol{D}_3 = 0$, $\mathbf{x} = \begin{bmatrix} x_1 \\ x_2 \end{bmatrix}$,
$\mathbf{y} = [y]$

$\Longrightarrow$

$$\boldsymbol{P}_{ISOR3}(z;\mathbf{x}_0) = \begin{bmatrix} \boldsymbol{T}_{ISOR3}(z;\mathbf{x}_0) & \boldsymbol{U}_{ISOR3}(z;\mathbf{x}_0) \\ -\boldsymbol{V}_{ISOR3}(z;\mathbf{x}_0) & \boldsymbol{W}_{ISOR3}(z;\mathbf{x}_0) \end{bmatrix} =$$

$$= \begin{bmatrix} z-1 & 0 & 0 & z & 0 \\ 0 & z+1 & 1 & 0 & z \\ 0 & -1 & 0 & 0 & 0 \end{bmatrix}, \ \mathbf{V}_{ISOR3}(z;\mathbf{x}_{0\mp}) = \begin{bmatrix} I(z) \\ \mathbf{x}_0 \end{bmatrix}.$$

4) $\mathbf{A}_4 = [-1]$, $\mathbf{B}_4 = [1]$, $\mathbf{C}_4 = [1]$, $\mathbf{D}_4 = 0$, $\mathbf{x} = [x]$, $\mathbf{y} = [y] \Longrightarrow$

$$\mathbf{P}_{ISOR4}(z;\mathbf{x}_0) = \begin{bmatrix} \mathbf{T}_{ISOR4}(z;\mathbf{x}_0) & \mathbf{U}_{ISOR4}(z;\mathbf{x}_0) \\ -\mathbf{V}_{ISOR4}(z;\mathbf{x}_0) & \mathbf{W}_{ISOR4}(z;\mathbf{x}_0) \end{bmatrix} =$$

$$= \begin{bmatrix} z+1 & 1 & z \\ -1 & 0 & 0 \end{bmatrix}, \ \mathbf{V}_{ISOR4}(z;\mathbf{x}_0) = \begin{bmatrix} I(s) \\ x_0 \end{bmatrix}.$$

Let for short $\mathbf{P}_{ISO}(z;\mathbf{x}_0) \equiv \mathbf{P}_{ISO}(z)$.

Definition 11.2 ***Strictly equivalent full system matrices***

Let $(n+N)\times(M+2n)$ *full system matrix* $\mathbf{P}_{ISO}(z)$ *be in its Rosenbrock form* $\mathbf{P}_{ISOR}(z)$ *(11.25). Let* $\mathbf{M}(z)$ *and* $\mathbf{N}(z)$ *be* $n\times n$ *unimodular polynomial matrices. Let* $\mathbf{X}(z)$ *and* $\mathbf{Y}(z)$ *be also polynomial matrices, respectively* $N\times n$ *and* $n\times(M+n)$*. If* $\mathbf{P}_{ISOR}(z)$ *and*

$$\mathbf{P}_1(z) = \begin{bmatrix} \mathbf{T}_1(z) & \mathbf{U}_1(z) \\ -\mathbf{V}_1(z) & \mathbf{W}_1(z) \end{bmatrix} \in \mathcal{C}^{(n+N)\times(M+n)} \tag{11.28}$$

are related by the transformation

$$\begin{bmatrix} \mathbf{M}(z) & \mathbf{O}_{n\times N} \\ \mathbf{X}(z) & \mathbf{I}_N \end{bmatrix} \begin{bmatrix} \mathbf{T}_{ISOR}(z) & \mathbf{U}_{ISOR}(z) \\ -\mathbf{V}_{ISOR}(z) & \mathbf{W}_{ISOR}(z) \end{bmatrix} \cdot$$

$$\cdot \begin{bmatrix} \mathbf{N}(z) & \mathbf{Y}(z) \\ \mathbf{O}_{(M+n)\times n} & \mathbf{I}_{M+n} \end{bmatrix} = \begin{bmatrix} \mathbf{T}_1(z) & \mathbf{U}_1(z) \\ -\mathbf{V}_1(z) & \mathbf{W}_1(z) \end{bmatrix} \tag{11.29}$$

then $\mathbf{P}_{ISOR}(z)$ *and* $\mathbf{P}_1(z)$ *are* ***strictly system equivalent****.*

Theorem 11.2 *Two full system matrices, which are strictly system equivalent, correspond to the same system full transfer function matrix* $\mathbf{F}_{ISO}(z)$*.*

Proof. We use Definition 11.2. Let $(n+N)\times(M+2n)$ system matrix $\mathbf{P}_{ISO}(z)$ be given in its Rosenbrock form $\mathbf{P}_{ISOR}(z)$ (11.25) and (11.26). Equation (11.29) can be set into the following form:

$$\begin{bmatrix} \mathbf{M}(z) & \mathbf{O}_{n\times N} \\ \mathbf{X}(z) & \mathbf{I}_N \end{bmatrix} \begin{bmatrix} \mathbf{T}_{ISOR}(z) & \mathbf{U}_{ISOR}(z) \\ -\mathbf{V}_{ISOR}(z) & \mathbf{W}_{ISOR}(z) \end{bmatrix} \cdot$$

$$\cdot \begin{bmatrix} \mathbf{N}(z) & \mathbf{Y}(z) \\ \mathbf{O}_{(M+n)\times n} & \mathbf{I}_{M+n} \end{bmatrix} = \begin{bmatrix} \mathbf{T}_1(z) & \mathbf{U}_1(z) \\ -\mathbf{V}_1(z) & \mathbf{W}_1(z) \end{bmatrix} =$$

$$= \begin{bmatrix} \mathbf{M}(z)\,\mathbf{T}_{ISOR}(z) & \mathbf{M}(z)\,\mathbf{U}_{ISOR}(z) \\ \mathbf{X}(z)\mathbf{T}_{ISOR}(z) - \mathbf{V}_{ISOR}(z) & \mathbf{X}(z)\mathbf{U}_{ISOR}(z) + \mathbf{W}_{ISOR}(z) \end{bmatrix} \cdot$$

$$\cdot \begin{bmatrix} \mathbf{N}(z) & \mathbf{Y}(z) \\ \mathbf{O}_{(M+n)\times n} & \mathbf{I}_{M+n} \end{bmatrix} =$$

$$= \begin{bmatrix} \boldsymbol{M}\boldsymbol{T}_{ISOR}\boldsymbol{N} & \boldsymbol{M}\boldsymbol{T}_{ISOR}\boldsymbol{Y} + \boldsymbol{M}\boldsymbol{U}_{ISOR} \\ \left(\boldsymbol{X}\boldsymbol{T}_{ISOR} - \boldsymbol{V}_{ISOR}\right)\boldsymbol{N} & \begin{matrix} \left(\boldsymbol{X}\boldsymbol{T}_{ISOR} - \boldsymbol{V}_{ISOR}\right)\boldsymbol{Y} + \\ + \boldsymbol{X}\boldsymbol{U}_{ISOR} + \boldsymbol{W}_{ISOR} \end{matrix} \end{bmatrix}.$$

This result, Equations (11.12):

$$\boldsymbol{F}_1(z) = \boldsymbol{V}_1(z)\boldsymbol{T}_1^{-1}(z)\boldsymbol{U}_1(z) + \boldsymbol{W}_1(z),$$

(11.29), (11.27) and the nonsingularity of the unimodular matrices imply the following:

$$\boldsymbol{F}_1(z) =$$
$$= -\left(\boldsymbol{X}\boldsymbol{T}_{ISOR} - \boldsymbol{V}_{ISOR}\right)\boldsymbol{N}\left(\boldsymbol{M}\boldsymbol{T}_{ISOR}\boldsymbol{N}\right)^{-1}\left(\boldsymbol{M}\boldsymbol{T}_{ISOR}\boldsymbol{Y} + \boldsymbol{M}\boldsymbol{U}_{ISOR}\right) +$$
$$+ \left(\boldsymbol{X}\boldsymbol{T}_{ISOR} - \boldsymbol{V}_{ISOR}\right)\boldsymbol{Y} + \boldsymbol{X}\boldsymbol{U}_{ISOR} + \boldsymbol{W}_{ISOR} =$$
$$= -\boldsymbol{X}\boldsymbol{T}_{ISOR}\boldsymbol{Y} - \boldsymbol{X}\boldsymbol{U}_{ISOR} + \boldsymbol{V}_{ISOR}\boldsymbol{Y} + \boldsymbol{V}_{ISOR}\boldsymbol{T}_{ISOR}^{-1}\boldsymbol{U}_{ISOR} +$$
$$+ \boldsymbol{X}\boldsymbol{T}_{ISOR}\boldsymbol{Y} - \boldsymbol{V}_{ISOR}\boldsymbol{Y} + \boldsymbol{X}\boldsymbol{U}_{ISOR} + \boldsymbol{W}_{ISOR} =$$
$$= \boldsymbol{V}_{ISOR}\boldsymbol{T}_{ISOR}^{-1}\boldsymbol{U}_{ISOR} + \boldsymbol{W}_{ISOR} = \boldsymbol{F}_{ISOD}^{-1}(z)\boldsymbol{F}_{ISON}(z) = \boldsymbol{F}_{ISO}(z),$$

which proves the theorem. ■

This theorem generalizes Rosenbrock's Theorem 3.1 [111, p. 52] for the *ISO* systems.

Let $\mathbf{r}(k) \equiv \mathbf{y}(k)$. Hence, $\mathbf{R}(z) \equiv \mathbf{Y}(z)$. We use this together with (8.31), (8.32), and (8.39) (in Section 8.2) in the *IO* form:

$$R(z) - \begin{bmatrix} \boldsymbol{C}(z\boldsymbol{I}_n - \boldsymbol{A})^{-1}\boldsymbol{B} + \boldsymbol{D} & z\boldsymbol{C}(z\boldsymbol{I}_n - \boldsymbol{A})^{-1} \end{bmatrix} V_{ISOIO}(z) = 0_N \Longrightarrow$$
$$\Delta(z)\mathbf{R}(z) +$$
$$+ \begin{bmatrix} \boldsymbol{C}\,\mathrm{adj}(z\boldsymbol{I}_n - \boldsymbol{A})\boldsymbol{B} + \boldsymbol{D}\Delta(z) & z\boldsymbol{C}\,\mathrm{adj}(z\boldsymbol{I}_n - \boldsymbol{A}) \end{bmatrix}\left(-\mathbf{V}_{ISOIO}(z)\right) = 0_N,$$
$$\mathbf{Y}(z) = \mathbf{R}(z), \text{ where } \mathbf{V}_{ISOIO}(z;\mathbf{x}_0) = \begin{bmatrix} \mathbf{I}(z) \\ \mathbf{x}_0 \end{bmatrix} \in \mathcal{C}^{M+n}. \tag{11.30}$$

The preceding equations imply the following *IO full system matrix* $\boldsymbol{P}_{ISOIO}(z)$ *of the ISO system*:

$$\overbrace{\begin{bmatrix} \Delta(z)\boldsymbol{I}_N & \boldsymbol{C}\,\mathrm{adj}(z\boldsymbol{I}_n - \boldsymbol{A})\boldsymbol{B} + \boldsymbol{D}\Delta(z) & z\boldsymbol{C}\,\mathrm{adj}(z\boldsymbol{I}_n - \boldsymbol{A}) \\ -\boldsymbol{I}_N & \boldsymbol{O}_{N\times M} & \boldsymbol{O}_{N\times n} \end{bmatrix}} \cdot$$
$$\cdot \begin{bmatrix} \mathbf{R}(z) \\ -\mathbf{V}_{ISOIO}(z) \end{bmatrix} = \begin{bmatrix} \mathbf{0}_N \\ -\mathbf{Y}(z) \end{bmatrix}. \tag{11.31}$$

We define *the IO full polynomial submatrices* $\boldsymbol{T}_{ISOIO}(z;\mathbf{x}_0)$, $\boldsymbol{U}_{ISOIO}(z;\mathbf{x}_0)$, $\boldsymbol{V}_{ISOIO}(z;\mathbf{x}_0)$, and $\boldsymbol{W}_{ISOIO}(z;\mathbf{x}_0)$ of $\boldsymbol{P}_{ISOIO}(z;\mathbf{x}_0)$, due to (8.31) (in Section 8.2), by

$$\begin{gathered}
\boldsymbol{T}_{ISOIO}(z;\mathbf{x}_0)=\Delta(z)\boldsymbol{I}_N=\boldsymbol{F}_{ISOIOD}(z),\\
\boldsymbol{U}_{ISOIO}(z;\mathbf{x}_0)=\left[\begin{array}{cc}\boldsymbol{C}\operatorname{adj}(z\boldsymbol{I}_n-\boldsymbol{A})\boldsymbol{B}+\boldsymbol{D}\Delta(z) & z\boldsymbol{C}\operatorname{adj}(z\boldsymbol{I}_n-\boldsymbol{A})\end{array}\right]=\\
=\boldsymbol{F}_{ISOION}(z),\\
\boldsymbol{V}_{ISOIO}(z;\mathbf{x}_0)=\boldsymbol{I}_N,\ \boldsymbol{W}_{ISOIO}(z;\mathbf{x}_0)=\boldsymbol{O}_{N\times(M+n)},
\end{gathered}\tag{11.32}$$

which and (11.31) imply

$$\boldsymbol{P}_{ISOIO}(z;\mathbf{x}_0)=\left[\begin{array}{cc}\boldsymbol{T}_{ISOIO}(z;\mathbf{x}_0) & \boldsymbol{U}_{ISOIO}(z;\mathbf{x}_0)\\ -\boldsymbol{V}_{ISOIO}(z;\mathbf{x}_0) & \boldsymbol{W}_{ISOIO}(z;\mathbf{x}_0)\end{array}\right].\tag{11.33}$$

Comment 11.1 *Equations (11.26), (11.32), and (11.33) prove that the full system matrices* $\boldsymbol{P}_{ISOR}(z;\mathbf{x}_0)$ *in (11.25) and* $\boldsymbol{P}_{ISOIO}(z;\mathbf{x}_0)$ *in (11.31), i.e., in (11.33), have the same form and structure as the general Rosenbrock system matrix (11.24) [111, p. 52].*

Definition 11.3 ***Strictly equivalent IO full system matrices in Rosenbrock form***

Let $2N\times(N+M+n)$ *IO full system matrix* $\boldsymbol{P}_{ISOIO}(z)$ *be given in the Rosenbrock form (11.25). Let* $\boldsymbol{M}(z)$ *and* $\boldsymbol{N}(z)$ *be* $N\times N$ *unimodular polynomial matrices. Let* $\boldsymbol{X}(z)$ *and* $\boldsymbol{Y}(z)$ *be also polynomial matrices, respectively* $N\times N$ *and* $N\times(M+n)$. *If* $\boldsymbol{P}_{ISOIO}(z)$ *and* $\boldsymbol{P}_1(z)$ *(11.28) are related by the transformation*

$$\begin{gathered}
\left[\begin{array}{cc}\boldsymbol{M}(z) & \boldsymbol{O}_{N\times N}\\ \boldsymbol{X}(z) & \boldsymbol{I}_N\end{array}\right]\left[\begin{array}{cc}\boldsymbol{T}_{ISOIOR}(z) & \boldsymbol{U}_{ISOIOR}(z)\\ -\boldsymbol{V}_{ISOIOR}(z) & \boldsymbol{W}_{ISOIOR}(z)\end{array}\right]\cdot\\
\cdot\left[\begin{array}{cc}\boldsymbol{N}(z) & \boldsymbol{Y}(z)\\ \boldsymbol{O}_{(M+n)\times N} & \boldsymbol{I}_{M+n}\end{array}\right]=\left[\begin{array}{cc}\boldsymbol{T}_1(z) & \boldsymbol{U}_1(z)\\ -\boldsymbol{V}_1(z) & \boldsymbol{W}_1(z)\end{array}\right]
\end{gathered}\tag{11.34}$$

then $\boldsymbol{P}_{ISOIO}(z)$ *and* $\boldsymbol{P}_1(z)$ *are* ***strictly system equivalent.***

Theorem 11.3 *Two full system matrices, which are strictly system equivalent, correspond to the same system full IO transfer function matrix* $\boldsymbol{F}_{ISOIO}(z)$.

The theorem is proved in the same way as Theorem 11.2.

Example 11.4 *Given* $\mathbf{G}_{ISO}(z) = \left(z^2-1\right)^{-1}(z-1) = (z+1)^{-1}$. *Four different (state space, i.e., ISO) realizations* $(\mathbf{A},\mathbf{B},\mathbf{C},\mathbf{D})$ *of* $\mathbf{G}_{ISO}(z)$ *are observed by analogy to the s−complex case determined in [5, p. 395], Example 8.7 (in Section 8.2). We use the full transfer function matrix* $\mathbf{F}_{ISO}(z)$ *determined for each in Example 8.7 in order to deduce from it the full system matrix* $\mathbf{P}_{ISOIO}(z;\mathbf{x}_0)$.

1) $\mathbf{A}_1 = \begin{bmatrix} 0 & 1 \\ 1 & 0 \end{bmatrix}$, $\mathbf{B}_1 = \begin{bmatrix} 0 \\ 1 \end{bmatrix}$, $\mathbf{C}_1 = \begin{bmatrix} -1 \vdots 1 \end{bmatrix}$, $\mathbf{D}_1 = 0$, $\mathbf{x} = \begin{bmatrix} x_1 \\ x_2 \end{bmatrix}$,
$\mathbf{y} = [y]$
$\Longrightarrow$

$$\mathbf{F}_{ISO1}(z) = \left(z^2-1\right)^{-1}\left[\begin{array}{ccc} z-1 & z(1-z) & z(z-1) \end{array}\right] \Longrightarrow$$

$$\mathbf{P}_{ISOIO1}(z;\mathbf{x}_0) = \begin{bmatrix} \mathbf{T}_{ISOIOR1}(z;\mathbf{x}_0) & \mathbf{U}_{ISOIOR1}(z;\mathbf{x}_0) \\ -\mathbf{V}_{ISOIOR1}(z;\mathbf{x}_0) & \mathbf{W}_{ISOIOR1}(z;\mathbf{x}_0) \end{bmatrix} =$$

$$= \begin{bmatrix} z^2-1 & z-1 & z(1-z) & z(z-1) \\ -1 & 0 & 0 & 0 \end{bmatrix} \neq \mathbf{P}_{ISOR1}(z;\mathbf{x}_0),$$

$$\mathbf{V}_{ISOIO1}(z;\mathbf{x}_{0\mp}) = \begin{bmatrix} I(z) \\ \mathbf{x}_0 \end{bmatrix}.$$

2) $\mathbf{A}_2 = \begin{bmatrix} 0 & 1 \\ 1 & 0 \end{bmatrix}$, $\mathbf{B}_2 = \begin{bmatrix} -1 \\ 1 \end{bmatrix}$, $\mathbf{C}_2 = \begin{bmatrix} 0 \vdots 1 \end{bmatrix}$, $\mathbf{D}_2 = 0$, $\mathbf{x} = \begin{bmatrix} x_1 \\ x_2 \end{bmatrix}$,
$\mathbf{y} = [y]$
$\Longrightarrow$

$$\mathbf{F}_{ISO2}(z) = \left(z^2-1\right)^{-1}\left[\begin{array}{ccc} z-1 & z & z^2 \end{array}\right] \Longrightarrow$$

$$\mathbf{P}_{ISOIO2}(z;\mathbf{x}_0) = \begin{bmatrix} \mathbf{T}_{ISOIOR2}(z;\mathbf{x}_0) & \mathbf{U}_{ISOIOR2}(z;\mathbf{x}_0) \\ -\mathbf{V}_{ISOIOR2}(z;\mathbf{x}_0) & \mathbf{W}_{ISOIOR2}(z;\mathbf{x}_0) \end{bmatrix} =$$

$$= \begin{bmatrix} z^2-1 & z-1 & z & z^2 \\ -1 & 0 & 0 & 0 \end{bmatrix} \neq \mathbf{P}_{ISOR2}(z;\mathbf{x}_0),$$

$$\mathbf{V}_{ISOIO2}(z;\mathbf{x}_0) = \begin{bmatrix} I(z) \\ \mathbf{x}_0 \end{bmatrix}.$$

3) $\mathbf{A}_3 = \begin{bmatrix} 1 & 0 \\ 0 & -1 \end{bmatrix}$, $\mathbf{B}_3 = \begin{bmatrix} 0 \\ 1 \end{bmatrix}$, $\mathbf{C}_3 = \begin{bmatrix} 0 \vdots 1 \end{bmatrix}$, $\mathbf{D}_3 = 0$, $\mathbf{x} = \begin{bmatrix} x_1 \\ x_2 \end{bmatrix}$,
$\mathbf{y} = [y]$
$\Longrightarrow$

$$\mathbf{F}_{ISO3}(z) = \left(z^2-1\right)^{-1}\left[\begin{array}{ccc} z-1 & 0 & z(z-1) \end{array}\right] \Longrightarrow$$

$$\boldsymbol{P}_{ISOIO3}(z;\mathbf{x}_0) = \begin{bmatrix} \boldsymbol{T}_{ISOIOR3}(z;\mathbf{x}_0) & \boldsymbol{U}_{ISOIOR3}(z;\mathbf{x}_0) \\ -\boldsymbol{V}_{ISOIOR3}(z;\mathbf{x}_0) & \boldsymbol{W}_{ISOIOR3}(z;\mathbf{x}_0) \end{bmatrix} =$$

$$= \begin{bmatrix} z^2-1 & z-1 & 0 & z(z-1) \\ -1 & 0 & 0 & 0 \end{bmatrix} \neq \boldsymbol{P}_{ISOR3}(z;\mathbf{x}_0),$$

$$\mathbf{V}_{ISOIO3}(z;\mathbf{x}_{0\mp}) = \begin{bmatrix} I(z) \\ \mathbf{x}_0 \end{bmatrix}.$$

4) $\boldsymbol{A}_4 = [-1]$, $\boldsymbol{B}_4 = [1]$, $\boldsymbol{C}_4 = [1]$, $\boldsymbol{D}_4 = 0$, $\mathbf{x} = [x]$, $\mathbf{y} = [y] \Longrightarrow$

$$\boldsymbol{F}_{ISO4}(z) = (z+1)^{-1} \begin{bmatrix} 1 & z \end{bmatrix} \Longrightarrow$$

$$\boldsymbol{P}_{ISOIO4}(z;\mathbf{x}_0) = \begin{bmatrix} \boldsymbol{T}_{ISOIOR4}(z;\mathbf{x}_0) & \boldsymbol{U}_{ISOIOR4}(z;\mathbf{x}_0) \\ -\boldsymbol{V}_{ISOIOR4}(z;\mathbf{x}_0) & \boldsymbol{W}_{ISOIOR4}(z;\mathbf{x}_0) \end{bmatrix} =$$

$$= \begin{bmatrix} z+1 & 1 & z \\ -1 & 0 & 0 \end{bmatrix} = \boldsymbol{P}_{ISOR4}(z;\mathbf{x}_0),$$

$$\mathbf{V}_{ISOIO4}(z;\mathbf{x}_{0\mp}) = \begin{bmatrix} I(z) \\ x_0 \end{bmatrix}.$$

Note 11.5 *The full system matrix functions* $\boldsymbol{P}_{ISOIO}(z;\mathbf{x}_0)$ *and* $\boldsymbol{P}_{ISOR}(z;\mathbf{x}_0)$ *are different in the cases 1) through 3) of Examples 11.3 and 11.4 because the subsidiary vectors* $\mathbf{r} = \mathbf{y} \in \mathcal{R}^N$ *and* $\mathbf{x} \in \mathcal{R}^n$ *are different in general due to* $\mathbf{x} \neq \mathbf{y}$.

Note 11.6 *The cases 1) through 3) of Examples 11.3 and 11.4 show that*

$$\boldsymbol{P}_{ISOR}(z;\mathbf{x}_0) \neq \boldsymbol{P}_{ISOIO}(z;\mathbf{x}_0) \text{ in general.}$$

However, the case 4) of the same Examples discovers that

$$\boldsymbol{P}_{ISOR}(z;\mathbf{x}_0) = \boldsymbol{P}_{ISOIO}(z;\mathbf{x}_0) \text{ is possible in special cases.}$$

11.3 System matrix of the *IIO* system

The $Z-$transform to the *IIO* system (3.64a) and (3.64b),

$$\boldsymbol{Q}^{(\alpha)}\mathbf{r}^{\alpha}(k) = \boldsymbol{P}^{(\beta)}\mathbf{i}^{\beta}(k),\ \boldsymbol{E}^{(\nu)}\mathbf{y}^{\nu}(k) = \boldsymbol{R}^{(\alpha)}\mathbf{r}^{\alpha}(k) + \boldsymbol{T}^{(\mu)}\mathbf{i}^{\mu}(k), \forall k \in \mathcal{N}_0,$$

reads

$$\left(\boldsymbol{Q}^{(\alpha)}\boldsymbol{S}_{\rho}^{(\alpha)}(z)\right)\mathbf{R}(z) = \boldsymbol{P}^{(\beta)}\boldsymbol{S}_{M}^{(\beta)}(z)\mathbf{I}(z) -$$
$$-\boldsymbol{P}^{(\beta)}\boldsymbol{Z}_{M}^{(\beta)}(z)\mathbf{i}_0^{\beta-1} + \boldsymbol{Q}^{(\alpha)}\boldsymbol{Z}_{\rho}^{(\alpha)}(z)\mathbf{r}_0^{\alpha-1},$$

$$\left(\boldsymbol{E}^{(\nu)}\boldsymbol{S}_N^{(\nu)}(z)\right)\mathbf{Y}(z)=\boldsymbol{R}^{(\alpha)}\boldsymbol{S}_\rho^{(\alpha)}(z)\mathbf{R}(z)+\boldsymbol{T}^{(\mu)}\boldsymbol{S}_M^{(\mu)}(z)\mathbf{I}(z)-$$
$$-\boldsymbol{R}^{(\alpha)}\boldsymbol{Z}_\rho^{(\alpha)}(z)\mathbf{r}_0^{\alpha-1}-\boldsymbol{T}^{(\mu)}\boldsymbol{Z}_M^{(\mu)}(z)\mathbf{i}_0^{\mu-1}+\boldsymbol{E}^{(\nu)}\boldsymbol{Z}_N^{(\nu)}(z)\mathbf{y}_0^{\nu-1}.$$

The equivalent form of these equations follows, in view (7.28) (in Section 7.4):

$$\boldsymbol{Q}^{(\alpha)}\boldsymbol{S}_\rho^{(\alpha)}(z)\mathbf{R}(z)-\boldsymbol{P}^{(\beta)}\boldsymbol{S}_M^{(\beta)}(z)\mathbf{I}(z)+$$
$$+\begin{bmatrix}\boldsymbol{P}^{(\beta)}\boldsymbol{Z}_M^{(\beta)}(z) & \boldsymbol{O}_{\rho\times[(\gamma-\beta)M]}\end{bmatrix}\mathbf{i}_0^{\gamma-1}-\boldsymbol{Q}^{(\alpha)}\boldsymbol{Z}_\rho^{(\alpha)}(z)\mathbf{r}_0^{\alpha-1}=\mathbf{0}_\rho,$$

$$-\boldsymbol{R}^{(\alpha)}\boldsymbol{S}_\rho^{(\alpha)}(z)\mathbf{R}(z)-\boldsymbol{T}^{(\mu)}\boldsymbol{S}_M^{(\mu)}(z)\mathbf{I}(z)+$$
$$+\begin{bmatrix}\boldsymbol{T}^{(\mu)}\boldsymbol{Z}_M^{(\mu)}(z) & \boldsymbol{O}_{N\times[(\gamma-\mu)M]}\end{bmatrix}\mathbf{i}_0^{\gamma-1}+\boldsymbol{R}^{(\alpha)}\boldsymbol{Z}_\rho^{(\alpha)}(z)\mathbf{r}_0^{\alpha-1}-$$
$$-\boldsymbol{E}^{(\nu)}\boldsymbol{Z}_N^{(\nu)}(z)\mathbf{y}_0^{\nu-1}=-\boldsymbol{E}^{(\nu)}\boldsymbol{S}_N^{(\nu)}(z)\mathbf{Y}(z)$$

These equations enable us to define ***the full system matrix function*** $P_{IIO}(.)$ ***of the IIO system*** by

$$\boldsymbol{P}_{IIO}(z;\mathbf{r}_0^{\alpha-1};\mathbf{y}_0^{\nu-1})\in\mathcal{C}^{(N+\rho)\times[(\gamma+1)M+(\alpha+1)\rho+\nu N]},$$
$$\boldsymbol{P}_{IIO}(z;\mathbf{r}_0^{\alpha-1};\mathbf{y}_0^{\nu-1})\equiv\boldsymbol{P}_{IIO}(z)=$$
$$\begin{bmatrix}
\left(\boldsymbol{Q}^{(\alpha)}\boldsymbol{S}_\rho^{(\alpha)}(z)\right)^T & \left(-\boldsymbol{R}^{(\alpha)}\boldsymbol{S}_\rho^{(\alpha)}(z)\right)^T\\
\left(\boldsymbol{P}^{(\beta)}\boldsymbol{S}_M^{(\beta)}(z)\right)^T & \left(\boldsymbol{T}^{(\mu)}\boldsymbol{S}_M^{(\mu)}(z)\right)^T\\
\begin{bmatrix}-\boldsymbol{P}^{(\beta)}\boldsymbol{Z}_M^{(\beta)}(z) & \boldsymbol{O}_A\end{bmatrix}^T & \begin{bmatrix}-\boldsymbol{T}^{(\mu)}\boldsymbol{Z}_M^{(\mu)}(z) & \boldsymbol{O}_C\end{bmatrix}^T\\
\left(\boldsymbol{Q}^{(\alpha)}\boldsymbol{Z}_\rho^{(\alpha)}(z)\right)^T & \left(-\boldsymbol{R}^{(\alpha)}\boldsymbol{Z}_\rho^{(\alpha)}(z)\right)^T\\
\boldsymbol{O}_B^T & \left(\boldsymbol{E}^{(\nu)}\boldsymbol{Z}_N^{(\nu)}(z)\right)^T
\end{bmatrix}^T$$
$$\boldsymbol{O}_A=\boldsymbol{O}_{\rho\times[(\gamma-\beta)M]},\ \boldsymbol{O}_B=\boldsymbol{O}_{\rho\times\nu N},\ \boldsymbol{O}_C=\boldsymbol{O}_{N\times[(\gamma-\mu)M]}$$
$$\Longrightarrow$$

$$\boldsymbol{P}_{IIO}(z)\begin{bmatrix}\mathbf{R}(z)\\ -\mathbf{V}_{IIO}(z;\mathbf{i}_0^{\gamma-1};\mathbf{r}_0^{\alpha-1};\mathbf{y}_0^{\nu-1})\end{bmatrix}=\begin{bmatrix}\mathbf{0}_\rho\\ -\boldsymbol{E}^{(\nu)}\boldsymbol{S}_N^{(\nu)}(z)\mathbf{Y}(z)\end{bmatrix},$$
$$\text{where }\mathbf{V}_{IIO}(z;\mathbf{i}_0^{\gamma-1};\mathbf{r}_0^{\alpha-1};\mathbf{y}_0^{\nu-1})=\begin{bmatrix}\mathbf{I}(z)\\ \mathbf{i}_0^{\gamma-1}\\ \mathbf{r}_0^{\alpha-1}\\ \mathbf{y}_0^{\nu-1}\end{bmatrix}. \tag{11.35}$$

This result determines the Rosenbrock type submatrices of the *IIO* system (3.64a), (3.64b), by

$$\boldsymbol{T}_{IIOR}(z) = \boldsymbol{Q}^{(\alpha)} \boldsymbol{S}_{\rho}^{(\alpha)}(z), \ \boldsymbol{T}_{IIOR}(z) \in \mathcal{C}^{\rho \times \rho},$$

$$\boldsymbol{U}_{IIOR}(z) =$$

$$= \begin{bmatrix} \boldsymbol{P}^{(\beta)} \boldsymbol{S}_{M}^{(\beta)}(z) & \begin{bmatrix} \left(-\boldsymbol{P}^{(\beta)} \boldsymbol{Z}_{M}^{(\beta)}(z)\right)^{T} \\ \left(\boldsymbol{O}_{\rho \times [(\gamma-\beta)M]}\right)^{T} \end{bmatrix}^{T} & \boldsymbol{Q}^{(\alpha)} \boldsymbol{Z}_{\rho}^{(\alpha)}(z) & \boldsymbol{O}_{\rho \times \nu N} \end{bmatrix},$$

$$\boldsymbol{U}_{IIOR}(z) \in \mathcal{C}^{\rho \times [\alpha\rho + (\gamma+1)M + \nu N]},$$

$$\boldsymbol{V}_{IIOR}(z) = \boldsymbol{R}^{(\alpha)} \boldsymbol{S}_{\rho}^{(\alpha)}(z), \ \boldsymbol{V}_{IIOR}(z) \in \mathcal{C}^{N \times \rho},$$

$$\boldsymbol{W}_{IIOR}(z) =$$

$$= \begin{bmatrix} \boldsymbol{T}^{(\mu)} \boldsymbol{S}_{M}^{(\mu)}(z) \vdots \begin{bmatrix} \left(-\boldsymbol{T}^{(\mu)} \boldsymbol{Z}_{M}^{(\mu)}(z)\right)^{T} \\ (\boldsymbol{O}_{C})^{T} \end{bmatrix}^{T} \vdots - \boldsymbol{R}^{(\alpha)} \boldsymbol{Z}_{\rho}^{(\alpha)}(z) \vdots \boldsymbol{E}^{(\nu)} \boldsymbol{Z}_{N}^{(\nu)}(z) \end{bmatrix},$$

$$\boldsymbol{O}_{C} = \boldsymbol{O}_{N \times [(\gamma-\mu)M]}, \ \boldsymbol{W}_{IIOR}(z) \in \mathcal{C}^{N \times [\alpha\rho + (\gamma+1)M + \nu N]}.$$

The so obtained Rosenbrock type submatrices of the *IIO* system (3.64a) and (3.64b), transform $\boldsymbol{P}_{IIO}(z)$ (11.35) into the Rosenbrock form $\boldsymbol{P}_{IIOR}(z)$,

$$\boldsymbol{P}_{IIOR}(z) = \begin{bmatrix} \boldsymbol{T}_{IIOR}(z) & \boldsymbol{U}_{IIOR}(z) \\ -\boldsymbol{V}_{IIOR}(z) & \boldsymbol{W}_{IIOR}(z) \end{bmatrix}. \tag{11.36}$$

Definition 11.4 *Let the full system matrix* $\boldsymbol{P}_{IIO}(z)$,

$$\boldsymbol{P}_{IIO}(z) \in \mathcal{C}^{(N+\rho) \times [(\alpha+1)\rho + (\gamma+1)M + \nu N]},$$

be given in its Rosenbrock form $\boldsymbol{P}_{IIOR}(z)$ *(11.36). Let* $\mathbf{M}(z)$ *and* $\mathbf{N}(z)$ *be* $\rho \times \rho$ *unimodular polynomial matrices. Let* $\mathbf{X}(z)$ *and* $\mathbf{Y}(z)$ *be also polynomial matrices,* $\mathbf{X}(z) \in \mathcal{C}^{N \times \rho}$ *and* $\mathbf{Y}(z) \in \mathcal{C}^{\rho \times [\alpha\rho + (\gamma+1)M + \nu N]}$. *If* $\boldsymbol{P}_{IIO}(z)$ *and*

$$\boldsymbol{P}_{1}(z) = \begin{bmatrix} \boldsymbol{T}_{1}(z) & \boldsymbol{U}_{1}(z) \\ -\boldsymbol{V}_{1}(z) & \boldsymbol{W}_{1}(z) \end{bmatrix} \tag{11.37}$$

are related by the transformation

$$\begin{bmatrix} \boldsymbol{M}(z) & \boldsymbol{O}_{\rho\times N} \\ \boldsymbol{X}(z) & \boldsymbol{I}_N \end{bmatrix} \begin{bmatrix} \boldsymbol{T}_{IIOR}(z) & \boldsymbol{U}_{IIOR}(z) \\ -\boldsymbol{V}_{IIOR}(z) & \boldsymbol{W}_{IIOR}(z) \end{bmatrix} \cdot$$
$$\cdot \begin{bmatrix} \boldsymbol{N}(z) & \boldsymbol{Y}(z) \\ \boldsymbol{O}_{[\rho\alpha+(\gamma+1)M+\nu N]\times\rho} & \boldsymbol{I}_{[\rho\alpha+(\gamma+1)M+\nu N]} \end{bmatrix} =$$
$$= \begin{bmatrix} \boldsymbol{T}_1(z) & \boldsymbol{U}_1(z) \\ -\boldsymbol{V}_1(z) & \boldsymbol{W}_1(z) \end{bmatrix} \tag{11.38}$$

then $\boldsymbol{P}_{IIO}(z)$ and $\boldsymbol{P}_1(z)$ are ***strictly system equivalent.***

Theorem 11.4 *Two full system matrices, which are strictly system equivalent, correspond to the same system full transfer function matrix $F_{IIO}(z)$.*

The proof of Theorem 11.1 (in Section 11.1) is to be only notationally adjusted to the system *IIO* system (3.64a) and (3.64b), in order to get the proof of Theorem 11.4.

Antsaklis and Michel [5, pp. 553, 554] determined the system matrix function $\boldsymbol{P}_{PMD}(\cdot)$ ($s-$complex case) of the *PMD* system under all zero initial conditions. By analogy the one can be obtained in the $z-$complex domain. It results in the Rosenbrock form for arbitrary initial conditions from $\boldsymbol{P}_{IIOR}(\cdot)$ as a special case,

$$\boldsymbol{P}_{PMDR}(z) = \begin{bmatrix} \boldsymbol{T}_{PMDR}(z) & \boldsymbol{U}_{PMDR}(z) \\ -\boldsymbol{V}_{PMDR}(z) & \boldsymbol{W}_{PMDR}(z) \end{bmatrix}, \tag{11.39}$$

$$\boldsymbol{T}_{PMDR}(z) = \boldsymbol{Q}^{(\alpha)}\boldsymbol{S}_{\rho}^{(\alpha)}(z),$$

$$\boldsymbol{U}_{PMDR}(z) = \begin{bmatrix} \boldsymbol{P}^{(\beta)}\boldsymbol{S}_M^{(\beta)}(z) & \begin{bmatrix} \left(-\boldsymbol{P}^{(\beta)}\boldsymbol{Z}_M^{(\beta)}(z)\right)^T \\ \left(\boldsymbol{O}_{\rho\times[(\gamma-\beta)M]}\right)^T \end{bmatrix}^T & \boldsymbol{Q}^{(\alpha)}\boldsymbol{Z}_{\rho}^{(\alpha)}(z) \end{bmatrix},$$

$$\boldsymbol{V}_{PMDR}(z) = \boldsymbol{R}^{(\alpha)}\boldsymbol{S}_{\rho}^{(\alpha)}(z),$$

$$\boldsymbol{W}_{PMDR}(z) = \begin{bmatrix} \left(\boldsymbol{T}^{(\mu)}\boldsymbol{S}_M^{(\mu)}(z)\right)^T \\ -\begin{bmatrix} \boldsymbol{T}^{(\mu)}\boldsymbol{Z}_M^{(\mu)}(z) & \boldsymbol{O}_{N\times[(\gamma-\mu)M]} \end{bmatrix}^T \\ -\left[\boldsymbol{R}^{(\alpha)}\boldsymbol{Z}_{\rho}^{(\alpha)}(z)\right]^T \end{bmatrix}^T, \tag{11.40}$$

$$\mathbf{V}_{PMDR}(z;\mathbf{i}_0^{\gamma-1};\mathbf{r}_0^{\alpha-1}) = \begin{bmatrix} \mathbf{I}(z) \\ \mathbf{i}_0^{\gamma-1} \\ \mathbf{r}_0^{\alpha-1} \end{bmatrix}. \tag{11.41}$$

Note 11.7 *Theorem 11.4 is directly applicable to $\boldsymbol{P}_{PMDR}(z)$ (11.39).*

Chapter 12

Realizations of $\boldsymbol{F}(z)$

12.1 Dynamical and least dimension of a system

The inherent differences between the system transfer function matrix $\boldsymbol{G}(z)$ and the system full transfer function matrix $\boldsymbol{F}(z)$, and the endeavor to escape a confusion or ambiguity, justify to present the following definitions [67].

Definition 12.1 ***The system characteristic polynomial*** *is the characteristic polynomial of the system full transfer function matrix* $\boldsymbol{F}(z)$.

Definition 12.2 ***The system minimal polynomial*** *is the minimal polynomial of the system full transfer function matrix* $\boldsymbol{F}(z)$.

Note 12.1 *The system minimal polynomial can be different from the minimal polynomial of the system transfer function matrix* $\boldsymbol{G}(z)$.

We will define *the dynamical dimension* and *the least dimension* of the system by following H. H. Rosenbrock, [111, pp. 30, 47, 48], and by noting that Rosenbrock used the term *order* in the sense of *dimension*. However, we accepted to distinguish *the dimension of the system* from *the order of the system* that we use in the classical mathematical sense of the order of a discrete equation that describes a physical dynamical system [Definition 2.2 and Definition 2.3 (in Section 2.1) for the IO system (2.1) (in Section 2.1), Definition 2.5 and Definition 2.6 for the ISO system (2.14a) and (2.14b) (in Section 2.2), Definition 2.7 and Definition 2.8 (in Section 2.3) for the IIO system (2.18a) and (2.18b) (in Section 2.3)]. Besides, we will define the (minimal) dynamical dimension of a system realization in the same sense as the (minimal) system dimension, respectively:

Definition 12.3 *(a)* ***The dynamical dimension of the system (realization)*** *denoted by* **ddim** *is, respectively, the number of initial conditions that determine uniquely the output response of the system (realization) to an arbitrary input vector function* $\mathbf{i}(\cdot)$ *and to arbitrary initial conditions; or equivalently,* ***the dynamical dimension of the system (realization)*** *is the degree of the characteristic polynomial of the system (realization).*

(b) ***The least (the minimal) dynamical dimension of the system (realization)*** *denoted by* **mddim** *is the number of independent initial conditions that must be known (i.e., the minimal number of the initial conditions that should be known) in order to determine uniquely the output response of the system (realization) to an arbitrary input vector function* $\mathbf{i}(\cdot)$ *and to arbitrary initial conditions; or equivalently,* ***the least (the minimal) dynamical dimension of the system (realization)*** *is the degree of the minimal polynomial of the system (realization).*

This definition is general. It is valid for the *IO* systems, for the *ISO* systems, and for the *IIO* systems.

Note 12.2 *The dimension of the basic vector, which together with its shifts, describes the complete system dynamics (which is the dimension* N *of the output vector* $\mathbf{y}$ *for the IO systems, the dimension* n *of the state vector* $\mathbf{x}$ *for the ISO systems, the dimension* $\rho + N$ *of the vectors* $\mathbf{r}$ *and* $\mathbf{y}$ *for the IIO systems) and the system order (which is* ν *for the IO systems, one for the ISO systems, and* $\alpha + \nu$ *for the IIO systems), determine jointly the dynamical dimension of the system (realization):*

$$\text{the dynamical dimension } ddim_{IO} \text{ of the } IO \text{ system obeys}$$
$$\nu \leq ddim_{IO} \leq \nu N,$$
$$\text{the dynamical dimension } ddim_{ISO} \text{ of the } ISO \text{ system is its dimension,}$$
$$ddim_{ISO} = n,$$
$$\text{the dynamical dimension } ddim_{IIO} \text{ of the } IIO \text{ system satisfies}$$
$$\rho + N \leq ddim_{IIO} \leq \alpha\rho + \nu N.$$

Example 12.1 *The dynamical dimension $ddim_{IO}$ of the three-dimensional second-order IO system (Example 2.2, in Section 2.1)*

$$\begin{bmatrix} 3 & 0 & 0 \\ 0 & 0 & 0 \\ 0 & 0 & 0 \end{bmatrix} E^2\mathbf{Y}(k) + \begin{bmatrix} 0 & 0 & 0 \\ 0 & 1 & 0 \\ 0 & 0 & 0 \end{bmatrix} E^1\mathbf{Y}(k) + \begin{bmatrix} 0 & 0 & 0 \\ 0 & 0 & 0 \\ 0 & 0 & 1 \end{bmatrix} E^0\mathbf{Y}(k) =$$
$$= \begin{bmatrix} 2 & 0 \\ 0 & 1 \\ 1 & 0 \end{bmatrix} E^0\mathbf{I}(k) + \begin{bmatrix} 1 & 0 \\ 0 & 1 \\ 1 & 1 \end{bmatrix} E^2\mathbf{I}(k)$$

is determined by the degree of the system characteristic polynomial,

$$\deg\left[\det\left(\sum_{r=0}^{r=\nu=2} \boldsymbol{A}_r z^r\right)\right] = \deg\begin{vmatrix} 3z^2 & 0 & 0 \\ 0 & z & 0 \\ 0 & 0 & 1 \end{vmatrix} = \deg\left(3z^3\right) = 3 = \eta > \nu = 2.$$

It equals 3, $ddim_{IO} = 3$. It is bigger than the system order (2), and equal to the system dimension ($dim_{IO} = 3$), hence less than their product $2 \times 3 = 6$.

Example 12.2 *The second-order three-dimensional IO system (Example 2.3, in Section 2.1)*

$$\begin{bmatrix} 3 & 0 & 1 \\ 2 & 0 & 0 \\ 0 & 1 & 1 \end{bmatrix} E^2\mathbf{Y}(k) + \begin{bmatrix} 0 & 0 & 0 \\ 0 & 1 & 0 \\ 0 & 0 & 0 \end{bmatrix} E^1\mathbf{Y}(k) + \begin{bmatrix} 0 & 0 & 0 \\ 0 & 0 & 0 \\ 0 & 0 & 1 \end{bmatrix} E^0\mathbf{Y}(k) =$$
$$= \begin{bmatrix} 2 & 0 \\ 0 & 1 \\ 1 & 0 \end{bmatrix} E^0\mathbf{I}(k) + \begin{bmatrix} 1 & 0 \\ 0 & 1 \\ 1 & 1 \end{bmatrix} E^2\mathbf{I}(k)$$

induces

$$\deg\left[\det\left(\sum_{r=0}^{r=\nu=2} \boldsymbol{A}_r z^r\right)\right] = \deg\begin{vmatrix} 3z^2 & 0 & z^2 \\ 2z^2 & z & 0 \\ 0 & z^2 & z^2+1 \end{vmatrix} =$$
$$= \deg\left(2z^6 + 3z^5 + 3z^3\right) = 6 = \eta = 2 \times 3 = \nu N.$$

Its dynamical dimension equals 6, $ddim_{IO} = 6$. In this case, the product of the system order (2) and of the system dimension ($dim_{IO} = 3$) equals the degree of the system characteristic polynomial; i.e., equals the system dynamical dimension, $\nu N = 2 \times 3 = 6 = \eta = ddim_{IO}$.

The Definition 12.3 implies:

Proposition 12.1 *[68] A system realization is* ***the minimal system realization*** *(or equivalently:* ***the irreducible system realization****) if and only if its characteristic polynomial is its minimal polynomial.*

We treat in more details this topic separately for the *IO* systems, the *ISO* systems, and the *IIO* systems in the sequel.

12.2 On realization and minimal realization

12.2.1 Minimal realization of the transfer function matrix

The realization of the system transfer function matrix $\mathbf{G}(z)$ is valid and useful only under nonzero input and under all zero initial conditions because it is defined and holds only for all zero initial conditions. C.-T. Chen noted correctly that *the dynamic equation realization gives only the same zero-state response of the system [18, p. 155]*. He proved the theorem [18, Theorem 4-10, p. 157] that it is valid in the framework of the *ISO* systems (see also [5, Theorem 3.3, p. 391]). Based on the proof of the theorem it can be verified that the following theorem is the discrete analogue of the former one:

Theorem 12.1 *[18, Theorem 4-10, p. 157] A transfer function matrix $\mathbf{G}(z)$ is realizable by a finite-dimensional linear time-invariant dynamical equation if and only if $\mathbf{G}(z)$ is a proper rational matrix (the degree of the numerator of every its entry $g_{ij}(z)$ does not exceed the degree of its denominator polynomial $\Delta(z)$, i.e., $lim[|g_{ij}(z)| : |z| \longrightarrow \infty] < \infty$, $\forall i, j$).*

We have used in the system and control theories the following definitions and the related criteria only in the framework of the *ISO* systems:

Definition 12.4 *A matrix quadruple ($\mathbf{A}$, $\mathbf{B}$, $\mathbf{C}$, $\mathbf{D}$) is* ***a realization of a given ISO system transfer function matrix*** $\mathbf{G}_{ISO}(z)$ *if and only if it is a realization of an ISO dynamical system, the transfer function matrix of which is $\mathbf{G}_{ISO}(z)$, i.e., if and only if*

$$\mathbf{C}(z\mathbf{I}-\mathbf{A})^{-1}\mathbf{B}+\mathbf{D}=\mathbf{G}_{ISO}(z).$$

Definition 12.5 *[5, Definition 3.2, p. 394], [18, p. 232]* ***The minimal (least-dimension, irreducible) ISO realization of a transfer function matrix*** $\mathbf{G}(z)$ *is its ISO realization with the least possible dynamical dimension.*

Theorem 12.2 *[5, Theorem 3.9, p. 395], [18, p. 233] For an n-dimensional ISO realization of $G(s)$ to be minimal, it is necessary and sufficient that it is both controllable and observable.*

Comment 12.1 *The dynamical dimension of the ISO system equals the dimension of the system state vector* $\mathbf{x}$ *because the order of the ISO system is minimal by the definition — it is equal to one.*

However, the order of the IO systems and the IIO systems can be greater than one. Hence, the minimum of the realization concerns both the dimension of the basic vector describing the system dynamics (which is the output vector $\mathbf{y}$ *for the IO systems) and the system order in general, i.e., the minimal realization is governed by the dimension and order of the system state vector (which is the extended output vector* $\mathbf{y}^{\nu-1}$ *for the IO systems).*

The actual linear control theory and the linear system theory have recognized only the *ISO* minimal realization of the transfer function matrix $\boldsymbol{G}(z)$ [5, Theorem 3.3, p. 391], [18] and [85]. We broaden it to the *IO* systems and the *IIO* systems, all for any initial conditions.

12.2.2 Realization and minimal realization of the full transfer function matrix and the system

Both initial conditions and the input vector act simultaneously on the system in reality. The full transfer function matrix $\boldsymbol{F}(z)$ is adequate to reflect the reality rather than the transfer function matrix $\boldsymbol{G}(z)$.

We need to distinguish the (minimal) realization of the system transfer function matrix $\boldsymbol{G}(z)$ from the (minimal) realization of the system itself and from the (minimal) realization of its full transfer function matrix $\boldsymbol{F}(z)$.

Definition 2.4 (in Section 2.2) and Definition 12.4 justify the following general definition:

Definition 12.6 ***A realization of the system full transfer function matrix*** $\boldsymbol{F}(z)$ *is a realization of a linear dynamical system if and only if its full transfer function matrix is the given* $\boldsymbol{F}(z)$.

This is the general definition. It holds for the *IO* systems, the *ISO* systems, and the *IIO* systems.

Comment 12.2 *A realization of the system full transfer function matrix* $\boldsymbol{F}(z)$ *is also a realization of the system itself, and vice versa.*

A realization of $\boldsymbol{G}(z)$ need not be a realization of the corresponding $\boldsymbol{F}(z)$. Theorem 12.1 is not valid for $\boldsymbol{F}(z)$. We will investigate this for different classes of the systems in the sequel. We refer to Definition 12.3 (in Section 12.1) in what follows.

Definition 12.7 *If and only if there is a realization of a linear dynamical system such that its full transfer function matrix is the nondegenerate form $\mathbf{F}_{nd}(z)$ of a given $\mathbf{F}(z)$, then such realization is* ***the minimal (the least dynamical dimensional, the least order, the irreducible) realization of the system full transfer function matrix*** $\mathbf{F}(z)$.

Comment 12.3 *The minimal (the least dynamical dimensional, the least order, the irreducible) realization of the system full transfer function matrix $\mathbf{F}(z)$ is also the minimal realization of the system itself, and vice versa.*

The preceding definition and comment together with Definition 12.3, in Section 12.1 lead to the following:

Comment 12.4 *The minimal system realization guarantees the minimal dynamical dimension of the system.*

Theorem 12.2 is not applicable to $\boldsymbol{F}(z)$ even in the framework of the *ISO* systems.

All submatrices of $\boldsymbol{F}(z)$ should be compatible, i.e., every submatrix of $\boldsymbol{F}(z)$ should be well related to the corresponding input action vector (either to the $Z-$transform of the input vector or to the corresponding initial vector). This condition is trivially satisfied for $\boldsymbol{G}(z)$ because it is its only one (single) submatrix related to the (single) input action (i.e., related to the $Z-$transform of the input vector) since all the initial conditions are then considered equal to zero vectors.

We will study the (minimal) realization of $\boldsymbol{F}(z)$ for all three classes of the systems treated in this book.

12.3 Realizations of $\boldsymbol{F}(z)$ of *IO* systems

The minimal realization of the system transfer function matrix $\boldsymbol{G}(z)$, which is the well-known notion, is crucially different from the minimal realization of the system full transfer function matrix $\boldsymbol{F}(z)$, which is defined in general in Definition 12.3, in Section 12.1, and in Definition 12.7, in Section 12.2. We specify it in more details for the *IO* system (3.55), i.e., (3.56) (in Subsection 3.5.2) in the sequel.

Definitions 12.3 (12.1), 12.6 and 12.7 (in Section 12.2) are the basis for what follows.

Definition 12.8 *If and only if there is a quadruple* $\left(\nu,\ \mu,\ \boldsymbol{A}^{(\nu)},\ \boldsymbol{B}^{(\mu)}\right)$ *that determines an IO realization of an IO dynamical linear system such that its full transfer function matrix is the nondegenerate form* $\boldsymbol{F}_{IOnd}(z)$ *of the given* $\boldsymbol{F}_{IO}(z)$ *then such IO realization is* ***the minimal (irreducible) IO realization of the system full transfer function matrix*** $\boldsymbol{F}_{IO}(z)$.

Comment 12.5 *The minimal IO system realization ensures the minimal dynamical dimension of the IO system.*

The conditions for the realizability of the full system transfer function matrix $\boldsymbol{F}_{IO}(z)$ are different from the realizability conditions of the system transfer function matrix $\boldsymbol{G}_{ISO}(z)$ (Theorem 12.2 in Section 12.2). Example 8.5 (in Section 8.1) illustrates that $\boldsymbol{F}_{IO}(z)$ can be improper, which violates the necessary and sufficient condition for the realizability of $\boldsymbol{G}(z)$ (Theorem 12.1, in Subsection 12.2.1).

Theorem 12.3 *In order for a quadruple* $\left(\nu,\ \mu,\ \boldsymbol{A}^{(\nu)},\ \boldsymbol{B}^{(\mu)}\right)$ *to determine the (minimal) input-output (IO) system realization (3.55), it is necessary and sufficient that*

$$\left(\boldsymbol{A}^{(\nu)}\boldsymbol{S}_N^{(\nu)}(z)\right)^{-1}\begin{cases}\left[\begin{array}{ccc}\boldsymbol{B}^{(\mu)}\boldsymbol{S}_M^{(\mu)}(z) & -\boldsymbol{B}^{(\mu)}\boldsymbol{Z}_M^{(\mu)}(z) & \boldsymbol{A}^{(\nu)}\boldsymbol{Z}_N^{(\nu)}(z)\end{array}\right], & \mu\geq 1\\ \left[\begin{array}{cc}\boldsymbol{B}^{(\mu)}\boldsymbol{S}_M^{(\mu)}(z) & \boldsymbol{A}^{(\nu)}\boldsymbol{Z}_N^{(\nu)}(z)\end{array}\right], & \mu=0\end{cases}$$

is equal to (the nondegenerate form of) the full system transfer function matrix $\boldsymbol{F}_{IO}(z)$, *respectively.*

Proof. *Necessity.* Let a quadruple $\left(\nu,\ \mu,\ \boldsymbol{A}^{(\nu)},\ \boldsymbol{B}^{(\mu)}\right)$ determine the (minimal) input-output system realization (3.55). Then, the following holds due to Definition 12.6 (in Section 12.2) (Definition 6.1, in Chapter 6, Definition 12.1 through Definition 12.3, Proposition 12.1, in Section 12.1, Definition 12.7 in Section 12.2, and Definition 12.8), respectively, and (8.6a) and (8.6b) (Theorem 8.1, in Section 8.1):

$$\left(\boldsymbol{A}^{(\nu)}\boldsymbol{S}_N^{(\nu)}(z)\right)^{-1}\begin{cases}\left[\begin{array}{ccc}\boldsymbol{B}^{(\mu)}\boldsymbol{S}_M^{(\mu)}(z) & -\boldsymbol{B}^{(\mu)}\boldsymbol{Z}_M^{(\mu)}(z) & \boldsymbol{A}^{(\nu)}\boldsymbol{Z}_N^{(\nu)}(z)\end{array}\right],\\ \mu\geq 1,\\ \left[\begin{array}{cc}\boldsymbol{B}^{(\mu)}\boldsymbol{S}_M^{(\mu)}(z) & \boldsymbol{A}^{(\nu)}\boldsymbol{Z}_N^{(\nu)}(z)\end{array}\right], \ \mu=0\end{cases} =$$

$$= \left[\det\left(\mathbf{A}^{(\nu)}\mathbf{S}_N^{(\nu)}(z)\right)\right]^{-1}\cdot\left\{\mathrm{adj}\left(\mathbf{A}^{(\nu)}\mathbf{S}_N^{(\nu)}(z)\right)\right\}\cdot$$

$$\cdot\left\{\begin{array}{l}\left[\begin{array}{ccc}\mathbf{B}^{(\mu)}\mathbf{S}_M^{(\mu)}(z) & -\mathbf{B}^{(\mu)}\mathbf{Z}_M^{(\mu)}(z) & \mathbf{A}^{(\nu)}\mathbf{Z}_N^{(\nu)}(z)\end{array}\right],\ \mu\geq 1\\ \left[\begin{array}{cc}\mathbf{B}^{(\mu)}\mathbf{S}_M^{(\mu)}(z) & \mathbf{A}^{(\nu)}\mathbf{Z}_N^{(\nu)}(z)\end{array}\right],\ \mu=0\end{array}\right. = \mathbf{F}_{IO}(z),$$

which proves necessity of the condition.

Sufficiency. Let the conditions be valid. The quadruple $\left(\nu,\ \mu,\ \mathbf{A}^{(\nu)},\ \mathbf{B}^{(\mu)}\right)$ determines (the minimal) input-output realization of $\mathbf{F}_{IO}(z)$, which is also (the minimal) IO realization of the IO system (3.55) due to Definition 12.6 (in Section 12.2), Definition 12.7 (in Section 12.2), and Definition 12.8, respectively. ■

Note 12.3 *The (full) IO transfer function matrix of the ISO system is, respectively, the system (full) transfer function matrix obtained from the IO system obtained from its ISO mathematical model (Note 7.7, in Section 7.3). It carries the subscript IO. However, the subscript ISO denotes the (full) transfer function matrix obtained from the ISO system.*

Example 12.3 *We will consider the ISO systems of Example 8.7 (in Section 8.2), which are analogy to the s−complex case given in [5, Example 3.8], in the framework of the IO systems. The given first three input-state-output (ISO) realizations* $(\mathbf{A}, \mathbf{B}, \mathbf{C}, \mathbf{D})$ *yield the degenerate transfer functions*

$$\mathbf{G}_{ISO1}(z) = \mathbf{G}_{ISO2}(z) = \mathbf{G}_{ISO3}(z) = \mathbf{G}_{ISO1-3}(z) = \frac{z-1}{z^2-1}$$

that are reducible to their nondegenerate forms

$$\mathbf{G}_{ISO1nd}(z) = \mathbf{G}_{ISO2nd}(z) = \mathbf{G}_{ISO3nd}(z) = \mathbf{G}_{ISO1-3nd}(z) = \frac{1}{z+1}.$$

The degenerate transfer functions imply the following IO system:

$$E^2y(k) - E^0y(k) = -E^0i(k) + E^1i(k), \tag{12.1}$$

and its IO realization (Definition 2.1, in Section 2.1)

$$(\nu, \mu, \mathbf{A}^{(\nu)}, \mathbf{B}^{(\mu)}) = \left(\begin{array}{cccc} 2 & 1 & \left[\begin{array}{ccc}-1 & 0 & 1\end{array}\right] & \left[\begin{array}{cc}-1 & 1\end{array}\right]\end{array}\right). \tag{12.2}$$

We can apply (8.6a) (in Section 8.1) to $\mathbf{F}_{IOISO1-3}(z)$ *by using the IO realization (12.2) of* $\mathbf{G}_{ISO1-3}(z)$ *in order to determine the full IO transfer function matrix*

$$\mathbf{F}_{IOISO1-3}(z) =$$

$$= \left(\mathbf{A}^{(\nu)} \mathbf{S}_N^{(\nu)}(z)\right)^{-1} \left[\begin{array}{ccc} \mathbf{B}^{(\mu)} \mathbf{S}_M^{(\mu)}(z) & -\mathbf{B}^{(\mu)} \mathbf{Z}_M^{(\mu)}(z) & \mathbf{A}^{(\nu)} \mathbf{Z}_N^{(\nu)}(z) \end{array} \right] =$$

$$= \left(\mathbf{A}^{(2)} \mathbf{S}_1^{(2)}(z)\right)^{-1} \left[\begin{array}{ccc} \mathbf{B}^{(1)} \mathbf{S}_1^{(1)}(z) & -\mathbf{B}^{(1)} \mathbf{Z}_1^{(1)}(z) & \mathbf{A}^{(2)} \mathbf{Z}_1^{(2)}(z) \end{array} \right],$$

$$\mathbf{A}^{(2)} \mathbf{S}_1^{(2)}(z) = \left[\begin{array}{ccc} -1 & 0 & 1 \end{array} \right] \left[\begin{array}{c} 1 \\ z \\ z^2 \end{array} \right] = \left[z^2 - 1\right],$$

$$\mathbf{B}^{(1)} \mathbf{S}_1^{(1)}(z) = \left[\begin{array}{cc} -1 & 1 \end{array} \right] \left[\begin{array}{c} 1 \\ z \end{array} \right] = [z - 1],$$

$$\mathbf{B}^{(1)} \mathbf{Z}_1^{(1)}(z) = \left[\begin{array}{cc} -1 & 1 \end{array} \right] \left[\begin{array}{c} 0 \\ z \end{array} \right] = [z],$$

$$\mathbf{A}^{(2)} \mathbf{Z}_1^{(2)}(z) = \left[\begin{array}{ccc} -1 & 0 & 1 \end{array} \right] \left[\begin{array}{cc} 0 & 0 \\ z & 0 \\ z^2 & z \end{array} \right] = \left[\begin{array}{cc} z^2 & z \end{array} \right] \Longrightarrow$$

$$\mathbf{F}_{IOISO1-3}(z) = \frac{\left[\begin{array}{cccc} z-1 & -z & z^2 & z \end{array} \right]}{z^2 - 1}. \tag{12.3}$$

The so obtained full IO transfer function matrix $\mathbf{F}_{IOISO1-3}(z)$ *(12.3), which resulted from the realization (12.2) of* $\mathbf{G}_{ISO1-3}(z)$*, is nondegenerate,*

$$\mathbf{F}_{IOISO1-3}(z) = \frac{\left[\begin{array}{cccc} z-1 & -z & z^2 & z \end{array} \right]}{z^2 - 1} = \mathbf{F}_{IOISO1-3nd}(z). \tag{12.4}$$

The IO system (12.1) induces directly $\mathbf{F}_{IO1-3}(z)$ *as follows:*

$$(z^2 - 1)Y(z) = (-1 + z)I(z) - zi_0 + z^2 y_0 + zy_1 \Longrightarrow$$

$$Y(z) = (z^2 - 1)^{-1} \left[\begin{array}{cccc} z-1 & -z & z^2 & z \end{array} \right] \left[\begin{array}{c} I(z) \\ i_0 \\ y_0 \\ Ey_0 = y_1 \end{array} \right] \Longrightarrow$$

$$\mathbf{F}_{IO1-3}(z) = (z^2 - 1)^{-1} \left[\begin{array}{cccc} z-1 & -z & z^2 & z \end{array} \right] = \mathbf{F}_{IOISO1-3}(z). \tag{12.5}$$

This verifies (12.3) and confirms the validity of (12.4) for $F_{IO1-3}(s)$*,*

$$\mathbf{F}_{IO1-3}(z) = (z^2 - 1)^{-1} \left[\begin{array}{cccc} z-1 & -z & z^2 & z \end{array} \right] = \mathbf{F}_{IO1-3nd}(z).$$

The nondegenerate form $\mathbf{G}_{ISO1-3nd}(z)$ *of* $\mathbf{G}_{ISO1-3}(z)$ *yields its minimal IO system in the form of the reduced differential equation*

$$Ey(k) + y(k) = i(k), \tag{12.6}$$

and its IO realization, which we will call the reduced IO realization (the subscript "rd"), reads

$$\left(\nu_{1-3rd}, \mu_{1-3rd}, \boldsymbol{A}_{1-3rd}^{(\nu_{1-3rd})}, \boldsymbol{B}_{1-3rd}^{(\mu_{\min})}\right) = \left(1 \quad 0 \quad \left[\begin{array}{cc} 1 & 1 \end{array}\right] \quad 1 \right). \tag{12.7}$$

The reduced IO system (12.6), i.e., its IO realization (12.7), determines the corresponding $\boldsymbol{F}_{IO1-3rd}(z)$*,*

$$\begin{gathered} Y(z) = (z+1)^{-1} \left[\begin{array}{cc} 1 & z \end{array}\right] \left[\begin{array}{c} I(z) \\ y_0 \end{array}\right] \Longrightarrow \\ \boldsymbol{F}_{IO1-3rd}(z) = (z+1)^{-1} \left[\begin{array}{cc} 1 & z \end{array}\right], \end{gathered} \tag{12.8}$$

which is nondegenerate, hence the irreducible, and different in this case from the nondegenerate form $\boldsymbol{F}_{IO1-3nd}(z)$ *of* $\boldsymbol{F}_{IO1-3}(z)$*,*

$$\begin{gathered} \boldsymbol{F}_{IO1-3rd}(z) = (z+1)^{-1} \left[\begin{array}{cc} 1 & z \end{array}\right] \neq \\ \neq (z^2-1)^{-1} \left[\begin{array}{cccc} z-1 & -z & z^2 & z \end{array}\right] = \boldsymbol{F}_{IO1-3nd}(z) = \boldsymbol{F}_{IO1-3}(z). \end{gathered}$$

The reducible $\boldsymbol{F}_{IO1-3rd}(z)$ *is related only to* y_0*, but nondegenerate* $\boldsymbol{F}_{IO1-3nd}(z)$ *corresponds to* i_0*,* y_0*, and* y_1*. This is the consequence of different orders of the IO models (12.6) and (12.1) to which they are related, respectively.*

However, the fourth ISO realization $(\boldsymbol{A}, \boldsymbol{B}, \boldsymbol{C}, \boldsymbol{D})$ *implies directly the nondegenerate transfer function*

$$\boldsymbol{G}_{ISO4}(z) = \boldsymbol{G}_{ISO4nd}(z) = \frac{1}{z+1} = \boldsymbol{G}_{ISOnd}(z) = \boldsymbol{G}_{ISO1-3nd}(z).$$

It induces the IO system (12.6) and the full IO transfer function matrix,

$$\boldsymbol{F}_{IOISO4}(z) = (z+1)^{-1} \left[\begin{array}{cc} 1 & z \end{array}\right] = \boldsymbol{F}_{IO4nd}(z) = \boldsymbol{F}_{IO1-3rd}(z). \tag{12.9}$$

Example 12.4 *Let*

$$\mathbf{i} \in \mathcal{R}^2 \Longrightarrow M = 2, \ \mathbf{y} \in \mathcal{R}^3 \Longrightarrow N = 3,$$

$$\boldsymbol{H}(z) = \left[\begin{array}{ccc} \frac{z^2+1}{3z^2+3z^4+2z^5} & \frac{z}{3z^2+2z^3+3} & -\frac{1}{3z^2+2z^3+3} \\ \frac{-2z^2-2}{3z+3z^3+2z^4} & \frac{3z^2+3}{3z+3z^3+2z^4} & 2\frac{z}{3z^2+2z^3+3} \\ 2\frac{z}{3z^2+2z^3+3} & -3\frac{s}{3z^2+2z^3+3} & \frac{3}{3z^2+2z^3+3} \end{array}\right] \cdot$$

$$\cdot \left[\begin{array}{cccccccccccc} 2+z^2 & 0 & z^2 & 0 & z & 0 & 3z^2 & 0 & 3+z^2 & z^2 & 3z & z \\ 0 & 1+z^2 & 0 & z^2 & 0 & z & 2z^2 & z & 0 & 2z & 0 & 0 \\ 1+z^2 & z^2 & z^2 & z^2 & z & z & 0 & z^2 & z^2 & 0 & z & z \end{array}\right].$$

We wish to test whether $\boldsymbol{H}(z)$ can be the full transfer function matrix $\boldsymbol{F}_{IO}(z)$ of an IO system. We set it at first in the form

$$\boldsymbol{H}(z)=\frac{\begin{bmatrix} z^2+1 & z^3 & -z^2 \\ -2z^3-2z & 3z^3+3z & 2z^3 \\ 2z^3 & -3z^3 & 3z^2 \end{bmatrix}}{2z^5+3z^4+3z^2}\cdot$$

$$\cdot\begin{bmatrix} 2+z^2 & 0 & z^2 & 0 & z & 0 & 3z^2 & 0 & 3+z^2 & z^2 & 3z & z \\ 0 & 1+z^2 & 0 & z^2 & 0 & z & 2z^2 & z & 0 & 2z & 0 & 0 \\ 1+z^2 & z^2 & z^2 & z^2 & z & z & 0 & z^2 & z^2 & 0 & z & z \end{bmatrix},$$

and now in the form

$$\boldsymbol{H}(z)=\begin{bmatrix} 3z^2 & 0 & z^2 \\ 2z^2 & z & 0 \\ 0 & z^2 & z^2+1 \end{bmatrix}^{-1}\cdot$$

$$\cdot\begin{bmatrix} 2+z^2 & 0 & z^2 & 0 & z & 0 & 3z^2 & 0 & 3+z^2 & z^2 & 3z & z \\ 0 & 1+z^2 & 0 & z^2 & 0 & z & 2z^2 & z & 0 & 2z & 0 & 0 \\ 1+z^2 & z^2 & z^2 & z^2 & z & z & 0 & z^2 & z^2 & 0 & z & z \end{bmatrix},$$

from which we deduce:

$$\sum_{r=0}^{r=\nu}\boldsymbol{A}_r z^r=\boldsymbol{A}^{(\nu)}\boldsymbol{S}_N^{(\nu)}(z)=\boldsymbol{A}^{(\nu)}\boldsymbol{S}_3^{(\nu)}(z)=\begin{bmatrix} 3z^2 & 0 & z^2 \\ 2z^2 & z & 0 \\ 0 & z^2 & z^2+1 \end{bmatrix}\Longrightarrow$$

$$\nu=2,\ \boldsymbol{A}_0=\begin{bmatrix} 0 & 0 & 0 \\ 0 & 0 & 0 \\ 0 & 0 & 1 \end{bmatrix},\ \boldsymbol{A}_1=\begin{bmatrix} 0 & 0 & 0 \\ 0 & 1 & 0 \\ 0 & 0 & 0 \end{bmatrix},\ \boldsymbol{A}_2=\begin{bmatrix} 3 & 0 & 1 \\ 2 & 0 & 0 \\ 0 & 1 & 1 \end{bmatrix}$$

$$\sum_{r=0}^{r=\mu}\boldsymbol{B}_r z^r=\boldsymbol{B}^{(\mu)}\boldsymbol{S}_M^{(\mu)}(z)=\boldsymbol{B}^{(\mu)}\boldsymbol{S}_2^{(\mu)}(z)=\begin{bmatrix} 2+z^2 & 0 \\ 0 & 1+z^2 \\ 1+z^2 & z^2 \end{bmatrix}\Longrightarrow$$

$$\mu=2,\ \boldsymbol{B}_0=\begin{bmatrix} 2 & 0 \\ 0 & 1 \\ 1 & 0 \end{bmatrix},\ \boldsymbol{B}_1=\begin{bmatrix} 0 & 0 \\ 0 & 0 \\ 0 & 0 \end{bmatrix},\ \boldsymbol{B}_2=\begin{bmatrix} 1 & 0 \\ 0 & 1 \\ 1 & 1 \end{bmatrix}.$$

This data yields

$$\boldsymbol{A}^{(2)}\boldsymbol{S}_3^{(2)}(z)=\begin{bmatrix} 3z^2 & 0 & z^2 \\ 2z^2 & z & 0 \\ 0 & z^2 & z^2+1 \end{bmatrix},\boldsymbol{B}^{(2)}\boldsymbol{S}_2^{(2)}(z)=\begin{bmatrix} 2+z^2 & 0 \\ 0 & 1+z^2 \\ 1+z^2 & z^2 \end{bmatrix},$$

$$\boldsymbol{A}^{(2)}\boldsymbol{Z}_3^{(2)}(z)=\begin{bmatrix}0&0&0&0&0&0&3&0&1\\0&0&0&0&1&0&2&0&0\\0&0&1&0&0&0&0&1&1\end{bmatrix}\begin{bmatrix}0&0&0&0&0&0\\0&0&0&0&0&0\\0&0&0&0&0&0\\z&0&0&0&0&0\\0&z&0&0&0&0\\0&0&z&0&0&0\\z^2&0&0&z&0&0\\0&z^2&0&0&z&0\\0&0&z^2&0&0&z\end{bmatrix}=$$

$$=\begin{bmatrix}3z^2&0&z^2&3z&0&z\\2z^2&z&0&2z&0&0\\0&z^2&z^2&0&z&z\end{bmatrix},$$

$$-\boldsymbol{B}^{(2)}\boldsymbol{Z}_2^{(2)}(z)=-\begin{bmatrix}2&0&0&0&1&0\\0&1&0&0&0&1\\1&0&0&0&1&1\end{bmatrix}\begin{bmatrix}0&0&0&0\\0&0&0&0\\z&0&0&0\\0&z&0&0\\z^2&0&z&0\\0&z^2&0&z\end{bmatrix}=$$

$$=-\begin{bmatrix}z^2&0&z&0\\0&z^2&0&z\\z^2&z^2&z&z\end{bmatrix}.$$

The obtained data determine the full transfer function matrix $\boldsymbol{F}_{IO}(z)$ of an IO system

$$\boldsymbol{F}_{IO}(z)=\left[\boldsymbol{A}^{(2)}\boldsymbol{S}_3^{(2)}(z)\right]^{-1}\left[\begin{array}{ccc}\boldsymbol{B}^{(2)}\boldsymbol{S}_2^{(2)}(z) & -\boldsymbol{B}^{(2)}\boldsymbol{Z}_2^{(2)}(z) & \boldsymbol{A}^{(2)}\boldsymbol{Z}_3^{(2)}(z)\end{array}\right]=$$

$$=\begin{bmatrix}3z^2&0&z^2\\2z^2&z&0\\0&z^2&z^2+1\end{bmatrix}^{-1}\cdot$$

$$\cdot\begin{bmatrix}2+z^2&0&-z^2&0&-z&0&3z^2&0&z^2&3z&0&z\\0&1+z^2&0&-z^2&0&-z&2z^2&z&0&2z&0&0\\1+z^2&z^2&-z^2&-z^2&-z&-z&0&z^2&z^2&0&z&z\end{bmatrix}\neq$$

$$\neq \begin{bmatrix} 3z^2 & 0 & z^2 \\ 2z^2 & z & 0 \\ 0 & z^2 & z^2+1 \end{bmatrix}^{-1} \cdot$$

$$\cdot \begin{bmatrix} 2+z^2 & 0 & z^2 & 0 & z & 0 & 3z^2 & 0 & 3+z^2 & z^2 & 3z & z \\ 0 & 1+z^2 & 0 & z^2 & 0 & z & 2z^2 & z & 0 & 2z & 0 & 0 \\ 1+z^2 & z^2 & z^2 & z^2 & z & z & 0 & z^2 & z^2 & 0 & z & z \end{bmatrix} = \mathbf{H}(z).$$

The given $\mathbf{H}(z)$ cannot be the full transfer function matrix $\mathbf{F}_{IO}(z)$ of any IO system because it does not obey Theorem 12.3. Notice that

$$\left[\mathbf{A}^{(2)}\mathbf{S}_3^{(2)}(z)\right]^{-1}\mathbf{B}^{(2)}\mathbf{S}_2^{(2)}(z) =$$

$$\begin{bmatrix} 3z^2 & 0 & z^2 \\ 2z^2 & z & 0 \\ 0 & z^2 & z^2+1 \end{bmatrix}^{-1} \begin{bmatrix} 2+z^2 & 0 \\ 0 & 1+z^2 \\ 1+z^2 & z^2 \end{bmatrix} =$$

$$= \frac{\begin{bmatrix} z^2+1 & z^3 & -z^2 \\ -2z^3-2z & 3z^3+3z & 2z^3 \\ 2z^3 & -3z^3 & 3z^2 \end{bmatrix} \begin{bmatrix} 2+z^2 & 0 \\ 0 & 1+z^2 \\ 1+z^2 & z^2 \end{bmatrix}}{2z^5+3z^4+3z^2}$$

is proper so that it can be the transfer function matrix $\mathbf{G}_{IO}(z)$ of an IO system in spite $\mathbf{H}(z)$ cannot be the full transfer function matrix $\mathbf{F}_{IO}(z)$ of any IO system. This warns us that the IO realizability of $\mathbf{G}_{IO}(z)$ does not guarantee the IO realizability of $\mathbf{H}(z)$.

Conclusion 12.1 *If an $\mathbf{H}(z)$ matrix is given, and a realizable transfer function matrix $\mathbf{G}_{IO}(z)$ of an IO system is its first submatrix, then it does not guarantee that $\mathbf{H}(z)$ can be the full transfer function matrix $\mathbf{F}_{IO}(z)$ of any IO system.*

However, if an $\mathbf{H}(z)$ matrix can be the full transfer function matrix $\mathbf{F}_{IO}(z)$ of an IO system, then its first submatrix is the transfer function matrix $\mathbf{G}_{IO}(z)$ of the same IO system.

Conclusion 12.2 *The full transfer function matrix $\mathbf{F}_{IO}(z)$ of an IO system can be improper, while its transfer function matrix $\mathbf{G}_{IO}(z)$ cannot (Example 8.5, in Section 8.1).*

Conclusion 12.3 *The preceding analysis shows the incompleteness of the IO system studies via its transfer function matrix $\mathbf{G}_{IO}(z)$. For the complete analysis or synthesis of the IO system in the complex domain, we should use its full transfer function matrix $\mathbf{F}_{IO}(z)$.*

12.4 Realizations of $\mathbf{F}(z)$ of *ISO* systems

We should distinguish the minimal realization of the system transfer function matrix $\mathbf{G}(z)$, which is a well-defined notion, from the minimal realization of the system, which we defined in general (Definition 12.3, in Section 12.1), and which we will specify in more details for the ISO system (3.60a) and (3.60b) (in Subsection 3.5.3) in the sequel.

We refer to Definitions 12.3 (in Section 12.1), 12.6 and 12.7 (in Section 12.2).

Definition 12.9 *If and only if there is a quintuple* $(n,\ \mathbf{A}, \mathbf{B}, \mathbf{C}, \mathbf{D})$ *that determines an* ISO *realization of an* ISO *dynamical linear system such that its full transfer function matrix is the nondegenerate form* $\mathbf{F}_{ISOnd}(z)$ *of the given* $\mathbf{F}_{ISO}(z)$*, then such* ISO *realization is* ***the minimal (irreducible) ISO realization of the full transfer function matrix*** $\mathbf{F}_{ISO}(z)$.

Note 12.4 *The dynamical dimension of the* ISO *system and its realization reduce to their dimensions since their orders are fixed to one by the definition.*

The conditions for the realizability of $\mathbf{F}_{ISO}(z)$ are different from the realizability conditions of $\mathbf{F}_{IO}(z)$ (Theorem 12.3 in Section 12.3).

Theorem 12.4 *In order for a quintuple* $(n,\ \mathbf{A}, \mathbf{B}, \mathbf{C}, \mathbf{D})$ *to determine an* ISO *system realization (3.60a) and (3.60b), it is necessary and sufficient that*

$$\left[\ \mathbf{C}(z\mathbf{I}_n - \mathbf{A})^{-1}\mathbf{B} + \mathbf{D} \quad z\mathbf{C}(z\mathbf{I}_n - \mathbf{A})^{-1}\ \right] \tag{12.10}$$

is proper and that it is the system full ISO *transfer function matrix* $\mathbf{F}_{ISO}(z)$,

$$\left[\ \mathbf{C}(z\mathbf{I}_n - \mathbf{A})^{-1}\mathbf{B} + \mathbf{D} \quad z\mathbf{C}(z\mathbf{I}_n - \mathbf{A})^{-1}\ \right] = \mathbf{F}_{ISO}(z). \tag{12.11}$$

Proof. *Necessity.* Let a quintuple $(n,\ \mathbf{A}, \mathbf{B}, \mathbf{C}, \mathbf{D})$ determine an ISO system realization (3.60a) and (3.60b). Then it is an ISO realization of the system full ISO transfer function matrix $\mathbf{F}_{ISO}(z)$, i.e., let (12.11) hold, due to Definition 12.6 and Comment 12.3 (in Section 12.2). The first submatrix in (12.10) is the system transfer function matrix $\mathbf{G}_{ISO}(z)$. It is proper (Theorem 12.1, in Section 12.2). The submatrix is proper. Let us set the matrix (12.10) into the following form:

$$\left[\frac{\mathbf{C}\,\mathrm{adj}(z\mathbf{I}_n - \mathbf{A})\mathbf{B} + \det(z\mathbf{I}_n - \mathbf{A})\mathbf{D} \quad z\mathbf{C}\,\mathrm{adj}(z\mathbf{I}_n - \mathbf{A})}{\det(z\mathbf{I}_n - \mathbf{A})}\right].$$

Since

$$\deg\left[\text{adj}(z\mathbf{I}_n - \mathbf{A})\right] < \deg\left[\det(z\mathbf{I}_n - \mathbf{A})\right]$$

then

$$\deg\left[\mathbf{C}\,\text{adj}(z\mathbf{I}_n - \mathbf{A})\mathbf{B}\right] < \deg\left[\det(z\mathbf{I}_n - \mathbf{A})\right],$$
$$\deg\left[z\mathbf{C}\,\text{adj}(z\mathbf{I}_n - \mathbf{A})\right] = \deg\left[\det(zI_n - A)\right]$$

and

$$\deg\left[\det(z\mathbf{I}_n - \mathbf{A})\mathbf{D}\right] = \deg\left[\det(z\mathbf{I}_n - \mathbf{A})\right].$$

Hence, the matrix (12.10),

$$\left[\begin{array}{cc} \mathbf{C}(z\mathbf{I}_n - \mathbf{A})^{-1}\mathbf{B} + \mathbf{D} & z\mathbf{C}(z\mathbf{I}_n - \mathbf{A})^{-1} \end{array}\right],$$

is proper, which verifies that $\mathbf{C}(z\mathbf{I}_n-\mathbf{A})^{-1}\mathbf{B}+\mathbf{D}$ obeys Theorem 12.1 (in Section 12.2).

Sufficiency. Let the conditions hold. Hence, $(n,\ \mathbf{A}, \mathbf{B}, \mathbf{C}, \mathbf{D})$ is the *ISO* system realization (3.60a) and (3.60b) in view of Definition 2.4 (in Section 2.2), Definition 12.6, Comment 12.3 (in Section 12.2) and Definition 12.9. ■

This theorem does not guarantee that the quintuple $(n,\ \mathbf{A}, \mathbf{B}, \mathbf{C}, \mathbf{D})$, which determines the matrix (12.10), is the minimal realization of the *ISO* system (3.60a) and (3.60b).

Theorem 12.5 *In order for a quintuple* $(n,\ \mathbf{A}, \mathbf{B}, \mathbf{C}, \mathbf{D})$ *to determine a minimal ISO system realization (3.60a) and (3.60b), it is necessary and sufficient that*

$$\left[\begin{array}{cc} \mathbf{C}(z\mathbf{I}_n - \mathbf{A})^{-1}\mathbf{B} + \mathbf{D} & z\mathbf{C}(z\mathbf{I}_n - \mathbf{A})^{-1} \end{array}\right] \tag{12.12}$$

is proper and that it is the nondegenerate form $\mathbf{F}_{ISOnd}(z)$ *of the system full ISO transfer function matrix* $\mathbf{F}_{ISO}(z)$,

$$\left[\begin{array}{cc} \mathbf{C}(z\mathbf{I}_n - \mathbf{A})^{-1}\mathbf{B} + \mathbf{D} & z\mathbf{C}(z\mathbf{I}_n - \mathbf{A})^{-1} \end{array}\right] = \mathbf{F}_{ISOnd}(z). \tag{12.13}$$

Proof. *Necessity.* Let a quintuple $(n,\ \mathbf{A}, \mathbf{B}, \mathbf{C}, \mathbf{D})$ determine a minimal *ISO* system realization (3.60a) and (3.60b). Hence, the matrix 12.12 is proper due to Theorem 12.4 and it is the nondegenerate form $F_{ISOnd}(s)$ of the system full *ISO* transfer function matrix $\mathbf{F}_{ISO}(z)$ due to Definition 12.7 (in Section 12.2) and Definition 12.9.

Sufficiency. Let the conditions hold. The conditions of Theorem 12.4 are satisfied. The matrix 12.12 is minimal *ISO* system realization (3.60a) and (3.60b) due to Definition 12.7 and Definition 12.9. ■

Example 12.5 *We analyze further the systems presented in Example 8.7 (in Section 8.2), and Example 12.3 (in Section 12.3), which are analogous to the s−complex cases given in [5, Example 3.8].*

Are the minimal ISO realizations of the transfer functions $\mathbf{G}_{ISO1-3}(z)$ and of $\mathbf{G}_{ISO4}(z)$ also the ISO minimal realizations of the full transfer functions $\boldsymbol{F}_{ISO1-4}(z)$?

Are the full ISO transfer functions $\boldsymbol{F}_{ISO1}(z)$ through $\boldsymbol{F}_{ISO4}(z)$ also the full IO transfer functions $\boldsymbol{F}_{ISOIO1}(z)$ through $\boldsymbol{F}_{ISOIO4}(z)$?

Are the nondegenerate forms $\boldsymbol{F}_{ISO1nd}(z)$ through $\boldsymbol{F}_{ISO4nd}(z)$ of $\boldsymbol{F}_{ISO1}(z)$ through $\boldsymbol{F}_{ISO4}(z)$ realizable?

Let us consider each case separately.

1) $\boldsymbol{A}_1 = \begin{bmatrix} 0 & 1 \\ 1 & 0 \end{bmatrix}$, $\boldsymbol{B}_1 = \begin{bmatrix} 0 \\ 1 \end{bmatrix}$, $\boldsymbol{C}_1 = \begin{bmatrix} -1 & 1 \end{bmatrix}$, $\boldsymbol{D}_1 = 0 \Longrightarrow$

$$\boldsymbol{F}_{ISO1}(z) = \left(z^2 - 1\right)^{-1} \begin{bmatrix} z-1 & z(1-z) & z(z-1) \end{bmatrix},$$
$$\mathbf{G}_{ISO1}(z) = \frac{z-1}{z^2-1}. \tag{12.14}$$

They follow from the ISO system realization. They are degenerate. Their nondegenerate forms read

$$\boldsymbol{F}_{ISO1nd}(z) = (z+1)^{-1} \begin{bmatrix} 1 & -z & z \end{bmatrix},$$
$$\mathbf{G}_{ISO1nd}(z) = \frac{1}{z+1}. \tag{12.15}$$

The transfer function $\mathbf{G}_{ISO1}(z)$ induces, via (12.1), Example 12.3 (in Section 12.3),

$$\nu_1 = 2,\ \mu_1 = 1,\ N_1 = 1,\ M_1 = 1,$$
$$\boldsymbol{A}_1^{(2)} = \begin{bmatrix} -1 & 0 & 1 \end{bmatrix},\ \boldsymbol{B}_1^{(1)} = \begin{bmatrix} -1 & 1 \end{bmatrix}.$$

The degenerate $\boldsymbol{F}_{ISO1}(z)$ (12.14) should obey the following IO condition due to (8.6a) and (8.6b) (in Section 8.1) and due to Theorem 12.3 in order to be also $\boldsymbol{F}_{ISOIO1}(z)$ that follows from the IO realization of the given ISO system, Example 12.3,

$$\boldsymbol{F}_{ISO1}(z) = \left(z^2-1\right)^{-1} \begin{bmatrix} z-1 & z(1-z) & z(z-1) \end{bmatrix} =$$
$$= \left(\boldsymbol{A}_i^{(\nu)} \boldsymbol{S}_N^{(\nu)}(z)\right)^{-1} \begin{bmatrix} \boldsymbol{B}_i^{(\mu)} \boldsymbol{S}_M^{(\mu)}(z) & -\boldsymbol{B}_i^{(\mu)} \boldsymbol{Z}_M^{(\mu)}(z) & \boldsymbol{A}_i^{(\nu)} \boldsymbol{Z}_N^{(\nu)}(z) \end{bmatrix} =$$
$$= \boldsymbol{F}_{ISOIO1}(z),$$

where

$$\boldsymbol{A}_1^{(2)}\boldsymbol{S}_1^{(2)}(z) = \begin{bmatrix} -1 & 0 & 1 \end{bmatrix}\begin{bmatrix} 1 \\ z \\ z^2 \end{bmatrix} = \left[-1+z^2\right] = \left[z^2-1\right],$$

$$\boldsymbol{B}_1^{(1)}\boldsymbol{S}_1^{(1)}(z) = \begin{bmatrix} -1 & 1 \end{bmatrix}\begin{bmatrix} 1 \\ z \end{bmatrix} = [-1+z] = [z-1],$$

$$-\boldsymbol{B}_1^{(1)}\boldsymbol{Z}_1^{(1)}(z) = -\begin{bmatrix} -1 & 1 \end{bmatrix}\begin{bmatrix} 0 \\ z \end{bmatrix} = [-z] \neq [z(1-z)],$$

$$\boldsymbol{A}_1^{(2)}\boldsymbol{Z}_1^{(2)}(z) = \begin{bmatrix} -1 & 0 & 1 \end{bmatrix}\begin{bmatrix} 0 & 0 \\ z & 0 \\ z^2 & z \end{bmatrix} = \begin{bmatrix} z^2 & z \end{bmatrix} \neq [z(z-1)]$$

$$\Longrightarrow$$

$$\boldsymbol{F}_{ISOIO1}(z) = (z^2-1)^{-1}\begin{bmatrix} z-1 & -z & z^2 & z \end{bmatrix}.$$

Since $-\boldsymbol{B}_1^{(1)}\boldsymbol{Z}_1^{(1)}(z) = [-z]$ *is different from the second term* $[z(1-z)]$ *of the given* $\boldsymbol{F}_{ISO1}(z)$*, and* $\boldsymbol{A}_1^{(2)}\boldsymbol{Z}_1^{(2)}(z) = \begin{bmatrix} z^2 & z \end{bmatrix}$ *is different from the third term* $[z(z-1)]$ *of the given* $\boldsymbol{F}_{ISO1}(z)$*, it follows that the degenerate* $\boldsymbol{F}_{ISO1}(z)$ *is IO unrealizable, i.e., there is not an IO realization. There is not an IO linear differential equation, which implies the given degenerate form of* $\boldsymbol{F}_{ISO1}(z)$*,*

$$\boldsymbol{F}_{ISO1}(z) = \left(z^2-1\right)^{-1}\begin{bmatrix} z-1 & z(1-z) & z(z-1) \end{bmatrix} \neq$$
$$\neq (z^2-1)^{-1}\begin{bmatrix} z-1 & -z & z^2 & z \end{bmatrix} = \boldsymbol{F}_{ISOIO1}(z) = \boldsymbol{F}_{IOISO1nd}(z).$$

We should remind ourselves that $\boldsymbol{F}_{ISO1}(z)$ *was obtained from the ISO system instead of the corresponding IO system.*

We can deduce the same result by comparing $\boldsymbol{F}_{ISO1}(z)$ *with* $\boldsymbol{F}_{IOISO1-3}(z)$ *(12.3), Example 12.3 (in Section 12.3), which is therein obtained directly from (12.1),*

$$\boldsymbol{F}_{ISO1}(z) = \left(z^2-1\right)^{-1}\begin{bmatrix} z-1 & z(1-z) & z(z-1) \end{bmatrix} \neq$$
$$\neq \left(z^2-1\right)^{-1}\begin{bmatrix} z-1 & -z & z^2 & z \end{bmatrix} =$$
$$= \boldsymbol{F}_{IOISO1-3}(z) = \boldsymbol{F}_{IOISO1-3nd}(z).$$

This verifies IO nonrealizability of $\boldsymbol{F}_{ISO1}(z)$*.*

The nondegenerate, hence the irreducible, form $\boldsymbol{G}_{ISO1nd}(z)$ *of* $\boldsymbol{G}_{ISO1}(z)$

$$\boldsymbol{G}_{ISO1nd}(z) = \frac{1}{z+1}$$

implies

$$\nu_{1\min} = 1,\ \mu_{1\min} = 0,\ N_{1\min} = 1,\ M_{1\min} = 1,$$
$$\boldsymbol{A}^{(1)}_{1\min} = \begin{bmatrix} 1 & 1 \end{bmatrix},\ \boldsymbol{B}^{(0)}_{1\min} = [1\ ].$$

These data and (8.6a), (8.6b) (in Section 8.1) yield

$$\boldsymbol{A}^{(1)}_{1\min}\boldsymbol{S}^{(1)}_{1}(z) = \begin{bmatrix} 1 & 1 \end{bmatrix}\begin{bmatrix} 1 \\ z \end{bmatrix} = [1+z] = [z+1],$$
$$\boldsymbol{B}^{(0)}_{1\min}\boldsymbol{S}^{(0)}_{1}(z) = [1]\,[1] = [1] = [1],$$
$$-\boldsymbol{B}^{(0)}_{1\min}\boldsymbol{Z}^{(0)}_{1}(z) \text{ does not exist,}$$
$$\boldsymbol{A}^{(1)}_{1\min}\boldsymbol{Z}^{(1)}_{1}(z) = \begin{bmatrix} 1 & 1 \end{bmatrix}\begin{bmatrix} 0 \\ z \end{bmatrix} = [z] = \ [z],$$

and,

$$\boldsymbol{F}_{IO1rd}(z) = \left[\boldsymbol{A}^{(1)}_{1\min}\boldsymbol{S}^{(1)}_{1}(z)\right]^{-1}\begin{bmatrix} \boldsymbol{B}^{(0)}_{1\min}\boldsymbol{S}^{(0)}_{1}(z) & \boldsymbol{A}^{(1)}_{1\min}\boldsymbol{Z}^{(1)}_{1}(z) \end{bmatrix} =$$
$$= (z+1)^{-1}\begin{bmatrix} 1 & z \end{bmatrix}.$$

Since $-\boldsymbol{B}^{(0)}_{1\min}\boldsymbol{Z}^{(0)}_{1}(z)$ *does not exist, then it is different from the second submatrix* $[-z]$ *of the nondegenerate form* $\boldsymbol{F}_{ISO1nd}(z)$ *of* $\boldsymbol{F}_{ISO1}(z)$*. We conclude that the nondegenerate form* $\boldsymbol{F}_{ISO1nd}(z)$ *(12.15) of* $\boldsymbol{F}_{ISO1}(z)$ *is not IO realizable,*

$$\boldsymbol{F}_{ISO1nd}(z) = (z+1)^{-1}\begin{bmatrix} 1 & -z & z \end{bmatrix} \neq$$
$$\neq (z+1)^{-1}\begin{bmatrix} 1 & z \end{bmatrix} = \boldsymbol{F}_{IO1rd}(z),$$

in spite the nondegenerate form $\boldsymbol{G}_{ISO1nd}(z)$ *of* $\boldsymbol{G}_{ISO1}(z)$ *is IO realizable,*

$$\boldsymbol{G}_{ISO1nd}(z) = \frac{1}{z+1} = \boldsymbol{G}_{IO1nd}(z).$$

However, $\boldsymbol{G}_{ISO1nd}(z)$ *yields the same quadruple* $(\nu_{1\min},\ \mu_{1\min},\ \boldsymbol{A}^{(1)}_{1\min},\ \boldsymbol{B}^{(0)}_{1\min})$ *that determines*

$$\boldsymbol{F}_{IOISO1rd}(z) = (z+1)^{-1}\begin{bmatrix} 1 & z \end{bmatrix}.$$

It is equal to the reduced full IO transfer function matrix $\boldsymbol{F}_{IO1-3rd}(z)$ *(12.9), Example 12.3 (in Section 12.3), which is obtained directly from the IO system (12.6), Example 12.3 (in Section 12.3),*

$$\boldsymbol{F}_{IOISO1rd}(z) = (z+1)^{-1}\begin{bmatrix} 1 & z \end{bmatrix} = \boldsymbol{F}_{IO1-3rd}(z),$$

but it is different from the nondegenerate full ISO *transfer function matrix* $\mathbf{F}_{ISO1nd}(z)$,

$$\mathbf{F}_{ISO1nd}(z) = (z+1)^{-1} \begin{bmatrix} 1 & -z & z \end{bmatrix} \neq$$
$$\neq (z+1)^{-1} \begin{bmatrix} 1 & z \end{bmatrix} = \mathbf{F}_{IOISO1rd}(z).$$

This shows that the IO *realization either of* $\mathbf{F}_{ISO}(z)$ *or of* $\mathbf{F}_{ISOnd}(z)$ *need not exist in spite an* IO *realization of* $\mathbf{G}_{ISO}(z)$ *exists, and in spite the* IO *model determined by* $\mathbf{G}_{ISO}(z)$ *implies the well-defined* $\mathbf{F}_{ISOIO}(z)$. *This is clear if we have in mind the different meaning of the initial conditions relative to which hold* $\mathbf{F}_{ISO}(z)$ *and* $\mathbf{F}_{ISOIO}(z)$. *However,* $\mathbf{F}_{ISO}(z)$ *can induce a reduced* IO *realization that gives the same* $\mathbf{F}_{IOISOrd}(z)$ *as* $\mathbf{F}_{IOrd}(z)$ *obtained from the corresponding* IO *system.*

2) $\mathbf{A}_2 = \begin{bmatrix} 0 & 1 \\ 1 & 0 \end{bmatrix}$, $\mathbf{B}_2 = \begin{bmatrix} -1 \\ 1 \end{bmatrix}$, $\mathbf{C}_2 = \begin{bmatrix} 0 & 1 \end{bmatrix}$, $\mathbf{D}_2 = 0 \Longrightarrow$

$$\mathbf{F}_{ISO2}(z) = \left(z^2-1\right)^{-1} \begin{bmatrix} z-1 & z & z^2 \end{bmatrix} = \mathbf{F}_{ISO2nd}(z),$$
$$\mathbf{G}_{ISO2}(z) = \frac{z-1}{z^2-1} = \mathbf{G}_{ISO1}(z) \Longrightarrow \mathbf{G}_{ISO2nd}(z) = \frac{1}{z+1}.$$

The transfer function matrix $\mathbf{G}_{ISO2}(z)$ *is degenerate, but the full transfer function matrix* $\mathbf{F}_{ISO2}(z)$ *is nondegenerate,* $\mathbf{F}_{ISO2}(z) = \mathbf{F}_{ISO2nd}(z)$.

The degenerate transfer function matrix $\mathbf{G}_{ISO2}(z)$ *implies the* IO *system (12.1), Example 12.3 (in Section 12.3), which holds only under all zero initial conditions.*

The nondegenerate form $\mathbf{G}_{ISO2nd}(z)$ *of* $\mathbf{G}_{ISO2}(z)$ *reads*

$$\mathbf{G}_{ISO2nd}(z) = \frac{1}{z+1} = \mathbf{G}_{IO2nd}(z).$$

It yields the IO *system (12.6) and (12.7), Example 12.3 (in Section 12.3), which is valid also only under all zero initial conditions.*

The transfer function matrix $\mathbf{G}_{ISO2}(z)$, *which is degenerate, furnishes*

$$\nu_2 = 2,\ \mu_2 = 1,\ N_2 = 1,\ M_2 = 1,$$
$$\mathbf{A}_2^{(2)} = \begin{bmatrix} -1 & 0 & 1 \end{bmatrix},\ \mathbf{B}_2^{(1)} = \begin{bmatrix} -1 & 1 \end{bmatrix},$$

which together with (8.6a), (8.6b) (in Section 8.1), imposes the following IO conditions:

$$\mathbf{A}_2^{(2)}\mathbf{S}_1^{(2)}(z) = \begin{bmatrix} -1 & 0 & 1 \end{bmatrix}\begin{bmatrix} 1 \\ z \\ z^2 \end{bmatrix} = \left[-1+z^2\right] = \left[z^2-1\right],$$

$$\mathbf{B}_2^{(1)}\mathbf{S}_1^{(1)}(z) = \begin{bmatrix} -1 & 1 \end{bmatrix}\begin{bmatrix} 1 \\ z \end{bmatrix} = [-1+z] = [z-1],$$

$$-\mathbf{B}_2^{(1)}\mathbf{Z}_1^{(1)}(z) = -\begin{bmatrix} -1 & 1 \end{bmatrix}\begin{bmatrix} 0 \\ z \end{bmatrix} = [-z] \neq [z],$$

$$\mathbf{A}_2^{(2)}\mathbf{Z}_1^{(2)}(z) = \begin{bmatrix} -1 & 0 & 1 \end{bmatrix}\begin{bmatrix} 0 & 0 \\ z & 0 \\ z^2 & z \end{bmatrix} = \begin{bmatrix} z^2 & z \end{bmatrix} \neq \left[z^2\right].$$

The result is

$$\mathbf{F}_{ISOIO2}(z) = (z^2-1)^{-1}\begin{bmatrix} z-1 & -z & z^2 & z \end{bmatrix} = \mathbf{F}_{IOISO2nd}(z),$$

which is nondegenerate and different from $\mathbf{F}_{ISO2}(z)$:

$$\mathbf{F}_{ISO2}(z) = \left(z^2-1\right)^{-1}\begin{bmatrix} z-1 & z & z^2 \end{bmatrix} \neq$$
$$\neq (z^2-1)^{-1}\begin{bmatrix} z-1 & -z & z^2 & z \end{bmatrix} = \mathbf{F}_{ISOIO2}(z).$$

The full transfer function matrix $\mathbf{F}_{ISO2}(z)$ *obtained from the ISO model is not IO realizable in spite* $\mathbf{G}_{ISO2}(z)$ *is IO realizable.*

The minimal IO realization of $\mathbf{G}_{ISO1}(z) = \mathbf{G}_{ISO2}(z)$ *was found in* 1) *to be*

$$\mathbf{A}_{1\min}^{(1)}\mathbf{S}_1^{(1)}(z) = \mathbf{A}_{2\min}^{(1)}\mathbf{S}_1^{(1)}(z) = \begin{bmatrix} 1 & 1 \end{bmatrix}\begin{bmatrix} 1 \\ z \end{bmatrix} = [1+z] = [z+1],$$

$$\mathbf{B}_{1\min}^{(0)}\mathbf{S}_1^{(0)}(z) = \mathbf{B}_{2\min}^{(0)}\mathbf{S}_1^{(0)}(z) = [1]\,[1] = [1] = [1],$$

$$-\mathbf{B}_{1\min}^{(0)}\mathbf{Z}_1^{(0)}(z) = -\mathbf{B}_{2\min}^{(0)}\mathbf{Z}_1^{(0)}(z) \text{ does not exist,}$$

$$\mathbf{A}_{1\min}^{(1)}\mathbf{Z}_1^{(1)}(z) = \mathbf{A}_{2\min}^{(1)}\mathbf{Z}_1^{(1)}(z) = \begin{bmatrix} 1 & 1 \end{bmatrix}\begin{bmatrix} 0 \\ z \end{bmatrix} = [z] = [z].$$

They imply

$$\mathbf{F}_{IO2rd}(z) = \left[\mathbf{A}_{2\min}^{(1)}\mathbf{S}_1^{(1)}(z)\right]^{-1}\begin{bmatrix} \mathbf{B}_{2\min}^{(0)}\mathbf{S}_1^{(0)}(z) & \mathbf{A}_{2\min}^{(1)}\mathbf{Z}_1^{(1)}(z) \end{bmatrix} =$$
$$= (z+1)^{-1}\begin{bmatrix} 1 & z \end{bmatrix},$$

which is different from

$$\mathbf{F}_{ISOIO2}(z) = (z^2-1)^{-1}\begin{bmatrix} z-1 & -z & z^2 & z \end{bmatrix} = \mathbf{F}_{IOISO2nd}(z).$$

The minimal IO realization of $\mathbf{G}_{ISO2}(z)$ is not the minimal realization of $\mathbf{F}_{ISO2}(z)$ and $\mathbf{F}_{ISO2}(z)$ is IO unrealizable in view of Theorem 12.3 (in Section 12.3).

3) $\mathbf{A}_3 = \begin{bmatrix} 1 & 0 \\ 0 & -1 \end{bmatrix}, \ \mathbf{B}_3 = \begin{bmatrix} 0 \\ 1 \end{bmatrix}, \ \mathbf{C}_3 = \begin{bmatrix} 0 & 1 \end{bmatrix}, \ \mathbf{D}_3 = 0$

$\Longrightarrow$

$$\mathbf{F}_{ISO3}(z) = \left(z^2-1\right)^{-1}\begin{bmatrix} z-1 & 0 & z(z-1) \end{bmatrix} \Longrightarrow$$
$$\mathbf{F}_{ISO3nd}(z) = (z+1)^{-1}\begin{bmatrix} 1 & 0 & z \end{bmatrix},$$
$$\mathbf{G}_{ISO3}(z) = \frac{z-1}{z^2-1} \Longrightarrow \mathbf{G}_{ISO3nd}(z) = \frac{1}{z+1}.$$

- *Both $\mathbf{F}_{ISO3}(z)$ and $\mathbf{G}_{ISO3}(z)$ are degenerate.*

  The conditions for the IO realization of the degenerate $\mathbf{F}_{ISO3}(z)$ read

$$\nu_3 = 2, \ \mu_3 = 1, \ N_3 = 1, \ M_3 = 1,$$
$$\mathbf{A}_3^{(2)} = \begin{bmatrix} -1 & 0 & 1 \end{bmatrix}, \ \mathbf{B}_3^{(1)} = \begin{bmatrix} -1 & 1 \end{bmatrix},$$

and

$$\mathbf{A}_3^{(2)}\mathbf{S}_1^{(2)}(z) = \begin{bmatrix} -1 & 0 & 1 \end{bmatrix}\begin{bmatrix} 1 \\ z \\ z^2 \end{bmatrix} = \left[-1+z^2\right] = \left[z^2-1\right],$$
$$\mathbf{B}_3^{(1)}\mathbf{S}_1^{(1)}(z) = \begin{bmatrix} -1 & 1 \end{bmatrix}\begin{bmatrix} 1 \\ z \end{bmatrix} = [-1+z] = [z-1],$$
$$-\mathbf{B}_3^{(1)}\mathbf{Z}_1^{(1)}(z) = -\begin{bmatrix} -1 & 1 \end{bmatrix}\begin{bmatrix} 0 \\ z \end{bmatrix} = [-z] \neq [0],$$
$$\mathbf{A}_3^{(2)}\mathbf{Z}_1^{(2)}(z) = \begin{bmatrix} -1 & 0 & 1 \end{bmatrix}\begin{bmatrix} 0 & 0 \\ z & 0 \\ z^2 & z \end{bmatrix} = \begin{bmatrix} z^2 & z \end{bmatrix} \neq [z(z-1)].$$

We determine now

$$\mathbf{F}_{ISOIO3}(z) = (z^2-1)^{-1}\begin{bmatrix} z-1 & -z & z^2 & z \end{bmatrix} = \mathbf{F}_{IOISO3nd}(z).$$

The last two conditions for the IO realizability of $\boldsymbol{F}_{ISO3}(z)$ are not fulfilled (Theorem 12.3, in Section 12.3). The degenerate $\boldsymbol{F}_{ISO3}(z)$ is not IO realizable

$$\boldsymbol{F}_{ISO3}(s) = \left(z^2-1\right)^{-1}\left[\begin{array}{ccc} z-1 & 0 & z(z-1) \end{array}\right] \neq$$
$$\neq \left(z^2-1\right)^{-1}\left[\begin{array}{cccc} z-1 & -z & z^2 & z \end{array}\right] = \boldsymbol{F}_{ISOIO3}(z).$$

However, the quadruple obtained from $\boldsymbol{G}_{ISO3}(z)$ determines the exact full transfer function matrix $\boldsymbol{F}_{ISOIO3}(z)$ of the corresponding IO model.

The minimal IO realization $\boldsymbol{G}_{ISO3nd}(z)$ of $\boldsymbol{G}_{ISO3}(z)$ is determined by (12.6) and (12.7), Example 12.3 (in Section 12.3).

Let us verify the conditions for the IO realization of the nondegenerate form $\boldsymbol{F}_{ISO3nd}(z)$ of $\boldsymbol{F}_{ISO3}(z)$,

$$\boldsymbol{F}_{ISO3nd}(z) = (z+1)^{-1}\left[\begin{array}{ccc} 1 & 0 & z \end{array}\right].$$

The conditions read

$$\nu_{3\min} = 1,\ \mu_{3\min} = 0,\ N_{3\min} = 1,\ M_{3\min} = 1,$$
$$\boldsymbol{A}_{3\min}^{(1)} = \left[\begin{array}{cc} 1 & 1 \end{array}\right],\ \boldsymbol{B}_{3\min}^{(0)} = [1],$$

and

$$\boldsymbol{A}_{3\min}^{(1)}\boldsymbol{S}_1^{(1)}(z) = \left[\begin{array}{cc} 1 & 1 \end{array}\right]\left[\begin{array}{c} 1 \\ z \end{array}\right] = [1+z] = [z+1],$$
$$\boldsymbol{B}_{3\min}^{(0)}\boldsymbol{S}_1^{(0)}(z) = [1]\,[1] = [1] = [1],$$
$$-\boldsymbol{B}_{3\min}^{(0)}\boldsymbol{Z}_1^{(0)}(s) \text{ does not exist,}$$
$$\boldsymbol{A}_{3\min}^{(1)}\boldsymbol{Z}_1^{(1)}(z) = \left[\begin{array}{cc} 1 & 1 \end{array}\right]\left[\begin{array}{c} 0 \\ z \end{array}\right] = [z] = [z].$$

Hence,

$$\boldsymbol{F}_{IOISO3nd}(z) = (z+1)^{-1}\left[\begin{array}{cc} 1 & z \end{array}\right] = \boldsymbol{F}_{IO1-3rd}(z).$$

In this case, the nondegenerate form $\boldsymbol{F}_{ISO3nd}(z)$ of $\boldsymbol{F}_{ISO3}(z)$ might seem IO realizable at first glance. Is that correct? Let us check this result by comparing $\boldsymbol{F}_{ISO3nd}(z)$ with $\boldsymbol{F}_{IOISO3nd}(z)$,

$$\boldsymbol{F}_{ISO3nd}(z) = (z+1)^{-1}\left[\begin{array}{ccc} 1 & 0 & z \end{array}\right] \neq$$
$$\neq (z+1)^{-1}\left[\begin{array}{cc} 1 & z \end{array}\right] = \boldsymbol{F}_{IOISO3nd}(z).$$

The nondegenerate form of the full transfer function matrix $\boldsymbol{F}_{ISO3nd}(z)$ *obtained from the* ISO *system corresponds to the input action vector*

$$\left[\begin{array}{lll} i(k) & x_{10}\delta_d(k) & x_{20}\delta_d(k) \end{array}\right]^T,$$

while the full transfer function matrix $\boldsymbol{F}_{IOISO3nd}(z)$ *obtained from the reduced* IO *system is related to the input action vector* $\left[\begin{array}{ll} i(k) & y_0\delta_d(k) \end{array}\right]^T$. *This explains why their dimensions are different,*

$$\boldsymbol{F}_{ISO3}(z) \in \mathcal{C}^{1\times 3}, \ \boldsymbol{F}_{IO}(z) \in \mathcal{C}^{1\times 2}.$$

The condition of Theorem 12.3, in Section 12.3, is not satisfied.

4) $\boldsymbol{A}_4 = [-1]$, $\boldsymbol{B}_4 = [1]$, $\boldsymbol{C}_4 = [1]$, $\boldsymbol{D}_4 = 0 \Longrightarrow$

$$\boldsymbol{F}_{ISO4}(z) = (z+1)^{-1}\left[\begin{array}{ll} 1 & z \end{array}\right] = \boldsymbol{F}_{ISO4nd}(z),$$
$$\boldsymbol{G}_{ISO4}(z) = \frac{1}{z+1} = \boldsymbol{G}_{ISO4nd}(z).$$

Both $\boldsymbol{F}_{ISO4}(z)$ *and* $\boldsymbol{G}_{ISO4}(z)$ *are nondegenerate. In this case*

$$\nu_4 = \nu_{4\min} = 1, \ \mu_4 = \mu_{4\min} = 0, \ N_4 = N_{4\min} = 1, \ M_4 = M_{4\min} = 1,$$
$$\boldsymbol{A}_4^{(1)} = \boldsymbol{A}_{4\min}^{(1)} = \left[\begin{array}{ll} 1 & 1 \end{array}\right], \ \boldsymbol{B}_4^{(0)} = \boldsymbol{B}_{4\min}^{(0)} = [1],$$

and the IO *conditions (Theorem 12.3, in Section 12.3) read*

$$\boldsymbol{A}_{4\min}^{(1)}\boldsymbol{S}_1^{(1)}(z) = \left[\begin{array}{ll} 1 & 1 \end{array}\right]\left[\begin{array}{c} 1 \\ z \end{array}\right] = [1+z] = [z+1],$$
$$\boldsymbol{B}_{4\min}^{(0)}\boldsymbol{S}_1^{(0)}(z) = [1]\,[1] = [1] = [1],$$
$$-\boldsymbol{B}_{4\min}^{(0)}\boldsymbol{Z}_1^{(0)}(z) \text{ does not exist},$$
$$\boldsymbol{A}_{4\min}^{(1)}\boldsymbol{Z}_1^{(1)}(z) = \left[\begin{array}{ll} 1 & 1 \end{array}\right]\left[\begin{array}{c} 0 \\ z \end{array}\right] = [z] = \ [z] \Longrightarrow$$
$$\boldsymbol{F}_{IO4}(z) = (z+1)^{-1}\left[\begin{array}{ll} 1 & z \end{array}\right] = \boldsymbol{F}_{ISO4}(z),$$
$$\boldsymbol{F}_{IO4nd}(z) = (z+1)^{-1}\left[\begin{array}{ll} 1 & z \end{array}\right] = \boldsymbol{F}_{ISO4nd}(s).$$

The full transfer function matrix $\boldsymbol{F}_{ISO4}(z)$ *obtained from the* ISO *model is* IO *realizable in this case and its* IO *realization is minimal.*

Equations (12.6), (12.7) and Example 12.3 (in Section 12.3), determine the minimal IO *model of both* $\boldsymbol{F}_{ISO4}(z)$ *and* $\boldsymbol{G}_{ISO4}(z)$. *It is the minimal* IO *model that realizes* $\boldsymbol{F}_{ISO4}(z)$.

Conclusion 12.4 *We conclude that the (non)degenerate form of $\boldsymbol{F}_{ISO}(z)$ can be IO unrealizable although the (non)degenerate form of $\mathbf{G}_{ISO}(s)$ is IO realizable, respectively. We should have in mind that $\boldsymbol{F}_{ISO}(z)$ follows from the ISO system model, for which the initial conditions are the initial values of the state variables rather than the initial values of output and its shifts that are related to the IO model. In other words, $\boldsymbol{F}_{ISO}(z)$ obtained from the ISO model can be different from $\boldsymbol{F}_{ISOIO}(z)$ obtained from the IO model (induced by the ISO model), in which case the former can be IO unrealizable although the latter is IO realizable. This is clearly expressed by necessity of the condition of Theorem 12.3.*

Example 12.6 *For the ISO system (presented in Example 8.9, in Section 8.2) which is discrete-time analogy to the continuous-time ISO system given by Kalman in [85, Example 8, pp. 188, 189]:*

$$\boldsymbol{A}=\begin{bmatrix}0 & 1 & 0\\ 5 & 0 & 2\\ -2 & 0 & -2\end{bmatrix},\ \boldsymbol{B}=\begin{bmatrix}0\\ 0\\ 0.5\end{bmatrix},\ \boldsymbol{C}=\begin{bmatrix}-2 & 1 & 0\end{bmatrix},\ \boldsymbol{D}=\boldsymbol{O},$$

we found

$$\boldsymbol{C}(z\boldsymbol{I}_n-\boldsymbol{A})^{-1}\boldsymbol{B}+\boldsymbol{D}\quad z\boldsymbol{C}(z\boldsymbol{I}_n-\boldsymbol{A})^{-1}$$

$$\boldsymbol{F}_{ISO}(z)=[\det(z\boldsymbol{I}_n-\boldsymbol{A})]^{-1}\cdot$$
$$\cdot\begin{bmatrix}\boldsymbol{C}\operatorname{adj}(z\boldsymbol{I}_n-\boldsymbol{A})\boldsymbol{B}+\boldsymbol{D}\det(z\boldsymbol{I}_n-\boldsymbol{A}) & z\boldsymbol{C}\operatorname{adj}(z\boldsymbol{I}_n-\boldsymbol{A})\end{bmatrix}=$$
$$=\frac{z-2}{z-2}\left[(z+1)(z+3)\right]^{-1}\begin{bmatrix}1 & z(-2z-3) & z(z+2) & 2z\end{bmatrix}\Longrightarrow$$
$$\boldsymbol{F}_{ISOnd}(z)=\frac{\begin{bmatrix}1 & z(-2z-3) & z(z+2) & 2z\end{bmatrix}}{(z+1)(z+3)},$$

$$\mathbf{G}_{ISO}(z)=[\det(z\boldsymbol{I}_n-\boldsymbol{A})]^{-1}\left[\boldsymbol{C}\operatorname{adj}(z\boldsymbol{I}_n-\boldsymbol{A})\boldsymbol{B}+\boldsymbol{D}\det(z\boldsymbol{I}_n-\boldsymbol{A})\right]=$$
$$=\left(z^3+2z^2-5z-6\right)^{-1}(z-2)=\frac{z-2}{z-2}\left[(z+1)(z+3)\right]^{-1}\Longrightarrow$$
$$\mathbf{G}_{ISOnd}(z)=\frac{1}{(z+1)(z+3)}.$$

Since $\mathbf{G}_{ISO}(z)$ is reducible to

$$\mathbf{G}_{ISOnd}(z)=\frac{1}{(z+1)(z+3)}$$

then the given ISO realization is not the minimal ISO realization of $\mathbf{G}_{ISO}(z)$. Its minimal ISO realization reads

$$\boldsymbol{A}_{\min G} = \begin{bmatrix} 0 & 1 \\ -3 & -4 \end{bmatrix}, \ \boldsymbol{B}_{\min G} = \begin{bmatrix} 0 \\ 1 \end{bmatrix},$$
$$\boldsymbol{C}_{\min G} = \begin{bmatrix} 1 & 0 \end{bmatrix}, \ \boldsymbol{D}_{\min G} = \boldsymbol{O}.$$

It yields

$$\frac{\begin{bmatrix} \left(\boldsymbol{C}_{\min G}\operatorname{adj}(z\boldsymbol{I}_n - \boldsymbol{A}_{\min G})\boldsymbol{B}_{\min G} + \boldsymbol{D}_{\min G}\det(z\boldsymbol{I}_n - \boldsymbol{A}_{\min G})\right)^T \\ \left(z\boldsymbol{C}_{\min G}\operatorname{adj}(z\boldsymbol{I}_n - \boldsymbol{A}_{\min G})\right)^T \end{bmatrix}^T}{\det(z\boldsymbol{I}_n - \boldsymbol{A}_{\min G})} =$$
$$= \frac{\begin{bmatrix} 1 & z(z+4) & z \end{bmatrix}}{(z+1)(z+3)},$$

which is different from the nondegenerate form $\boldsymbol{F}_{ISOnd}(z)$ of $\boldsymbol{F}_{ISO}(z)$,

$$\frac{\begin{bmatrix} 1 & z(z+4) & z \end{bmatrix}}{(z+1)(z+3)} \neq \frac{\begin{bmatrix} 1 & z(-2z-3) & z(z+2) & 2z \end{bmatrix}}{(z+1)(z+3)} = \boldsymbol{F}_{ISOnd}(z).$$

The minimal ISO realization $(\boldsymbol{A}_{\min G}, \boldsymbol{B}_{\min G}, \boldsymbol{C}_{\min G}, \boldsymbol{D}_{\min G})$ of $\mathbf{G}_{ISO}(z)$ is not the minimal ISO realization of $\boldsymbol{F}_{ISO}(z)$.

Does the nondegenerate form $\boldsymbol{F}_{ISOnd}(z)$ of $\boldsymbol{F}_{ISO}(z)$ imply an ISO realization of $\boldsymbol{F}_{ISO}(z)$ with a lower dimension than that of the given ISO realization $(\boldsymbol{A}, \boldsymbol{B}, \boldsymbol{C}, \boldsymbol{D})$?

We start with

$$\boldsymbol{F}_{ISOnd}(z) = \frac{\begin{bmatrix} \begin{pmatrix} \boldsymbol{C}_{ISOnd}\operatorname{adj}(z\boldsymbol{I}_n - \boldsymbol{A}_{ISOnd})\boldsymbol{B}_{ISOnd} + \\ +\boldsymbol{D}_{ISOnd}\det(z\boldsymbol{I}_n - \boldsymbol{A}_{ISOnd}) \end{pmatrix}^T \\ \left(\boldsymbol{C}_{ISOnd}\operatorname{adj}(z\boldsymbol{I}_n - \boldsymbol{A}_{ISOnd})\right)^T \end{bmatrix}^T}{\det(z\boldsymbol{I}_n - \boldsymbol{A}_{ISOnd})} =$$
$$= \frac{\begin{bmatrix} 1 & z(-2z-3) & z(z+2) & 2z \end{bmatrix}}{(z+1)(z+3)} \Longrightarrow$$
$$\deg \begin{bmatrix} 1 & z(-2z-3) & z(z+2) & 2z \end{bmatrix} = 2,$$
$$\deg (z+1)(z+3) = 2.$$

This implies

$$\deg \begin{bmatrix} \begin{pmatrix} \boldsymbol{C}_{ISOnd}\operatorname{adj}(z\boldsymbol{I}_n - \boldsymbol{A}_{ISOnd})\boldsymbol{B}_{ISOnd} + \\ +\boldsymbol{D}_{ISOnd}\det(z\boldsymbol{I}_n - \boldsymbol{A}_{ISOnd}) \end{pmatrix}^T \\ \left(z\boldsymbol{C}_{ISOnd}\operatorname{adj}(z\boldsymbol{I}_n - \boldsymbol{A}_{ISOnd})\right)^T \end{bmatrix}^T = 2,$$

$$\deg\left[\det(z\boldsymbol{I}_n - \boldsymbol{A}_{ISOnd})\right] = 2$$

$$\Longrightarrow$$

$$\boldsymbol{D}_{ISOnd} = 0,\ n = 2,$$

$$\Longrightarrow$$

$$\begin{aligned} \boldsymbol{C}_{ISOnd}\,\mathrm{adj}(z\boldsymbol{I}_n - \boldsymbol{A}_{ISOnd})\boldsymbol{B}_{ISOnd} &= \left[\begin{array}{cc} 1 & z(-2z-3) \end{array}\right], \\ z\boldsymbol{C}_{ISOnd}\,\mathrm{adj}(z\boldsymbol{I}_n - \boldsymbol{A}_{ISOnd}) &= \left[\begin{array}{cc} z(z+2) & 2z \end{array}\right]. \\ \det(z\boldsymbol{I}_n - \boldsymbol{A}_{ISOnd}) &= (z+1)(z+3) \end{aligned}$$

$$\boldsymbol{A}_{ISOnd} = \left[\begin{array}{cc} a_{11} & a_{12} \\ a_{21} & a_{22} \end{array}\right],\ \boldsymbol{B}_{ISOnd} = \left[\begin{array}{c} b_1 \\ b_2 \end{array}\right],\ \boldsymbol{C}_{ISOnd} = \left[\begin{array}{cc} c_1 & c_2 \end{array}\right]$$

$$\Longrightarrow$$

$$\left[\begin{array}{cc} c_1 & c_2 \end{array}\right]\left[\begin{array}{cc} z-a_{22} & a_{12} \\ a_{21} & z-a_{11} \end{array}\right]\left[\begin{array}{c} b_1 \\ b_2 \end{array}\right] = \left[\begin{array}{cc} 1 & z(-2z-3) \end{array}\right],$$

$$z\left[\begin{array}{cc} c_1 & c_2 \end{array}\right]\left[\begin{array}{cc} z-a_{22} & a_{12} \\ a_{21} & z-a_{11} \end{array}\right] = \left[\begin{array}{cc} z(z+2) & 2z \end{array}\right],$$

$$\det\left[\begin{array}{cc} z-a_{11} & -a_{12} \\ -a_{21} & z-a_{22} \end{array}\right] = (z+1)(z+3) = z^2+4z+3.$$

The last three equations have many solutions. There exist many ISO systems realizations that imply their full transfer function matrices equal to the nondegenerate form $\boldsymbol{F}_{ISOnd}(z)$ of the given $\boldsymbol{F}_{ISO}(z)$. There exists many ISO realizations $(\boldsymbol{A}, \boldsymbol{B}, \boldsymbol{C}, \boldsymbol{D})$ of the nondegenerate form $\boldsymbol{F}_{ISOnd}(z)$ of the given $\boldsymbol{F}_{ISO}(z)$. The dimension n of the given ISO realization $(\boldsymbol{A}, \boldsymbol{B}, \boldsymbol{C}, \boldsymbol{D})$ is not the minimal dimension of the ISO realization of the given $\boldsymbol{F}_{ISO}(z)$.

Conclusion 12.5 *The preceding analysis shows that the minimal ISO realization of the transfer function matrix $\boldsymbol{G}_{ISO}(z)$ can be, but need not be, the minimal ISO realization of the full transfer function matrix $\boldsymbol{F}_{ISO}(z)$.*

Conclusion 12.6 *The nondegenerate form $\boldsymbol{F}_{ISOnd}(z)$ of the full transfer function matrix $\boldsymbol{F}_{ISO}(z)$ can have an ISO realization.*

12.5 Realizations of $\boldsymbol{F}(z)$ of IIO systems

We refer to Definitions 12.3 (in Section 12.1), 12.6 and 12.7 (in Section 12.2) for the following:

Definition 12.10 *If and only if there is*

$$\left(\alpha, \beta, \mu, \nu, \boldsymbol{E}^{(\nu)}, \boldsymbol{P}^{(\beta)}, \boldsymbol{Q}^{(\alpha)}, \boldsymbol{R}^{(\alpha)}, \boldsymbol{T}^{(\mu)}\right)$$

that determines an IIO system realization (3.64a) and (3.64b) (in Subsection 3.5.4) such that its full transfer function matrix is equal to the nondegenerate form $\boldsymbol{F}_{IIOnd}(z)$ *of a given* $\boldsymbol{F}_{IIO}(z)$*, then the realization is* ***the minimal (irreducible) IIO realization of the given full transfer function matrix*** $\boldsymbol{F}_{IIO}(z)$.

Theorem 12.6 *In order for* $\left(\alpha, \beta, \mu, \nu, \boldsymbol{E}^{(\nu)}, \boldsymbol{P}^{(\beta)}, \boldsymbol{Q}^{(\alpha)}, \boldsymbol{R}^{(\alpha)}, \boldsymbol{T}^{(\mu)}\right)$ *to determine the (minimal) IIO system realization (3.64a) and (3.64b) (in Subsection 3.5.4), it is necessary and sufficient that*

$$\left(\boldsymbol{E}^{(\nu)} \boldsymbol{S}_N^{(\nu)}(z)\right)^{-1} \cdot$$

$$\cdot \left[\boldsymbol{R}^{(\alpha)} \boldsymbol{S}_\rho^{(\alpha)}(z) \left(\boldsymbol{Q}^{(\alpha)} \boldsymbol{S}_\rho^{(\alpha)}(z)\right)^{-1} \boldsymbol{P}^{(\beta)} \boldsymbol{S}_M^{(\beta)}(z) + \boldsymbol{T}^{(\mu)} \boldsymbol{S}_M^{(\mu)}(z)\right] = \boldsymbol{G}_{IIO}(z),$$

$$\left(\boldsymbol{E}^{(\nu)} \boldsymbol{S}_N^{(\nu)}(z)\right)^{-1} \cdot$$

$$\cdot \left\{\left[\begin{array}{cc} \boldsymbol{R}^{(\alpha)} \boldsymbol{S}_\rho^{(\alpha)}(z) \left(\boldsymbol{Q}^{(\alpha)} \boldsymbol{S}_\rho^{(\alpha)}(z)\right)^{-1} \boldsymbol{P}^{(\beta)} \boldsymbol{Z}_M^{(\beta)}(z) & \boldsymbol{O}_{N,(\gamma-\beta)M} \end{array}\right] + \right.$$

$$\left. + \left[\begin{array}{cc} \boldsymbol{T}^{(\mu)} \boldsymbol{Z}_M^{(\mu)}(z) & \boldsymbol{O}_{N,(\gamma-\mu)M} \end{array}\right]\right\} = -\ \boldsymbol{G}_{IIOi_0}(z),$$

$$\left(\boldsymbol{E}^{(\nu)} \boldsymbol{S}_N^{(\nu)}(z)\right)^{-1} \cdot$$

$$\cdot \left[\boldsymbol{R}^{(\alpha)} \boldsymbol{S}_\rho^{(\alpha)}(z) \left(\boldsymbol{Q}^{(\alpha)} \boldsymbol{S}_\rho^{(\alpha)}(z)\right)^{-1} \boldsymbol{Q}^{(\alpha)} \boldsymbol{Z}_\rho^{(\alpha)}(z) - \boldsymbol{R}^{(\alpha)} \boldsymbol{Z}_\rho^{(\alpha)}(z)\right] = \boldsymbol{G}_{IIOr_0}(z),$$

$$\left(\boldsymbol{E}^{(\nu)} \boldsymbol{S}_N^{(\nu)}(z)\right)^{-1} \boldsymbol{E}^{(\nu)} \boldsymbol{Z}_N^{(\nu)}(z) = \boldsymbol{G}_{IIOy_0}(z), \tag{12.16}$$

and that they determine the nondegenerate form $\boldsymbol{F}_{IIOnd}(z)$ *of the full IIO system transfer function matrix* $\boldsymbol{F}_{IIO}(z)$ *as follows, respectively,*

$$\left[\begin{array}{cccc} \boldsymbol{G}_{IIO}(z) & \boldsymbol{G}_{IIOi_0}(z) & \boldsymbol{G}_{IIOr_0}(z) & \boldsymbol{G}_{IIOy_0}(z) \end{array}\right] = \boldsymbol{F}_{IIO}(z) \tag{12.17}$$

$$\left(\left[\begin{array}{cccc} \boldsymbol{G}_{IIO}(z) & \boldsymbol{G}_{IIOi_0}(z) & \boldsymbol{G}_{IIOr_0}(z) & \boldsymbol{G}_{IIOy_0}(z) \end{array}\right] = \boldsymbol{F}_{IIOnd}(z)\right). \tag{12.18}$$

Proof. *Necessity.* Let $(\alpha,\ \beta,\ \mu,\ \nu,\ \boldsymbol{E}^{(\nu)},\ \boldsymbol{P}^{(\beta)},\ \boldsymbol{Q}^{(\alpha)},\ \boldsymbol{R}^{(\alpha)},\ \boldsymbol{T}^{(\mu)})$ determine the (minimal) *IIO* system realization (3.64a) and (3.64b). Then, Equations 12.16 through 12.18 hold due to Definition 6.1 (in Chapter 6),

Definition 12.1 through Definition 12.3, Proposition 12.1 (in Section 12.1), Definition 12.6 (in Section 12.2), Definition 12.7 (in Section 12.2), Definition 12.10, (8.74) and (8.78) (Theorem 8.5, in Section 8.3), respectively, which proves necessity of the conditions.

Sufficiency. Let the conditions hold. Hence, $(\alpha,\ \beta,\ \mu,\ \nu,\ \boldsymbol{E}^{(\nu)},\ \boldsymbol{P}^{(\beta)},\ \boldsymbol{Q}^{(\alpha)},\ \boldsymbol{R}^{(\alpha)},\ \boldsymbol{T}^{(\mu)})$ determines (the minimal) an input-output realization of $\boldsymbol{F}_{IIO}(z)$ in view of (8.74) and (8.78) (Theorem 8.5, in Section 8.3), which is also (the minimal) an *IIO* realization of the *IIO* system (3.64a) and (3.64b) due to Definition 12.6 (in Section 12.2), Definition 12.7 (in Section 12.2), and Definition 12.10, respectively. ■

Example 12.7 *Let the full IIO system transfer function matrix be*

$$\boldsymbol{F}_{IIO}(z) = \left[(z-2)^2(z+0.2)(z+0.5)\right]^{-1} \cdot$$
$$\cdot \begin{bmatrix} (z-2)^2(z-6)(17z+1) \\ -z(z-2)(z-8)(17z+1) \\ -z(z-2)(17z+1) \\ 5z(z-2) \\ -2.5z(z-2) \\ z(z-2)(z+0.5)(z-1.8) \\ z(z-2)(z+0.5) \end{bmatrix}^T . \tag{12.19}$$

Its IIO system realization $\left(\alpha,\ \beta,\ \mu,\ \nu,\ \boldsymbol{E}^{(\nu)},\ \boldsymbol{P}^{(\beta)},\ \boldsymbol{Q}^{(\alpha)},\ \boldsymbol{R}^{(\alpha)},\ \boldsymbol{T}^{(\mu)}\right)$ *is determined (Example 8.10, in Section 8.3) by:*

$$\alpha = 2,\ \beta = 2, \mu = 2,\ \nu = 2,$$
$$\boldsymbol{E}^{(\nu)} = \boldsymbol{E}^{(2)} = \begin{bmatrix} -0.4 & -1.8 & 1 \end{bmatrix},\ \boldsymbol{P}^{(\beta)} = \boldsymbol{P}^{(2)} = \begin{bmatrix} 36 & -24 & 3 \end{bmatrix},$$
$$\boldsymbol{Q}^{(\alpha)} = \boldsymbol{Q}^{(2)} = \begin{bmatrix} -1 & -1.5 & 1 \end{bmatrix},\ \boldsymbol{R}^{(\alpha)} = \boldsymbol{R}^{(2)} = \begin{bmatrix} 2 & -10 & 5 \end{bmatrix},$$
$$\boldsymbol{T}^{(\mu)} = \boldsymbol{T}^{(2)} = \begin{bmatrix} 24 & -16 & 2 \end{bmatrix}.$$

These data determine the IIO system realization in the vector form:

$$\boldsymbol{Q}^{(\alpha)}\mathbf{r}^{\alpha}(t) = \boldsymbol{P}^{(\beta)}\mathbf{i}^{\beta}(t),$$
$$\boldsymbol{E}^{(\nu)}\mathbf{y}^{\nu}(t) = \boldsymbol{R}^{(\alpha)}\mathbf{r}^{\alpha}(t) + \boldsymbol{T}^{(\mu)}\mathbf{i}^{\mu}(t), \tag{12.20}$$

and, equivalently, in the scalar form:

$$E^2r(k) - 1.5E^1r(k) - r(k) = 3E^2i(k) - 24E^1i(k) + 36i(k),$$
$$E^2y(k) - 1.8E^1y(k) - 0.4y(k) =$$
$$= 5E^2r(k) - 10E^1r(k) + 2E^2i(k) - 16E^1i(k) + 24i(k). \tag{12.21}$$

Part III

STABILITY STUDY

Chapter 13

Lyapunov stability

13.1 Lyapunov stability concept

The stability concept of Lyapunov [93] has the following main qualitative characteristics:

- It concerns the internal dynamical behavior of the system.
- It concerns the system behavior in the nominal regime in terms of total coordinates $\mathbf{I}$, i.e., in the free regime in terms of deviations $\mathbf{i}$, $\mathbf{i} = \mathbf{I} - \mathbf{I}_N$, Subsection 3.5.1,

$$\mathbf{I}(k) = \mathbf{I}_N(k), \text{ i.e., } \mathbf{i}(k) = \mathbf{0}_M, \ \forall k \in \mathcal{N}_0.$$

- It deals only with the influence of nonzero initial conditions on the system dynamical behavior.
- It concerns the system dynamical behavior over the unbounded time set $\mathcal{N}_0$. If the system dynamic behavior is satisfactory over $\mathcal{N}_0$, then it is satisfactory also on any subset of $\mathcal{N}_0$.
- It allows any permitted upper bound ε, $\varepsilon \in \mathcal{R}^+$, of the norm of the deviation $\mathbf{x}(k)$, or $\mathbf{y}(k)$ at discrete moment k, $k \in \mathbb{Z}$, of real total systems discrete behavior $\mathbf{X}(k)$, or $\mathbf{Y}(k)$, from the total discrete nominal behavior $\mathbf{X}_d(k)$, or $\mathbf{Y}_d(k)$, respectively.
- It demands the existence of the appropriate upper bound δ, $\delta \in \mathcal{R}^+$, of the norm of the initial deviation $\mathbf{x}_0$, or $\mathbf{y}_0$, of the total initial system behavior $\mathbf{X}_0$, or $\mathbf{Y}_0$, from the total initial desired behavior $\mathbf{X}_{d0}$, or $\mathbf{Y}_{d0}$.

- Analogously, a positive real number α specifies the arbitrarily requested lower bound of the closeness at discrete moment k of real system discrete behavior $\mathbf{X}(k)$, or $\mathbf{Y}(k)$, to the desired discrete behavior $\mathbf{X}_d(k)$, or $\mathbf{Y}_d(k)$, where $\alpha = \varepsilon^{-1} \in \mathcal{R}^+$.

- A positive real number β specifies the lower bound of the initial closeness of $\mathbf{X}_0$, or $\mathbf{Y}_0$, to $\mathbf{X}_{d0}$, or $\mathbf{Y}_{d0}$, which corresponds to α, where $\beta = \delta^{-1} \in \mathcal{R}^+$.

- If and only if for arbitrarily chosen permitted upper bound ε, there exists an appropriate upper bound δ of the initial deviation $\mathbf{x}_0$, or $\mathbf{y}_0$, such that the norm $\|\mathbf{x}_0\|$, or $\|\mathbf{y}_0\|$ less than δ guarantees that the norm $\|\mathbf{x}(k)\|$, or $\|\mathbf{y}(k)\|$, is less than the chosen permitted upper bound ε at every $k \in \mathcal{N}_0$, then **the desired behavior $\mathbf{X}_d(k)$**, or $\mathbf{Y}_d(k)$, respectively, is **stable** (**the linear system is limiting stable**, in other words **the linear system is on the boundary of stability,** equivalently: **it is critically stable**; see Fig. 13.1. Equivalently, **the**

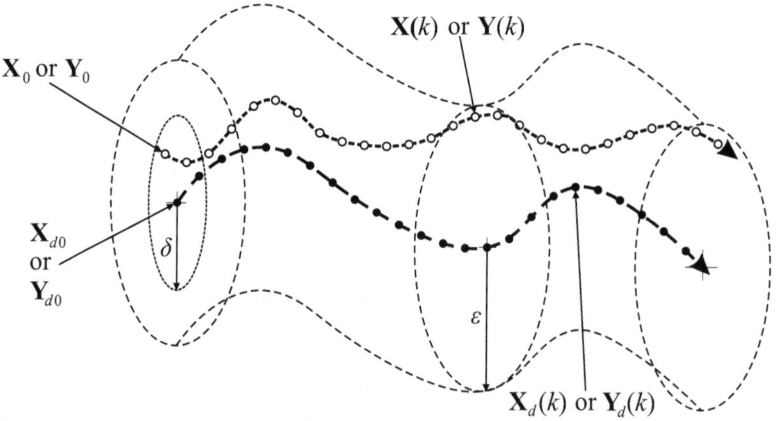

Figure 13.1: A stable total desired behavior $\mathbf{X}_d(k)$, or $\mathbf{Y}_d(k)$.

zero vector deviation $\mathbf{x} = \mathbf{0}_n$, or $\mathbf{y} = \mathbf{0}_N$, **is stable**; see Fig. 13.2.

- If and only if there exists a Δ-neighborhood of the desired behavior $\mathbf{X}_d(k)$, or $\mathbf{Y}_d(k)$, such that for every initial condition $\mathbf{X}_0$, or $\mathbf{Y}_0$, from the Δ-neighborhood, the corresponding system dynamical behavior asymptotically approaches the desired behavior $\mathbf{X}_d(k)$, or $\mathbf{Y}_d(k)$, as $k \to \infty$, then the desired behavior $\mathbf{X}_d(k)$, or $\mathbf{Y}_d(k)$, is **attractive**;

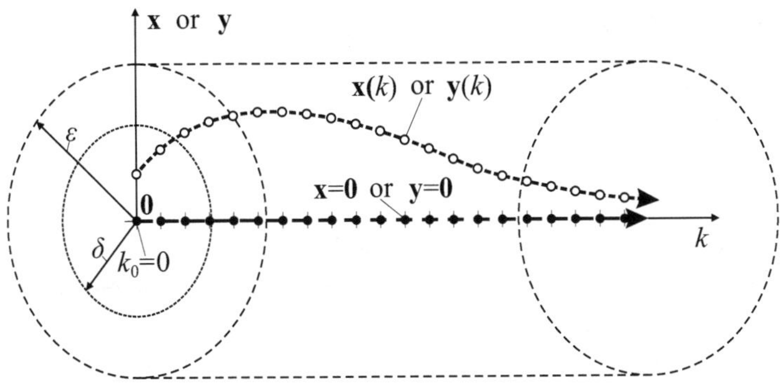

Figure 13.2: The zero vector deviation $\mathbf{x} = \mathbf{0}_n$, or $\mathbf{y} = \mathbf{0}_N$, is stable.

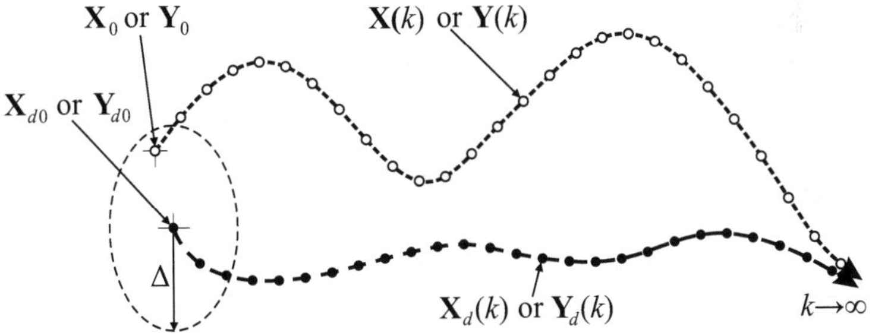

Figure 13.3: The total desired behavior $\mathbf{X}_d(k)$, or $\mathbf{Y}_d(k)$, is attractive.

see Fig. 13.3. Equivalently, then, and only then, **the zero vector deviation** $\mathbf{x} = \mathbf{0}_n$, or $\mathbf{y} = \mathbf{0}_N$, **is attractive**; see Fig. 13.4. If and only if this holds for any initial conditions then the attraction is **global**, i.e., **in the whole**.

- If the desired behavior $\mathbf{X}_d(k)$, or $\mathbf{Y}_d(k)$, is both stable and (globally) attractive, then it is **(globally) asymptotically stable** (**asymptotically stable in the whole, the system is stable**), respectively. Equivalently, then, and only then, **the zero vector deviation** $\mathbf{x} = \mathbf{0}_n$, or $\mathbf{y} = \mathbf{0}_N$, **is (globally) asymptotically stable.**

- If and only if additionally the desired behavior $\mathbf{X}_d(k)$, or $\mathbf{Y}_d(k)$, is globally stable, then the desired behavior $\mathbf{X}_d(k)$ or $\mathbf{Y}_d(k)$ is **strictly globally asymptotically stable.** Equivalently, then, and only then, **the zero vector deviation** $\mathbf{x} = \mathbf{0}_n$, or $\mathbf{y} = \mathbf{0}_N$, **is strictly globally asymptotically stable**, respectively.

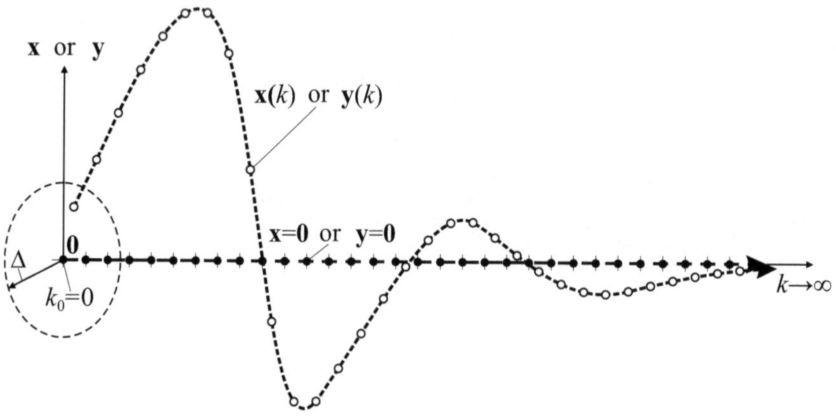

Figure 13.4: The zero vector deviation $\mathbf{x} = \mathbf{0}_n$, or $\mathbf{y} = \mathbf{0}_N$, is attractive.

13.2 Definitions

13.2.1 *IO* systems

Stability of the zero equilibrium vector

We start with the definition in Lyapunov sense of *stability* of the desired behavior $\mathbf{Y}_d^{\nu-1}(k)$ of the *IO* system (2.1) (in Section 2.1), in terms of the total coordinates, i.e., of the desired behavior deviation $\mathbf{y}_d^{\nu-1}(k) \equiv \mathbf{0}_{\nu N}$ of the system (3.56) (in Subsection 3.5.2), although Lyapunov gave it only for the general *ISO* continuous nonlinear systems that are in the Cauchy (normal) form, [93].

Although $\|\cdot\| : \mathcal{R}^r \rightarrow \mathcal{R}_+$ can be *any norm on* $\mathcal{R}^r$,

$$r \in \{1, 2, 3, \cdots, M, \cdots, N, \cdots, n, \cdots, \nu N, \cdots\},$$

usually it is accepted to be the *Euclidean norm on* $\mathcal{R}^r$:

$$\mathbf{w} = \begin{bmatrix} w_1 & w_2 & \cdots & w_r \end{bmatrix}^T \in \mathcal{R}^r \Longrightarrow \|\mathbf{w}\| = \sqrt{\mathbf{w}^T\mathbf{w}} = \sqrt{\sum_{i=1}^{i=r} w_i^2},$$

$\mathcal{B}_\xi(\mathbf{z})$ is an open hyperball with the radius ξ centered at the point $\mathbf{z}$ also in the space $\mathcal{R}^r$,

$$\mathcal{B}_\xi(\mathbf{z}) = \{\mathbf{w} : \ \|\mathbf{w} - \mathbf{z}\| < \xi\}.$$

We omit $(\mathbf{0})$ from the notation $\mathcal{B}_\xi(\mathbf{0})$ if and only if $\mathbf{z} = \mathbf{0}_r$,

$$\mathcal{B}_\xi = \mathcal{B}_\xi(\mathbf{0}).$$

Notice that in what follows, $r = \nu N$, $\mathcal{R}^r = \mathcal{R}^{\nu N}$.

As it was pointed out, Lyapunov defined stability properties of the desired behavior by accepting a priory that the input vector $\mathbf{I}$ is nominal, what in the discrete time case means,

$$\mathbf{I}(k) = \mathbf{I}_N(k), \ \forall k \in \mathcal{N}_0.$$

In terms of the input deviation vector $\mathbf{i}(k)$,

$$\mathbf{i}(k) = \mathbf{I}(k) - \mathbf{I}_N(k) \ \forall k \in \mathcal{N}_0,$$

it means that we treat the system in the free regime,

$$\mathbf{i}(k) = \mathbf{I}_N(k) - \mathbf{I}_N(k) = \mathbf{0}_M, \ \forall k \in \mathcal{N}_0.$$

Definition 13.1 *a) The desired behavior $\mathbf{Y}_d^{\nu-1}(k)$ of the IO system (2.1) is **stable** if and only if for every positive real number ε there exists a positive real number δ, the value of which depends on ε, such that*

$$\forall \varepsilon \in \mathcal{R}^+, \ \exists \delta \in \mathcal{R}^+, \ \delta = \delta(\varepsilon), \left\| \mathbf{Y}_{d0}^{\nu-1} - \mathbf{Y}_0^{\nu-1} \right\| < \delta \Longrightarrow$$
$$\left\| \mathbf{Y}_d^{\nu-1}(k) - \mathbf{Y}^{\nu-1}(k; \mathbf{Y}_0^{\nu-1}; \mathbf{I}_N) \right\| < \varepsilon, \ \forall k \in \mathcal{N}_0. \qquad (13.1)$$

*The zero equilibrium vector $\mathbf{y}_e^{\nu-1} = \mathbf{0}_{\nu N}$ of the IO system (3.56) is **stable** if and only if*

$$\forall \varepsilon \in \mathcal{R}^+, \ \exists \delta \in \mathcal{R}^+, \ \delta = \delta(\varepsilon),$$
$$\left\| \mathbf{y}_0^{\nu-1} \right\| < \delta \Longrightarrow \left\| \mathbf{y}^{\nu-1}(k; \mathbf{y}_0^{\nu-1}; \mathbf{0}_M) \right\| < \varepsilon, \ \forall k \in \mathcal{N}_0. \qquad (13.2)$$

*b) The desired behavior $\mathbf{Y}_d^{\nu-1}(k)$ of the IO system (2.1) is **globally stable** (i.e., **stable in the whole**) if and only if a) holds so that the maximal $\delta(\varepsilon)$, denoted by $\delta_{\max}(\varepsilon)$, diverges to infinity as ε goes to infinity,*

$$\varepsilon \to \infty \Longrightarrow \delta_{\max}(\varepsilon) \to \infty. \qquad (13.3)$$

*The zero equilibrium vector $\mathbf{y}_e^{\nu-1} = \mathbf{0}_{\nu N}$ of the IO system (3.56) is **globally stable** (i.e., **stable in the whole**) if and only if a) holds so that the maximal $\delta(\varepsilon)$, denoted by $\delta_{\max}(\varepsilon)$, diverges to infinity as ε goes to infinity, that is that (13.3) is valid.*

Conclusion 13.1 ***Relationship between stability of $\mathbf{Y}_d^{\nu-1}(t)$ and of $\mathbf{y}_e^{\nu-1} = \mathbf{0}_{\nu N}$***

The zero equilibrium vector $\mathbf{y}_e^{\nu-1} = \mathbf{0}_{\nu N}$ of the IO system (3.56) is (globally) stable, if and only if the desired behavior $\mathbf{Y}_d^{\nu-1}(k)$ of the IO system (2.1) is (globally) stable, respectively.

Note 13.1 *Geometrical interpretation*

Geometrically considering the sense of this definition is that the zero equilibrium vector $\mathbf{y}_e^{\nu-1} = \mathbf{0}_{\nu N}$ of the IO system (3.56) is ***stable*** *if and only if for every open hyperball $\mathcal{B}_\varepsilon \in \mathcal{R}^{N\nu}$ centered at the origin with the radius ε, there exists an open hyperball $\mathcal{B}_\delta \in \mathcal{R}^{N\nu}$ centered at the origin with the radius δ, which depends on $\mathcal{B}_\varepsilon$, such that $\mathbf{y}_0^{\nu-1} \in \mathcal{B}_\delta$ guarantees that $\mathbf{y}^{\nu-1}(k;\mathbf{y}_0^{\nu-1};\mathbf{0}_M)$ stays in $\mathcal{B}_\varepsilon$ all the discrete time from $k=0$ on, Fig. 13.2 (in Section 13.1),*

$$\begin{gathered}\forall \mathcal{B}_\varepsilon \subseteq \mathcal{R}^{N\nu},\ \exists \mathcal{B}_\delta \subseteq \mathcal{R}^{N\nu},\ \mathcal{B}_\delta = \mathcal{B}_\delta(\mathcal{B}_\varepsilon),\ \Longrightarrow \\ \mathbf{y}_0^{\nu-1} \in \mathcal{B}_\delta \Longrightarrow \mathbf{y}^{\nu-1}(k;\mathbf{y}_0^{\nu-1};\mathbf{0}_M) \in \mathcal{B}_\varepsilon,\ \forall k \in \mathcal{N}_0.\end{gathered} \tag{13.4}$$

If and only if additionally $\mathcal{B}_\varepsilon \longrightarrow \mathcal{R}^n$ implies $\mathcal{B}_\delta \longrightarrow \mathcal{R}^n$, then $\mathbf{y}_e^{\nu-1} = \mathbf{0}_n$ is ***globally stable*** *(i.e.,* ***stable in the whole****).*

Equation 10.4 (in Section 10.1) determines the response of the *IO* system (3.56) in the free regime by

$$\mathbf{y}(k;\mathbf{y}_0^{\nu-1};\mathbf{0}_M) = \mathbf{\Gamma}_{IOy0}(k)\mathbf{y}_0^{\nu-1}, \tag{13.5}$$

which implies

$$E^r\mathbf{y}(k;\mathbf{y}_0^{\nu-1};\mathbf{0}_M) = E^r\mathbf{\Gamma}_{IOy0}(k)\mathbf{y}_0^{\nu-1},\ r = 0,1,\cdots,\nu,$$

and

$$\mathbf{y}^{\nu-1}(k;\mathbf{y}_0^{\nu-1};\mathbf{0}_M) = \mathbf{\Gamma}_{IOy0}^{\nu-1}(k)\mathbf{y}_0^{\nu-1}, \tag{13.6}$$

where

$$\mathbf{\Gamma}_{IOy0}^{\nu-1}(k) = \begin{bmatrix} \mathbf{\Gamma}_{IOy0}(k) \\ E^1\mathbf{\Gamma}_{IOy0}(k) \\ \vdots \\ E^{\nu-1}\mathbf{\Gamma}_{IOy0}(k) \end{bmatrix} \in \mathcal{R}^{\nu N \times \nu N}. \tag{13.7}$$

Theorem 13.1 *In order for the zero equilibrium vector $\mathbf{y}_e^{\nu-1} = \mathbf{0}_{\nu N}$ of the IO system (3.56) to be globally stable, it is necessary and sufficient to be stable.*

Proof. *Necessity.* Let the zero equilibrium vector $\mathbf{y}_e^{\nu-1} = \mathbf{0}_{\nu N}$ of the *IO* system (3.56) be globally stable. Then it is stable due to *b*) of Definition 13.1.

Sufficiency. Let the zero equilibrium vector $\mathbf{y}_e^{\nu-1} = \mathbf{0}_{\nu N}$ of the *IO* system (3.56) be stable. Then, Definition 13.1 holds. Let $\varepsilon \in \mathcal{R}^+$ be arbitrarily chosen. Let $\delta = \delta(\varepsilon) \in \mathcal{R}^+$ be the corresponding δ obeying (13.1) so that, due to Equation 13.5 through Equation 13.7,

$$\left\|\mathbf{y}_0^{\nu-1}\right\| < \delta \Longrightarrow \left\|\mathbf{y}^{\nu-1}(k;\mathbf{y}_0^{\nu-1};\mathbf{0}_M)\right\| = \left\|\mathbf{\Gamma}_{IOy_0}^{\nu-1}(k)\mathbf{y}_0^{\nu-1}\right\| < \varepsilon,\ \forall k \in \mathcal{N}_0.$$

Let δ_{ij} be the Kronecker delta, $\delta_{ij} = 0$ for $i \neq j$, and $\delta_{ij} = 1$ for $i = j$, and let $\mathbf{w}_l^{\nu N} = \begin{bmatrix} \delta_{1l} & \delta_{2l} & \cdots & \delta_{\nu N,l} \end{bmatrix}^T \in \mathcal{R}^{\nu N}$, $l \in \{1, 2, ..., \nu N\}$, and

$$\begin{gathered}
\mathbf{y}_{0l}^{\nu-1} = \frac{\delta}{2}\mathbf{w}_l^{\nu N} \Longrightarrow \left\|\mathbf{y}_{0l}^{\nu-1}\right\| = \frac{\delta}{2} < \delta \Longrightarrow \\
\left\|\mathbf{y}^{\nu-1}(k;\mathbf{y}_{0l}^{\nu-1};\mathbf{0}_M)\right\| = \left\|\mathbf{\Gamma}_{IOy_0}^{\nu-1}(k)\mathbf{y}_{0l}^{\nu-1}\right\| = \\
= \frac{\delta}{2}\left\|\mathbf{\Gamma}_{IOy_0}^{\nu-1}(k)\mathbf{w}_l^{\nu N}\right\| < \varepsilon,\ \forall k \in \mathcal{N}_0, \\
\forall l \in \{1, 2, \cdots, \nu N\} \Longrightarrow
\end{gathered}$$

$$\begin{gathered}
\exists \gamma \in \mathcal{R}^+ \Longrightarrow \left\|\mathbf{\Gamma}_{IOy_0}^{\nu-1}(k)\right\| < \gamma,\ \forall k \in \mathcal{N}_0 \Longrightarrow \\
0 \leq \left\|\mathbf{y}^{\nu-1}(k;\mathbf{y}_0^{\nu-1};\mathbf{0}_M)\right\| = \left\|\mathbf{\Gamma}_{IOy_0}^{\nu-1}(k)\mathbf{y}_0^{\nu-1}\right\| \leq \gamma\left\|\mathbf{y}_0^{\nu-1}\right\|.
\end{gathered}$$

Hence,

$$\begin{gathered}
\forall k \in \mathcal{N}_0,\ \forall \mathbf{y}_0^{\nu-1} \in \mathcal{R}^{\nu N},\ \left\|\mathbf{y}_0^{\nu-1}\right\| < \delta \text{ for } \delta = \delta(\varepsilon) = \frac{\varepsilon}{2\gamma} \Longrightarrow \\
\left\|\mathbf{y}^{\nu-1}(k;\mathbf{y}_0^{\nu-1};\mathbf{0}_M)\right\| = \left\|\mathbf{\Gamma}_{IOy_0}^{\nu-1}(k)\mathbf{y}_0^{\nu-1}\right\| \leq \gamma\left\|\mathbf{y}_0^{\nu-1}\right\| < \\
< \gamma\delta(\varepsilon) = \frac{\varepsilon}{2} < \varepsilon,\ \forall k \in \mathcal{N}_0, \\
\delta_{\max}(\varepsilon) \geq \delta(\varepsilon) = \frac{\varepsilon}{2\gamma} \to \infty \text{ as } \varepsilon \to \infty.
\end{gathered}$$

This proves global stability of the zero equilibrium vector $\mathbf{y}_e^{\nu-1} = \mathbf{0}_{\nu N}$ of the *IO* system (3.56). ■

Conclusion 13.1 and this theorem imply the following:

Theorem 13.2 *In order for the desired behavior $\mathbf{Y}_d^{\nu-1}(t)$ of the IO system (2.1) to be globally stable it is necessary and sufficient to be stable.*

Definition 13.2 *The desired behavior $\mathbf{Y}_d^{\nu-1}(k)$ of the IO system (2.1) is* ***unstable*** *if and only if there exists a positive real number ε such that for every positive real number δ there exist an initial vector $\mathbf{Y}_0^{\nu-1}$ satisfying*

$$\left\|\mathbf{Y}_{d0}^{\nu-1}-\mathbf{Y}_0^{\nu-1}\right\|<\delta,$$

and a discrete moment $k_\tau \in \mathcal{N}_0$, which imply

$$\left\|\mathbf{Y}_d^{\nu-1}(k)-\mathbf{Y}^{\nu-1}(k_\tau;\mathbf{y}_0^{\nu-1};\mathbf{I}_N)\right\|\geq\varepsilon.$$

The zero equilibrium vector $\mathbf{y}_e^{\nu-1}=\mathbf{0}_{\nu N}$ of the IO system (3.56) is ***unstable*** *if and only if*

$$\exists\varepsilon\in\mathcal{R}^+\longrightarrow\forall\delta\in\mathcal{R}^+,\ \exists(\mathbf{y}_0^{\nu-1},k_\tau)\in\mathcal{R}^{N\nu}\times\mathcal{N}_0,$$
$$\left\|\mathbf{y}_0^{\nu-1}\right\|<\delta\Longrightarrow\left\|\mathbf{y}^{\nu-1}(k_\tau;\mathbf{y}_0^{\nu-1};\mathbf{0}_M)\right\|\geq\varepsilon.$$

Conclusion 13.2 ***Relationship between instability of $\mathbf{Y}_d^{\nu-1}(k)$ and of $\mathbf{y}_e^{\nu-1}=\mathbf{0}_{\nu N}$***

The zero equilibrium vector $\mathbf{y}_e^{\nu-1}=\mathbf{0}_{\nu N}$ of the IO system (3.56) is unstable if and only if the desired behavior $\mathbf{Y}_d^{\nu-1}(k)$ of the IO system (2.1) is unstable.

Note 13.2 ***Geometrical interpretation***

The zero equilibrium vector $\mathbf{y}_e^{\nu-1}=\mathbf{0}_{\nu N}$ of the IO system (3.56) is ***unstable*** *if and only if there exists an open hyperball $\mathcal{B}_\varepsilon\subset\mathcal{R}^{N\nu}$ centered at the origin with the radius ε, such that for every open hyperball $\mathcal{B}_\delta\subset\mathcal{R}^{N\nu}$ there are a moment $k_\tau\in\mathcal{N}_0$ and $\mathbf{y}_0^{\nu-1}\in\mathcal{B}_\delta$ implying in the free regime that $\mathbf{y}^{\nu-1}(k;\mathbf{y}_0^{\nu-1};\mathbf{0}_M)$ escapes $\mathcal{B}_\varepsilon$ at the instant k_τ,*

$$\exists\mathcal{B}_\varepsilon\subset\mathcal{R}^{N\nu},\ \forall\mathcal{B}_\delta\subset\mathcal{R}^{N\nu}\Longrightarrow\exists(\mathbf{y}_0^{\nu-1},k_\tau)\in\mathcal{B}_\delta\times\mathcal{N}_0\Longrightarrow$$
$$\mathbf{y}^{\nu-1}(k_\tau;\mathbf{y}_0^{\nu-1};\mathbf{0}_M)\in\mathcal{R}^{N\nu}\backslash\mathcal{B}_\varepsilon,$$
$$\mathcal{R}^{N\nu}\backslash\mathcal{B}_\varepsilon=\left\{\mathbf{y}^{\nu-1}:\ \mathbf{y}^{\nu-1}\in\mathcal{R}^{N\nu},\ \mathbf{y}^{\nu-1}\notin\mathcal{B}_\varepsilon\right\}.$$

Attraction of the zero equilibrium vector

Another important qualitative dynamical system property is *attraction* of the desired motion and of the equilibrium vector. We present their definitions now in the framework of the *IO* systems.

Definition 13.3 *a) The desired behavior* $\mathbf{Y}_d^{\nu-1}(k)$ *of the IO system (2.1) is* ***attractive*** *if and only if there exists a positive real number* Δ, *or* $\Delta = \infty$, *such that*

$$\left\| \mathbf{Y}_{d0}^{\nu-1} - \mathbf{Y}_0^{\nu-1} \right\| < \Delta$$

implies

$$\lim_{k \to \infty} \left\| \mathbf{Y}_d^{\nu-1}(k) - \mathbf{Y}^{\nu-1}(k; \mathbf{Y}_0^{\nu-1}; \mathbf{I}_N) \right\| = 0. \tag{13.8}$$

The zero equilibrium vector $\mathbf{y}_e^{\nu-1} = \mathbf{0}_{\nu N}$ *of the IO system (3.56) is* ***attractive*** *if and only if*

$$\exists \Delta > 0 \Longrightarrow$$
$$\left\| \mathbf{y}_0^{\nu-1} \right\| < \Delta \Longrightarrow \lim_{k \to \infty} \mathbf{y}^{\nu-1}(k; \mathbf{y}_0^{\nu-1}; \mathbf{0}_M) = \mathbf{0}_{\nu N}. \tag{13.9}$$

b) If and only if a) holds for $\Delta = \infty$ *then the desired behavior* $\mathbf{Y}_d^{\nu-1}(k)$ *of the IO system (2.1), i.e., the zero equilibrium vector* $\mathbf{y}_e^{\nu-1} = \mathbf{0}_{\nu N}$ *of the IO system (3.56), is* ***globally attractive (attractive in the whole)****.*

Conclusion 13.3 ***Relationship between attraction of*** $\mathbf{Y}_d^{\nu-1}(k)$ ***and of*** $\mathbf{y}_e^{\nu-1} = \mathbf{0}_{\nu N}$

The desired behavior $\mathbf{Y}_d^{\nu-1}(k)$ *of the IO system (2.1) is (globally) attractive if and only if the zero equilibrium vector* $\mathbf{y}_e^{\nu-1} = \mathbf{0}_{\nu N}$ *of the IO system (3.56) is (globally) attractive, respectively.*

Note 13.3 ***Geometrical interpretation***

The zero equilibrium vector $\mathbf{y}_e^{\nu-1} = \mathbf{0}_{\nu N}$ *of the IO system (3.56) is* ***attractive*** *if and only if there exists an open neighborhood* $\mathcal{B}_\Delta \subseteq \mathcal{R}^{N\nu}$ *of* $\mathbf{0}_{\nu N}$ *such that every initial vector* $\mathbf{y}_0^{\nu-1}$ *in* $\mathcal{B}_\Delta$, $\mathbf{y}_0^{\nu-1} \in \mathcal{B}_\Delta$, *ensures that the system motion* $\mathbf{y}^{\nu-1}(k; \mathbf{y}_0^{\nu-1}; \mathbf{0}_M)$ *approaches asymptotically the equilibrium state* $\mathbf{y}_e^{\nu-1} = \mathbf{0}_{\nu N}$ *as* $k \to \infty$, *Fig. 13.4 (in Section 13.1).*

The meaning of the global attraction of the zero equilibrium vector $\mathbf{y}_e^{\nu-1} = \mathbf{0}_{\nu N}$ of the *IO* system (3.56) is that for every initial vector $\mathbf{y}_0^{\nu-1} \in \mathcal{R}^{N\nu}$ the system motion $\mathbf{y}^{\nu-1}(k; \mathbf{y}_0^{\nu-1}; \mathbf{0}_M)$ approaches asymptotically the equilibrium state $\mathbf{y}_e^{\nu-1} = \mathbf{0}_{\nu N}$ as $k \to \infty$.

Global attraction of the zero equilibrium vector $\mathbf{y}_e^{\nu-1} = \mathbf{0}_{\nu N}$ of the *IO* system (3.56) implies its attraction. The inverse also holds as the following theorem explains.

Theorem 13.3 *In order for the zero equilibrium vector* $\mathbf{y}_e^{\nu-1} = \mathbf{0}_{\nu N}$ *of the IO system (3.56) to be globally attractive, it is necessary and sufficient to be attractive.*

Proof. *Necessity.* Let the zero equilibrium vector $\mathbf{y}_e^{\nu-1} = \mathbf{0}_{\nu N}$ of the *IO* system (3.56) be globally attractive. Let Δ be any positive real number. Since $\mathbf{y}^{\nu-1}(k;\mathbf{y}_0^{\nu-1};\mathbf{0}_M)$ approaches asymptotically the equilibrium state $\mathbf{y}_e^{\nu-1} = \mathbf{0}_{\nu N}$ as $k \to \infty$ for every $\mathbf{y}_0^{\nu-1} \in \mathcal{R}^{\nu N}$, *b*) of Definition 13.3, then $\left\|\mathbf{y}_0^{\nu-1}\right\| < \Delta < \infty$ implies in the free regime $\lim_{k\to\infty} \mathbf{y}^{\nu-1}(k;\mathbf{y}_0^{\nu-1};\mathbf{0}_M) = \mathbf{0}_{\nu N}$. The zero equilibrium vector $\mathbf{y}_e^{\nu-1} = \mathbf{0}_{\nu N}$ of the *IO* system (3.56) is attractive in view of *a*) of Definition 13.3.

Sufficiency. Let the zero equilibrium vector $\mathbf{y}_e^{\nu-1} = \mathbf{0}_{\nu N}$ of the *IO* system (3.56) be attractive. Its response in the free regime is given by Equation 13.5 through Equation 13.7. Since

$$\mathbf{y}^{\nu-1}(k;\mathbf{y}_0^{\nu-1};\mathbf{0}_M) = \mathbf{\Gamma}_{IOy_0}^{\nu-1}(k)\mathbf{y}_0^{\nu-1} \to \mathbf{0}_{\nu N} \text{ as } k \to \infty,$$
$$\forall \mathbf{y}_0^{\nu-1} \in \mathcal{R}^{\nu N} \text{ obeying } \left\|\mathbf{y}_0^{\nu-1}\right\| < \Delta,\ a) \text{ of Definition 13.3,}$$

then

$$\mathbf{\Gamma}_{IOy_0}^{\nu-1}(k) \to \boldsymbol{O}_{\nu N,\, \nu N} \text{ as } k \to \infty.$$

This implies

$$\lim_{k\to\infty} \mathbf{y}^{\nu-1}(k;\mathbf{y}_0^{\nu-1};\mathbf{0}_M) = \mathbf{0}_{\nu N},\ \forall \mathbf{y}_0^{\nu-1} \in \mathcal{R}^{\nu N},$$

which shows that the zero equilibrium vector $\mathbf{y}_e^{\nu-1} = \mathbf{0}_{\nu N}$ of the *IO* system (3.56) is globally attractive (part *b*) of Definition 13.3). ∎

Theorem 13.3 enables us to establish the relationship between the uniqueness of the zero equilibrium vector and its attraction.

Theorem 13.4 *In order for the zero equilibrium vector* $\mathbf{y}_e^{\nu-1} = \mathbf{0}_{\nu N}$ *of the IO system (3.56) to be attractive, it is necessary (but not sufficient) to be the unique equilibrium vector of the system.*

Proof. *Necessity.* Let the zero equilibrium vector $\mathbf{y}_e^{\nu-1} = \mathbf{0}_{\nu N}$ of the *IO* system (3.56) be nonunique and attractive. Then there is another equilibrium vector $\mathbf{y}_{e2}^{\nu-1} \neq \mathbf{0}_{\nu N}$ of the *IO* system (3.56). The zero equilibrium vector $\mathbf{y}_e^{\nu-1} = \mathbf{0}_{\nu N}$ of the *IO* system is globally attractive due to Theorem 13.3. However,

$$\lim_{k\to\infty} \mathbf{y}^{\nu-1}(k;\mathbf{y}_{e2}^{\nu-1};\mathbf{0}_M) = \mathbf{y}_{e2}^{\nu-1} \neq \mathbf{0}_{\nu N}$$

due to Definition 3.16 (in Subsection 3.7.2), of the equilibrium vector of the IO system (3.56), which contradicts global attraction of $\mathbf{y}_e^{\nu-1} = \mathbf{0}_{\nu N}$. Hence, there does not exist another equilibrium vector $\mathbf{y}_{e2}^{\nu-1} \neq \mathbf{0}_{\nu N}$ of the IO system (3.56), i.e., the zero equilibrium vector $\mathbf{y}_e^{\nu-1} = \mathbf{0}_{\nu N}$ of the IO system is unique.

Insufficiency. Let the IO system (3.56) be defined by

$$y(k+1) - 2y(k) = i(k).$$

Its equilibrium point $y_e = 0$ is unique. Its response in the free regime reads $y(k; y_0; 0) = 2^k y_0$, $\forall k \in \mathcal{N}_0$, so that $\lim_{k\to\infty} y(k; y_0; 0) = \infty \text{ sign } y_0$, $\forall\, (y_0 \neq 0) \in \mathcal{R}$. The equilibrium point $y_e = 0$ is not attractive. ■

This theorem and Theorem 3.8 (in Subsection 3.7.2), imply

Theorem 13.5 *In order for the zero equilibrium vector* $\mathbf{y}_e^{\nu-1} = \mathbf{0}_{\nu N}$ *of the IO system (3.56) to be attractive, it is necessary (but not sufficient) that the matrix* $(A_0 + A_1 + \cdots + A_\nu)$ *is nonsingular.*

Asymptotic stability of the zero equilibrium vector

Definition 13.4 *The desired behavior* $\mathbf{Y}_d^{\nu-1}(k)$ *of the IO system (2.1), i.e., the zero equilibrium vector* $\mathbf{y}_e^{\nu-1} = \mathbf{0}_{\nu N}$ *of the IO system (3.56) is:*

a) ***asymptotically stable*** *if and only if it is both stable and attractive,*

b) ***globally asymptotically stable*** *if and only if it is both stable and globally attractive,*

c) ***strictly (completely, fully) globally asymptotically stable*** *if and only if it is both globally stable and globally attractive.*

This definition, Theorem 13.1, and Theorem 13.3 imply,

Theorem 13.6 *In order for the zero equilibrium vector* $\mathbf{y}_e^{\nu-1} = \mathbf{0}_{\nu N}$ *of the IO system (3.56) to be strictly globally asymptotically stable, it is necessary and sufficient to be asymptotically stable.*

Exponential stability of the zero equilibrium vector

The exponential stability is introduced in order to ensure a higher convergence rate of system behaviors to the equilibrium vector than what its asymptotic stability can assure.

Definition 13.5 *a) The desired behavior* $\mathbf{Y}_d^{\nu-1}(k)$ *of the IO system (2.1) is* ***exponentially stable*** *if and only if there exist positive real numbers* $\alpha \geq 1$, β, *and* Δ, *or* $\Delta = \infty$, *such that*

$$\left\|\mathbf{Y}_{d0}^{\nu-1} - \mathbf{Y}_0^{\nu-1}\right\| < \Delta$$

implies

$$\left\|\mathbf{Y}_d^{\nu-1}(k) - \mathbf{Y}^{\nu-1}(k;\mathbf{Y}_0^{\nu-1};\mathbf{I}_N)\right\| \leq \alpha \exp(-\beta k)\left\|\mathbf{Y}_{d0}^{\nu-1} - \mathbf{Y}_0^{\nu-1}\right\|, \quad \forall k \in \mathcal{N}_0. \tag{13.10}$$

The zero equilibrium vector $\mathbf{y}_e^{\nu-1} = \mathbf{0}_{\nu N}$ *of the IO system (3.56) is* ***exponentially stable*** *if and only if*

$$\exists \alpha \in \mathcal{R}^+,\ \alpha \geq 1,\ \exists \beta \in \mathcal{R}^+,\ \text{and}\ \exists \Delta > 0 \Longrightarrow \left\|\mathbf{y}_0^{\nu-1}\right\| < \Delta \Longrightarrow \left\|\mathbf{y}^{\nu-1}(k;\mathbf{y}_0^{\nu-1};\mathbf{0}_M)\right\| \leq \alpha \exp(-\beta k)\left\|\mathbf{y}_0^{\nu-1}\right\|,\ \forall k \in \mathcal{N}_0. \tag{13.11}$$

b) If and only if a) holds for $\Delta = \infty$, *then the desired behavior* $\mathbf{Y}_d^{\nu-1}(k)$ *of the IO system (2.1), i.e., the zero equilibrium vector* $\mathbf{y}_e^{\nu-1} = \mathbf{0}_{\nu N}$ *of the IO system (3.56), is* ***globally exponentially stable (exponentially stable in the whole)****.*

Conclusion 13.4 ***Relationship between exponential stability of*** $\mathbf{Y}_d^{\nu-1}(k)$ ***and of*** $\mathbf{y}_e^{\nu-1} = \mathbf{0}_{\nu N}$

The zero equilibrium vector $\mathbf{y}_e^{\nu-1} = \mathbf{0}_{\nu N}$ *of the IO system (3.56) is (globally) exponentially stable if and only if the desired behavior* $\mathbf{Y}_d^{\nu-1}(k)$ *of the IO system (2.1) is (globally) exponentially stable, respectively.*

The number α determines the upper bound of $\left\|\mathbf{y}^{\nu-1}(k;\mathbf{y}_0^{\nu-1};\mathbf{0}_M)\right\|$ on $\mathcal{N}_0$ for fixed $\mathbf{y}_0^{\nu-1}$, which is a measure of the maximal deviation of $\mathbf{y}^{\nu-1}(k;\mathbf{y}_0^{\nu-1};\mathbf{0}_M)$ from the zero equilibrium vector $\mathbf{y}_e^{\nu-1} = \mathbf{0}_{\nu N}$. The number β signifies the minimal convergence rate of $\left\|\mathbf{y}^{\nu-1}(k;\mathbf{y}_0^{\nu-1};\mathbf{0}_M)\right\|$ to zero, i.e., of the system state $\mathbf{y}^{\nu-1}(k;\mathbf{y}_0^{\nu-1};\mathbf{0}_M)$ to the zero equilibrium vector $\mathbf{y}_e^{\nu-1} = \mathbf{0}_{\nu N}$.

Theorem 13.7 *In order for the zero equilibrium vector* $\mathbf{y}_e^{\nu-1} = \mathbf{0}_{\nu N}$ *of the IO system (3.56) to be globally exponentially stable, it is necessary and sufficient to be exponentially stable.*

The proof of this theorem is analogous to the proof of Theorem 13.3.

Theorem 13.8 *In order for the zero equilibrium vector* $\mathbf{y}_e^{\nu-1} = \mathbf{0}_{\nu N}$ *of the IO system (3.56) to be exponentially stable, it is necessary and sufficient to be asymptotically stable.*

Proof. *Necessity.* Let the zero equilibrium vector $\mathbf{y}_e^{\nu-1} = \mathbf{0}_{\nu N}$ of the *IO* system (3.56) be exponentially stable. Then, it is attractive because (Definition 13.5),

$$\left\|\mathbf{y}_0^{\nu-1}\right\| < \Delta \Longrightarrow$$
$$\left\|\mathbf{y}^{\nu-1}(k;\mathbf{y}_0^{\nu-1};\mathbf{0}_M)\right\| \le \alpha \exp(-\beta k)\left\|\mathbf{y}_0^{\nu-1}\right\|, \ \forall k \in \mathcal{N}_0,$$

implies

$$\mathbf{y}^{\nu-1}(k;\mathbf{y}_0^{\nu-1};\mathbf{0}_M) \to \mathbf{0}_{\nu N} \text{ as } k \longrightarrow \infty.$$

Let $\delta(\varepsilon) = \varepsilon\alpha^{-1}$, $\forall \varepsilon \in \mathcal{R}^+$. Hence,

$$\left\|\mathbf{y}_0^{\nu-1}\right\| < \delta(\varepsilon) = \varepsilon\alpha^{-1}$$

and

$$\left\|\mathbf{y}^{\nu-1}(k;\mathbf{y}_0^{\nu-1};\mathbf{0}_M)\right\| \le \alpha \exp(-\beta k)\left\|\mathbf{y}_0^{\nu-1}\right\|, \ \forall k \in \mathcal{N}_0,$$

guarantee

$$\left\|\mathbf{y}^{\nu-1}(k;\mathbf{y}_0^{\nu-1};\mathbf{0}_M)\right\| \le \alpha \exp(-\beta k)\left\|\mathbf{y}_0^{\nu-1}\right\| <$$
$$< \alpha \exp(-\beta k)\varepsilon\alpha^{-1} \le \varepsilon, \ \forall \varepsilon \in \mathcal{R}^+, \ \forall k \in \mathcal{N}_0.$$

This proves stability of $\mathbf{y}_e^{\nu-1} = \mathbf{0}_{\nu N}$. Since it is also attractive, then it is asymptotically stable.

Sufficiency. Notice that the *IO* system (3.56) can be written in the free regime in the *ISO* form:

$$E^1\mathbf{y}^{\nu-1} = \begin{bmatrix} E^1\mathbf{y} \\ E^2\mathbf{y} \\ \vdots \\ E^{\nu-1}\mathbf{y} \\ E^{\nu}\mathbf{y} \end{bmatrix} =$$

$$= \underbrace{\begin{bmatrix} O_N & I_N & O_N & \cdots & O_N & O_N \\ O_N & O_N & I_N & \cdots & O_N & O_N \\ \vdots & \vdots & \vdots & \vdots & \vdots & \vdots \\ O_N & O_N & O_N & \cdots & O_N & I_N \\ -\mathbf{A}_\nu^{-1}\mathbf{A}_0 & -\mathbf{A}_\nu^{-1}\mathbf{A}_1 & -\mathbf{A}_\nu^{-1}\mathbf{A}_2 & \cdots & -\mathbf{A}_\nu^{-1}\mathbf{A}_{\nu-2} & -\mathbf{A}_\nu^{-1}\mathbf{A}_{\nu-1} \end{bmatrix}}_{\mathbf{A}_{IO}} \cdot$$

$$\cdot \mathbf{y}^{\nu-1}, \tag{13.12}$$

i.e.,

$$E^1\mathbf{w} = \boldsymbol{A}_{IO}\mathbf{w}, \text{ for } \mathbf{w} = \mathbf{y}^{\nu-1}, \ \boldsymbol{A}_{IO} \in \mathcal{R}^{\nu N \times \nu N}. \tag{13.13}$$

The solution $\mathbf{w}(k;\mathbf{w}_0)$ of (13.13) reads

$$\mathbf{w}(k;\mathbf{w}_0) = \boldsymbol{A}_{IO}^k\mathbf{w}_0,$$

i.e.,

$$\mathbf{y}^{\nu-1}(k;\mathbf{y}_0^{\nu-1};\mathbf{0}_M) = \boldsymbol{A}_{IO}^k\mathbf{y}_0^{\nu-1}. \tag{13.14}$$

Let the zero equilibrium vector $\mathbf{y}_e^{\nu-1} = \mathbf{0}_{\nu N}$ of the IO system (3.56) be asymptotically stable. Then, Definition 13.1, Definition 13.4 and (13.14) yield

$$\forall\varepsilon \in \mathcal{R}^+, \ \exists\delta \in \mathcal{R}^+, \ \delta = \delta(\varepsilon), \implies$$
$$\left\|\mathbf{y}_0^{\nu-1}\right\| < \delta \implies \left\|\boldsymbol{A}_{IO}^k\mathbf{y}_0^{\nu-1}\right\| < \varepsilon, \ \forall k \in \mathcal{N}_0. \tag{13.15}$$

Definition 13.3 and Definition 13.4 furnish

$$\exists\Delta > 0 \implies$$
$$\left\|\mathbf{y}_0^{\nu-1}\right\| < \Delta \implies \lim_{k\to\infty} \boldsymbol{A}_{IO}^k\mathbf{y}_0^{\nu-1} = \mathbf{0}_{\nu N}. \tag{13.16}$$

From (13.15) and (13.16) follows that

$$\exists\gamma \in \mathcal{R}^+ \implies \left\|\boldsymbol{A}_{IO}^k\right\| < \gamma, \ \forall k \in \mathcal{N}_0, \tag{13.17}$$

and

$$\lim_{k\to\infty} \boldsymbol{A}_{IO}^k = \boldsymbol{O}_{\nu N}, \tag{13.18}$$

which, together with (13.14), implies, for $\alpha \geq \gamma$, $\alpha \in \mathcal{R}^+$, and for some $\beta \in \mathcal{R}^+$, which exists due to (13.14),

$$\left\|\mathbf{y}^{\nu-1}(k;\mathbf{y}_0^{\nu-1};\mathbf{0}_M)\right\| = \left\|\boldsymbol{A}_{IO}^k\mathbf{y}_0^{\nu-1}\right\| \leq \left\|\boldsymbol{A}_{IO}^k\right\| \left\|\mathbf{y}_0^{\nu-1}\right\| \leq$$
$$\leq \alpha\exp(-\beta k)\left\|\mathbf{y}_0^{\nu-1}\right\|, \ \forall\mathbf{y}_0^{\nu-1} \in \mathcal{R}^{N\nu}, \ \forall k \in \mathcal{N}_0. \tag{13.19}$$

The zero equilibrium vector $\mathbf{y}_e^{\nu-1} = \mathbf{0}_{\nu N}$ of the IO system (3.56) is globally exponentially stable (Definition 13.5), hence exponentially stable (Theorem 13.7). ■

This proof permits the proof of the following:

Theorem 13.9 *In order for the zero equilibrium vector* $\mathbf{y}_e^{\nu-1} = \mathbf{0}_{\nu N}$ *of the IO system (3.56) to be exponentially stable, it is necessary and sufficient to be attractive.*

Proof. *Necessity.* Let the zero equilibrium vector $y_e^{\nu-1} = 0_{\nu N}$ of the *IO* system (3.56) be exponentially stable. Then it is attractive that has been proved in the proof of the necessity of the condition of Theorem 13.8.

Sufficiency. Let the zero equilibrium vector $y_e^{\nu-1} = 0_{\nu N}$ of the *IO* system (3.56) be attractive. The proof of Theorem 13.8 provides (13.14) through (13.16), which imply (13.17) through (13.19), hence prove exponential stability of the zero equilibrium vector $y_e^{\nu-1} = 0_{\nu N}$ of the *IO* system (3.56). It is also global in view of Theorem 13.7. ■

This theorem, Theorem 13.4, Theorem 13.5 and Definition 13.4 imply directly the following:

Theorem 13.10 *In order for the zero equilibrium vector* $\mathbf{y}_e^{\nu-1} = \mathbf{0}_{\nu N}$ *of the IO system (3.56) to be exponentially stable, it is necessary (but not sufficient) that* $\mathbf{y}_e^{\nu-1} = \mathbf{0}_{\nu N}$ *is the unique equilibrium vector of the system, i.e., that the matrix* $(A_0 + A_1 + \cdots + A_\nu)$ *is nonsingular.*

Conclusions 13.1, 13.2, 13.3, and 13.4 result in:

Conclusion 13.5 ***Stability properties of the total desired behavior and of the zero deviation vector***

The zero deviation vector $\mathbf{y}_e^{\nu-1} = \mathbf{0}_{\nu N}$ *of the IO system (3.56) and the total desired behavior* $\mathbf{Y}_d^{\nu-1}(k)$ *of the IO system (2.1) possess the same stability properties.*

13.2.2 *ISO* systems

Stability of the zero equilibrium state

A. M. Lyapunov [93] defined stability of the zero equilibrium vector of the *ISO* continuous-time nonlinear systems, which are in Cauchy (normal, *IS* systems) form. We will broaden it for the discrete-time *ISO* system (2.14a) and (2.14b) (in Section 2.2); i.e., for (3.60a) and (3.60b) (in Subsection 3.5.3).

Definition 13.6 *a) The desired motion* $\mathbf{X}_d(k)$ *of the ISO system (2.14a) and (2.14b) is* ***stable*** *if and only if for every positive real number* ε *there exists a positive real number* δ*, the value of which depends on* $\varepsilon, \delta = \delta(\varepsilon)$*, such that* $\|\mathbf{X}_{d0} - \mathbf{X}_0\| < \delta$ *implies* $\|\mathbf{X}_d(k) - \mathbf{X}(k; \mathbf{X}_0; \mathbf{I}_N)\| < \varepsilon$ *for all* $k \in \mathcal{N}_0$,

The zero equilibrium vector $\mathbf{x}_e = \mathbf{0}_n$ *of the ISO system (3.60a) and (3.60b) is* ***stable*** *if and only if*

$$\forall \varepsilon \in \mathcal{R}^+,\ \exists \delta \in \mathcal{R}^+,\ \delta = \delta(\varepsilon),\ \Longrightarrow$$
$$\|\mathbf{x}_0\| < \delta \Longrightarrow \|\mathbf{x}(k;\mathbf{x}_0;\mathbf{0}_M)\| < \varepsilon,\ \forall k \in \mathcal{N}_0. \tag{13.20}$$

b) *The desired motion* $\mathbf{X}_d(k)$ *of the ISO system (2.14a) and (2.14b), i.e., the zero equilibrium vector* $\mathbf{x}_e = \mathbf{0}_n$ *of the ISO system (3.60a) and (3.60b), is* ***globally stable*** *(i.e.,* ***stable in the whole*** *) if and only if a) holds so that the maximal* $\delta(\varepsilon)$*, denoted by* $\delta_{\max}(\varepsilon)$*, diverges to infinity as* ε *goes to infinity,*

$$\varepsilon \to \infty \Longrightarrow \delta_{\max}(\varepsilon) \to \infty. \tag{13.21}$$

Conclusion 13.6 ***Relationship between stability of*** $\mathbf{X}_d(k)$ ***and of*** $\mathbf{x}_e = \mathbf{0}_n$

For the desired motion $\mathbf{X}_d(k)$ *of the ISO system (2.14a) and (2.14b) to be* ***stable,*** *it is necessary and sufficient that the zero equilibrium vector* $\mathbf{x}_e = \mathbf{0}_n$ *of the ISO system (3.60a) and (3.60b) is stable.*

Note 13.4 ***Geometrical interpretation***

This definition geometrically means that the zero equilibrium vector $\mathbf{x}_e = \mathbf{0}_n$ *of the ISO system (3.60a) and (3.60b) is* ***stable*** *if and only if for every open hyperball* $\mathcal{B}_\varepsilon \subseteq \mathcal{R}^n$ *centered at the origin with the radius* ε*, there exists an open hyperball* $\mathcal{B}_\delta \subseteq \mathcal{R}^n$ *centered at the origin* $\mathbf{x} = \mathbf{0}_n$ *with the radius* δ*, which depends on* $\mathcal{B}_\varepsilon$*, such that* $\mathbf{x}_0 \in \mathcal{B}_\delta$ *guarantees that the system state* $\mathbf{x}(k) = \mathbf{x}(k;\mathbf{x}_0;\mathbf{0}_M)$ *at discrete instant* k *stays in* $\mathcal{B}_\varepsilon$ *for any* k *at and after* $k = k_0 = 0$*, Fig. 13.2 (in Section 13.1),*

$$\forall \mathcal{B}_\varepsilon \subseteq \mathcal{R}^n,\ \exists \mathcal{B}_\delta \subseteq \mathcal{R}^n,\ \mathcal{B}_\delta = \mathcal{B}_\delta(\mathcal{B}_\varepsilon),\ \Longrightarrow$$
$$\mathbf{x}_0 \in \mathcal{B}_\delta \Longrightarrow \mathbf{x}(k;\mathbf{x}_0;\mathbf{0}_M) \in \mathcal{B}_\varepsilon,\ \forall k \in \mathcal{N}_0. \tag{13.22}$$

If and only if additionally $\mathcal{B}_\varepsilon \longrightarrow \mathcal{R}^n$ *implies* $\mathcal{B}_\delta \longrightarrow \mathcal{R}^n$*, then* $\mathbf{x}_e = \mathbf{0}_n$ *is* ***globally stable*** *(i.e.,* ***stable in the whole****).*

The following is well-known in the linear systems theory (which is easy to verify or see Section 10.2) that the motion $\mathbf{x}(.;\mathbf{x}_0;\mathbf{0}_M)$ of the *ISO* system (3.60a) and (3.60b) in the free regime is given by

$$\mathbf{x}(k;\mathbf{x}_0;\mathbf{0}_M) = \boldsymbol{A}^k\mathbf{x}_0,\ \forall k \in \mathcal{N}_0. \tag{13.23}$$

Theorem 13.11 *In order for the zero equilibrium vector* $\mathbf{x}_e = \mathbf{0}_n$ *of the ISO system (3.60a) and (3.60b) to be globally stable, it is necessary and sufficient to be stable.*

Proof. *Necessity.* Let the zero equilibrium vector $\mathbf{x}_e = \mathbf{0}_n$ of the *ISO* system (3.60a) and (3.60b) be globally stable. Then, it is stable due to *b)* of Definition 13.6.

Sufficiency. Let the zero equilibrium vector $\mathbf{x}_e = \mathbf{0}_n$ of the *ISO* system (3.60a) and (3.60b) be stable. Then, Definition 13.6 holds. Let $\varepsilon \in \mathcal{R}^+$ be arbitrarily chosen. Let $\delta = \delta(\varepsilon) \in \mathcal{R}^+$ be the corresponding δ obeying Definition 13.6 so that, due to Equation 13.23,

$$\|\mathbf{x}_0\| < \delta \Longrightarrow \|\mathbf{x}(k;\mathbf{x}_0;\mathbf{0}_M)\| = \left\|\boldsymbol{A}^k\mathbf{x}_0\right\| < \varepsilon, \ \forall k \in \mathcal{N}_0.$$

Let δ_{ij} be the Kronecker delta,

$$\mathbf{w}_l^n = \begin{bmatrix} \delta_{1l} & \delta_{2l} & \cdots & \delta_{nl} \end{bmatrix}^T \in \mathcal{R}^n, l \in \{1, 2, ..., n\},$$

and

$$\mathbf{x}_0 = \frac{\delta}{2}\mathbf{w}_l^n \Longrightarrow \|\mathbf{x}_0\| = \frac{\delta}{2} < \delta \Longrightarrow$$
$$\|\mathbf{x}(t;\mathbf{x}_0;\mathbf{0}_M)\| = \left\|\boldsymbol{A}^k\mathbf{x}_0\right\| = \frac{\delta}{2}\left\|\boldsymbol{A}^k\mathbf{w}_l^n\right\| < \varepsilon, \ \forall k \in \mathcal{N}_0 \Longrightarrow$$
$$\exists \gamma \in \mathcal{R}^+ \Longrightarrow \left\|\boldsymbol{A}^k\right\| < \gamma, \ \forall k \in \mathcal{N}_0.$$

Therefore,

$$\forall k \in \mathcal{N}_0, \ \forall \mathbf{x}_0 \in \mathcal{R}^n, \ \|\mathbf{x}_0\| < \delta \text{ for } \delta = \delta(\varepsilon) = \frac{\varepsilon}{2\gamma} \Longrightarrow$$
$$\|\mathbf{x}(k;\mathbf{x}_0;\mathbf{0}_M)\| = \left\|\boldsymbol{A}^k\mathbf{x}_0\right\| \le \gamma\|\mathbf{x}_0\| < \gamma\delta = \frac{\varepsilon}{2} < \varepsilon, \ \forall k \in \mathcal{N}_0, \Longrightarrow$$
$$\delta_{\max}(\varepsilon) \ge \delta(\varepsilon) = \frac{\varepsilon}{2\gamma} \to \infty \text{ as } \varepsilon \to \infty.$$

This proves global stability of the zero equilibrium vector $\mathbf{x}_e = \mathbf{0}_n$ of the *ISO* system (3.60a) and (3.60b). ■

Definition 13.7 *The zero equilibrium vector* $\mathbf{x}_e = \mathbf{0}_n$ *of the ISO system (3.60a) and (3.60b) is* ***unstable*** *if and only if there exists a positive real number* ε *such that for every positive real number* δ *there exist an initial* *vector* $\mathbf{x}_0$ *satisfying* $\|\mathbf{x}_0\| < \delta$ *and a discrete moment* $k_\tau \in \mathcal{N}_0$*, which imply* $\|\mathbf{x}_0(k_\tau; \mathbf{x}_0; \mathbf{0}_M)\| \geq \varepsilon$ *in the free regime,*

$$\exists \varepsilon \in \mathcal{R}^+ \longrightarrow \forall \delta \in \mathcal{R}^+, \ \exists (\mathbf{x}_0, k_\tau) \in \mathcal{R}^n \times \mathcal{N}_0, \\ \|\mathbf{x}_0\| < \delta \Longrightarrow \|\mathbf{x}_0(k_\tau; \mathbf{x}_0; \mathbf{0}_M)\| \geq \varepsilon.$$

Note 13.5 *Geometrical explanation is the following: the zero equilibrium vector* $\mathbf{x}_e = \mathbf{0}_n$ *of the ISO system (3.60a) and (3.60b) is* ***unstable*** *if and only if there exists an open hyperball* $\mathcal{B}_\varepsilon \subseteq \mathcal{R}^n$ *centered at the origin with the radius* ε*, such that for every open hyperball* $\mathcal{B}_\delta \subseteq \mathcal{R}^n$ *there are a discrete moment* $k_\tau \in \mathcal{N}_0$ *and* $\mathbf{x}_0 \in \mathcal{B}_\delta$ *implying in the free regime that* $\mathbf{x}(k; \mathbf{x}_0; \mathbf{0}_M)$ *escapes* $\mathcal{B}_\varepsilon$ *at the instant* k_τ*,*

$$\exists \mathcal{B}_\varepsilon \subseteq \mathcal{R}^n, \ \forall \mathcal{B}_\delta \subseteq \mathcal{R}^n \Longrightarrow \exists (\mathbf{x}_0, k_\tau) \in \mathcal{B}_\delta \times \mathcal{N}_0 \Longrightarrow \\ \mathbf{x}_0(k_\tau; \mathbf{x}_0; \mathbf{0}_M) \notin \mathcal{B}_\varepsilon.$$

Attraction of the zero equilibrium state

Attraction of the equilibrium state $\mathbf{x} = \mathbf{0}_n$ is another crucial qualitative dynamical system property. It is independent of the stability property. We present its definition now in the framework of the *ISO* systems described by (2.14a) and (2.14b), i.e., (3.60a) and (3.60b).

Definition 13.8 *a) The desired motion* $\mathbf{X}_d(k)$ *of the ISO system (2.14a) and (2.14b) is* ***attractive*** *if and only if there exists a positive real number* Δ*, or* $\Delta = \infty$*, such that* $\|\mathbf{X}_{d0} - \mathbf{X}_0\| < \Delta$ *implies in the unperturbed regime*

$$\lim_{k \to \infty} [\mathbf{X}_d(k) - \mathbf{X}(k; \mathbf{X}_0; \mathbf{I}_N)] = \mathbf{0}_n.$$

The zero equilibrium vector $\mathbf{x}_e = \mathbf{0}_n$ *of the ISO system (3.60a) and (3.60b) is* ***attractive*** *if and only if*

$$\exists \Delta > 0 \Longrightarrow \|\mathbf{x}_0\| < \Delta \Longrightarrow \lim_{k \to \infty} \mathbf{x}(k; \mathbf{x}_0; \mathbf{0}_M) = \mathbf{0}_n.$$

b) If and only if a) holds for $\Delta = \infty$*, then the desired motion* $\mathbf{X}_d(k)$ *of the ISO system (2.14a) and (2.14b), i.e., the zero equilibrium vector* $\mathbf{x}_e = \mathbf{0}_n$ *of the ISO system (3.60a) and (3.60b) is* ***globally attractive (attractive in the whole)****.*

Conclusion 13.7 ***Relationship between attraction of*** $\mathbf{X}_d(k)$ ***and of*** $\mathbf{x}_e = \mathbf{0}_n$

For the desired motion $\mathbf{X}_d(k)$ *of the ISO system (2.14a) and (2.14b) to be* ***(globally) attractive,*** *it is necessary and sufficient that the zero equilibrium vector* $\mathbf{x}_e = \mathbf{0}_n$ *of the ISO system (3.60a) and (3.60b) is (globally) attractive, respectively.*

Note 13.6 ***Geometrical interpretation***

The zero equilibrium vector $\mathbf{x}_e = \mathbf{0}_n$ *of the ISO system (3.60a) and (3.60b) is* **attractive** *if and only if there exists an open hyperball* $\mathcal{B}_\Delta \subseteq \mathcal{R}^n$ *such that for every initial vector* $\mathbf{x}_0$ *in* $\mathcal{B}_\Delta$, $\mathbf{x}_0 \in \mathcal{B}_\Delta$, *the system state* $\mathbf{x}(k;\mathbf{x}_0;\mathbf{0}_M)$ *approaches asymptotically the equilibrium state* $\mathbf{x} = \mathbf{0}_n$ *as* $k \to \infty$, *Fig. 13.4 (in Section 13.1).*

The global attraction of the zero equilibrium state $\mathbf{x} = \mathbf{0}_n$ *of the ISO system (3.60a) and (3.60b) means that for every initial vector* $\mathbf{x}_0 \in \mathcal{R}^n$ *the system state* $\mathbf{x}(k;\mathbf{x}_0;\mathbf{0}_M)$ *approaches asymptotically the equilibrium state* $\mathbf{x}_e = \mathbf{0}_n$ *as* $k \to \infty$.

Global attraction of the zero equilibrium vector $\mathbf{x}_e = \mathbf{0}_n$ of the *ISO* system (3.60a) and (3.60b) implies its attraction. The inverse also holds as the following theorem explains.

Theorem 13.12 *In order for the zero equilibrium vector* $\mathbf{x}_e = \mathbf{0}_n$ *of the ISO system (3.60a) and (3.60b) to be globally attractive, it is necessary and sufficient to be attractive.*

Proof. *Necessity.* Let the zero equilibrium vector $\mathbf{x}_e = \mathbf{0}_n$ of the *ISO* system (3.60a) and (3.60b) be globally attractive. The equilibrium state $\mathbf{x} = \mathbf{0}_n$ is attractive due to *b*) of Definition 13.8.

Sufficiency. Let the zero equilibrium vector $\mathbf{x}_e = \mathbf{0}_n$ of the *ISO* system (3.60a) and (3.60b) be attractive. Its motion in the free regime is given by Equation 13.23. Since

$$\mathbf{x}(k;\mathbf{x}_0;\mathbf{0}_M) = \boldsymbol{A}^k\mathbf{x}_0 \to 0 \text{ as } k \to \infty,$$
$$\forall \mathbf{x}_0 \in \mathcal{R}^n \text{ obeying } \|\mathbf{x}_0\| < \Delta,\ a) \text{ of Definition 13.8},$$

then

$$\boldsymbol{A}^k \to \boldsymbol{O}_n \text{ as } k \to \infty.$$

This implies

$$\lim_{k\to\infty} \mathbf{x}(k;\mathbf{x}_0;\mathbf{0}_M) = \lim_{k\to\infty} \boldsymbol{A}^k\mathbf{x}_0 = \mathbf{0}_n,\ \forall \mathbf{x}_0 \in \mathcal{R}^n,$$

which shows that the zero equilibrium vector $\mathbf{x}_e = \mathbf{0}_n$ of the ISO system (3.60a) and (3.60b) is globally attractive (part b) of Definition 13.8). ■

This theorem enables us to state the relationship between the uniqueness of the zero equilibrium vector and its attraction.

Theorem 13.13 *In order for the zero equilibrium vector* $\mathbf{x}_e = \mathbf{0}_n$ *of the ISO system (3.60a) and (3.60b) to be attractive, it is necessary (but not sufficient) that it is the unique equilibrium vector of the system.*

Proof. *Necessity.* Let the zero equilibrium vector $\mathbf{x}_e = \mathbf{0}_n$ of the ISO system (3.60a) and (3.60b) be nonunique and attractive. Then there is another equilibrium vector $\mathbf{x}_{e2} \neq \mathbf{0}_n$ of the ISO system (3.60a) and (3.60b). The zero equilibrium vector $\mathbf{x}_e = \mathbf{0}_n$ of the ISO system is globally attractive due to Theorem 13.12. However,

$$\lim_{k \to \infty} \mathbf{x}(k; \mathbf{x}_{e2}; \mathbf{0}_M) = \mathbf{x}_{e2} \neq \mathbf{0}_n$$

due to Definition 3.17 (in Subsection 3.7.3), of the equilibrium vector of the ISO system (3.60a) and (3.60b), which contradicts global attraction of $\mathbf{x}_e = \mathbf{0}_n$. Hence, there does not exist another equilibrium vector $\mathbf{x_{e2}} \neq \mathbf{0}_n$ of the ISO system (3.60a) and (3.60b). The zero equilibrium vector $\mathbf{x}_e = \mathbf{0}_n$ of the ISO system is unique.

Insufficiency. Let the ISO system (3.60a) and (3.60b) be defined by

$$x\,(k+1) = 2x(k) + i(k),\ y = x.$$

Its equilibrium point $x_e = 0$ is unique. Its motion in the free regime reads $x(k; x_0; 0) = 2^k x_0$ so that $\lim_{k \to \infty} x(k; x_0; 0) = \infty \operatorname{sign} x_0,\ \forall\,(x_0 \neq 0) \in \mathcal{R}$. The equilibrium point $x_e = 0$ is not attractive. ■

This theorem and Theorem 3.10 (in Subsection 3.7.3), imply,

Theorem 13.14 *In order for the zero equilibrium vector* $\mathbf{x}_e = \mathbf{0}_n$ *of the ISO system (3.60a) and (3.60b) to be attractive, it is necessary (but not sufficient) that the matrix* $(\boldsymbol{I} - \boldsymbol{A})$ *is nonsingular.*

Asymptotic stability of the zero equilibrium state

Definition 13.9 *The desired motion* $\mathbf{X}_d\,(k)$ *of the ISO system (2.14a) and (2.14b), i.e., the zero equilibrium vector* $\mathbf{x}_e = \mathbf{0}_n$ *of the ISO system (3.60a) and (3.60b), is:*

a) ***asymptotically stable*** *if and only if it is both stable and attractive,*

b) ***globally asymptotically stable*** *if and only if it is both stable and globally attractive,*

c) ***strictly (completely, fully) globally asymptotically stable*** *if and only if it is both globally stable and globally attractive.*

This definition, Theorem 13.11, and Theorem 13.12 imply directly,

Theorem 13.15 *In order for the zero equilibrium vector* $\mathbf{x}_e = \mathbf{0}_n$ *of the ISO system (3.60a) and (3.60b) to be strictly globally asymptotically stable, it is necessary and sufficient to be asymptotically stable.*

Exponential stability of the zero equilibrium state

The exponential stability is originally defined only in the framework of the *ISO* systems.

Definition 13.10 *a) The desired motion* $\mathbf{X}_d(k)$ *of the ISO system (2.14a) and (2.14b) is* ***exponentially stable*** *if and only if there exist positive real numbers* $\alpha \geq 1$, β, *and* Δ, *or* $\Delta = \infty$, *such that* $\|\mathbf{X}_{d0} - \mathbf{X}_0\| < \Delta$ *implies*

$$\|\mathbf{X}_d(k) - \mathbf{X}(k;\mathbf{x}_0;\mathbf{I}_N)\| \leq \alpha \exp(-\beta k) \|\mathbf{X}_{d0} - \mathbf{X}_0\|$$

for all $k \in \mathcal{N}_0$.

The zero equilibrium state $\mathbf{x}_e = \mathbf{0}_n$ *of the ISO system (3.60a) and (3.60b) is* ***exponentially stable*** *if and only if there exist positive real numbers* $\alpha \geq 1$, β, *and* Δ, *or* $\Delta = \infty$, *such that* $\|\mathbf{x}_0\| < \Delta$ *implies in the free regime* $\|\mathbf{x}(k;\mathbf{x}_0;\mathbf{0}_M)\| \leq \alpha \exp(-\beta k) \|\mathbf{x}_0\|$ *for all* $k \in \mathcal{N}_0$,

$$\begin{gathered}\exists \alpha \in \mathcal{R}^+,\ \alpha \geq 1,\ \exists \beta \in \mathcal{R}^+,\ \text{and}\ \exists \Delta > 0 \Longrightarrow \\ \|\mathbf{x}_0\| < \Delta \Longrightarrow \\ \|\mathbf{x}(k;\mathbf{x}_0;\mathbf{0}_M)\| \leq \alpha \exp(-\beta k) \|\mathbf{x}_0\|,\ \forall k \in \mathcal{N}_0.\end{gathered} \tag{13.24}$$

b) *If and only if a) holds for* $\Delta = \infty$, *then the desired motion* $\mathbf{X}_d(k)$ *of the ISO system (2.14a) and (2.14b), i.e., the zero equilibrium state* $\mathbf{x}_e = \mathbf{0}_n$ *of the ISO system (3.60a) and (3.60b), is* ***globally exponentially stable (exponentially stable in the whole).***

The number α specifies the upper bound of $\|\mathbf{x}(k;\mathbf{x}_0;\mathbf{0}_M)\|$. It serves as a measure of the maximal deviation of $\mathbf{x}(k;\mathbf{x}_0;\mathbf{0}_M)$ from the zero equilibrium state $\mathbf{x}_e = \mathbf{0}_n$. The number β expresses the minimal convergence rate of $\|\mathbf{x}(k;\mathbf{x}_0;\mathbf{0}_M)\|$ to zero, i.e., $\mathbf{x}(k;\mathbf{x}_0;\mathbf{0}_M)$ to the zero equilibrium state $\mathbf{x}_e = \mathbf{0}_n$.

Conclusion 13.8 ***Exponential stability of*** $\mathbf{X}_d(k)$ ***and of*** $\mathbf{x}_e = \mathbf{0}_n$

For the desired motion $\mathbf{X}_d(k)$ *of the ISO system (2.14a) and (2.14b) to be* ***(globally) exponentially stable,*** *it is necessary and sufficient that the zero equilibrium vector* $\mathbf{x}_e = \mathbf{0}_n$ *of the ISO system (3.60a) and (3.60b) is (globally) exponentially stable, respectively.*

Theorem 13.16 *In order for the zero equilibrium state* $\mathbf{x}_e = \mathbf{0}_n$ *of the ISO system (3.60a) and (3.60b) to be globally exponentially stable, it is necessary and sufficient to be exponentially stable.*

The analogy between the proof of this theorem and the proof of Theorem 13.12 is complete.

Theorem 13.17 *In order for the zero equilibrium state* $\mathbf{x}_e = \mathbf{0}_n$ *of the ISO system (3.60a) and (3.60b) to be exponentially stable, it is necessary and sufficient to be asymptotically stable.*

Proof. *Necessity.* Let the zero equilibrium state $\mathbf{x}_e = \mathbf{0}_n$ of the *ISO* system (3.60a) and (3.60b) be exponentially stable. Then it is attractive because (Definition 13.10),

$$\|\mathbf{x}_0\| < \Delta \Longrightarrow \|\mathbf{x}(k;\mathbf{x}_0;\mathbf{0}_M)\| \leq \alpha \exp(-\beta k) \|\mathbf{x}_0\|, \ \forall k \in \mathcal{N}_0,$$

which implies

$$k \longrightarrow \infty \Longrightarrow \mathbf{x}(k;\mathbf{x}_0;\mathbf{0}_M) \longrightarrow \mathbf{0}_n.$$

Let $\delta(\varepsilon) = \varepsilon\alpha^{-1}$, $\forall \varepsilon \in \mathcal{R}^+$. Hence,

$$\|\mathbf{x}_0\| < \delta(\varepsilon) = \varepsilon\alpha^{-1}$$

and

$$\|\mathbf{x}(k;\mathbf{x}_0;\mathbf{0}_M)\| \leq \alpha \exp(-\beta k) \|\mathbf{x}_0\|, \ \forall k \in \mathcal{N}_0,$$

guarantee

$$\|\mathbf{x}(k;\mathbf{x}_0;\mathbf{0}_M)\| \leq \alpha \exp(-\beta k) \|\mathbf{x}_0\| <$$
$$< \alpha \exp(-\beta k)\varepsilon\alpha^{-1} \leq \varepsilon, \ \forall \varepsilon \in \mathcal{R}^+, \ \forall k \in \mathcal{N}_0.$$

This proves stability of $\mathbf{x}_e = \mathbf{0}_n$. Since it is also attractive, then it is asymptotically stable.

Sufficiency. Let the zero equilibrium state $\mathbf{x}_e = \mathbf{0}_n$ of the ISO system (3.60a) and (3.60b) be asymptotically stable. The system solutions are found in the form

$$\mathbf{x}(k;\mathbf{x}_0;\mathbf{0}_M) = \boldsymbol{A}^k\mathbf{x}_0,\ \forall k \in \mathcal{N}_0.$$

This yields

$$\|\mathbf{x}_0\| < \Delta \Longrightarrow \|\mathbf{x}(k;\mathbf{x}_0;\mathbf{0}_M)\| = \left\|\boldsymbol{A}^k\mathbf{x}_0\right\| \leq$$
$$\leq \left\|\boldsymbol{A}^k\right\| \|\mathbf{x}_0\|,\ \forall k \in \mathcal{N}_0.$$

By repeating the procedure of the proof of sufficiency of Theorem 13.1 (in Subsection 13.2.1), we prove that

$$\exists \beta, \gamma \in \mathcal{R}^+ \Longrightarrow$$
$$\left\|\boldsymbol{A}^k\right\| < \gamma \exp(-\beta k),\ \forall k \in \mathcal{N}_0.$$

This further implies, for $\alpha \geq \gamma$, $\alpha \in \mathcal{R}^+$,

$$\|\mathbf{x}(k;\mathbf{x}_0;\mathbf{0}_M)\| = \left\|\boldsymbol{A}^k\mathbf{x}_0\right\| \leq \left\|\boldsymbol{A}^k\right\| \|\mathbf{x}_0\| \leq$$
$$\leq \alpha \exp(-\beta k) \|\mathbf{x}_0\|,\ \forall \mathbf{x}_0 \in \mathcal{R}^n,\ \forall k \in \mathcal{N}_0.$$

The zero equilibrium state $\mathbf{x}_e = \mathbf{0}_n$ of the IO system (3.60a) and (3.60b) is globally exponentially stable (Definition 13.10), hence exponentially stable. ■

This theorem and Theorem 13.14 imply directly the following:

Theorem 13.18 *In order for the zero equilibrium state* $\mathbf{x}_e = \mathbf{0}_n$ *of the ISO system (3.60a) and (3.60b) to be exponentially stable, it is necessary (but not sufficient) that* $\mathbf{x}_e = \mathbf{0}_n$ *is the unique equilibrium state of the system, i.e., that the matrix* $(\boldsymbol{I} - \boldsymbol{A})$ *is nonsingular.*

We summarize Conclusions 13.6, 13.7 and 13.8:

Conclusion 13.9 ***Stability properties of*** $\mathbf{X}_d(k)$ ***and of*** $\mathbf{x}_e = \mathbf{0}_n$

The zero equilibrium state $\mathbf{x}_e = \mathbf{0}_n$ *of the ISO system (3.60a) and (3.60b) and the desired motion* $\mathbf{X}_d(k)$ *of the ISO system (2.14a) and (2.14b) have the same stability properties.*

13.2.3 *IIO* systems

Stability of the zero equilibrium vector

Notice that the definitions of the stability properties of $\mathbf{r}_e^{\alpha-1} = \mathbf{0}_{\alpha\rho}$ are the same as the definitions of the corresponding stability properties of $\mathbf{y}_e^{\nu-1} = \mathbf{0}_{\nu N}$ of the *IO* system (in Subsection 13.2.1).

We introduce the precise definition of stability of the zero equilibrium vector $\left[\begin{array}{cc} (\mathbf{r}_e^{\alpha-1})^T & (\mathbf{y}_e^{\nu-1})^T \end{array} \right]^T$ of the *IIO* system (3.64a) and (3.64b) (in Subsection 3.5.4), [93].

Definition 13.11 *a) The zero equilibrium vector* $\left[(\mathbf{r}_e^{\alpha-1})^T \ (\mathbf{y}_e^{\nu-1})^T \right]^T = \mathbf{0}_{\alpha\rho+\nu N}$ *of the IIO system (3.64a) and (3.64b), in which $\nu \geq 1$, is* ***stable*** *if and only if for every positive real number ε there exists a positive real number δ, the value of which depends on ε, $\delta = \delta(\varepsilon)$, such that* $\left\| \left[\begin{array}{cc} (\mathbf{r}_0^{\alpha-1})^T & (\mathbf{y}_0^{\nu-1})^T \end{array} \right]^T \right\| < \delta$ *implies in the free regime*

$$\left\| \begin{array}{c} \mathbf{r}^{\alpha-1}(k; \mathbf{r}_0^{\alpha-1}; \mathbf{0}_M) \\ \mathbf{y}^{\nu-1}(k; \mathbf{r}_0^{\alpha-1}; \mathbf{y}_0^{\nu-1}; \mathbf{0}_M) \end{array} \right\| < \varepsilon$$

for all $k \in \mathcal{N}_0$,

$$\begin{gathered} \forall \varepsilon \in \mathcal{R}^+, \ \exists \delta \in \mathcal{R}^+, \ \delta = \delta(\varepsilon), \implies \\ \left\| \left[\begin{array}{c} \mathbf{r}_0^{\alpha-1} \\ \mathbf{y}_0^{\nu-1} \end{array} \right] \right\| < \delta \implies \\ \left\| \left[\begin{array}{c} \mathbf{r}^{\alpha-1}(k; \mathbf{r}_0^{\alpha-1}; \mathbf{0}_M) \\ \mathbf{y}^{\nu-1}(k; \mathbf{r}_0^{\alpha-1}; \mathbf{y}_0^{\nu-1}; \mathbf{0}_M) \end{array} \right] \right\| < \varepsilon, \ \forall k \in \mathcal{N}_0. \end{gathered} \tag{13.25}$$

b) The zero equilibrium vector $\left[\begin{array}{cc} (\mathbf{r}_e^{\alpha-1})^T & (\mathbf{y}_e^{\nu-1})^T \end{array} \right]^T = \mathbf{0}_{\alpha\rho+\nu N}$ *of the IIO system (3.64a) and (3.64b) in which $\nu \geq 1$, is* ***globally stable*** *(i.e.,* ***stable in the whole****) if and only if a) holds so that the maximal $\delta(\varepsilon)$, denoted by $\delta_{\max}(\varepsilon)$, diverges to infinity as ε goes to infinity,*

$$\varepsilon \to \infty \Longrightarrow \delta_{\max}(\varepsilon) \to \infty. \tag{13.26}$$

Note 13.7 ***Geometrical interpretation***

The geometrical sense of this definition is the same as for the IO systems and for the ISO systems, i.e., that the zero equilibrium vector

$\left[\begin{array}{cc}\left(\mathbf{r}_e^{\alpha-1}\right)^T & \left(\mathbf{y}_e^{\nu-1}\right)^T\end{array}\right]^T = \mathbf{0}_{\alpha\rho+\nu N}$ *of the IIO system (3.64a) and (3.64b), in which $\nu \geq 1$, is* ***stable*** *if and only if for every open hyperball $\mathcal{B}_\varepsilon \subseteq \mathcal{R}^{\alpha\rho+\nu N}$ centered at the origin with the radius ε, there exists an open hyperball $\mathcal{B}_\delta \subseteq \mathcal{R}^{\alpha\rho+\nu N}$ centered at the origin with the radius δ, which depends on $\mathcal{B}_\varepsilon$, such that $\left[\begin{array}{cc}\left(\mathbf{r}_0^{\alpha-1}\right)^T & \left(\mathbf{y}_0^{\nu-1}\right)^T\end{array}\right]^T \in \mathcal{B}_\delta$ guarantees that $\left[\begin{array}{cc}\left(\mathbf{r}^{\alpha-1}(k;\mathbf{r}_0^{\alpha-1};\mathbf{0}_M)\right)^T & \left(\mathbf{y}^{\nu-1}(k;\mathbf{r}_0^{\alpha-1};\mathbf{y}_0^{\nu-1};\mathbf{0}_M)\right)^T\end{array}\right]^T$ stays in $\mathcal{B}_\varepsilon$ all the time from $k = k_0 = 0$ on, Fig. 13.2 (in Section 13.1),*

$$\forall \mathcal{B}_\varepsilon \subseteq \mathcal{R}^{\alpha\rho+\nu N},\ \exists \mathcal{B}_\delta \subseteq \mathcal{R}^{\alpha\rho+\nu N},\ \mathcal{B}_\delta = \mathcal{B}_\delta(\mathcal{B}_\varepsilon),\ \Longrightarrow$$
$$\left[\begin{array}{c}\mathbf{r}_0^{\alpha-1}\\ \mathbf{y}_0^{\nu-1}\end{array}\right] \in \mathcal{B}_\delta \Longrightarrow$$
$$\left[\begin{array}{c}\mathbf{r}^{\alpha-1}(k;\mathbf{r}_0^{\alpha-1};\mathbf{0}_M)\\ \mathbf{y}^{\nu-1}(k;\mathbf{r}_0^{\alpha-1};\mathbf{y}_0^{\nu-1};\mathbf{0}_M)\end{array}\right] \in \mathcal{B}_\varepsilon,\ \forall k \in \mathcal{N}_0. \tag{13.27}$$

Equation (10.33) and (Section 10.3), determines the dynamic behavior of the *IIO* system (3.64a) and (3.64b), which we apply to the *IIO* system in the free regime,

$$\mathbf{r}(k;\mathbf{r}_0^{\alpha-1};\mathbf{0}_M) = \mathbf{\Gamma}_{IIOr_0IS}(k)\mathbf{r}_0^{\alpha-1},\ \mathbf{\Gamma}_{IIOr_0IS}(k) \in \mathcal{R}^{\rho\times\alpha\rho}. \tag{13.28}$$

This implies

$$E^r\mathbf{r}(k;\mathbf{r}_0^{\alpha-1};\mathbf{0}_M) = E^r\mathbf{\Gamma}_{IIOr_0IS}(k)\mathbf{r}_0^{\alpha-1},$$
$$\forall k \in \mathcal{N}_0,\ r = 0,1,...,\alpha-1, \tag{13.29}$$

where

$$\mathbf{\Gamma}_{IIOr_0IS}^{\alpha-1}(k) = \left[\begin{array}{c}\mathbf{\Gamma}_{IIOr_0IS}(k)\\ E^1\mathbf{\Gamma}_{IIOr_0IS}(k)\\ \vdots\\ E^{\alpha-1}\mathbf{\Gamma}_{IIOr_0IS}(k)\end{array}\right] \in \mathcal{R}^{\alpha\rho\times\alpha\rho}. \tag{13.30}$$

Hence,

$$\mathbf{r}^{\alpha-1}(k;\mathbf{r}_0^{\alpha-1};\mathbf{0}_M) = \mathbf{\Gamma}_{IIOr_0IS}^{\alpha-1}(k)\mathbf{r}_0^{\alpha-1} \in \mathfrak{R}^{\alpha\rho}. \tag{13.31}$$

and due to Equation 10.47 (in Section 10.3),

$$\mathbf{y}(k;\mathbf{r}_0^{\alpha-1};\mathbf{y}_0^{\nu-1};\mathbf{i}) = \left(\sum_{j=0}^{j=k}\mathbf{\Gamma}_{IIO}(j)\mathbf{i}(k-j)\right) + \mathbf{\Gamma}_{IIOi_0}(k)\mathbf{i}_0^{\gamma-1} +$$
$$+\mathbf{\Gamma}_{IIOr_0}(k)\mathbf{r}_0^{\alpha-1} + \mathbf{\Gamma}_{IIOy_0}(k)\mathbf{y}_0^{\nu-1},\ \forall k \in \mathcal{N}_0, \tag{13.32}$$

which becomes for the free regime

$$E^r\mathbf{y}(k;\mathbf{r}_0^{\alpha-1};\mathbf{y}_0^{\nu-1};\mathbf{0}_M) = E^r\mathbf{\Gamma}_{IIOr_0}(k)\mathbf{r}_0^{\alpha-1} + E^r\mathbf{\Gamma}_{IIOy_0}(k)\mathbf{y}_0^{\nu-1},$$
$$\forall k \in \mathcal{N}_0,\ r = 0, 1, ..., \nu - 1.$$

This furnishes

$$\mathbf{y}^{\nu-1}(k;\mathbf{r}_0^{\alpha-1};\mathbf{y}_0^{\nu-1};\mathbf{0}_M) = \mathbf{\Gamma}_{IIOr_0}^{\nu-1}(k)\mathbf{r}_0^{\alpha-1} + \mathbf{\Gamma}_{IIOy_0}^{\nu-1}(k)\mathbf{y}_0^{\nu-1} =$$
$$= \mathbf{\Gamma}_{IIO(ry)_0}^{\nu-1}(k)\begin{bmatrix}\mathbf{r}_0^{\alpha-1}\\ \mathbf{y}_0^{\nu-1}\end{bmatrix},\ \forall k \in \mathcal{N}_0, \tag{13.33}$$

where

$$\mathbf{\Gamma}_{IIOr_0}^{\nu-1}(k) = \begin{bmatrix}\mathbf{\Gamma}_{IIOr_0}(k)\\ E^1\mathbf{\Gamma}_{IIOr_0}(k)\\ \vdots\\ E^{\nu-1}\mathbf{\Gamma}_{IIOr_0}(k)\end{bmatrix} \in \mathcal{R}^{\nu N\times\alpha\rho},$$

$$\mathbf{\Gamma}_{IIOy_0}^{\nu-1}(k) = \begin{bmatrix}\mathbf{\Gamma}_{IIOy_0}(k)\\ E^1\mathbf{\Gamma}_{IIOy_0}(k)\\ \vdots\\ E^{\nu-1}\mathbf{\Gamma}_{IIOy_0}(k)\end{bmatrix} \in \mathcal{R}^{\nu N\times\nu N},$$
$$\mathbf{\Gamma}_{IIO(ry)_0}^{\nu-1}(k) = \begin{bmatrix}\mathbf{\Gamma}_{IIOr_0}^{\nu-1}(k) & \mathbf{\Gamma}_{IIOy_0}^{\nu-1}(k)\end{bmatrix} \in \mathcal{R}^{\nu N\times(\alpha\rho+\nu N)}. \tag{13.34}$$

The following notation simplifies the exposition:

$$\mathbf{\Gamma}_{IIO0}(k) =$$
$$= \begin{bmatrix}\mathbf{\Gamma}_{IIOr_0IS}^{\alpha-1}(k) & \boldsymbol{O}_{\rho,\nu N}\\ \mathbf{\Gamma}_{IIOr_0}^{\nu-1}(k) & \mathbf{\Gamma}_{IIOy_0}^{\nu-1}(k)\end{bmatrix} \in \mathcal{R}^{(\alpha\rho+\nu N)\times(\alpha\rho+\nu N)}, \tag{13.35}$$

so that

$$\begin{bmatrix}\mathbf{r}^{\alpha-1}(k;\mathbf{r}_0^{\alpha-1};\mathbf{0}_M)\\ \mathbf{y}^{\nu-1}(k;\mathbf{r}_0^{\alpha-1};\mathbf{y}_0^{\nu-1};\mathbf{0}_M)\end{bmatrix} = \mathbf{\Gamma}_{IIO0}(k)\begin{bmatrix}\mathbf{r}_0^{\alpha-1}\\ \mathbf{y}_0^{\nu-1}\end{bmatrix} \in \mathcal{R}^{\alpha\rho+\nu N}. \tag{13.36}$$

Theorem 13.19 *In order for the zero equilibrium vector* $\begin{bmatrix}\mathbf{r}_e^{\alpha-1}\\ \mathbf{y}_e^{\nu-1}\end{bmatrix} = \mathbf{0}_{\alpha\rho+\nu N}$ *of the IIO system (3.64a) and (3.64b), in which* $\nu \geq 1$, *to be globally stable, it is necessary and sufficient to be stable.*

Proof. *Necessity.* Let the zero equilibrium vector $\begin{bmatrix} \mathbf{r}_e^{\alpha-1} \\ \mathbf{y}_e^{\nu-1} \end{bmatrix} = \mathbf{0}_{\alpha\rho+\nu N}$ of the IIO system (3.64a) and (3.64b), in which $\nu \geq 1$, be globally stable. Then, it is stable due to $b)$ of Definition 13.11.

Sufficiency. Let the zero equilibrium vector $\begin{bmatrix} (\mathbf{r}_e^{\alpha-1})^T & (\mathbf{y}_e^{\nu-1})^T \end{bmatrix}^T = \mathbf{0}_{\alpha\rho+\nu N}$ of the IIO system (3.64a) and (3.64b), in which $\nu \geq 1$, be stable. Then, Definition (13.11) holds. Let $\varepsilon \in \mathcal{R}^+$ be arbitrarily chosen. Let $\delta = \delta(\varepsilon) \in \mathcal{R}^+$ be the corresponding δ obeying Definition 13.11 so that, due to Equation 13.33 through Equation 13.36,

$$\left\| \begin{bmatrix} \mathbf{r}_0^{\alpha-1} \\ \mathbf{y}_0^{\nu-1} \end{bmatrix} \right\| < \delta \Longrightarrow$$

$$\left\| \begin{bmatrix} \mathbf{r}^{\alpha-1}(k;\mathbf{r}_0^{\alpha-1};\mathbf{0}_M) \\ \mathbf{y}^{\nu-1}(k;\mathbf{r}_0^{\alpha-1};\mathbf{y}_0^{\nu-1};\mathbf{0}_M) \end{bmatrix} \right\| =$$

$$= \left\| \mathbf{\Gamma}_{IIO0}(k) \begin{bmatrix} \mathbf{r}_0^{\alpha-1} \\ \mathbf{y}_0^{\nu-1} \end{bmatrix} \right\| < \varepsilon, \ \forall k \in \mathcal{N}_0.$$

Let δ_{ij} be the Kronecker delta, $L = \alpha\rho + \nu N$,

$$\mathbf{w}_l^L = \begin{bmatrix} \delta_{1l} & \delta_{2l} & \cdots & \delta_{Ll} \end{bmatrix}^T \in \mathcal{R}^L, l \in \{1, 2, ..., L\},$$

and

$$\begin{bmatrix} \mathbf{r}_{0l}^{\alpha-1} \\ \mathbf{y}_{0l}^{\nu-1} \end{bmatrix} = \frac{\delta}{2}\mathbf{w}_l^L \Longrightarrow$$

$$\left\| \begin{bmatrix} \mathbf{r}_{0l}^{\alpha-1} \\ \mathbf{y}_{0l}^{\nu-1} \end{bmatrix} \right\| = \frac{\delta}{2} < \delta \Longrightarrow$$

$$\left\| \begin{bmatrix} \mathbf{r}^{\alpha-1}(k;\mathbf{r}_{0l}^{\alpha-1};\mathbf{0}_M) \\ \mathbf{y}^{\nu-1}(k;\mathbf{r}_{0l}^{\alpha-1};\mathbf{y}_{0l}^{\nu-1};\mathbf{0}_M) \end{bmatrix} \right\| =$$

$$= \left\| \mathbf{\Gamma}_{IIO0}(k) \begin{bmatrix} \mathbf{r}_0^{\alpha-1} \\ \mathbf{y}_0^{\nu-1} \end{bmatrix} \right\| =$$

$$= \frac{\delta}{2} \left\| \mathbf{\Gamma}_{IIO0}(k)\mathbf{w}_k^L \right\| < \varepsilon, \ \forall k \in \mathcal{N}_0 \Longrightarrow$$

$$\exists \gamma \in \mathcal{R}^+ \Longrightarrow \left\| \mathbf{\Gamma}_{IIO0}(k) \right\| < \gamma, \ \forall k \in \mathcal{N}_0 \Longrightarrow$$

$$\forall \begin{bmatrix} \mathbf{r}_0^{\alpha-1} \\ \mathbf{y}_0^{\nu-1} \end{bmatrix} \in \mathcal{R}^{\alpha\rho+\nu N},$$

$$\left\| \begin{bmatrix} \mathbf{r}_0^{\alpha-1} \\ \mathbf{y}_0^{\nu-1} \end{bmatrix} \right\| < \delta \text{ for } \delta = \delta(\varepsilon) = \frac{\varepsilon}{2\gamma} \Longrightarrow$$

$$\left\| \begin{bmatrix} \mathbf{r}^{\alpha-1}(k;\mathbf{r}_0^{\alpha-1};\mathbf{0}_M) \\ \mathbf{y}^{\nu-1}(k;\mathbf{r}_0^{\alpha-1};\mathbf{y}_0^{\nu-1};\mathbf{0}_M) \end{bmatrix} \right\| =$$
$$= \left\| \mathbf{\Gamma}_{IIO0}(k) \begin{bmatrix} \mathbf{r}_0^{\alpha-1} \\ \mathbf{y}_0^{\nu-1} \end{bmatrix} \right\| \leq$$
$$\leq \gamma \left\| \begin{bmatrix} \mathbf{r}_0^{\alpha-1} \\ \mathbf{y}_0^{\nu-1} \end{bmatrix} \right\| < \delta\gamma = \frac{\varepsilon}{2} < \varepsilon, \ \forall k \in \mathcal{N}_0,$$

$$\delta_{\max}(\varepsilon) \geq \delta(\varepsilon) = \frac{\varepsilon}{2\gamma} \to \infty \text{ as } \varepsilon \to \infty.$$

This proves global stability of the zero equilibrium vector $\begin{bmatrix} \mathbf{r}_e^{\alpha-1} \\ \mathbf{y}_e^{\nu-1} \end{bmatrix} = \mathbf{0}_{\alpha\rho+\nu N}$ of the IIO system (3.64a) and (3.64b), in which $\nu \geq 1$. ■

Definition 13.12 *The zero equilibrium vector* $\begin{bmatrix} (\mathbf{r}_e^{\alpha-1})^T & (\mathbf{y}_e^{\nu-1})^T \end{bmatrix}^T = \mathbf{0}_{\alpha\rho+\nu N}$ *of the IIO system (3.64a) and (3.64b), in which* $\nu \geq 1$, *is **unstable** if and only if there exists a positive real number* ε *such that for every positive real number* μ *there exist an initial vector* $\begin{bmatrix} (\mathbf{r}_0^{\alpha-1})^T & (\mathbf{y}_0^{\nu-1})^T \end{bmatrix}^T$ *satisfying*

$$\left\| \begin{bmatrix} (\mathbf{r}_0^{\alpha-1})^T & (\mathbf{y}_0^{\nu-1})^T \end{bmatrix}^T \right\| < \mu$$

and a moment $k_\tau \in \mathcal{N}_0$, *which imply*

$$\left\| \begin{bmatrix} \mathbf{r}^{\alpha-1}(k_\tau;\mathbf{r}_0^{\alpha-1};\mathbf{0}_M) \\ \mathbf{y}^{\nu-1}(k_\tau;\mathbf{r}_0^{\alpha-1};\mathbf{y}_0^{\nu-1};\mathbf{0}_M) \end{bmatrix} \right\| \geq \varepsilon$$

in the free regime,

$$\exists \varepsilon \in R^+ \longrightarrow \forall \mu \in R^+, \ \exists(\begin{bmatrix} (\mathbf{r}_0^{\alpha-1})^T & (\mathbf{y}_0^{\nu-1})^T \end{bmatrix}^T, k_\tau) \in R^{\alpha\rho+N\nu} \times N_0, \Rightarrow$$
$$\left\| \begin{bmatrix} \mathbf{r}_0^{\alpha-1} \\ \mathbf{y}_0^{\nu-1} \end{bmatrix} \right\| < \mu \Longrightarrow$$
$$\left\| \begin{bmatrix} \mathbf{r}^{\alpha-1}(k_\tau;\mathbf{r}_0^{\alpha-1};\mathbf{0}_M) \\ \mathbf{y}^{\nu-1}(k_\tau;\mathbf{r}_0^{\alpha-1};\mathbf{y}_0^{\nu-1};\mathbf{0}_M) \end{bmatrix} \right\| \geq \varepsilon.$$

Note 13.8 ***Geometrical interpretation***

The geometrical sense of this is the following: the zero equilibrium vector $\begin{bmatrix} (\mathbf{r}_e^{\alpha-1})^T & (\mathbf{y}_e^{\nu-1})^T \end{bmatrix}^T = \mathbf{0}_{\alpha\rho+\nu N}$ *of the IIO system (3.64a) and (3.64b), in which* $\nu \geq 1$, *is **unstable** if and only if there exists an open hyperball*

$\mathcal{B}_\varepsilon \subseteq \mathcal{R}^{\alpha\rho+N\nu}$ *centered at the origin with the radius* ε*, such that for every open hyperball* $\mathcal{B}_\mu \subseteq \mathcal{R}^{\alpha\rho+N\nu}$ *there is a moment* $k_\tau \in \mathcal{N}_0$ *and* $\begin{bmatrix} \mathbf{r}_0^{\alpha-1} \\ \mathbf{y}_0^{\nu-1} \end{bmatrix} \in \mathcal{B}_\mu$ *implying in the free regime that* $\begin{bmatrix} \mathbf{r}^{\alpha-1}(k_\tau;\mathbf{r}_0^{\alpha-1};\mathbf{0}_M) \\ \mathbf{y}^{\nu-1}(k_\tau;\mathbf{r}_0^{\alpha-1};\mathbf{y}_0^{\nu-1};\mathbf{0}_M) \end{bmatrix}$ *escapes* $\mathcal{B}_\varepsilon$ *at the instant* k_τ*,*

$$\exists\mathcal{B}_\varepsilon \subseteq \mathcal{R}^{\alpha\rho+N\nu},\ \forall\mathcal{B}_\mu \subseteq \mathcal{R}^{\alpha\rho+N\nu} \Longrightarrow$$
$$\exists\left(\begin{bmatrix} \left(\mathbf{r}_0^{\alpha-1}\right)^T & \left(\mathbf{y}_0^{\nu-1}\right)^T \end{bmatrix}^T, k_\tau\right) \in \mathcal{B}_\mu \times \mathcal{N}_0 \Longrightarrow$$
$$\begin{bmatrix} \left(\mathbf{r}^{\alpha-1}(k_\tau;\mathbf{r}_0^{\alpha-1};\mathbf{0}_M)\right)^T & \left(\mathbf{y}^{\nu-1}(k_\tau;\mathbf{r}_0^{\alpha-1};\mathbf{y}_0^{\nu-1};\mathbf{0}_M)\right)^T \end{bmatrix}^T \in \mathcal{R}^{\alpha\rho+\nu N}\backslash\mathcal{B}_\varepsilon,$$

where

$$\mathcal{R}^{\alpha\rho+\nu N}\backslash\mathcal{B}_\varepsilon = \left\{\begin{bmatrix} \left(\mathbf{r}_0^{\alpha-1}\right)^T & \left(\mathbf{y}_0^{\nu-1}\right)^T \end{bmatrix}^T : \right.$$
$$\left. : \left(\begin{array}{c} \begin{bmatrix} \left(\mathbf{r}_0^{\alpha-1}\right)^T & \left(\mathbf{y}_0^{\nu-1}\right)^T \end{bmatrix}^T \in \mathcal{R}^{\alpha\rho+\nu N}, \\ \begin{bmatrix} \left(\mathbf{r}_0^{\alpha-1}\right)^T & \left(\mathbf{y}_0^{\nu-1}\right)^T \end{bmatrix}^T \notin \mathcal{B}_\varepsilon \end{array} \right)\right\}.$$

Attraction of the zero equilibrium vector

The definition of attraction of the equilibrium vector in the framework of the *IIO* systems described by (3.64a) and (3.64b), follows:

Definition 13.13 *a) The zero equilibrium vector* $\begin{bmatrix} \mathbf{r}_e^{\alpha-1} \\ \mathbf{y}_e^{\nu-1} \end{bmatrix} = \mathbf{0}_{\alpha\rho+\nu N}$ *of the IIO system (3.64a) and (3.64b), in which* $\nu \geq 1$*, is* ***attractive*** *if and only if there exists a positive real number* Δ*, or* $\Delta = \infty$*, such that* $\left\|\begin{bmatrix} \mathbf{r}_0^{\alpha-1} \\ \mathbf{y}_0^{\nu-1} \end{bmatrix}\right\| < \Delta$ *implies in the free regime that*

$$\begin{bmatrix} \left(\mathbf{r}^{\alpha-1}(k;\mathbf{r}_0^{\alpha-1};\mathbf{0}_M)\right)^T & \left(\mathbf{y}^{\nu-1}(k;\mathbf{r}_0^{\alpha-1};\mathbf{y}_0^{\nu-1};\mathbf{0}_M)\right)^T \end{bmatrix}^T$$

tends to $\mathbf{0}_{\alpha\rho+\nu N}$ *as* k *escapes to infinity,*

$$\exists\Delta > 0 \Longrightarrow \left\|\begin{bmatrix} \mathbf{r}_0^{\alpha-1} \\ \mathbf{y}_0^{\nu-1} \end{bmatrix}\right\| < \Delta \Longrightarrow$$
$$\lim_{k\to\infty}\begin{bmatrix} \mathbf{r}^{\alpha-1}(k;\mathbf{r}_0^{\alpha-1};\mathbf{0}_M) \\ \mathbf{y}^{\nu-1}(k;\mathbf{r}_0^{\alpha-1};\mathbf{y}_0^{\nu-1};\mathbf{0}_M) \end{bmatrix} = \mathbf{0}_{\alpha\rho+\nu N}.$$

b) If and only if a) holds for $\Delta = \infty$, *then the zero equilibrium vector* $\begin{bmatrix} \mathbf{r}_e^{\alpha-1} \\ \mathbf{y}_e^{\nu-1} \end{bmatrix} = \mathbf{0}_{\alpha\rho+\nu N}$ *of the IIO system (3.64a) and (3.64b), in which* $\nu \geq 1$, *is* ***globally attractive*** *(**attractive in the whole**).*

Note 13.9 ***Geometrical interpretation***

The zero equilibrium vector $\begin{bmatrix} (\mathbf{r}_e^{\alpha-1})^T & (\mathbf{y}_e^{\nu-1})^T \end{bmatrix}^T = \mathbf{0}_{\alpha\rho+\nu N}$ *of the IIO system (3.64a) and (3.64b), in which* $\nu \geq 1$, *is* ***attractive*** *if and only if there exists its open neighborhood* $\mathcal{B}_\Delta \subseteq \mathcal{R}^{\alpha\rho+N\nu}$ *such that every initial vector* $\begin{bmatrix} (\mathbf{r}_0^{\alpha-1})^T & (\mathbf{y}_0^{\nu-1})^T \end{bmatrix}^T$ *in* $\mathcal{B}_\Delta$, $\begin{bmatrix} (\mathbf{r}_0^{\alpha-1})^T & (\mathbf{y}_0^{\nu-1})^T \end{bmatrix}^T \in \mathcal{B}_\Delta$, *ensures that the system behavior* $\begin{bmatrix} \mathbf{r}^{\alpha-1}(k;\mathbf{r}_0^{\alpha-1};\mathbf{0}_M) \\ \mathbf{y}^{\nu-1}(k;\mathbf{r}_0^{\alpha-1};\mathbf{y}_0^{\nu-1};\mathbf{0}_M) \end{bmatrix}$ *in the free regime approaches asymptotically the equilibrium state* $\begin{bmatrix} \mathbf{r}_e^{\alpha-1} \\ \mathbf{y}_e^{\nu-1} \end{bmatrix} = \mathbf{0}_{\alpha\rho+\nu N}$ *as* $k \to \infty$, *Fig. 13.4 (in Section 13.1).*

Global attraction of the zero equilibrium vector $\begin{bmatrix} \mathbf{r}_e^{\alpha-1} \\ \mathbf{y}_e^{\nu-1} \end{bmatrix} = \mathbf{0}_{\alpha\rho+\nu N}$ *of the IIO system (3.64a) and (3.64b), means that for every* $\begin{bmatrix} \mathbf{r}_0^{\alpha-1} \\ \mathbf{y}_0^{\nu-1} \end{bmatrix} \in \mathcal{R}^{\alpha\rho+N\nu}$ *the system behavior* $\begin{bmatrix} \mathbf{r}^{\alpha-1}(k;\mathbf{r}_0^{\alpha-1};\mathbf{0}_M) \\ \mathbf{y}^{\nu-1}(k;\mathbf{r}_0^{\alpha-1};\mathbf{y}_0^{\nu-1};\mathbf{0}_M) \end{bmatrix}$ *approaches asymptotically the equilibrium state* $\begin{bmatrix} (\mathbf{r}_e^{\alpha-1})^T & (\mathbf{y}_e^{\nu-1})^T \end{bmatrix}^T = \mathbf{0}_{\alpha\rho+\nu N}$ *as* $k \to \infty$.

The following theorem expresses the relationship between global attraction of the zero equilibrium vector $\begin{bmatrix} (\mathbf{r}_e^{\alpha-1})^T & (\mathbf{y}_e^{\nu-1})^T \end{bmatrix}^T = \mathbf{0}_{\alpha\rho+\nu N}$ of the *IIO* system (3.64a) and (3.64b), and its attraction:

Theorem 13.20 *In order for the zero equilibrium vector* $\begin{bmatrix} \mathbf{r}_e^{\alpha-1} \\ \mathbf{y}_e^{\nu-1} \end{bmatrix} = \mathbf{0}_{\alpha\rho+\nu N}$ *of the IIO system (3.64a) and (3.64b), in which* $\nu \geq 1$, *to be globally attractive, it is necessary and sufficient to be attractive.*

Proof. *Necessity.* Let the zero equilibrium vector $\begin{bmatrix} \mathbf{r}_e^{\alpha-1} \\ \mathbf{y}_e^{\nu-1} \end{bmatrix} = \mathbf{0}_{\alpha\rho+\nu N}$ of the *IIO* system (3.64a) and (3.64b), in which $\nu \geq 1$, be globally attractive. The zero equilibrium vector $\begin{bmatrix} \mathbf{r}_e^{\alpha-1} \\ \mathbf{y}_e^{\nu-1} \end{bmatrix} = \mathbf{0}_{\alpha\rho+\nu N}$ of the *IIO* system

(3.64a) and (3.64b), in which $\nu \geq 1$, is attractive in view of $a)$ of Definition 13.13.

Sufficiency. Let the zero equilibrium vector $\left[\begin{array}{cc} (\mathbf{r}_e^{\alpha-1})^T & (\mathbf{y}_e^{\nu-1})^T \end{array} \right]^T = \mathbf{0}_{\alpha\rho+\nu N}$ of the *IIO* system (3.64a) and (3.64b), in which $\nu \geq 1$, be attractive. Its response in the free regime is given by Equation (13.28) through Equation (13.36). Since

$$\left[\begin{array}{c} \mathbf{r}^{\alpha-1}(k;\mathbf{r}_0^{\alpha-1};\mathbf{0}_M) \\ \mathbf{y}^{\nu-1}(k;\mathbf{r}_0^{\alpha-1};\mathbf{y}_0^{\nu-1};\mathbf{0}_M) \end{array} \right] =$$

$$= \mathbf{\Gamma}_{IIO0}(k) \left[\begin{array}{c} \mathbf{r}_0^{\alpha-1} \\ \mathbf{y}_0^{\nu-1} \end{array} \right] \rightarrow \mathbf{0}_{\alpha\rho+\nu N} \text{ as } k \rightarrow \infty,$$

$$\forall \left[\begin{array}{c} \mathbf{r}_0^{\alpha-1} \\ \mathbf{y}_0^{\nu-1} \end{array} \right] \in \mathcal{R}^{\alpha\rho+N\nu} \text{ obeying } \left\| \left[\begin{array}{c} \mathbf{r}_0^{\alpha-1} \\ \mathbf{y}_0^{\nu-1} \end{array} \right] \right\| < \Delta,$$

$$\text{due to } a) \text{ of Definition 13.13,}$$

then

$$\mathbf{\Gamma}_{IIO0}(k) \rightarrow \boldsymbol{O}_{(\alpha\rho+\nu N)\times(\alpha\rho+\nu N)} \text{ as } k \rightarrow \infty.$$

This implies

$$\lim_{k\rightarrow\infty} \left[\begin{array}{c} \mathbf{r}^{\alpha-1}(k;\mathbf{r}_0^{\alpha-1};\mathbf{0}_M) \\ \mathbf{y}^{\nu-1}(k;\mathbf{r}_0^{\alpha-1};\mathbf{y}_0^{\nu-1};\mathbf{0}_M) \end{array} \right] = \mathbf{0}_{\alpha\rho+N\nu},\ \forall \left[\begin{array}{c} \mathbf{r}_0^{\alpha-1} \\ \mathbf{y}_0^{\nu-1} \end{array} \right] \in \mathcal{R}^{\alpha\rho+N\nu},$$

which shows that the zero equilibrium vector $\left[\begin{array}{cc} (\mathbf{r}_e^{\alpha-1})^T & (\mathbf{y}_e^{\nu-1})^T \end{array} \right]^T = \mathbf{0}_{\alpha\rho+\nu N}$ of the *IIO* system (3.64a) and (3.64b), in which $\nu \geq 1$, is globally attractive (part b) of Definition 13.13). ■

This theorem permits us to discover the link between the uniqueness of the zero equilibrium vector and its attraction in the framework of the *IIO* system described by (3.64a) and (3.64b).

Theorem 13.21 *In order for the zero equilibrium vector* $\left[\begin{array}{c} \mathbf{r}_e^{\alpha-1} \\ \mathbf{y}_e^{\nu-1} \end{array} \right] = \mathbf{0}_{\alpha\rho+\nu N}$ *of the IIO system (3.64a) and (3.64b), in which* $\nu \geq 1$, *to be attractive, it is necessary (but not sufficient) to be the unique equilibrium vector of the system.*

Proof. *Necessity.* Let the zero equilibrium vector $\left[\begin{array}{c} \mathbf{r}_e^{\alpha-1} \\ \mathbf{y}_e^{\nu-1} \end{array} \right] = \mathbf{0}_{\alpha\rho+\nu N}$ of the *IIO* system (3.64a) and (3.64b), in which $\nu \geq 1$, be nonunique and attractive. Then, there is another equilibrium vector $\begin{bmatrix} \mathbf{r}_{e2}^{\alpha-1} \\ \mathbf{y}_{e2}^{\nu-1} \end{bmatrix} \neq \mathbf{0}_{\alpha\rho+\nu N}$ of the *IIO* system (3.64a) and (3.64b), in which $\nu \geq 1$. The zero equilibrium vector $\begin{bmatrix} (\mathbf{r}_e^{\alpha-1})^T & (\mathbf{y}_e^{\nu-1})^T \end{bmatrix}^T = \mathbf{0}_{\alpha\rho+\nu N}$ of the *IIO* system is globally attractive due to Theorem 13.20. However,

$$\lim_{k\to\infty} \begin{bmatrix} \mathbf{r}^{\alpha-1}(k; \mathbf{r}_{e2}^{\alpha-1}; \mathbf{0}_M) \\ \mathbf{y}^{\nu-1}(k; \mathbf{r}_{e2}^{\alpha-1}; \mathbf{y}_{e2}^{\nu-1}; \mathbf{0}_M) \end{bmatrix} = \begin{bmatrix} \mathbf{r}_{e2}^{\alpha-1} \\ \mathbf{y}_{e2}^{\nu-1} \end{bmatrix} \neq \mathbf{0}_{\alpha\rho+\nu N}$$

due to Definition 3.18, (in Subsection 3.7.4), of the equilibrium vector of the *IIO* system (3.64a) and (3.64b), in which $\nu \geq 1$. This contradicts global attraction of $\begin{bmatrix} (\mathbf{r}_e^{\alpha-1})^T & (\mathbf{y}_e^{\nu-1})^T \end{bmatrix}^T = \mathbf{0}_{\alpha\rho+\nu N}$. Hence, there does not exist another equilibrium vector $\begin{bmatrix} (\mathbf{r}_{e2}^{\alpha-1})^T & (\mathbf{y}_{e2}^{\nu-1})^T \end{bmatrix}^T \neq \mathbf{0}_{\alpha\rho+\nu N}$ of the *IIO* system (3.64a) and (3.64b), in which $\nu \geq 1$, i.e., the zero equilibrium vector $\begin{bmatrix} (\mathbf{r}_e^{\alpha-1})^T & (\mathbf{y}_e^{\nu-1})^T \end{bmatrix}^T = \mathbf{0}_{\alpha\rho+\nu N}$ of the *IIO* system is unique.

Insufficiency. Let the *IIO* system (3.64a) and (3.64b), be defined by

$$r(k+1) - 2r(k) = i(k)$$
$$y(k+1) - 2y(k) = r(k) + i(k).$$

Its equilibrium point $\begin{bmatrix} r_e & y_e \end{bmatrix}^T = \mathbf{0}_2$ is unique. Its dynamic behavior in the free regime reads for $|r_0| > 0$ and $|y_0| > 0$:

$$\begin{bmatrix} r(k; |r_0|; 0) \\ y(k; |r_0|; |y_0|; 0) \end{bmatrix} =$$
$$= \begin{bmatrix} 2^k & 0 \\ k2^{k-1} & 2^k \end{bmatrix} \begin{bmatrix} |r_0| \\ |y_0| \end{bmatrix} \longrightarrow \begin{bmatrix} \infty \\ \infty \end{bmatrix}, \ k \longrightarrow \infty.$$

The equilibrium vector $\begin{bmatrix} r_e & y_e \end{bmatrix}^T = \mathbf{0}_2$ is not attractive. ■

This theorem and Theorem 3.12 (in Subsection 3.7.4), yield directly the following:

Theorem 13.22 *In order for the zero equilibrium vector* $\begin{bmatrix} \mathbf{r}_e^{\alpha-1} \\ \mathbf{y}_e^{\nu-1} \end{bmatrix} = \mathbf{0}_{\alpha\rho+\nu N}$ *of the IIO system (3.64a) and (3.64b), in which* $\nu \geq 1$*, to be attractive, it is necessary (but not sufficient) that the matrix*

$$\begin{bmatrix} \boldsymbol{Q}_0 + \boldsymbol{Q}_1 + \cdots + \boldsymbol{Q}_\alpha & \boldsymbol{O}_{\rho N} \\ -(\boldsymbol{R}_0 + \boldsymbol{R}_1 + \cdots + \boldsymbol{R}_\alpha) & \boldsymbol{E}_0 + \boldsymbol{E}_1 + \cdots + \boldsymbol{E}_\nu \end{bmatrix}$$

is nonsingular.

Asymptotic stability of the zero equilibrium vector

Definition 13.14 *The zero equilibrium vector* $\left[\begin{array}{cc} \left(\mathbf{r}_e^{\alpha-1}\right)^T & \left(\mathbf{y}_e^{\nu-1}\right)^T \end{array} \right]^T = \mathbf{0}_{\alpha\rho+\nu N}$ *of the IIO system (3.64a) and (3.64b), in which* $\nu \geq 1$*, is*

a) ***asymptotically stable*** *if and only if it is both stable and attractive,*

b) ***globally asymptotically stable*** *if and only if it is both stable and globally attractive,*

c) ***strictly (completely, fully) globally asymptotically stable*** *if and only if it is both globally stable and globally attractive.*

This definition, Theorem 13.19 and Theorem 13.20 imply:

Theorem 13.23 *In order for the zero equilibrium vector* $\left[\begin{array}{c} \mathbf{r}_e^{\alpha-1} \\ \mathbf{y}_e^{\nu-1} \end{array} \right] = \mathbf{0}_{\alpha\rho+\nu N}$ *of the IIO system (3.64a) and (3.64b), in which* $\nu \geq 1$*, to be strictly globally asymptotically stable, it is necessary and sufficient that it is asymptotically stable.*

Definition 13.15 *The zero equilibrium vector* $\mathbf{r}_e^{\alpha-1} = \mathbf{0}_{\alpha\rho}$ *of the first subsystem of the GISO system (3.68a) and (3.68b) is*

a) ***asymptotically stable*** *if and only if it is both stable and attractive,*

b) ***globally asymptotically stable*** *if and only if it is both stable and globally attractive,*

c) ***strictly (completely, fully) globally asymptotically stable*** *if and only if it is both globally stable and globally attractive.*

Exponential stability of the zero equilibrium vector

The definition of exponential stability to the *IIO* systems described by (3.64a) and (3.64b), reads:

Definition 13.16 *a) The zero equilibrium vector* $\left[\begin{array}{c} \mathbf{r}_e^{\alpha-1} \\ \mathbf{y}_e^{\nu-1} \end{array} \right] = \mathbf{0}_{\alpha\rho+\nu N}$ *of the IIO system (3.64a) and (3.64b), in which* $\nu \geq 1$*, is* ***exponentially stable*** *if and only if there exist positive real numbers* $\xi \geq 1$*,* β*, and* Δ*, or* $\Delta = \infty$*, such that* $\left\| \left[\begin{array}{cc} \left(\mathbf{r}_0^{\alpha-1}\right)^T & \left(\mathbf{y}_0^{\nu-1}\right)^T \end{array} \right]^T \right\| < \Delta$ *implies* *in the free regime*

$$\left\| \begin{bmatrix} \mathbf{r}^{\alpha-1}(k;\mathbf{r}_0^{\alpha-1};\mathbf{0}_M) \\ \mathbf{y}^{\nu-1}(k;\mathbf{r}_0^{\alpha-1};\mathbf{y}_0^{\nu-1};\mathbf{0}_M) \end{bmatrix} \right\| \leq$$

$$\leq \xi \exp(-\beta k) \left\| \begin{bmatrix} \mathbf{r}_0^{\alpha-1} \\ \mathbf{y}_0^{\nu-1} \end{bmatrix} \right\| \text{ for all } k \in \mathcal{N}_0,$$

i.e.,

$$\exists \xi \in \mathcal{R}^+,\ \xi \geq 1, \exists \beta \in \mathcal{R}^+, \text{ and } \exists \Delta > 0 \Longrightarrow$$

$$\left\| \begin{bmatrix} \mathbf{r}_0^{\alpha-1} \\ \mathbf{y}_0^{\nu-1} \end{bmatrix} \right\| < \Delta \Longrightarrow$$

$$\left\| \begin{bmatrix} \mathbf{r}^{\alpha-1}(k;\mathbf{r}_0^{\alpha-1};\mathbf{0}_M) \\ \mathbf{y}^{\nu-1}(k;\mathbf{r}_0^{\alpha-1};\mathbf{y}_0^{\nu-1};\mathbf{0}_M) \end{bmatrix} \right\| \leq$$

$$\leq \xi \exp(-\beta k) \left\| \begin{bmatrix} \mathbf{r}_0^{\alpha-1} \\ \mathbf{y}_0^{\nu-1} \end{bmatrix} \right\|,\ \forall k \in \mathcal{N}_0.$$

b) If and only if a) holds for $\Delta = \infty$ *then the zero equilibrium vector* $\begin{bmatrix} (\mathbf{r}_e^{\alpha-1})^T & (\mathbf{y}_e^{\nu-1})^T \end{bmatrix}^T = \mathbf{0}_{\alpha\rho+\nu N}$ *of the IIO system (3.64a) and (3.64b), in which* $\nu \geq 1$*, is* ***globally exponentially stable*** *(****exponentially stable in the whole****).*

The number ξ determines the upper bound of

$$\left\| \begin{bmatrix} \mathbf{r}^{\alpha-1}(k;\mathbf{r}_0^{\alpha-1};\mathbf{0}_M) \\ \mathbf{y}^{\nu-1}(k;\mathbf{r}_0^{\alpha-1};\mathbf{y}_0^{\nu-1};\mathbf{0}_M) \end{bmatrix} \right\|,$$

which is a measure of the maximal deviation of

$$\begin{bmatrix} \mathbf{r}^{\alpha-1}(k;\mathbf{r}_0^{\alpha-1};\mathbf{0}_M) \\ \mathbf{y}^{\nu-1}(k;\mathbf{r}_0^{\alpha-1};\mathbf{y}_0^{\nu-1};\mathbf{0}_M) \end{bmatrix}$$

from the zero equilibrium vector $\begin{bmatrix} (\mathbf{r}_e^{\alpha-1})^T & (\mathbf{y}_e^{\nu-1})^T \end{bmatrix}^T = \mathbf{0}_{\alpha\rho+\nu N}$. The number β signifies the minimal convergence rate of

$$\left\| \begin{bmatrix} \mathbf{r}^{\alpha-1}(k;\mathbf{r}_0^{\alpha-1};\mathbf{0}_M) \\ \mathbf{y}^{\nu-1}(k;\mathbf{r}_0^{\alpha-1};\mathbf{y}_0^{\nu-1};\mathbf{0}_M) \end{bmatrix} \right\|$$

to zero, i.e., of

$$\begin{bmatrix} \mathbf{r}^{\alpha-1}(k;\mathbf{r}_0^{\alpha-1};\mathbf{0}_M) \\ \mathbf{y}^{\nu-1}(k;\mathbf{r}_0^{\alpha-1};\mathbf{y}_0^{\nu-1};\mathbf{0}_M) \end{bmatrix}$$

to the zero equilibrium vector $\begin{bmatrix} (\mathbf{r}_e^{\alpha-1})^T & (\mathbf{y}_e^{\nu-1})^T \end{bmatrix}^T = \mathbf{0}_{\alpha\rho+\nu N}$.

Theorem 13.24 *In order for the zero equilibrium vector* $\begin{bmatrix} \mathbf{r}_e^{\alpha-1} \\ \mathbf{y}_e^{\nu-1} \end{bmatrix} = \mathbf{0}_{\alpha\rho+\nu N}$ *of the IIO system (3.64a) and (3.64b), in which $\nu \geq 1$, to be globally exponentially stable, it is necessary and sufficient to be exponentially stable.*

The proof of Theorem 13.20 is to be repeated with the obvious change of the notation in order to prove Theorem 13.24.

Theorem 13.25 *In order for the zero equilibrium vector* $\begin{bmatrix} \mathbf{r}_e^{\alpha-1} \\ \mathbf{y}_e^{\nu-1} \end{bmatrix} = \mathbf{0}_{\alpha\rho+\nu N}$ *of the IIO system (3.64a) and (3.64b), in which $\nu \geq 1$, to be exponentially stable, it is necessary and sufficient to be asymptotically stable.*

Proof. *Necessity.* Let the zero equilibrium vector $\begin{bmatrix} \mathbf{r}_e^{\alpha-1} \\ \mathbf{y}_e^{\nu-1} \end{bmatrix} = \mathbf{0}_{\alpha\rho+\nu N}$ of the *IIO* system (3.64a) and (3.64b), in which $\nu \geq 1$, be exponentially stable. Then it is attractive because (Definition 13.16)

$$\left\| \begin{bmatrix} \mathbf{r}_0^{\alpha-1} \\ \mathbf{y}_0^{\nu-1} \end{bmatrix} \right\| < \Delta \Longrightarrow$$

$$\left\| \begin{bmatrix} \mathbf{r}^{\alpha-1}(k; \mathbf{r}_0^{\alpha-1}; \mathbf{0}_M) \\ \mathbf{y}^{\nu-1}(k; \mathbf{r}_0^{\alpha-1}; \mathbf{y}_0^{\nu-1}; \mathbf{0}_M) \end{bmatrix} \right\| \leq$$
$$\leq \xi \exp(-\beta k) \left\| \begin{bmatrix} \mathbf{r}_0^{\alpha-1} \\ \mathbf{y}_0^{\nu-1} \end{bmatrix} \right\|, \ \forall k \in \mathcal{N}_0, \ \xi \in \mathcal{R}^+, \ \xi \geq 1$$

implies

$$\begin{bmatrix} \mathbf{r}^{\alpha-1}(k; \mathbf{r}_0^{\alpha-1}; \mathbf{0}_M) \\ \mathbf{y}^{\nu-1}(k; \mathbf{r}_0^{\alpha-1}; \mathbf{y}_0^{\nu-1}; \mathbf{0}_M) \end{bmatrix} \longrightarrow \mathbf{0}_{\alpha\rho+\nu N}, \ k \to \infty.$$

Let $\delta(\varepsilon) = \varepsilon\xi^{-1}$, $\forall \varepsilon \in \mathcal{R}^+$. Hence,

$$\left\| \begin{bmatrix} \mathbf{r}_0^{\alpha-1} \\ \mathbf{y}_0^{\nu-1} \end{bmatrix} \right\| < \delta(\varepsilon) = \varepsilon\xi^{-1}$$

and

$$\left\| \begin{bmatrix} \mathbf{r}^{\alpha-1}(k; \mathbf{r}_0^{\alpha-1}; \mathbf{0}_M) \\ \mathbf{y}^{\nu-1}(k; \mathbf{r}_0^{\alpha-1}; \mathbf{y}_0^{\nu-1}; \mathbf{0}_M) \end{bmatrix} \right\| \leq$$
$$\leq \xi \exp(-\beta k) \left\| \begin{bmatrix} \mathbf{r}_0^{\alpha-1} \\ \mathbf{y}_0^{\nu-1} \end{bmatrix} \right\|, \ \forall k \in \mathcal{N}_0,$$

guarantee

$$\left\| \begin{bmatrix} \mathbf{r}^{\alpha-1}(k;\mathbf{r}_0^{\alpha-1};\mathbf{0}_M) \\ \mathbf{y}^{\nu-1}(k;\mathbf{r}_0^{\alpha-1};\mathbf{y}_0^{\nu-1};\mathbf{0}_M) \end{bmatrix} \right\| \leq$$
$$\leq \xi \exp(-\beta k) \left\| \begin{bmatrix} \mathbf{r}_0^{\alpha-1} \\ \mathbf{y}_0^{\nu-1} \end{bmatrix} \right\| <$$
$$< \xi \exp(-\beta k)\varepsilon\xi^{-1} \leq \varepsilon, \ \forall\varepsilon \in \mathcal{R}^+, \ \forall k \in \mathcal{N}_0.$$

This proves stability of $\begin{bmatrix} (\mathbf{r}_e^{\alpha-1})^T & (\mathbf{y}_e^{\nu-1})^T \end{bmatrix}^T = \mathbf{0}_{\alpha\rho+\nu N}$. Since it is also attractive, then it is asymptotically stable.

Sufficiency. Let the zero equilibrium vector $\begin{bmatrix} (\mathbf{r}_e^{\alpha-1})^T & (\mathbf{y}_e^{\nu-1})^T \end{bmatrix}^T = \mathbf{0}_{\alpha\rho+\nu N}$ of the *IIO* system (3.64a) and (3.64b), in which $\nu \geq 1$, be asymptotically stable. In the free regime, the system takes the following form:

$$Q^{(\alpha)}\mathbf{r}^{\alpha}(k) = \mathbf{0}_{\rho}, \ \forall k \in \mathcal{N}_0,$$
$$R^{(\alpha)}\mathbf{r}^{\alpha}(k) - E^{(\nu)}\mathbf{y}^{\nu}(k) = \mathbf{0}_N, \ k \in \mathcal{N}_0. \tag{13.37}$$

or equivalently

$$E^1\mathbf{w} = \boldsymbol{A}_w\mathbf{w}, \tag{13.38}$$

where

$$p = \alpha\rho + \nu N, \ \mathbf{w} = \begin{bmatrix} (\mathbf{r}^{\alpha-1})^T & (\mathbf{y}^{\nu-1})^T \end{bmatrix}^{\mathbf{T}} \in \mathcal{R}^p,$$

and $\boldsymbol{A}_w$ is defined by (13.39) through (13.42),

$$\boldsymbol{A}_{IIO1} =$$
$$= \begin{bmatrix} \boldsymbol{O}_\rho & \boldsymbol{I}_\rho & \boldsymbol{O}_\rho & \cdots & \boldsymbol{O}_\rho & \boldsymbol{O}_\rho \\ \boldsymbol{O}_\rho & \boldsymbol{O}_\rho & \boldsymbol{I}_\rho & \cdots & \boldsymbol{O}_\rho & \boldsymbol{O}_\rho \\ \vdots & \vdots & \vdots & \vdots & \vdots & \vdots \\ \boldsymbol{O}_\rho & \boldsymbol{O}_\rho & \boldsymbol{O}_\rho & \cdots & \boldsymbol{O}_\rho & \boldsymbol{I}_\rho \\ -\boldsymbol{Q}_\alpha^{-1}\boldsymbol{Q}_0 & -\boldsymbol{Q}_\alpha^{-1}\boldsymbol{Q}_1 & -\boldsymbol{Q}_\alpha^{-1}\boldsymbol{Q}_2 & \cdots & -\boldsymbol{Q}_\alpha^{-1}\boldsymbol{Q}_{\alpha-2} & -\boldsymbol{Q}_\alpha^{-1}\boldsymbol{Q}_{\alpha-1} \end{bmatrix} \in$$
$$\in \mathcal{R}^{\alpha\rho\times\alpha\rho}, \tag{13.39}$$

$$\boldsymbol{A}_{IIO2} = \begin{bmatrix} \boldsymbol{O}_{N,\rho} & \boldsymbol{O}_{N,\rho} & \boldsymbol{O}_{N,\rho} & \cdots & \boldsymbol{O}_{N,\rho} & \boldsymbol{O}_{N,\rho} \\ \boldsymbol{O}_{N,\rho} & \boldsymbol{O}_{N,\rho} & \boldsymbol{O}_{N,\rho} & \cdots & \boldsymbol{O}_{N,\rho} & \boldsymbol{O}_{N,\rho} \\ \vdots & \vdots & \vdots & \vdots & \vdots & \vdots \\ \boldsymbol{O}_{N,\rho} & \boldsymbol{O}_{N,\rho} & \boldsymbol{O}_{N,\rho} & \cdots & \boldsymbol{O}_{N,\rho} & \boldsymbol{O}_{N,\rho} \\ \boldsymbol{E}_\nu^{-1}\boldsymbol{R}_0 & \boldsymbol{E}_\nu^{-1}\boldsymbol{R}_1 & \boldsymbol{E}_\nu^{-1}\boldsymbol{R}_2 & \cdots & \boldsymbol{E}_\nu^{-1}\boldsymbol{R}_{\alpha-2} & \boldsymbol{E}_\nu^{-1}\boldsymbol{R}_{\alpha-1} \end{bmatrix} \in$$
$$\in \mathcal{R}^{\nu N\times\alpha\rho}, \tag{13.40}$$

$$\boldsymbol{A}_{IIO3} =$$

$$= \begin{bmatrix} \boldsymbol{O}_N & \boldsymbol{I}_N & \boldsymbol{O}_N & \cdots & \boldsymbol{O}_N & \boldsymbol{O}_N \\ \boldsymbol{O}_N & \boldsymbol{O}_N & \boldsymbol{I}_N & \cdots & \boldsymbol{O}_N & \boldsymbol{O}_N \\ \vdots & \vdots & \vdots & \vdots & \vdots & \vdots \\ \boldsymbol{O}_N & \boldsymbol{O}_N & \boldsymbol{O}_N & \cdots & \boldsymbol{O}_N & \boldsymbol{I}_N \\ -\boldsymbol{E}_\nu^{-1}\boldsymbol{E}_0 & -\boldsymbol{E}_\nu^{-1}\boldsymbol{E}_1 & -\boldsymbol{E}_\nu^{-1}\boldsymbol{E}_2 & \cdots & -\boldsymbol{E}_\nu^{-1}\boldsymbol{E}_{\nu-2} & -\boldsymbol{E}_\nu^{-1}\boldsymbol{E}_{\nu-1} \end{bmatrix} \in$$

$$\in \mathcal{R}^{\nu N \times \nu N}, \tag{13.41}$$

$$\boldsymbol{A}_w = \boldsymbol{A}_{IIOry} = \begin{bmatrix} \boldsymbol{A}_{IIO1} & \boldsymbol{O}_{\alpha\rho,\nu N} \\ \boldsymbol{A}_{IIO2} & \boldsymbol{A}_{IIO3} \end{bmatrix}, \ \nu \geq 1,. \tag{13.42}$$

The solution of (13.38) reads for $\nu > 0$:

$$\mathbf{w}(k;\mathbf{w}_0) = \begin{bmatrix} \mathbf{r}^{\alpha-1}(k;\mathbf{r}_0^{\alpha-1};\mathbf{0}_M) \\ \mathbf{y}^{\nu-1}(k;\mathbf{r}_0^{\alpha-1};\mathbf{y}_0^{\nu-1};\mathbf{0}_M) \end{bmatrix} = \boldsymbol{A}_w^k \begin{bmatrix} \mathbf{r}_0^{\alpha-1} \\ \mathbf{y}_0^{\nu-1} \end{bmatrix}.$$

Asymptotic stability of $\begin{bmatrix} (\mathbf{r}_e^{\alpha-1})^T & (\mathbf{y}_e^{\nu-1})^T \end{bmatrix}^T = \mathbf{0}_{\alpha\rho+\nu N}$ implies both

$$\exists \gamma \in \mathcal{R}^+ \Longrightarrow \left\| \boldsymbol{A}_w^k \right\| < \gamma, \ \forall k \in \mathcal{N}_0,$$

and

$$k \to \infty \Longrightarrow \boldsymbol{A}_w^k \to \boldsymbol{O}_{p.}$$

Hence,

$$\exists\, (\zeta > 1, \xi) \in \mathcal{R}^+ \times \mathcal{R}^+$$

such that

$$\left\| \begin{bmatrix} \mathbf{r}^{\alpha-1}(k;\mathbf{r}_0^{\alpha-1};\mathbf{0}_M) \\ \mathbf{y}^{\nu-1}(k;\mathbf{r}_0^{\alpha-1};\mathbf{y}_0^{\nu-1};\mathbf{0}_M) \end{bmatrix} \right\| = \left\| \boldsymbol{A}_w^k \begin{bmatrix} \mathbf{r}_0^{\alpha-1} \\ \mathbf{y}_0^{\nu-1} \end{bmatrix} \right\| \leq$$

$$\leq \zeta e^{-\xi k} \left\| \begin{bmatrix} \mathbf{r}_0^{\alpha-1} \\ \mathbf{y}_0^{\nu-1} \end{bmatrix} \right\|, \ \forall k \in \mathbb{Z}, \forall \begin{bmatrix} \mathbf{r}_0^{\alpha-1} \\ \mathbf{y}_0^{\nu-1} \end{bmatrix} \in \mathcal{R}^{\alpha\rho+\nu N}.$$

The zero equilibrium vector $\begin{bmatrix} (\mathbf{r}_e^{\alpha-1})^T & (\mathbf{y}_e^{\nu-1})^T \end{bmatrix}^T = \mathbf{0}_{\alpha\rho+\nu N}$ of the *IIO* system (3.64a) and (3.64b), is (globally) exponentially stable. ■

This theorem, Theorem 13.22 and Definition 13.14 imply directly the following:

Theorem 13.26 *In order for the zero equilibrium vector* $\begin{bmatrix} \mathbf{r}_e^{\alpha-1} \\ \mathbf{y}_e^{\nu-1} \end{bmatrix} = \mathbf{0}_{\alpha\rho+\nu N}$ *of the IIO system (3.64a) and (3.64b), in which* $\nu \geq 1$, *to be exponentially* *stable, it is necessary (but not sufficient) that* $\left[\begin{array}{cc} \left(\mathbf{r}_e^{\alpha-1}\right)^T & \left(\mathbf{y}_e^{\nu-1}\right)^T \end{array} \right]^T =$ $\mathbf{0}_{\alpha\rho+\nu N}$ *is the unique equilibrium vector of the system, i.e., that the matrix*

$$\left[\begin{array}{cc} \boldsymbol{Q}_0 + \boldsymbol{Q}_1 + \cdots + \boldsymbol{Q}_\alpha & \boldsymbol{O}_{\rho,N} \\ -\left(\boldsymbol{R}_0 + \boldsymbol{R}_1 + \cdots + \boldsymbol{R}_\alpha\right) & \boldsymbol{E}_0 + \boldsymbol{E}_1 + \cdots + \boldsymbol{E}_\nu \end{array} \right]$$

is nonsingular.

13.3 Lyapunov method and theorems

13.3.1 Outline of Lyapunov's original theory

Let us explain the meaning of some basic notions of Lyapunov stability theory.

Lyapunov established his method, which he originally called *the second method*, known also as *the direct Lyapunov method, for the continuous-time ISO systems.* We broaden it to discrete time *IO*, *ISO* and *IIO* systems. The method means the study of properties of the system nominal motion under the influence of any initial conditions via the sign properties of a subsidiary function $V(\cdot)$ and of its total *time* difference along system motions without using any information about the motions themselves, hence without solving the system mathematical model (the system discrete-time equation).

The *Lyapunov theorems* determine the conditions on the subsidiary function $V(\cdot)$ called the *Lyapunov function* and on its total *time* difference along system motions in order for the system motions to have some qualitative properties.

The *Lyapunov methodology* results from the Lyapunov theorems and it determines how to apply the Lyapunov method: should we start with a choice of a subsidiary function $V(\cdot)$ or with its total *time* difference along system motions. The Lyapunov himself determined one methodology for *time*-invariant linear systems and another one for *time*-varying linear and all nonlinear systems.

For *time*-invariant linear systems, Lyapunov methodology demands to begin with a choice of the total *time* difference of the subsidiary function $V(\cdot)$ along system motions, and then to test the properties of the function $V(\cdot)$. This methodology resolves stability problems completely.

For *time*-varying linear systems and for all nonlinear systems, Lyapunov's methodology requires to start with a choice of the function $V(\cdot)$ and then to test the properties of its total *time* difference along system motions, which is inverse to his methodology for *time*-invariant linear systems.

We cite at first Lyapunov's original theorems for *time*-invariant continuous-*time* linear systems, [93, p. 67 in the Russian edition], which A. M. Lyapunov himself described by (13.43),

$$\frac{dx_s}{dt} = p_{s1}x_1 + p_{s2}x_2 + ... + p_{sn}x_n, \; s = 1, 2, ..., n, \tag{13.43}$$

or in the vector form (13.44),

$$\frac{d\mathbf{x}}{dt} = \boldsymbol{P}_L\mathbf{x}, \; \boldsymbol{P}_L = \begin{bmatrix} p_{11} & \cdots & p_{1n} \\ \vdots & \vdots & \vdots \\ p_{n1} & \cdots & p_{nn} \end{bmatrix}. \tag{13.44}$$

The zeros (the roots) of the system characteristic equation, denoted by $\varkappa_i$ in them, and the natural numbers m and m_i also in them, obey

$$m_1 + m_2 + ... + m_n = m.$$

Theorem 13.27 ***Original Lyapunov's Theorem I on the linear systems in the free regime [93, pp. 75, 76 in the Russian edition]***

When the roots $\varkappa_1$, $\varkappa_2$, ... , $\varkappa_n$, of the characteristic equation are such that for a given positive integer m the relationships of the following forms:

$$m_1\varkappa_1 + m_2\varkappa_2 + ... + m_n\varkappa_n = 0,$$

are impossible for them, in which all m_s are nonnegative integers, the sum of which is m, then it is possible to find and then unique fully homogeneous function V of the power m of the variables x_s, which satisfies the equation

$$\sum_{s=1}^{n} (p_{s1}x_1 + p_{s2}x_2 + ... + p_{sn}x_n) \frac{\partial V}{\partial x_s} = U \tag{13.45}$$

for an arbitrarily given fully homogeneous function U of quantities x_s with the same power m.

The vector form of (13.45) reads

$$(gradV)^T \boldsymbol{P}_L\mathbf{x} = \mathbf{U}. \tag{13.46}$$

Remark 13.1 ***Lyapunov's remark on his Theorem I 13.27, [93, p. 76 in the Russian edition]***
The conditions, considered in the theorem, will be, for example, fulfilled and then for every m, when the real parts of all quantities $\varkappa_s$ are different from zero and have the same sign.

Comment 13.1 *Lyapunov's theorem I 13.27 for the linear systems establishes only necessary conditions for the existence of the unique solution function $V(\cdot)$ of (13.45), but not sufficiency for roots $\varkappa_1, \varkappa_2, \ldots, \varkappa_n$, of the characteristic equation to obey the conditions of the theorem.*

Theorem 13.28 ***Original Lyapunov's Theorem II on the linear systems in the free regime [93, p. 76 in the Russian edition]***
When the real parts of all roots $\varkappa_s$ are negative and when in (13.45) the function U is sign definite form of any even power m, then the form V of the power m satisfying that equation is also sign definite and additionally with the sign opposite to the sign of U.

Comment 13.2 *The conditions in this Lyapunov's theorem 13.28 are presented as necessary, but not as both necessary and sufficient for the real parts of all roots $\varkappa_s$ to be negative. The sufficient conditions follow from Lyapunov's following statements related to stability properties of the unperturbed motion of the nonlinear systems.*

Theorem 13.29 ***Original Lyapunov's Theorem I on the nonlinear systems in the free regime [93, pp. 59 in the Russian edition]***
If the differential equations of the perturbed motions are such that it is possible to find a sign definite function V, the derivative V' of which due to those equations is either a sign semidefinite function with the sign inverse to the sign of V, or identically equal to zero, the unperturbed motion is stable.

Note 13.10 ***Original Lyapunov's Note II on the nonlinear systems in the free regime [93, p. 61 in the Russian edition]***
If the function V, which satisfies the conditions of Theorem 13.29, and simultaneously permits the infinitesimally small upper bound, and its derivative is sign definite function, then it is possible to prove that every perturbed motion, sufficiently close to the unperturbed one, will converge to it asymptotically.

Comment 13.3 *Every positive definite time-independent function permits the infinitesimally small upper bound.*

13.3.2 Lyapunov method, theorems and methodology for the linear systems

We will broaden Lyapunov's original theory to be directly applicable in the same, unified, manner to all classes of the systems treated herein.

Let $\mathbf{w}$ be *a subsidiary real-valued vector,*

$$\mathbf{w} = \begin{bmatrix} w_1 & w_2 & \cdots & w_p \end{bmatrix}^T \in \mathcal{R}^p,$$
$$\mathbf{w} \in \left\{ \begin{bmatrix} \left(\mathbf{r}^{\alpha-1}\right)^T & \left(\mathbf{y}^{\nu-1}\right)^T \end{bmatrix}^T, \; \mathbf{r}^{\alpha-1}, \mathbf{x}, \; \mathbf{y}^{\nu-1} \right\},$$
$$p \in \{ n, \; \nu N, \; \alpha\rho, \; \alpha\rho + \nu N \}. \tag{13.47}$$

The vector $\mathbf{w}$ becomes

- $\mathbf{y}^{\nu-1}$,

$$\mathbf{w} = \mathbf{y}^{\nu-1} \in \mathcal{R}^{\nu N}, \tag{13.48}$$

for the IO systems (3.56),(in Subsection 3.5.2),

- $\mathbf{x}$,

$$\mathbf{w} = \mathbf{x} \in \mathcal{R}^n, \tag{13.49}$$

for the ISO systems (3.60a) and(3.60b) (in Subsection 3.5.3),

- $\begin{bmatrix} \left(\mathbf{r}^{\alpha-1}\right)^T & \left(\mathbf{y}^{\nu-1}\right)^T \end{bmatrix}^T$ if $\nu \geq 1$, or $\mathbf{r}^{\alpha-1}$ if $\nu = 0$,

$$\mathbf{w} = \left\{ \begin{array}{c} \begin{bmatrix} \left(\mathbf{r}^{\alpha-1}\right)^T & \left(\mathbf{y}^{\nu-1}\right)^T \end{bmatrix}^T \in \mathcal{R}^{\alpha\rho+\nu N}, \; \nu \geq 1, \\ \mathbf{r}^{\alpha-1} \in \mathcal{R}^{\alpha\rho}, \; \nu = 0 \end{array} \right\} \tag{13.50}$$

for the IIO systems (3.63a) and (3.63b) (in Subsection 3.5.4).

The properties of the symmetric matrix $\boldsymbol{W}$ of the quadratic form $v(\mathbf{w})$,

$$v(\mathbf{w}) = \mathbf{w}^T \boldsymbol{W} \mathbf{w}, \; \boldsymbol{W} = \boldsymbol{W}^T \in \mathcal{R}^{p \times p}, \tag{13.51}$$

determine some particular properties of the quadratic form itself, where

$$\boldsymbol{W} = \boldsymbol{W}^T \in \left\{ \boldsymbol{G} = \boldsymbol{G}^T \in \mathcal{R}^{q \times q}, \; \boldsymbol{H} = \boldsymbol{H}^T \in \mathcal{R}^{q \times q} \right\}, \tag{13.52}$$

and

$$q \in \{ n, \; N, \; \nu N, \; \alpha\rho, \; \rho + N, \; \alpha\rho + \nu N \}. \tag{13.53}$$

The following is well known in the matrix theory:

Claim 13.1 *Let $\lambda_m(\boldsymbol{W})$ and $\lambda_M(\boldsymbol{W})$ be the minimal and the maximal eigenvalue of the matrix $\boldsymbol{W}$. They are real numbers because the matrix $\boldsymbol{W}$ is symmetric. The quadratic form $v(\mathbf{w}) = \mathbf{w}^T \boldsymbol{W}\mathbf{w}$ obeys the following estimates:*

$$\lambda_m(\boldsymbol{W}) \|\mathbf{w}\|^2 \leq \mathbf{w}^T \boldsymbol{W}\mathbf{w} \leq \lambda_M(\boldsymbol{W}) \|\mathbf{w}\|^2, \forall \mathbf{w} \in \mathcal{R}^p. \tag{13.54}$$

It is easy to verify the following properties of the quadratic form:

Property 13.1 *The quadratic form (13.51)*

$$v(\mathbf{w}) = \mathbf{w}^T \boldsymbol{W}\mathbf{w}, \;\; \boldsymbol{W} = \boldsymbol{W}^T \in \mathcal{R}^{p\times p}$$

is

1) *everywhere defined and continuous on $\mathcal{R}^p$,*

$$v(\mathbf{w}) \in \mathbb{C}(\mathcal{R}^p),$$

and

2) *zero valued at the origin,*

$$v(\mathbf{0}_p) = 0.$$

Definition 13.17 *The quadratic form (13.51):*

– *is **positive definite**, denoted by $v > 0$, if and only if it is both*

  a) *zero valued only at the origin of $\mathcal{R}^p$,*

$$v(\mathbf{w}) = 0 \Leftrightarrow \mathbf{w} = \mathbf{0}_p,$$

  and

  b) *positive valued out of the origin on $\mathcal{R}^p$,*

$$v(\mathbf{w}) > 0, \; \forall\, (\mathbf{w} \neq \mathbf{0}_p) \in \mathcal{R}^p,$$

– *is **negative definite** if and only if $-v(\mathbf{w})$ is positive definite.*

Definition 13.18 *The matrix $\boldsymbol{W}$, $\boldsymbol{W} = \boldsymbol{W}^T$, of the quadratic form $v(\mathbf{w})$ is positive (negative) definite, denoted respectively by $\boldsymbol{W} > \boldsymbol{O}$ ($\boldsymbol{W} < \boldsymbol{O}$), if and only if the quadratic form itself is positive (negative) definite, respectively.*

The well-known **Sylvester criterion,** [27], [28], represents the necessary and sufficient conditions for positive definiteness of the matrix $\boldsymbol{W}$, $\boldsymbol{W} = [w_{ij}]$. It reads:

Criterion 13.1 ***Sylvester criterion***

In order for the symmetric matrix $\boldsymbol{W}$, $\boldsymbol{W} = \boldsymbol{W}^T = [w_{ij}] \in \mathcal{R}^{p\times p}$, *to be positive definite it is:*

– *necessary (but not sufficient) that*

$$w_{k,k} > 0, \ \forall k = 1, \ 2, \ \ldots \ , \ p,$$

– *both necessary and sufficient that*

$$\begin{vmatrix} w_{11} & w_{12} & \cdots & w_{1,k} \\ w_{21} & w_{22} & \cdots & w_{2,k} \\ - & - & \cdots & - \\ w_{k,1} & w_{k,2} & \cdots & w_{k,k} \end{vmatrix} > 0, \ \forall k = 1, \ 2, \ \ldots \ , \ p.$$

For the quadratic form (13.51) which is defined everywhere and continuous on $\mathcal{R}^p$, the first forward difference $\boldsymbol{\Delta}_w v(\mathbf{w})$ according to the vector $\mathbf{w}$ is:

$$\boldsymbol{\Delta}_w v(\mathbf{w}) = v(\mathbf{w} + \boldsymbol{\Delta}\mathbf{w}) - v(\mathbf{w}) =$$

$$= (\mathbf{w} + \boldsymbol{\Delta}\mathbf{w})^T \boldsymbol{W} (\mathbf{w} + \boldsymbol{\Delta}\mathbf{w}) - \mathbf{w}^T \boldsymbol{W}\mathbf{w} =$$
$$= 2\mathbf{w}^T \boldsymbol{W} \Delta\mathbf{w} + \boldsymbol{\Delta}\mathbf{w}^T \boldsymbol{W} \Delta\mathbf{w}, \tag{13.55}$$

so that the total *time* difference $\Delta v[\mathbf{w}(k)]$ of $v(\mathbf{w})$ can be further expressed in terms of $\Delta\mathbf{w}(k) = \mathbf{w}(k+1) - \mathbf{w}(k)$,

$$\Delta v[\mathbf{w}(k)] = 2\mathbf{w}(k)^T \boldsymbol{W}[\mathbf{w}(k+1) - \mathbf{w}(k)] +$$
$$+ [\mathbf{w}(k+1) - \mathbf{w}(k)]^T \boldsymbol{W}[\mathbf{w}(k+1) - \mathbf{w}(k)] =$$
$$= 2\mathbf{w}(k)^T \boldsymbol{W}\mathbf{w}(k+1) - 2\mathbf{w}(k)^T \boldsymbol{W}\mathbf{w}(k) + \mathbf{w}^T(k+1)\boldsymbol{W}\mathbf{w}(k+1) -$$

$$-\mathbf{w}^T(k)\boldsymbol{W}\mathbf{w}(k+1) - \mathbf{w}^T(k+1)\boldsymbol{W}\mathbf{w}(k) + \mathbf{w}^T(k)\boldsymbol{W}\mathbf{w}(k). \tag{13.56}$$

Since

$$\mathbf{w}^T(k+1)\boldsymbol{W}\mathbf{w}(k)$$

is scalar and $\boldsymbol{W} = \boldsymbol{W}^T$ (13.52) then,

$$\mathbf{w}^T(k+1)\boldsymbol{W}\mathbf{w}(k) = \mathbf{w}^T(k)\boldsymbol{W}^T\mathbf{w}(k+1) = \mathbf{w}^T(k)\boldsymbol{W}\mathbf{w}(k+1),$$

and, due to (13.56),

$$\Delta v\left[\mathbf{w}(k)\right] = \mathbf{w}^T(k+1)\,\boldsymbol{W}\mathbf{w}(k+1) - \mathbf{w}^T(k)\,\boldsymbol{W}\mathbf{w}(k). \tag{13.57}$$

Let

$$E^1\mathbf{w} = \boldsymbol{A}_w\mathbf{w}. \tag{13.58}$$

This, transform (13.57) into

$$\Delta v\left[\mathbf{w}(k)\right] = \mathbf{w}^T(k)\left(\boldsymbol{A}_w^T\boldsymbol{W}\boldsymbol{A}_w - \boldsymbol{W}\right)\mathbf{w}(k). \tag{13.59}$$

Definition 13.19 *A square matrix* $\boldsymbol{A}_w \in \mathcal{R}^{p\times p}$ *is* **discrete *stable*** *(or* **discrete *stability*,** *or* ***Schur*****)** ***matrix*** *if and only if the modulus of all its eigenvalues are less than one.*

Remark 13.2 *The word* **discrete** *in Definition 13.19 in denomination* **discrete *stable matrix*** *or* **discrete *stability matrix*** *does not mean the matrix* $\boldsymbol{A}_w$ *dependance on discrete time* k *but the* ***stable matrix*** *or* ***stability matrix*** *related to the discrete time systems.*

The following is well known in the matrix theory [27] and [28]:

Claim 13.2 *If and only if* $\boldsymbol{W}$ *is positive definite, then its minimal and maximal eigenvalues are positive,*

$$\boldsymbol{W} > \boldsymbol{O} \Leftrightarrow \lambda_M(\boldsymbol{W}) \geq \lambda_m(\boldsymbol{W}) > 0. \tag{13.60}$$

We will present the complete proof of the following theorem which is well known as *Lyapunov theorem* in the stability theory. The proof is taken from [41, p. 293].

Theorem 13.30 ***Lyapunov theorem for the system (13.58)***

In order for the zero equilibrium state $\mathbf{w}_e = \mathbf{0}_p$ *of the system (13.58) to be asymptotically stable, it is necessary and sufficient that for an arbitrary positive definite quadratic form* $\mathbf{w}^T\boldsymbol{G}\mathbf{w}$, $\boldsymbol{G} = \boldsymbol{G}^T \in \mathcal{R}^{p\times p}$, $\boldsymbol{G} > \boldsymbol{O}$, *the solution function* $v(\cdot)$, $v(\cdot): \mathcal{R}^p \to \mathcal{R}$, *of the difference equation*

$$\Delta v(\mathbf{w}) = -\mathbf{w}^T\boldsymbol{G}\mathbf{w} \tag{13.61}$$

is also positive definite, and unique, quadratic form,

$$v(\mathbf{w}) = \mathbf{w}^T\boldsymbol{H}\mathbf{w},\ \boldsymbol{H} = \boldsymbol{H}^T \in \mathcal{R}^{p\times p}.\ \boldsymbol{H} > \boldsymbol{O}. \tag{13.62}$$

Proof. *Necessity. Let* $\boldsymbol{G}$ *be an arbitrary symmetric positive definite matrix,* $\boldsymbol{G} = \boldsymbol{G}^T \in \mathcal{R}^{p\times p}$, $\boldsymbol{G} > \boldsymbol{O}$, *what by the definition means that the quadratic form* $\mathbf{w}^T\boldsymbol{G}\mathbf{w}$ *is also positive definite. Let the zero equilibrium state* $\mathbf{w}_e = \mathbf{0}_p$ *of the system (13.58) be asymptotically stable. Theorem 13.13 and Definition 13.9 (in Subsection 13.2.2), ensure that the zero equilibrium state* $\mathbf{w}_e = \mathbf{0}_p$ *is unique. It follows that the modulus of all, matrix* $\boldsymbol{A}_w$, *eigenvalues are less than one. This implies [7, Theorem 2-9-5, p. 38]*

$$\left(\boldsymbol{A}_w^T\right)^N \boldsymbol{G}\boldsymbol{A}_w^N \rightarrow \boldsymbol{O} \text{ when } N \rightarrow \infty, \tag{13.63}$$

and existence of a matrix $\boldsymbol{H} \in \mathcal{R}^{p\times p}$ *such that the sum* $\sum_{r=0}^{N}\left(\boldsymbol{A}_w^T\right)^r \boldsymbol{G}\boldsymbol{A}_w^r$ *converges as N infinitely grows [7, Exercise 4-1-4, p. 60]:*

$$\sum_{r=0}^{N}\left(\boldsymbol{A}_w^T\right)^r \boldsymbol{G}\boldsymbol{A}_w^r \rightarrow \boldsymbol{H} \text{ as } N \rightarrow \infty, \tag{13.64}$$

i.e., geometric matrix series $\sum_{r=0}^{\infty}\left(\boldsymbol{A}_w^T\right)^r \boldsymbol{G}\boldsymbol{A}_w^r$ *is absolutely convergent, so it is convergent, with the matrix limit* $\boldsymbol{H}$,

$$\sum_{r=0}^{\infty}\left(\boldsymbol{A}_w^T\right)^r \boldsymbol{G}\boldsymbol{A}_w^r = \boldsymbol{H}. \tag{13.65}$$

By the definition, the convergence of the matrix series (13.65), means convergence of the sequence of the partial sums:

$$\boldsymbol{X}(N) = \sum_{r=0}^{N}\left(\boldsymbol{A}_w^T\right)^r \boldsymbol{G}\boldsymbol{A}_w^r,\ \forall N = 0, 1, 2, \cdots \tag{13.66}$$

where the matrix $\boldsymbol{X}(N)$ *is well defined since matrices* $\boldsymbol{A}_w$ *and* $\boldsymbol{G}$ ***are*** *defined, symmetric and positive definite* $\forall N = 0, 1, 2, \cdots$ *[7, Theorem 2-8-7, p. 35]. From (13.65) and (13.66) follows:*

$$\boldsymbol{X}(\infty) = \boldsymbol{H}. \tag{13.67}$$

Based on (13.66) we may write:

$$\boldsymbol{A}_w^T\boldsymbol{X}(N)\boldsymbol{A}_w - \boldsymbol{X}(N) = \sum_{r=0}^{N}\left(\boldsymbol{A}_w^T\right)^{r+1} \boldsymbol{G}\boldsymbol{A}_w^{r+1} -$$

$$-\sum_{r=0}^{N}\left(\boldsymbol{A}_w^T\right)^r \boldsymbol{G}\boldsymbol{A}_w^r = \left(\boldsymbol{A}_w^T\right)^{N+1} \boldsymbol{G}\boldsymbol{A}_w^{N+1} - \boldsymbol{G}. \tag{13.68}$$

When $N \to \infty$, (13.68), by means (13.67) and (13.63), becomes

$$\boldsymbol{A}_w^T \boldsymbol{H} \boldsymbol{A}_w - \boldsymbol{H} = -\boldsymbol{G}. \tag{13.69}$$

It follows that the symmetric and positive definite matrix $\boldsymbol{H}$ is the solution of matrix Equation (13.69), i.e., there exists matrix solution $\boldsymbol{H}$ of (13.69). We will prove the uniqueness of $\boldsymbol{H}$ by contradiction. Suppose that $\boldsymbol{H}$ is not unique solution of (13.69), i.e., that there exists another solution $\boldsymbol{P} \neq \boldsymbol{H}$,

$$\exists \boldsymbol{P} \in \mathcal{R}^{p \times p},\ \boldsymbol{P} \neq \boldsymbol{H} \text{ such that } \boldsymbol{A}_w^T \boldsymbol{P} \boldsymbol{A}_w - \boldsymbol{P} = -\boldsymbol{G}. \tag{13.70}$$

From (13.70) it follows (see: [92]):

$$\left(\boldsymbol{A}_w^T\right)^{N+1} \boldsymbol{P} \boldsymbol{A}_w^{N+1} = \boldsymbol{P} - \sum_{r=0}^{N} \left(\boldsymbol{A}_w^T\right)^r \boldsymbol{G} \boldsymbol{A}_w^r, \tag{13.71}$$

that can be easily proved by the mathematical induction. Equation (13.71) is transformed by means of (13.66) in

$$\left(\boldsymbol{A}_w^T\right)^{N+1} \boldsymbol{P} \boldsymbol{A}_w^{N+1} = \boldsymbol{P} - \boldsymbol{X}(N). \tag{13.72}$$

When $N \to \infty$ (13.72), by using (13.63) and (13.67), becomes

$$\boldsymbol{O} = \boldsymbol{P} - \boldsymbol{H} \Longrightarrow \boldsymbol{P} = \boldsymbol{H}.$$

There does not exist another matrix solution $\boldsymbol{P}$ of (13.69) different from $\boldsymbol{H}$. The unique matrix solution of (13.69) is symmetric positive definite matrix $\boldsymbol{H}$. Let Equation (13.69) be premultiplied by $\mathbf{w}^T$ and postmuliplied by $\mathbf{w}$

$$\mathbf{w}^T \boldsymbol{A}_w^T \boldsymbol{H} \boldsymbol{A}_w \mathbf{w} - \mathbf{w}^T \boldsymbol{H} \mathbf{w} \equiv -\mathbf{w}^T \boldsymbol{G} \mathbf{w}, \tag{13.73}$$

and let the quadratic form $\mathbf{w}^T \boldsymbol{H} \mathbf{w}$ be denoted by $v(\mathbf{w})$

$$v(\mathbf{w}) = \mathbf{w}^T \boldsymbol{H} \mathbf{w}. \tag{13.74}$$

Since matrix $\boldsymbol{H}$ is positive definite, so is the quadratic form $v(\mathbf{w})$. The first forward difference $\Delta v(\mathbf{w})$ of $v(\mathbf{w})$ related to the (13.58) reads:

$$\Delta v(\mathbf{w}) = \mathbf{w}^T \boldsymbol{A}_w^T \boldsymbol{H} \boldsymbol{A}_w \mathbf{w} - \mathbf{w}^T \boldsymbol{H} \mathbf{w},$$

what with (13.73) give

$$\Delta v(\mathbf{w}) \equiv -\mathbf{w}^T \boldsymbol{G} \mathbf{w}.$$

This confirms that the $v(\mathbf{w})$ is unique positive definite quadratic form which is solution of Equation (13.61).

Sufficiency: Let all the conditions of the theorem statement hold. Arbitrary $\varepsilon > 0$ is adopted, which determines hyperball $\mathcal{K}_\varepsilon$ with the boundary hypersphere $\partial\mathcal{K}_\varepsilon$,

$$\partial\mathcal{K}_\varepsilon = \{\mathbf{w} : \|\mathbf{w}\| = \varepsilon\}.$$

Let ς_1 be the smallest value of the function $v(\mathbf{w})$ on the hypersphere $\partial\mathcal{K}_\varepsilon$,

$$\varsigma_1 = \inf\left[v(\mathbf{w}) : \mathbf{w} \in \partial\mathcal{K}_\varepsilon\right].$$

V_{ς_1} denotes the greatest connected set with the following properties:

a) $\mathbf{0}_w \in V_{\varsigma_1}$

b) $v(\mathbf{w}) \in [0, \varsigma_1[, \forall \mathbf{w} \in V_{\varsigma_1}$,

c) $v(\mathbf{w}) = \varsigma_1, \forall \mathbf{w} \in \partial V_{\varsigma_1}$,

where ∂V_{ς_1} is boundary of the set V_{ς_1}.

Let δ be the smallest Euclidean distance between a point which belongs to the boundary ∂V_{ς_1} of the set V_{ς_1}, and the origin $\mathbf{0}_w$:

$$\delta = \inf\left(\|\mathbf{w}\| : \mathbf{w} \in \partial V_{\varsigma_1}\right).$$

It follows that

$$\mathcal{K}_\delta \subseteq V_{\varsigma_1} \subseteq \mathcal{K}_\varepsilon. \tag{13.75}$$

If $\mathbf{w}_0 \in \mathcal{K}_\delta \Longrightarrow$ then from (13.75) $\Longrightarrow$

$$v(\mathbf{w}_0) < \varsigma_1. \tag{13.76}$$

Due to the conditions of the theorem statement, $\Delta v(\mathbf{w}) < 0$, $\forall (\mathbf{w} \neq \mathbf{0}_w) \in \mathcal{R}^p$ which implies:

$$\begin{gathered} v\left[\mathbf{w}(k+1; \mathbf{w}_0; \mathbf{0}_M)\right] < v\left[\mathbf{w}(k; \mathbf{w}_0; \mathbf{0}_M)\right], \\ \forall k = 0, 1, 2, \cdots, \forall (\mathbf{w}_0 \neq \mathbf{0}_w) \in \mathcal{R}^p. \end{gathered} \tag{13.77}$$

As $v(\mathbf{w})$ is positive definite quadratic form, the following is valid for it:

$$v(\alpha_i \mathbf{w}) = \alpha_i^2 v(\mathbf{w}),$$

what implies,

$$v(\alpha_1 \mathbf{w}) < v(\alpha_2 \mathbf{w}), \forall \alpha_1, \alpha_2 \in \mathcal{R},\ 0 \leq \alpha_1 < \alpha_2, \forall \mathbf{w} \in \mathcal{R}^p. \tag{13.78}$$

(13.78) proves the truth of the following statement:

$$0 \leq \varsigma_1 < \varsigma_2 \Longrightarrow V_{\varsigma_1} \subset V_{\varsigma_2},\ \partial V_{\varsigma_1} \cap \partial V_{\varsigma_2} = \emptyset, \tag{13.79}$$

where $\emptyset$ *denotes empty set. Connection (13.77) and (13.79) gives:*

$$\mathbf{w}_0 \in V_{\varsigma_1} \Longrightarrow \mathbf{w}(k; \mathbf{w}_0; \mathbf{0}_M) \in V_{\varsigma_1}, \forall k = 0, 1, 2, \cdots,$$

and further with (13.75) leads to:

$$\mathbf{w}_0 \in \mathcal{K}_\delta \Longrightarrow \mathbf{w}(k; \mathbf{w}_0; \mathbf{0}_M) \in \mathcal{K}_\varepsilon, \forall k = 0, 1, 2, \cdots$$

Stability of the zero equilibrium state, $\mathbf{w}_e = \mathbf{0}_w$, *related to the system (13.58), is proved.*

Attraction of the zero equilibrium state, $\mathbf{w}_e = \mathbf{0}_w$, *related to the system (13.58), is proved by contradiction. Let suppose that equilibrium state,* $\mathbf{w}_e = \mathbf{0}_w$, *related to the system (13.58), is not attractive. As a consequence of this assumption there exists* $\hat{\mathbf{w}}_0 \in \mathcal{R}^p$, *and* $\gamma > 0$, $\gamma = \gamma(\hat{\mathbf{w}}_0)$, *such that*

$$\|\mathbf{w}(k; \hat{\mathbf{w}}_0; \mathbf{0}_M)\| \to \gamma \text{ when } k \to \infty. \tag{13.80}$$

This implies that there does not exist $i \in \{0, 1, 2, \cdots\}$, *such that*

$$\|\mathbf{w}(i; \hat{\mathbf{w}}_0; \mathbf{0}_M)\| = 0. \tag{13.81}$$

If (13.81) would be true then it would imply

$$\|\mathbf{w}(k; \hat{\mathbf{w}}_0; \mathbf{0}_M)\| = 0, \forall k = i, i+1, i+2, \cdots \tag{13.82}$$

for (13.81) means that $\mathbf{w}(i) = \mathbf{0}_w$, *i.e., the system is in equilibrium state at* $k = i$, *and (13.82) would be valid. It is directly in collision with (13.80) what means that (13.81) and (13.82) would not be valid if* $\mathbf{w}_e = \mathbf{0}_w$ *were not attractive. Then there would exist* $\theta > 0$ *such that*

$$\|\mathbf{w}(k; \hat{\mathbf{w}}_0; \mathbf{0}_M)\| > \theta, \forall k = 0, 1, 2, \cdots \tag{13.83}$$

Positive definiteness of $v(\mathbf{w})$ *and* $\mathbf{w}^T \boldsymbol{G}\mathbf{w}$, *and (13.83) would imply that there existed* $\varsigma > 0$ *and* $\xi > 0$ *such that,*

$$v[\mathbf{w}(k; \hat{\mathbf{w}}_0; \mathbf{0}_M)] > \varsigma, \forall k = 0, 1, 2, \cdots, \tag{13.84}$$

and

$$g[\mathbf{w}(k; \hat{\mathbf{w}}_0; \mathbf{0}_M)] > \xi, \forall k = 0, 1, 2, \cdots, \text{ where } g(\mathbf{w}) = \mathbf{w}^T \boldsymbol{G}\mathbf{w}. \tag{13.85}$$

Based on the theorem condition (13.61) and (13.85) it would follow,

$$\Delta v[\mathbf{w}(k; \hat{\mathbf{w}}_0; \mathbf{0}_M)] < -\xi, \forall k = 0, 1, 2, \cdots \tag{13.86}$$

By summing up the inequality (13.86) due to $k = 0, 1, 2, \cdots, N$ we would get:

$$v\left[\mathbf{w}\left(N; \hat{\mathbf{w}}_0; \mathbf{0}_M\right)\right] - v\left(\hat{\mathbf{w}}_0\right) < -N\xi, \ \forall N = 1, 2, 3, \cdots \tag{13.87}$$

From (13.87) it is clear that,

$$v\left[\mathbf{w}\left(N; \hat{\mathbf{w}}_0; \mathbf{0}_M\right)\right] < 0, \ \forall N \geq v\left(\hat{\mathbf{w}}_0\right)\xi^{-1}, \tag{13.88}$$

what would mean, for $0 < v\left(\hat{\mathbf{w}}_0\right)\xi^{-1} < +\infty$, that the function $v\left(\mathbf{w}\right)$ would become negative (13.88) after finite number of discrete instants. This would be in collision with positive definiteness of $v\left(\mathbf{w}\right)$. The collision would appear because of the wrong assumption (13.80). Therefore,

$$v\left[\mathbf{w}\left(k; \mathbf{w}_0; \mathbf{0}_M\right)\right] \to 0 \ \textit{when} \ k \to \infty, \ \forall \mathbf{w}_0 \in \mathcal{R}^p, \ \Longrightarrow$$
$$\Longrightarrow \mathbf{w}\left(k; \mathbf{w}_0; \mathbf{0}_M\right) \to \mathbf{0}_M \ \textit{as} \ k \to \infty,$$

due to positive definiteness of $v\left(\cdot\right)$. This means that $\mathbf{w}_e = \mathbf{0}_w$ of the system (13.58) is attractive in the whole and proves its global asymptotic stability, (Definition 13.9, in Subsection 13.2.1). ■

Definition 13.20 ***Lyapunov function of the system (13.58)***
The quadratic form (13.62), which is the solution of (13.61), is ***a Lyapunov function of the system (13.58)****.*

Comment 13.4 ***The physical meaning of the Lyapunov theorem***
Let a mathematical model of a physical system be the linear time-invariant discrete-time system of the form (13.58). If the energy $e(\cdot)$ and power $p(\cdot)$ of the physical system are quadratic forms such that the power is negative definite quadratic form,

$$p(\mathbf{w}) = \mathbf{w}^T \boldsymbol{P}\mathbf{w} = -\mathbf{w}^T \boldsymbol{G}\mathbf{w} \ \textit{for} \ \boldsymbol{P} = \boldsymbol{P}^T = - \ \boldsymbol{G}, \ \boldsymbol{G} > O,$$

then for the zero equilibrium state of the system to be asymptotically (hence, exponentially) stable, it is both necessary and sufficient that the system energy $e(\cdot)$, $e(\mathbf{w}) = \mathbf{w}^T \boldsymbol{E}\mathbf{w}$, $\boldsymbol{E} = \boldsymbol{E}^T$, is positive definite quadratic form since the power $p(\cdot)$ is approximately the first forward difference of the energy when sampling period T is equal to one, $T = 1$,

$$\Delta e\left(\mathbf{w}\right) = p(\mathbf{w}).$$

For $\boldsymbol{H} = \boldsymbol{E} = \boldsymbol{E}^T = \boldsymbol{H}^T$,

$$\Delta e\left(\mathbf{w}\right) = \mathbf{w}^T\left(\boldsymbol{A}_w^T\boldsymbol{E}\boldsymbol{A}_w - \boldsymbol{E}\right)\mathbf{w} = \mathbf{w}^T\left(\boldsymbol{A}_w^T\boldsymbol{H}\boldsymbol{A}_w - \boldsymbol{H}\right)\mathbf{w} =$$
$$= -\mathbf{w}^{\mathbf{T}}\boldsymbol{G}\mathbf{w} = \mathbf{w}^{\mathbf{T}}\boldsymbol{P}\mathbf{w} = p(\mathbf{w}),\ -\boldsymbol{G} = \boldsymbol{P} = \left(\boldsymbol{A}_w^T\boldsymbol{H}\boldsymbol{A}_w - \boldsymbol{H}\right) \Longrightarrow$$
$$\Delta e\left(\mathbf{w}\right) = -\mathbf{w}^T\boldsymbol{G}\mathbf{w}, \tag{13.89}$$

which is (13.61).

Theorem 13.31 ***Lyapunov matrix theorem for the system (13.58)***
In order for the matrix $\boldsymbol{A}_w$ *to be stable matrix, it is necessary and sufficient that for any positive definite symmetric matrix* $\boldsymbol{G}_w$, $\boldsymbol{G}_w = \boldsymbol{G}_w^T \in \mathcal{R}^{p\times p}$, *the matrix solution* $\boldsymbol{H}_w$ *of the Lyapunov matrix equation*

$$\boldsymbol{A}_w^T\boldsymbol{H}\boldsymbol{A}_w - \boldsymbol{H}_w = -\ \boldsymbol{G}_w \tag{13.90}$$

is also positive definite symmetric matrix and the unique solution to (13.90).

Proof. It is well known that for the zero equilibrium vector $\mathbf{w}_e = \mathbf{0}_p$ of the system (13.58) to be asymptotically stable, it is both necessary and sufficient that the modulus of all eigenvalues of the system matrix $\boldsymbol{A}_w$ (13.58) are less than one. This, and Definition 13.19 imply that for the zero equilibrium vector $\mathbf{w}_e = \mathbf{0}_p$ of the system (13.58) to be asymptotically stable, it is both necessary and sufficient that the system matrix $\boldsymbol{A}_w$ (13.58) is stable (i.e., Schur) matrix.

Let $\boldsymbol{G}_w = \boldsymbol{G}_w^T \in \mathcal{R}^{p\times p}$ be any positive definite symmetric matrix, which determines the quadratic form

$$\mathbf{w}^T\boldsymbol{G}_w\mathbf{w}.$$

Let $\boldsymbol{H}_w = \boldsymbol{H}_w^T \in \mathcal{R}^{p\times p}$ be a symmetric matrix of the quadratic form

$$v(\mathbf{w}) = \mathbf{w}^T\boldsymbol{H}_w\mathbf{w},$$

which is the solution of the Lyapunov difference Equation (13.61),

$$\Delta v\left(\mathbf{w}\right) \equiv -\mathbf{w}^T\boldsymbol{G}_w\mathbf{w}. \tag{13.91}$$

The Lyapunov Theorem 13.30 implies that the matrix $\boldsymbol{H}_w = \boldsymbol{H}_w^T$ is also positive definite and that its quadratic form $v(\mathbf{w}) = \mathbf{w}^T\boldsymbol{H}_w\mathbf{w}$ is the unique solution of (13.91) if and only if the zero equilibrium vector $\mathbf{w}_e = \mathbf{0}_p$ of the system (13.58) is asymptotically stable, i.e., if and only if the matrix $\boldsymbol{A}_w$

is stable matrix. The first forward difference of $v(\mathbf{w})$ along the motions of the system (13.58) reads

$$\Delta v\left(\mathbf{w}\right)=\Delta\left(\mathbf{w}^T\boldsymbol{H}_w\mathbf{w}\right)=\mathbf{w}^T\left(k+1\right)\boldsymbol{H}_w\mathbf{w}\left(k+1\right)-\mathbf{w}\left(k\right)^T\boldsymbol{H}_w^T\mathbf{w}\left(k\right).$$

This, and (13.58) permit

$$\Delta v\left(\mathbf{w}\right)=\mathbf{w}^T\boldsymbol{A}_w^T\boldsymbol{H}_w\boldsymbol{A}_w\mathbf{w}-\mathbf{w}^T\boldsymbol{H}_w\mathbf{w}.$$

We replace $\Delta v\left(\mathbf{w}\right)$ by $\mathbf{w}^T\boldsymbol{A}_w^T\boldsymbol{H}_w\boldsymbol{A}_w\mathbf{w}-\mathbf{w}^T\boldsymbol{H}_w\mathbf{w}$ in (13.91),

$$\mathbf{w}^T\boldsymbol{A}_w^T\boldsymbol{H}_w\boldsymbol{A}_w\mathbf{w}-\mathbf{w}^T\boldsymbol{H}_w\mathbf{w}=-\mathbf{w}^T\boldsymbol{G}_w\mathbf{w},$$

i.e.,

$$\mathbf{w}^T\left(\boldsymbol{A}_w^T\boldsymbol{H}_w\boldsymbol{A}_w-\boldsymbol{H}_w\right)\mathbf{w}=-\,\mathbf{w}^T\boldsymbol{G}_w\mathbf{w}.$$

Since this holds for every $\mathbf{w}\in\mathcal{R}^p$, then

$$\boldsymbol{A}_w^T\boldsymbol{H}_w\boldsymbol{A}_w-\boldsymbol{H}_w=-\,\boldsymbol{G}_w.$$

This completes the proof. ■

Note 13.11 ***The fundamental matrix theorem***

Theorem 13.31 is the well known fundamental matrix theorem of the stability theory of the linear time-invariant discrete-time systems.

In order to show how we can effectively solve Lyapunov matrix Equation (13.90) for $\boldsymbol{H}_w$ we explain the Kronecker matrix product of two matrices $\boldsymbol{M}=[m_{ij}]\in\mathcal{R}^{\mu\times s}$ and $\boldsymbol{U}=[u_{ij}]\in\mathcal{R}^{\upsilon\times\sigma}$, which is denoted by $\otimes$,

$$\boldsymbol{M}\otimes\boldsymbol{U}=\begin{bmatrix} m_{11}\boldsymbol{U} & m_{12}\boldsymbol{U} & \cdots & m_{1s}\boldsymbol{U}\\ m_{21}\boldsymbol{U} & m_{21}\boldsymbol{U} & \cdots & m_{2s}\boldsymbol{U}\\ \vdots & \vdots & \vdots & \vdots\\ m_{\mu 1}\boldsymbol{U} & m_{\mu 2}\boldsymbol{U} & \cdots & m_{\mu s}\boldsymbol{U}\end{bmatrix}\in\mathcal{R}^{\mu\upsilon\times s\sigma}. \tag{13.92}$$

For the same purpose, we explain also application of operator $\mathrm{Vec}\,(\cdot)$ to the matrix $\boldsymbol{M}=[m_{ij}]\in\mathcal{R}^{\mu\times s}$,

$$\begin{aligned}\mathrm{Vec}\,(\boldsymbol{M})=\big[\; m_{11}\;\; m_{21}\;\; \cdots\;\; m_{\mu 1}\;\; m_{12}\;\; m_{22}\;\; \cdots\;\; m_{\mu 2}\;\; \cdots\\ \cdots\;\; m_{1s}\;\; m_{2s}\;\; \cdots\;\; m_{\mu s}\;\big]^T\in\mathcal{R}^{\mu s}.\end{aligned} \tag{13.93}$$

Note 13.12 *Well-known properties of the Kronecker product and the operator* $\mathrm{Vec}\,(\cdot)$ *are [79], [116]:*

1. *if* $\boldsymbol{A} \in \mathcal{R}^{n\times m}$, $\boldsymbol{B} \in \mathcal{R}^{m\times l}$, *and* $\boldsymbol{C} \in \mathcal{R}^{l\times r}$, *then*

$$\mathrm{Vec}\,(\boldsymbol{ABC}) = \left(\boldsymbol{C}^T \otimes \boldsymbol{A}\right)\mathrm{Vec}\,(\boldsymbol{B}), \tag{13.94}$$

2. *if* $\boldsymbol{A} \in \mathcal{R}^{n\times n}$, $\boldsymbol{B} \in \mathcal{R}^{m\times m}$, *and* $\lambda_i(\boldsymbol{A})$, $i \in \{1, 2, \cdots, n\}$, $\lambda_j(\boldsymbol{B})$, $j \in \{1, 2, \cdots, m\}$ *are eigenvalues of matrices* $\boldsymbol{A}$ *and* $\boldsymbol{B}$, *respectively, then eigenvalues* $\lambda_r(\boldsymbol{A} \otimes \boldsymbol{B})$, $r \in \{1, 2, \cdots, nm\}$, *of matrix* $\boldsymbol{A} \otimes \boldsymbol{B}$ *are:*

$$\lambda_r(\boldsymbol{A} \otimes \boldsymbol{B}) = \lambda_i(\boldsymbol{A})\,\lambda_j(\boldsymbol{B}),\ r \in \{1, 2, \cdots, nm\},$$
$$i \in \{1, 2, \cdots, n\},\ j \in \{1, 2, \cdots, m\}. \tag{13.95}$$

We define the vectors $\mathbf{h}_w$ and $\mathbf{g}_w$ by application of the operator Vec $(\cdot)$ to the matrices $\boldsymbol{H}_w = \boldsymbol{H}_w^T = [h_{ij}] \in \mathcal{R}^{p\times p}$ and $\boldsymbol{G}_w = \boldsymbol{G}_w^T = [g_{ij}] \in \mathcal{R}^{p\times p}$, respectively,

$$\mathrm{Vec}\,(\boldsymbol{H}_w) = \mathbf{h}_w = \begin{bmatrix} h_{11} & h_{21} & \cdots & h_{p1} & h_{12} & h_{22} & \cdots & h_{p2} & \cdots \end{bmatrix}$$
$$\begin{bmatrix} \cdots & h_{1p} & h_{2p} & \cdots & h_{pp} \end{bmatrix}^T \in \mathcal{R}^{pp}, \tag{13.96}$$

$$\mathrm{Vec}\,(\boldsymbol{G}_w) = \mathbf{g}_w = \begin{bmatrix} g_{11} & g_{21} & \cdots & g_{p1} & g_{12} & g_{22} & \cdots & g_{p2} & \cdots \end{bmatrix}$$
$$\begin{bmatrix} \cdots & g_{1p} & g_{2p} & \cdots & g_{pp} \end{bmatrix}^T \in \mathcal{R}^{pp}. \tag{13.97}$$

These vectors, and application of (13.93) and (13.94) to Equation (13.90) enable us

- to determine in the straightforward procedure the Lyapunov function of the system (13.58),

  and

- to set the Lyapunov matrix Equation (13.90) in the vector form (13.98) by using the Kronecker matrix product of the matrix $\boldsymbol{A}_w^T$ with itself,

$$\left(\boldsymbol{A}_w^T \otimes \boldsymbol{A}_w^T - \boldsymbol{I}_{pp}\right)\mathbf{h}_w = -\mathbf{g}_w. \tag{13.98}$$

Theorem 13.32 *For Equation (13.98) to be solvable in* $\mathbf{h}_w$,

$$\mathbf{h}_w = -\left(\boldsymbol{A}_w^T \otimes \boldsymbol{A}_w^T - \boldsymbol{I}_p\right)^{-1}\mathbf{g}_w, \tag{13.99}$$

it is necessary and sufficient that the eigenvalues $\lambda_i(\boldsymbol{A}_w)$ *of the matrix* $\boldsymbol{A}_w$ *obey*

$$\lambda_i(\boldsymbol{A}_w)\lambda_j(\boldsymbol{A}_w) \neq 1, \ \forall i, j = 1, 2, ..., p. \tag{13.100}$$

Proof. *According to the Kronecker-Capelli Theorem, Equation (13.98) has got unique solution if and only if*

$$\text{rank}\left(\boldsymbol{A}_w^T \otimes \boldsymbol{A}_w^T - \boldsymbol{I}_{pp}\right) = \text{rank}\left[\begin{array}{cc} \boldsymbol{A}_w^T \otimes \boldsymbol{A}_w^T - \boldsymbol{I}_{pp} & -\mathbf{g}_w \end{array}\right] = pp \Longleftrightarrow$$
$$\det\left(\boldsymbol{A}_w^T \otimes \boldsymbol{A}_w^T - \boldsymbol{I}_{pp}\right) \neq 0. \tag{13.101}$$

Characteristic equation of the matrix $\left(\boldsymbol{A}_w^T \otimes \boldsymbol{A}_w^T - \boldsymbol{I}_{pp}\right)$ *reads*

$$\det\left[\lambda \boldsymbol{I}_{pp} - \left(\boldsymbol{A}_w^T \otimes \boldsymbol{A}_w^T - \boldsymbol{I}_{pp}\right)\right] = 0, \tag{13.102}$$

which implies that λ *must be different from zero,* $\lambda \neq 0$*, in order for the condition (13.101) to be fulfilled, i.e., in order for the unique solution of Equation (13.98) to exist. After small arrangement of (13.102) we get*

$$\det\left[\hat{\lambda}\boldsymbol{I}_{pp} - \boldsymbol{A}_w^T \otimes \boldsymbol{A}_w^T\right] = \det\left[\hat{\lambda}\boldsymbol{I}_{pp}^T - \left(\boldsymbol{A}_w \otimes \boldsymbol{A}_w\right)^T\right] =$$
$$= \det\left[\hat{\lambda}\boldsymbol{I}_{pp} - \boldsymbol{A}_w \otimes \boldsymbol{A}_w\right]^T = \det\left[\hat{\lambda}\boldsymbol{I}_{pp} - \boldsymbol{A}_w \otimes \boldsymbol{A}_w\right] = 0,$$
$$\hat{\lambda} = \lambda + 1. \tag{13.103}$$

As Equation (13.103) is the characteristic equation of matrix $\boldsymbol{A}_w \otimes \boldsymbol{A}_w$*, and as the condition* $\lambda \neq 0$ *implies* $\hat{\lambda} \neq 1$*, it means that eigenvalues of matrix* $\boldsymbol{A}_w \otimes \boldsymbol{A}_w$ *must be different from one.*

Necessity: Let Equation (13.98) be uniquely solvable in $\mathbf{h}_w$*. Due to (13.101), (13.103) and (13.95) the eigenvalues* $\lambda_i(\boldsymbol{A}_w)\lambda_j(\boldsymbol{A}_w)$*,* $\forall i, j = 1, 2, ..., p$ *of matrix* $\boldsymbol{A}_w \otimes \boldsymbol{A}_w$ *are different from one what is the condition (13.100) of the theorem.*

Sufficiency: Let the conditions of the theorem statement hold. Due to (13.95)
$\lambda_i(\boldsymbol{A}_w)\lambda_j(\boldsymbol{A}_w)$*,* $\forall i, j = 1, 2, ..., p$ *are the eigenvalues of the matrix* $\boldsymbol{A}_w \otimes \boldsymbol{A}_w$*. Taking into account (13.103) and (13.101) it follows that Equation (13.98) is uniquely solvable in* $\mathbf{h}_w$*.*

In this way the proof is ended. ■

Equation (13.99) determines the vector $\mathbf{h}_w$, which, together with (13.96) and (13.97), defines completely the matrix $\boldsymbol{H}_w$, $\boldsymbol{H}_w = \boldsymbol{H}_w^T \in \mathcal{R}^{p \times p}$, and its quadratic form $v(\mathbf{w}) = \mathbf{w}^T \boldsymbol{H}_w \mathbf{w}$ being Lyapunov function of the system (13.58).

The requirement (13.100) opens the problem of the conditions under which (13.100) holds.

Theorem 13.33 *In order for the condition (13.100) to be fulfilled, it is necessary and sufficient, that the following relationships are obeyed,*

$$\operatorname{Re}\lambda_j(\boldsymbol{A}_w)\operatorname{Im}\lambda_i(\boldsymbol{A}_w)+\operatorname{Re}\lambda_i(\boldsymbol{A}_w)\operatorname{Im}\lambda_j(\boldsymbol{A}_w)=0, \tag{13.104a}$$

$$\operatorname{Re}\lambda_i(\boldsymbol{A}_w)\operatorname{Re}\lambda_j(\boldsymbol{A}_w)-\operatorname{Im}\lambda_i(\boldsymbol{A}_w)\operatorname{Im}\lambda_j(\boldsymbol{A}_w)\neq 1. \tag{13.104b}$$

Proof. *Necessity:* Let the condition (13.100) holds. If we use rectangular form of eigenvalues $\lambda_i(\boldsymbol{A}_w)$ and $\lambda_j(\boldsymbol{A}_w)$,

$$\lambda_i(\boldsymbol{A}_w)=\sigma_i+j\omega_i,\ \sigma_i=\operatorname{Re}\lambda_i(\boldsymbol{A}_w),\ \omega_i=\operatorname{Im}\lambda_i(\boldsymbol{A}_w),$$
$$\lambda_j(\boldsymbol{A}_w)=\sigma_j+j\omega_j,\ \sigma_j=\operatorname{Re}\lambda_j(\boldsymbol{A}_w),\ \omega_j=\operatorname{Im}\lambda_j(\boldsymbol{A}_w),$$

the condition (13.100) becomes

$$\begin{aligned}\lambda_i(\boldsymbol{A}_w)\lambda_j(\boldsymbol{A}_w)&=(\sigma_i+j\omega_i)(\sigma_j+j\omega_j)=\\&=(\sigma_i\sigma_j-\omega_i\omega_j)+j(\sigma_j\omega_i+\sigma_i\omega_j)\neq 1,\end{aligned} \tag{13.105}$$

what implies

$$(\sigma_j\omega_i+\sigma_i\omega_j)=\operatorname{Re}\lambda_j(\boldsymbol{A}_w)\operatorname{Im}\lambda_i(\boldsymbol{A}_w)+\operatorname{Re}\lambda_i(\boldsymbol{A}_w)\operatorname{Im}\lambda_j(\boldsymbol{A}_w)=0,$$
$$(\sigma_i\sigma_j-\omega_i\omega_j)=\operatorname{Re}\lambda_i(\boldsymbol{A}_w)\operatorname{Re}\lambda_j(\boldsymbol{A}_w)-\operatorname{Im}\lambda_i(\boldsymbol{A}_w)\operatorname{Im}\lambda_j(\boldsymbol{A}_w)\neq 1.$$

Sufficiency: Let the relations (13.104a) and (13.104b) hold. Let form a complex number whose real part is left side of the relation (13.104b) and imaginary part is left side of the relation (13.104a). According to the relations (13.104a) and (13.104b), the complex number is different from one,

$$\begin{aligned}&(\sigma_i\sigma_j-\omega_i\omega_j)+j(\sigma_j\omega_i+\sigma_i\omega_j)=\\&=[\operatorname{Re}\lambda_i(\boldsymbol{A}_w)\operatorname{Re}\lambda_j(\boldsymbol{A}_w)-\operatorname{Im}\lambda_i(\boldsymbol{A}_w)\operatorname{Im}\lambda_j(\boldsymbol{A}_w)]+\\&+j[\operatorname{Re}\lambda_j(\boldsymbol{A}_w)\operatorname{Im}\lambda_i(\boldsymbol{A}_w)+\operatorname{Re}\lambda_i(\boldsymbol{A}_w)\operatorname{Im}\lambda_j(\boldsymbol{A}_w)]\neq 1.\end{aligned} \tag{13.106}$$

As the formed complex number is obviously product of two complex numbers,

$$\sigma_i+j\omega_i,\ \sigma_j+j\omega_j,$$

which are eigenvalues of matrix $\boldsymbol{A}_w$, it implies that

$$\begin{aligned}(\sigma_i+j\omega_i)(\sigma_j+j\omega_j)&=[\operatorname{Re}\lambda_i(\boldsymbol{A}_w)+j\operatorname{Im}\lambda_i(\boldsymbol{A}_w)]\cdot\\&\cdot[\operatorname{Re}\lambda_j(\boldsymbol{A}_w)+j\operatorname{Im}\lambda_j(\boldsymbol{A}_w)]=\lambda_i(\boldsymbol{A}_w)\lambda_j(\boldsymbol{A}_w)\neq 1,\end{aligned}$$

by which the theorem is proved. ■

The next theorem explains in more details the meaning of the conditions of the previous theorem.

Theorem 13.34 *In order for the conditions (13.104a) and (13.104b) of Theorem (13.33) to be valid, it is necessary and sufficient, that the following is obeyed,*

a) the complex eigenvalues $\lambda_i(\boldsymbol{A}_w)$ and $\lambda_j(\boldsymbol{A}_w)$ of the matrix $\boldsymbol{A}_w$ whose real parts are equal to zero,

$$\operatorname{Re}\lambda_i(\boldsymbol{A}_w) = \operatorname{Re}\lambda_j(\boldsymbol{A}_w) = 0, \tag{13.107}$$

must not have imaginary parts with opposite sign and reciprocal modulus,

$$\operatorname{Im}\lambda_i(\boldsymbol{A}_w) \neq -\frac{1}{\operatorname{Im}\lambda_j(\boldsymbol{A}_w)}, \tag{13.108}$$

b) the real eigenvalues $\lambda_i(\boldsymbol{A}_w)$ and $\lambda_j(\boldsymbol{A}_w)$ of the matrix $\boldsymbol{A}_w$, different from zero, $\lambda_i(\boldsymbol{A}_w) \neq 0$, $\lambda_j(\boldsymbol{A}_w) \neq 0$ (their imaginary parts are equal to zero, $\operatorname{Im}\lambda_i(\boldsymbol{A}_w) = \operatorname{Im}\lambda_j(\boldsymbol{A}_w) = 0$) whose signs are equal, must not have reciprocal modulus,

$$\operatorname{Re}\lambda_i(\boldsymbol{A}_w) \neq \frac{1}{\operatorname{Re}\lambda_j(\boldsymbol{A}_w)}. \tag{13.109}$$

Proof. *Necessity:* Let the conditions (13.104a) and (13.104b) hold.
a) In this case relation (13.104b) is reduced to

$$\lambda_i(\boldsymbol{A}_w)\lambda_j(\boldsymbol{A}_w) = j^2\omega_i\omega_j = -\omega_i\omega_j = -\operatorname{Im}\lambda_i(\boldsymbol{A}_w)\operatorname{Im}\lambda_j(\boldsymbol{A}_w) \neq 1,$$

which implies

$$\operatorname{Im}\lambda_i(\boldsymbol{A}_w) \neq -\frac{1}{\operatorname{Im}\lambda_j(\boldsymbol{A}_w)},$$

b) In this case relation (13.104b) is reduced to

$$\lambda_i(\boldsymbol{A}_w)\lambda_j(\boldsymbol{A}_w) = \sigma_i\sigma_j = \operatorname{Re}\lambda_i(\boldsymbol{A}_w)\operatorname{Re}\lambda_j(\boldsymbol{A}_w) \neq 1$$

which implies

$$\operatorname{Re}\lambda_i(\boldsymbol{A}_w) \neq \frac{1}{\operatorname{Re}\lambda_j(\boldsymbol{A}_w)}.$$

Sufficiency: *a*) Let the relations (13.107) and (13.108) hold. By using them in forming, in this special case, the relations (13.104a) and (13.104b), we get,

$$0\cdot\operatorname{Im}\lambda_i(\boldsymbol{A}_w) + 0\cdot\operatorname{Im}\lambda_j(\boldsymbol{A}_w) = 0,$$
$$0\cdot 0 - \operatorname{Im}\lambda_i(\boldsymbol{A}_w)\operatorname{Im}\lambda_j(\boldsymbol{A}_w) \neq 1,$$

which directly prove *a*) of the theorem.

b) Let the relation (13.109) hold and $\operatorname{Im}\lambda_i(\boldsymbol{A}_w) = \operatorname{Im}\lambda_j(\boldsymbol{A}_w) = 0$. By using this in forming, in this special case, the relations (13.104a) and (13.104a), we get,

$$\operatorname{Re}\lambda_j(\boldsymbol{A}_w)\cdot 0 + \operatorname{Re}\lambda_i(\boldsymbol{A}_w)\cdot 0 = 0,$$
$$\operatorname{Re}\lambda_i(\boldsymbol{A}_w)\operatorname{Re}\lambda_j(\boldsymbol{A}_w) - 0\cdot 0 \neq 1,$$

which directly prove *b*) of the theorem. ■

The following theorem is induced by Theorems (13.32), (13.33) and (13.34).

Theorem 13.35 *The Equation (13.98) to be solvable in* $\mathbf{h}_w$, *it is necessary and sufficient that conditions of Theorem (13.33), that is, conditions under a) and b) of Theorem (13.34), are fulfilled.*

If the matrix $\boldsymbol{A}_w$ is stability matrix, then it satisfies all these conditions, i.e., then Equation (13.98) is solvable in $\mathbf{h}_w$. Equation (13.99) determines the solution $\mathbf{h}_w$ that induces directly the matrix $\boldsymbol{H}_w$ via (13.96).

13.3.3 Lyapunov theorem for the *IO* systems

In the framework of the *IO* systems described by (3.56) (in Subsection 3.5.2),

$$p = \nu N,\ \mathbf{w} = \mathbf{y}^{\nu-1},\ \boldsymbol{W} = \boldsymbol{G}_{IO} = \boldsymbol{G}_{IO}^T \in \mathcal{R}^{\nu N\times\nu N},$$
$$\boldsymbol{H}_w = \boldsymbol{H}_{IO} = \boldsymbol{H}_{IO}^T \in \mathcal{R}^{\nu N\times\nu N\nu},\ v(\mathbf{w}) = v_{IO}\left(\mathbf{y}^{\nu-1}\right). \tag{13.110}$$

In the sequel the total *time* difference, (13.56), of a function $v_{IO}(\cdot)$, $v_{IO}(\cdot) : \mathcal{R}^{\nu N} \to \mathcal{R}$, along motions of the *IO* system (3.56), i.e., (13.111)

$$\boldsymbol{A}^{(\nu)}\mathbf{y}^{\nu}(k) = \boldsymbol{B}^{(\mu)}\mathbf{i}^{\mu}(k),\ \forall k \in \mathcal{N}_0, \tag{13.111}$$

in the free regime,

$$\boldsymbol{A}^{(\nu)}\mathbf{y}^{\nu}(k) = \sum_{r=0}^{r=\nu}\boldsymbol{A}_r E^r\mathbf{y}(k) = \mathbf{0}_{\nu N},\ \forall k \in \mathcal{N}_0, \tag{13.112}$$

is denoted by $\Delta v_{IO}\left[\mathbf{y}^{\nu-1}(k)\right]$,

$$\Delta v_{IO}\left[\mathbf{y}^{\nu-1}(k)\right] = \left[\mathbf{y}^{\nu-1}(k+1)\right]^T \boldsymbol{H}_{IO}\mathbf{y}^{\nu-1}(k+1) -$$
$$- \left[\mathbf{y}^{\nu-1}(k)\right]^T \boldsymbol{H}_{IO}\mathbf{y}^{\nu-1}(k). \tag{13.113}$$

Theorem 13.36 ***Lyapunov theorem for the IO system (13.111) in the free regime***

In order for the zero equilibrium state $\mathbf{y}_e^{\nu-1} = \mathbf{0}_{\nu N}$ *of the IO system (13.111) to be asymptotically stable, it is necessary and sufficient that for an arbitrary positive definite quadratic form* $\left(\mathbf{y}^{\nu-1}\right)^T \boldsymbol{G}_{IO}\mathbf{y}^{\nu-1}$, $\boldsymbol{G}_{IO} = \boldsymbol{G}_{IO}^T \in \mathcal{R}^{\nu N\times\nu N\nu}$, $\boldsymbol{G}_{IO} > \boldsymbol{O}$, *the solution function* $v_{IO}(\cdot)$, $v_{IO}(\cdot) : \mathcal{R}^{\nu N} \to \mathcal{R}$, *of the difference equation*

$$\Delta v\left(\mathbf{y}^{\nu-1}\right) = -\left(\mathbf{y}^{\nu-1}\right) \boldsymbol{G}_{IO}\mathbf{y}^{\nu-1} \tag{13.114}$$

is also positive definite, and unique, quadratic form,

$$v_{IO}\left(\mathbf{y}^{\nu-1}\right) =$$
$$= \left(\mathbf{y}^{\nu-1}\right) \boldsymbol{H}_{IO}\mathbf{y}^{\nu-1},\ \boldsymbol{H}_{IO} = \boldsymbol{H}_{IO}^T \in \mathcal{R}^{\nu N\times\nu N\nu},\ \boldsymbol{H}_{IO} > \boldsymbol{O}. \tag{13.115}$$

Proof. We use $\mathbf{w}$ defined by (13.110) so that along the behavior of the *IO* system (13.111) in the free regime, i.e., (13.112), we find the following since A_ν^{-1} exists due to Condition 2.1 (in Section 2.1):

$$E^1\mathbf{w} = E^1\mathbf{y}^{\nu-1} =$$
$$= \left[\ \left(E^1\mathbf{y}\right)^T\ \ \left(E^2\mathbf{y}\right)^T\ \ \cdots\ \ \left(E^{\nu-1}\mathbf{y}\right)^T\ \ \left(E^{\nu}\mathbf{y}\right)^T\ \right]^T = \boldsymbol{A}_w\mathbf{w}, \tag{13.116}$$

$$\boldsymbol{A}_w = \underbrace{\begin{bmatrix} \boldsymbol{O}_N & \boldsymbol{I}_N & \boldsymbol{O}_N & \cdots & \boldsymbol{O}_N & \boldsymbol{O}_N \\ \boldsymbol{O}_N & \boldsymbol{O}_N & \boldsymbol{I}_N & \cdots & \boldsymbol{O}_N & \boldsymbol{O}_N \\ \vdots & \vdots & \vdots & \vdots & \vdots & \vdots \\ \boldsymbol{O}_N & \boldsymbol{O}_N & \boldsymbol{O}_N & \cdots & \boldsymbol{O}_N & \boldsymbol{I}_N \\ -\boldsymbol{A}_\nu^{-1}\boldsymbol{A}_0 & -\boldsymbol{A}_\nu^{-1}\boldsymbol{A}_1 & -\boldsymbol{A}_\nu^{-1}\boldsymbol{A}_2 & \cdots & -\boldsymbol{A}_\nu^{-1}\boldsymbol{A}_{\nu-2} & -\boldsymbol{A}_\nu^{-1}\boldsymbol{A}_{\nu-1} \end{bmatrix}}. \tag{13.117}$$

Equations (13.116), (13.117) and (13.110) give

$$E^1\mathbf{w} = \boldsymbol{A}_w\mathbf{w},\ \text{for } p = \nu N,\ \boldsymbol{A}_w = \boldsymbol{A}_{IO} \in \mathcal{R}^{\nu N\times\nu N}. \tag{13.118}$$

Equation (13.118) is the *ISO* form of the *IO* system (13.111) in the free regime, i.e., the *ISO* form of (13.112). Theorem 13.30 is therefore applicable to the *IO* system (13.111) via (13.110), and (13.118) that is (13.58). This proves the statement of the theorem. ■

From this theorem we will deduce in Subsection 13.4.2, in the framework of the *IO* system (13.111), the well-known Lyapunov matrix theorem.

Definition 13.21 ***Lyapunov function of the*** IO ***system (13.111)***
The quadratic form (13.115) that is the solution of (13.114) is the ***Lyapunov function of the*** IO ***system (13.111).***

When we set

$$\boldsymbol{A}_w = \boldsymbol{A}_{IO},\ \mathbf{h}_w = \mathbf{h}_{IO},\ \mathbf{g}_w = \mathbf{g}_{IO},$$

in Equation (13.99) then it determines the vector $\mathbf{h}_{IO}$, which, together with (13.96) and (13.97), defines completely the matrix $\boldsymbol{H}_{IO}$, $\boldsymbol{H}_{IO} = \boldsymbol{H}_{IO}^T \in \mathcal{R}^{p\times p}$, and its quadratic form $v_{IO}\left(\mathbf{y}^{\nu-1}\right) = \left(\mathbf{y}^{\nu-1}\right)^T \boldsymbol{H}_{IO}\mathbf{y}^{\nu-1}$ being the Lyapunov function of the IO system (13.111).

13.3.4 Lyapunov theorem for the ISO systems

Theorem 13.37 ***Lyapunov theorem for the*** ISO ***system (3.60a) and (3.60b), (in Subsection 3.5.3), in the free regime***
In order for the zero equilibrium state $\mathbf{x}_e = \mathbf{0}_n$ *of the* ISO *system (3.60a) and (3.60b) to be asymptotically stable, it is necessary and sufficient that for an arbitrary positive definite quadratic form* $\mathbf{x}^T \boldsymbol{G}_w\mathbf{x}$, $\boldsymbol{G}_w = \boldsymbol{G}_w^T \in \mathcal{R}^{n\times n}$, $\boldsymbol{G}_w > \boldsymbol{O}$, *the solution function* $v(\cdot)$, $v(\cdot) : \mathcal{R}^n \to \mathcal{R}$, *of the difference equation*

$$\Delta v(\mathbf{x}) = -\mathbf{x}^T \boldsymbol{G}_w\mathbf{x} \tag{13.119}$$

is also positive definite, and unique, quadratic form,

$$v(\mathbf{x}) = \mathbf{x}^T \boldsymbol{H}_w\mathbf{x},\ \boldsymbol{H}_w = \boldsymbol{H}_w^T \in \mathcal{R}^{n\times n},\ \boldsymbol{H}_w > \boldsymbol{O}. \tag{13.120}$$

Proof. When we set

$$p = n,\ \mathbf{w} = \mathbf{x}, \tag{13.121}$$

in (13.58) (in Section 13.3), then it becomes (3.60a) and (3.60b) in the free regime, and vice versa. Besides we set

$$v(\mathbf{w}) = v(\mathbf{x}) \tag{13.122}$$

in (13.61) and (13.99), so that Theorem 13.30 takes the form of this theorem. ■

Definition 13.22 ***Lyapunov function of the*** ISO ***system (3.60a) and***
(3.60b)
The quadratic form (13.120) that is the solution of (13.119) is the ***Lyapunov function of the*** ISO ***system (3.60a) and (3.60b).***

When we set

$$v(\mathbf{w}) = v_{ISO}(\mathbf{x}), \ \boldsymbol{H}_w = \boldsymbol{H}_{ISO}, \ \boldsymbol{G}_w = \boldsymbol{G}_{ISO},$$

in (13.61), (13.99), and

$$\boldsymbol{A}_w = \boldsymbol{A},$$

in Equation (13.99) then it determines the vector $\mathbf{h}_{ISO}$, which, together with (13.96) and (13.97), defines completely the matrix $\boldsymbol{H}_{ISO}$, $\boldsymbol{H}_{ISO} = \boldsymbol{H}_{ISO}^T \in \mathcal{R}^{n\times n}$, and its quadratic form $v_{ISO}(\mathbf{x}) = \mathbf{x}^T \boldsymbol{H}_{ISO}\mathbf{x}$ that is the Lyapunov function of the ISO system (3.60a) and (3.60b).

13.3.5 Lyapunov theorem for the IIO systems

Theorem 13.38 ***Lyapunov theorem for the*** IIO ***system*** *(2.31a) and (2.31b),*

a) *Let $\nu \geq 1$ in (2.31a) and (2.31b). In order for the zero equilibrium state $\left[\begin{array}{cc} (\mathbf{r}_e^{\alpha-1})^T & (\mathbf{y}_e^{\nu-1})^T \end{array} \right]^T = \mathbf{0}_{\alpha\rho+\nu N}$ of the IIO system (2.31a) and (2.31b), to be asymptotically stable, it is necessary and sufficient that for an arbitrary positive definite quadratic form*

$$\left[\begin{array}{cc} (\mathbf{r}^{\alpha-1})^T & (\mathbf{y}^{\nu-1})^T \end{array} \right] \boldsymbol{G}_{IIO} \left[\begin{array}{cc} (\mathbf{r}^{\alpha-1})^T & (\mathbf{y}^{\nu-1})^T \end{array} \right]^T,$$
$$\boldsymbol{G}_{IIO} = \boldsymbol{G}_{IIO}^T \in \mathcal{R}^{(\alpha\rho+\nu N)\times(\alpha\rho+\nu N)}, \ \boldsymbol{G}_{IIO} > \boldsymbol{O}$$

the solution function $v(\cdot)$, $v(\cdot) : \mathcal{R}^{\alpha\rho+\nu N} \rightarrow \mathcal{R}$, of the difference equation

$$\Delta v\left(\left[\begin{array}{cc} (\mathbf{r}^{\alpha-1})^T & (\mathbf{y}^{\nu-1})^T \end{array} \right]^T\right) =$$
$$= -\left[\begin{array}{cc} (\mathbf{r}^{\alpha-1})^T & (\mathbf{y}^{\nu-1})^T \end{array} \right] \boldsymbol{G}_{IIO} \left[\begin{array}{cc} (\mathbf{r}^{\alpha-1})^T & (\mathbf{y}^{\nu-1})^T \end{array} \right]^T \quad (13.123)$$

is also positive definite, and unique, quadratic form,

$$v\left(\left[\begin{array}{cc} (\mathbf{r}^{\alpha-1})^T & (\mathbf{y}^{\nu-1})^T \end{array} \right]^T\right) =$$
$$= \left[\begin{array}{cc} (\mathbf{r}^{\alpha-1})^T & (\mathbf{y}^{\nu-1})^T \end{array} \right] \boldsymbol{H}_{IIO} \left[\begin{array}{cc} (\mathbf{r}^{\alpha-1})^T & (\mathbf{y}^{\nu-1})^T \end{array} \right]^T,$$
$$\boldsymbol{H}_{IIO} = \boldsymbol{H}_{IIO}^T \in \mathcal{R}^{(\alpha\rho+\nu N)\times(\alpha\rho+\nu N)}, \ \boldsymbol{H}_{IIO} > \boldsymbol{O}. \quad (13.124)$$

b) Let $\nu = 0$ in (2.31a) and (2.31b). In order for the zero equilibrium state $\mathbf{r}_e^{\alpha-1} = \mathbf{0}_\rho$ of the IIO system (2.31a) and (2.31b), the PMD system (2.32a) and (2.32b) the GISO system (2.35a) and (2.35b) (in Section 2.3), to be asymptotically stable, it is necessary and sufficient that for an arbitrary positive definite quadratic form $\mathbf{r}^{\alpha-1^T}\boldsymbol{G}_{IIOr}\mathbf{r}^{\alpha-1}$, $\boldsymbol{G}_{IIOr} = \boldsymbol{G}_{IIOr}^T \in \mathcal{R}^{\rho\times\rho}$, the solution function $v(\cdot)$, $v(\cdot) : \mathcal{R}^\rho \to \mathcal{R}$, of the difference equation

$$\Delta v\left(\mathbf{r}^{\alpha-1}\right) = -\mathbf{r}^{\alpha-1^T}\boldsymbol{G}_{IIOr}\mathbf{r}^{\alpha-1} \tag{13.125}$$

is also positive definite, and unique, quadratic form,

$$v\left(\mathbf{r}^{\alpha-1}\right) =$$

$$= \mathbf{r}^{\alpha-1^T}\boldsymbol{H}_{IIOr}\mathbf{r}^{\alpha-1},\ \boldsymbol{H}_{IIOr} = \boldsymbol{H}_{IIOr}^T \in \mathcal{R}^{\rho\times\rho},\ \boldsymbol{H}_{IIOr} > \boldsymbol{O}. \tag{13.126}$$

Proof. *a*) Let $\nu \geq 1$ in (2.31a) and (2.31b). The matrices $\boldsymbol{Q}_\alpha$ and $\boldsymbol{E}_\nu$ are nonsingular. They have inverses $\boldsymbol{Q}_\alpha^{-1}$ and $\boldsymbol{E}_\nu^{-1}$, respectively. When we set

$$p = \alpha\rho + \nu N,\ \mathbf{w} = \left[\begin{array}{cc} \left(\mathbf{r}^{\alpha-1}\right)^T & \left(\mathbf{y}^{\nu-1}\right)^T \end{array}\right]^{\mathbf{T}},$$

$$v(\mathbf{w}) = v_{IIO}\left(\left[\begin{array}{cc} \left(\mathbf{r}^{\alpha-1}\right)^T & \left(\mathbf{y}^{\nu-1}\right)^T \end{array}\right]^T\right),$$

$$\begin{aligned} \boldsymbol{H} &= \boldsymbol{H}_{IIO} = \boldsymbol{H}_{IIO}^T \in \mathcal{R}^{(\alpha\rho+\nu N)\times(\alpha\rho+\nu N)}, \\ \boldsymbol{G} &= \boldsymbol{G}_{IIO} = \boldsymbol{G}_{IIO}^T \in \mathcal{R}^{(\alpha\rho+\nu N)\times(\alpha\rho+\nu N)}, \end{aligned}$$

and set (13.39)-(13.42) (in Subsection 13.2.3), in (13.58) then it becomes (2.31a) and (2.31b), and vice versa, so that Theorem 13.30 takes the form of the statement under *a*) of this theorem.

b) Let $\nu = 0$ in (2.31a) and (2.31b). We set

$$p = \alpha\rho,\ \mathbf{w} = \mathbf{r}^{\boldsymbol{\alpha}-\mathbf{1}},\ v(\mathbf{w}) = v_{IIOr}\left(\mathbf{r}^{\boldsymbol{\alpha}-\mathbf{1}}\right),$$

$$\boldsymbol{H} = \boldsymbol{H}_{IIOr} = \boldsymbol{H}_{IIOr}^T \in \mathcal{R}^{\alpha\rho\times\alpha\rho},\ \boldsymbol{G} = \boldsymbol{G}_{IIOr} = \boldsymbol{G}_{IIOr}^T \in \mathcal{R}^{\alpha\rho\times\alpha\rho},$$

$$\boldsymbol{A}_w = \boldsymbol{A}_{IIO1} = \boldsymbol{A}_{IIOr} =$$

$$= \left[\begin{array}{cccccc} \boldsymbol{O}_\rho & \boldsymbol{I}_\rho & \boldsymbol{O}_\rho & \cdots & \boldsymbol{O}_\rho & \boldsymbol{O}_\rho \\ \boldsymbol{O}_\rho & \boldsymbol{O}_\rho & \boldsymbol{I}_\rho & \cdots & \boldsymbol{O}_\rho & \boldsymbol{O}_\rho \\ \vdots & \vdots & \vdots & \vdots & \vdots & \vdots \\ \boldsymbol{O}_\rho & \boldsymbol{O}_\rho & \boldsymbol{O}_\rho & \cdots & \boldsymbol{O}_\rho & \boldsymbol{I}_\rho \\ -\boldsymbol{Q}_\alpha^{-1}\boldsymbol{Q}_0 & -\boldsymbol{Q}_\alpha^{-1}\boldsymbol{Q}_1 & -\boldsymbol{Q}_\alpha^{-1}\boldsymbol{Q}_2 & \cdots & -\boldsymbol{Q}_\alpha^{-1}\boldsymbol{Q}_{\alpha-2} & -\boldsymbol{Q}_\alpha^{-1}\boldsymbol{Q}_{\alpha-1} \end{array}\right] \in$$

$$\in \mathcal{R}^{\alpha\rho\times\alpha\rho}, \tag{13.127}$$

in (13.58) that takes the form of (2.31a) and (2.31b), i.e., (2.35a) and (2.35b), in the free regime, and vice versa. Hence, Theorem 13.30 becomes the statement under b) of this theorem. ■

Definition 13.23 ***Lyapunov function of the IIO system (2.31a) and (2.31b)***

The quadratic form (13.124) that is the solution of (13.123) is the ***Lyapunov function of the IIO system (2.31a) and (2.31b).***

13.4 Conditions via $\boldsymbol{F}(z)$

13.4.1 Generating theorem

A complex-valued matrix function $\boldsymbol{F}(\cdot) : \mathcal{C} \rightarrow \mathcal{C}^{m \times n}$ is real rational matrix function if and only if every its entry is a quotient of two polynomials in z and it becomes a real-valued matrix for the real value of the complex variable z, i.e., for $z = \sigma_z \in \mathcal{R}$.

Let $\boldsymbol{F}(z)$ have μ different poles denoted by z_i^*, $i = 1, 2, \cdots, \mu$. The multiplicity of the pole z_i^* is designated by ν_i^*. We denote its real and imaginary part by $\operatorname{Re} z_i^*$ and $\operatorname{Im} z_i^*$, respectively.

Theorem 13.39 *Generating theorem*

Let $\boldsymbol{F}(\cdot) : \mathcal{R} \rightarrow \mathcal{R}^{p \times p}$, $\boldsymbol{F}(k) = [F_{ij}(k)]$, *have the Z−transform* $\boldsymbol{F}(\cdot) : \mathcal{C} \rightarrow \mathcal{C}^{p \times p}$, $\boldsymbol{F}(z) = [F_{ij}(z)]$, *which is real rational matrix function. In order for the norm* $\|\boldsymbol{F}(k)\|$ *of the original* $\boldsymbol{F}(k)$*:*

a) to be ***bounded****, i.e.,*

$$\exists \alpha \in \mathcal{R}^+ \Longrightarrow \|\boldsymbol{F}(k)\| < \alpha, \ \forall k \in \mathcal{N}_0,$$

it is necessary and sufficient that:

1. *the modulus of all poles of* $\boldsymbol{F}(z)$ *are less or equal to one, see Fig. 13.5,*

$$|z_i^*| \leq 1, \ \forall i = 1, 2, ..., \ \mu,$$

2. *all poles with modulus equal to one of* $\boldsymbol{F}(z)$ *are simple (i.e., with the multiplicity* ν_i^* *that is equal to one),*

$$|z_i^*| = 1, \nu_i^* = 1,$$

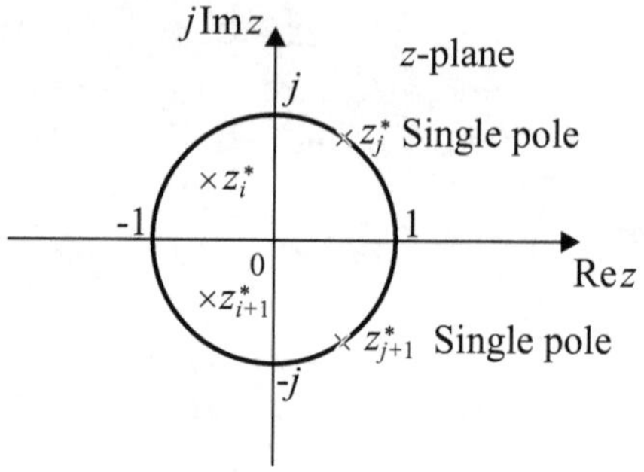

Figure 13.5: Poles with modulus less or equal to one.

3. $\boldsymbol{F}(z)$ is either proper or strictly proper,

b) and in order for $\|\boldsymbol{F}(k)\|$ to ***vanish asymptotically****, i.e., in order for the following condition to hold:*

$$\lim\left[\|\boldsymbol{F}(k)\|: \ k \longrightarrow \infty\right] = 0,$$

it is necessary and sufficient that

1. the modulus of all poles of $\boldsymbol{F}(z)$ are less than one, see Fig. 13.6,

$$|z_i^*| < 1, \ \forall i = 1, 2, ..., \ \mu,$$

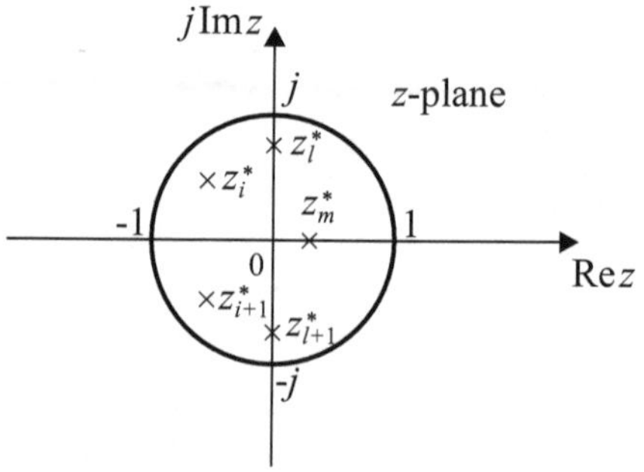

Figure 13.6: Poles with modulus less than one.

2. $\boldsymbol{F}(z)$ is either proper or strictly proper.

Proof. Let the $Z-$transform $\boldsymbol{F}(z)$ of $\boldsymbol{F}(k)$ have μ different poles denoted by z_i^* with the multiplicity ν_i^*, $i = 1, 2, \ldots, \mu$. We know (from the Heaviside expansion of $\boldsymbol{F}(z)$) that the original $\boldsymbol{F}(k)$ and its $Z-$transform $\boldsymbol{F}(z)$ are interrelated by the following formulae [41]:

- in the matrix form (13.128) and (13.129),

$$\boldsymbol{F}(k) = \mathcal{Z}^{-1}\left\{\boldsymbol{F}(z)\right\} = \delta_d(k)\boldsymbol{R}_0 +$$
$$+\sum_{i=1}^{i=\mu}(z_i^*)^{k-1}\left[\sum_{r=1}^{r=\nu_i^*}\frac{1}{(z_i^*)^{r-1}(r-1)!}\frac{(k-1)!}{(k-r)!}\boldsymbol{R}_{ir}\right],$$
$$\boldsymbol{R}_{ir} \in \mathcal{R}^{p\times n}, \forall k \in \mathcal{N}_0, \tag{13.128}$$

$$\boldsymbol{F}(z) = \mathcal{Z}\left\{\boldsymbol{F}(k)\right\} =$$
$$= \mathcal{Z}\left\{\delta_d(k)\boldsymbol{R}_0 + \sum_{i=1}^{i=\mu}(z_i^*)^{k-1}\left[\sum_{r=1}^{r=\nu_i^*}\frac{1}{(z_i^*)^{r-1}(r-1)!}\frac{(k-1)!}{(k-r)!}\boldsymbol{R}_{ir}\right]\right\}, \tag{13.129}$$

- in the scalar form (13.130) and (13.131), where z_i^{lm*} is one of poles of the $\mu_{lm} - th$ entry $F_{lm}(z)$ of $\boldsymbol{F}(z)$, $i \in \{1, 2, \ldots, \mu\}$, the multiplicity of which is denoted by ν_i^{lm*}:

$$F_{lm}(k) = \mathcal{Z}^{-1}\left\{F_{lm}(z)\right\} = \delta_d(k)R_0^{lm} +$$
$$+\sum_{i=1}^{i=\mu_{lm}}\left(z_i^{lm*}\right)^{k-1}\left[\sum_{r=1}^{r=\nu_i^{lm*}}\frac{1}{(z_i^{lm*})^{r-1}(r-1)!}\frac{(k-1)!}{(k-r)!}R_{ir}^{lm}\right],$$
$$R_{ir}^{lm} \in \mathcal{R}, \forall k \in \mathcal{N}_0, \tag{13.130}$$

$$F_{lm}(z) = \mathcal{Z}\left\{F_{lm}(k)\right\} =$$
$$= \mathcal{Z}\left\{\delta_d(k)R_0^{lm} + \sum_{i=1}^{i=\mu_{lm}}\left(z_i^{lm*}\right)^{k-1}\cdot\right.$$
$$\left.\cdot\left[\sum_{r=1}^{r=\nu_i^{lm*}}\frac{1}{(z_i^{lm*})^{r-1}(r-1)!}\frac{(k-1)!}{(k-r)!}R_{ir}^{lm}\right]\right\}. \tag{13.131}$$

Necessity. *a*) Let $\boldsymbol{F}(k)$ be bounded, i.e.,

$$\exists\alpha \in \mathcal{R}^{+} \Longrightarrow \|\boldsymbol{F}(k)\| < \alpha, \ \forall k \in \mathcal{N}_0. \tag{13.132}$$

We will primarily apply the method of contradiction to complete the proof of the necessity. Let us assume that condition *a*-1) does not hold, i.e.,

$$\exists z_i^{lm*} = \left|z_i^{lm*}\right| e^{j\arg z_i^{lm*}} \in \mathcal{C} \Longrightarrow \left|z_i^{lm*}\right| > 1. \tag{13.133}$$

This, $\delta_d(k)R_0^{lm} = 0$ for $k \neq 0$, $\delta_d(0)R_0^{lm} = R_0^{lm}$, and (13.130) imply

$$\left|\left(\left|z_i^{lm*}\right| e^{j\arg z_i^{lm*}}\right)^{k-1}\left[\sum_{r=1}^{r=\nu_i^{lm*}} \frac{1}{(z_i^{lm*})^{r-1}(r-1)!}\frac{(k-1)!}{(k-r)!}R_{ir}^{lm}\right]\right| =$$

$$= \left|z_i^{lm*}\right|^{k-1}\left|\left(e^{j\arg z_i^{lm*}}\right)^{k-1}\right|\left|\left[\sum_{r=1}^{r=\nu_i^{lm*}} \frac{1}{(z_i^{lm*})^{r-1}(r-1)!}\frac{(k-1)!}{(k-r)!}R_{ir}^{lm}\right]\right| =$$

$$= \left|z_i^{lm*}\right|^{k-1}\left|\left[\sum_{r=1}^{r=\nu_i^{lm*}} \frac{1}{(z_i^{lm*})^{r-1}(r-1)!}\frac{(k-1)!}{(k-r)!}R_{ir}^{lm}\right]\right| \longrightarrow \infty \text{ as } k \to \infty,$$

$$\Longrightarrow$$

$$\lim_{k\to\infty}\left|F_{lm}(k)\right| = \lim_{k\to\infty}\left|\delta_d(k)R_0^{lm} + \sum_{i=1}^{i=\mu_{lm}}\left(z_i^{lm*}\right)^{k-1}\cdot\right.$$

$$\left.\cdot\left[\sum_{r=1}^{r=\nu_i^{lm*}} \frac{1}{(z_i^{lm*})^{r-1}(r-1)!}\frac{(k-1)!}{(k-r)!}R_{ir}^{lm}\right]\right| = \infty \Longrightarrow$$

$$\lim_{k\to\infty}\|\boldsymbol{F}(k)\| = \infty\ .$$

It follows that $\boldsymbol{F}(k)$ is not bounded, which contradicts (13.132). The contradiction is a consequence of (13.133) implying that (13.133) is incorrect. This proves necessity of *a*-1).

We continue with the method of contradiction. Let us suppose that the condition *a*-2) does not hold, i.e.,

$$\exists z_i^{lm*} = \left|z_i^{lm*}\right| e^{j\arg z_i^{lm*}} \in \mathcal{C} \Longrightarrow \left|z_i^{lm*}\right| = 1 \text{ and } \nu_i^{lm*} \geq 2. \tag{13.134}$$

Now, $\delta_d(k)R_0^{lm} = 0$ for $k \neq 0$, $\delta_d(0)R_0^{lm} = R_0^{lm}$, (13.130) and (13.134) imply:

$$\lim_{k\to\infty}\left\|\left(\left|z_i^{lm*}\right|e^{j\arg z_i^{lm*}}\right)^{k-1}\left[\sum_{r=1}^{r=\nu_i^{lm*}\geq 2}\frac{1}{(z_i^{lm*})^{r-1}(r-1)!}\frac{(k-1)!}{(k-r)!}R_{ir}^{lm}\right]\right\| =$$

$$= \lim_{k\to\infty}\left|\left(e^{j\arg z_i^{lm*}}\right)^{k-1}\right|\left\|\left[\sum_{r=1}^{r=\nu_i^{lm*}\geq 2}\frac{1}{(z_i^{lm*})^{r-1}(r-1)!}\frac{(k-1)!}{(k-r)!}R_{ir}^{lm}\right]\right\| =$$

$$= \lim_{k\to\infty}\left\|\left[\sum_{r=1}^{r=\nu_i^{lm*}\geq 2}\frac{1}{(z_i^{lm*})^{r-1}(r-1)!}\frac{(k-1)!}{(k-r)!}R_{ir}^{lm}\right]\right\| = \infty\ ,$$

$$\Longrightarrow$$

$$\lim_{k\to\infty}|F_{lm}(k)| = \lim_{k\to\infty}\left|\delta_d(k)R_0^{lm} + \right.$$

$$\left. + \sum_{i=1}^{i=\mu_{lm}}\left(z_i^{lm*}\right)^{k-1}\left[\sum_{r=1}^{r=\nu_i^{lm*}}\frac{1}{(z_i^{lm*})^{r-1}(r-1)!}\frac{(k-1)!}{(k-r)!}R_{ir}^{lm}\right]\right| = \infty \Longrightarrow$$

$$\lim_{k\to\infty}\|\boldsymbol{F}(k)\| = \infty.$$

It follows that $\boldsymbol{F}(k)$ is unbounded, which contradicts (13.132). The contradiction is a consequence of (13.134), which implies that (13.134) is not correct. This proves necessity of the condition a-2).

$|F_{lm}(k)|$, for any $l \in \{1,2,\cdots,p\}$ and $m \in \{1,2,\cdots,n\}$, may either contain or not discrete impulse component $\delta_d(k)$ for it is bounded and does not disrupt the boundedness of the norm $\|\boldsymbol{F}(k)\|$ of $\boldsymbol{F}(k)$. This proves necessity of a-3)

b) We keep on using the contradiction method. Let

$$\lim\left[\|\boldsymbol{F}(k)\| :\ k\to\infty\right] = 0 \tag{13.135}$$

be true and let us suppose that the condition b-1) does not hold. If (13.133) were valid, then $\boldsymbol{F}(k)$ would be unbounded as shown above in the proof of the necessity of a-1), which would contradict (13.135). If (13.134) were valid, then $\boldsymbol{F}(k)$ would be unbounded as shown above in the proof of the necessity of a-2), which would again contradict (13.135). If

$$\exists z_i^{lm*} = \left|z_i^{lm*}\right|e^{j\arg z_i^{lm*}} \in \mathcal{C} \Longrightarrow \left|z_i^{lm*}\right| = 1 \text{ and } \nu_i^{lm*} = 1 \tag{13.136}$$

then

$$\lim_{k\to\infty} |F_{lm}(k)| = \lim_{k\to\infty} \left| \delta_d(k) R_0^{lm} + \sum_{i=1}^{i=\mu_{lm}} \left(z_i^{lm*}\right)^{k-1} \cdot \right.$$

$$\left. \cdot \left[\sum_{r=1}^{r=\nu_i^{lm*}=1} \frac{1}{(z_i^{lm*})^{r-1}(r-1)!} \frac{(k-1)!}{(k-r)!} R_{ir}^{lm} \right] \right| \geq$$

$$\geq \lim_{k\to\infty} \left| \left(e^{j \arg z_i^{lm*}}\right)^{k-1} \right| \left| \left[R_{ir}^{lm} \right] \right| \in \mathcal{R}^+ ,$$

which would also contradict (13.135). Altogether, the validity of (13.135) proves the validity of the condition *b*-1).

$|F_{lm}(k)|$, for any $l \in \{1, 2, \cdots, p\}$ and $m \in \{1, 2, \cdots, n\}$, may either contain or not discrete impulse component $\delta_d(k)$ whose value is bounded and different from zero only for $k = 0$, $\delta_d(0) = 1 \neq 0$, such that it does not influence the limit value $\lim_{k\to\infty} |F_{lm}(k)|$ of $|F_{lm}(k)|$. This proves necessity of *b*-2).

Sufficiency. *a*) Let the conditions under *a*) hold. Then $\boldsymbol{F}(z)$ is either proper or strictly proper. Hence,

$$R_0^{lm} = 0 \vee R_0^{lm} \neq 0,\ \forall l \in \{1, 2, \cdots, p\},\ \forall m \in \{1, 2, \cdots, n\}. \tag{13.137}$$

We recall the following facts:

1) If (13.136) holds, then

$$\lim_{k\to\infty} |F_{lm}(k)| \in \mathcal{R}^+. \tag{13.138}$$

2) If $\left|z_i^{lm*}\right| < 1$, then

$$\lim_{k\to\infty} |F_{lm}(k)| = \lim_{k\to\infty} \left| \delta_d(k) R_0^{lm} + \sum_{i=1}^{i=\mu_{lm}} \left(z_i^{lm*}\right)^{k-1} \cdot \right.$$

$$\left. \cdot \left[\sum_{r=1}^{r=\nu_i^{lm*}} \frac{1}{(z_i^{lm*})^{r-1}(r-1)!} \frac{(k-1)!}{(k-r)!} R_{ir}^{lm} \right] \right|$$

$$\lim_{k\to\infty} \left| \left(z_i^{lm*}\right)^{k-1} \left[\sum_{r=1}^{r=\nu_i^{lm*}} \frac{1}{(z_i^{lm*})^{r-1}(r-1)!} \frac{(k-1)!}{(k-r)!} R_{ir}^{lm} \right] \right| =$$

$$= \lim_{k\to\infty} \left| \left|z_i^{lm*}\right|^{k-1} \left[\sum_{r=1}^{r=\nu_k^{im*}} \frac{1}{(r-1)!} t^{r-1} R_{kr}^{im} \right] \right| = 0. \tag{13.139}$$

3) The results (13.137) through (13.139) prove boundedness of $\|\boldsymbol{F}(k)\|$, i.e.,

$$\exists \alpha \in \mathcal{R}^{+} \Longrightarrow \|\boldsymbol{F}(k)\| < \alpha, \ \forall k \in \mathcal{N}_0.$$

b) Let the condition under b) hold. Now, $\left|z_i^{lm*}\right| < 1$, $\forall l \in \{1, 2, \cdots, p\}$, $\forall m \in \{1, 2, \cdots, n\}$, so that (13.139) holds $\forall l \in \{1, 2, \cdots, p\}$, $\forall m \in \{1, 2, \cdots, n\}$, which proves that $\|\boldsymbol{F}(k)\|$ vanishes asymptotically, i.e.,

$$\lim \left[\|\boldsymbol{F}(k)\| : \ k \to \infty\right] = 0.$$

This completes the proof. ■

Comment 13.5 ***Importance of the Generating theorem***

Qualitative system properties (e.g., controllability, observability, optimality, stability, trackability) concern families of dynamical behaviors of a dynamical system, which are caused by sets of initial conditions and/or by sets of external actions. They take place in time. Their definitions are given in the time domain. It is impractical in the framework of linear systems (practically impossible in the framework of nonlinear systems) to use their definitions directly in order to test whether a given system possesses a requested qualitative dynamical property. It is preferable to establish conditions and criteria for them in the algebraic and/or in the complex domain, which enable us to test them without knowing individual system behaviors, i.e., without solving system mathematical models for every initial condition and for every external action. The Generating Theorem 13.39 is the basis to establish such conditions and criteria in the complex domain for stability properties of the discrete-time time-invariant linear dynamical systems.

13.4.2 *IO* systems

We say usually that the modulus less than one of all poles of the transfer function matrix $\boldsymbol{G}_{IO}(z)$ of the IO system (3.56) (in Subsection 3.5.2), is necessary and sufficient for asymptotic stability of the zero equilibrium vector. The following counterexample illustrates the need to refine this statement.

Example 13.1 *The SISO IO system of Example 8.1 (Section 8.1),*

$$E^2 y(k) - E^1 y(k) - 0.75 E^0 y(k) = E^2 i(k) - 7.5 E^1 i(k) + 9 E^0 i(k)$$

has the following transfer function $G_{IO}(z)$:

$$G_{IO}(z) = \frac{z^2 - 7.5z + 9}{z^2 - z - 0.75} = \frac{(z - 1.5)(z - 6)}{(z - 1.5)(z + 0.5)}.$$

It has the same zero z^0 *and pol* z^*, $z^0 = z^* = 1.5$. *They do not influence the impulse response of the system under all zero initial conditions. The nondegenerate form* $G_{IOnd}(z)$ *of* $G_{IO}(z)$ *reads:*

$$G_{IOnd}(z) = \frac{z-6}{z+0.5}.$$

It has only one pole $z^* = -0.5 < 0$. *It is real and with modulus less than one. If we used it to conclude about the stability properties of the system, then we would conclude (wrongly) that its zero state is asymptotically stable equilibrium. Let us verify this. The system discrete equation yields the system transfer function matrix relative to* $\mathbf{y}_0^1$*:*

$$\boldsymbol{G}_{IOyo}(z) = \left[\begin{array}{cc} \frac{z(z-1)}{(z-1.5)(z+0.5)} & \frac{z}{(z-1.5)(z+0.5)} \end{array} \right],$$

so that in the free regime

$$\mathbf{y}^1(k;\mathbf{y}_0^1;0) = \frac{1}{4}\left[\begin{array}{cc} 1.5^k + 3\left(-0.5\right)^k & 3\left(1.5\right)^k + \left(-0.5\right)^k \\ 1.5^{k+1} + 3\left(-0.5\right)^{k+1} & 3\left(1.5\right)^{k+1} + \left(-0.5\right)^{k+1} \end{array} \right] \mathbf{y}_0^1 \Longrightarrow$$
$$\mathbf{y}_0^1 \neq \mathbf{0}_2 \Longrightarrow \left\|\mathbf{y}^1(t;\mathbf{y}^1;0)\right\| \rightharpoonup \infty \text{ as } k \rightharpoonup \infty.$$

The zero equilibrium state is unstable. If we concluded that it is asymptotically stable by referring to the poles of the nondegenerate form $G_{IOnd}(z)$ *of* $G_{IO}(z)$, *then we would make a cardinal mistake. It would be the consequence of using the nondegenerate form* $G_{IOnd}(z)$ *of* $G_{IO}(z)$ *without testing the system full transfer function matrix* $\boldsymbol{F}_{IO}(z)$ *for the cancellation of the pole* $z^* = 1.5$ *with modulus greater than one.*

The system full transfer function matrix $\boldsymbol{F}_{IO}(z)$ *(Example 10.1, in Section 10.1),*

$$\boldsymbol{F}_{IO}(z) = \frac{1}{(z-1.5)(z+0.5)}\cdot$$
$$\cdot\left[\begin{array}{ccccc} (z-1.5)\,(z-6) & -z\,(z-7.5) & -z & z\,(z-1) & z \end{array} \right]$$

shows that it is nondegenerate. The cancellation of $(z-1.5)$ *in the numerator and in the denominator is neither possible in all entries of* $\boldsymbol{F}_{IO}(z)$ *nor in all entries of any row of* $\boldsymbol{G}_{IOyo}(z)$, *which means in the single row of* $\boldsymbol{G}_{IOyo}(z)$ *because it is* 1×2 *matrix, i.e., the row vector,*

$$\boldsymbol{G}_{IOy0}(z) = \left[\begin{array}{cc} \frac{z(z-1)}{(z-1.5)(z+0.5)} & \frac{z}{(z-1.5)(z+0.5)} \end{array} \right],$$

so that $(z-1.5)$ *may not be cancelled in* $G_{IO}(z)$ *when it is to be used for the Lyapunov stability test.*

Comment 13.6 *There was discussion in [11] on the problem of the correctness of the pole-zero cancellation in the system transfer function related to the internal, i.e., Lyapunov stability of the system. The preceding example (Example 13.1) gives the clear, correct and unique explanation of the problem, the answer and the solution to the problem, related to the discrete-time systems. The pole-zero cancellation from Lyapunov stability point of view is permitted only in the IO system transfer function (matrix)* $\boldsymbol{G}_{IOyo}(z)$, *and from the point of view of the IO system complete response it is allowed only in the system full transfer function matrix* $\boldsymbol{F}_{IO}(z)$.

Lemma 13.1 *In order for* $\mathbf{y}^{\nu-1}(k;\mathbf{y}_0^{\nu-1};\mathbf{0}_M)$ *of the IO system (3.56) in the free regime*

a) to be bounded on $\mathcal{N}_0$,

$$\exists\zeta \in \mathcal{R}^+ \Longrightarrow \left\|\mathbf{y}^{\nu-1}(k;\mathbf{y}_0^{\nu-1};\mathbf{0}_M)\right\| < \zeta,\ \forall k \in \mathcal{N}_0,$$

it is necessary and sufficient that $\mathbf{y}(k;\mathbf{y}_0^{\nu-1};\mathbf{0}_M)$ *is bounded on* $\mathcal{N}_0$,

$$\exists\xi \in \mathcal{R}^+ \Longrightarrow \left\|\mathbf{y}(k;\mathbf{y}_0^{\nu-1};\mathbf{0}_M)\right\| < \xi,\ \forall k \in \mathcal{N}_0,$$

b) to vanish as $k \to \infty$,

$$k \to \infty \Longrightarrow \mathbf{y}^{\nu-1}(k;\mathbf{y}_0^{\nu-1};\mathbf{0}_M) \to \mathbf{0}_{\nu N},$$

it is necessary and sufficient that $\mathbf{y}(k;0;\mathbf{y}_0^{\nu-1})$ *vanishes as* $k \to \infty$,

$$k \to \infty \Longrightarrow \mathbf{y}(k;\mathbf{y}_0^{\nu-1};\mathbf{0}_M) \to \mathbf{0}_N.$$

Proof. *Necessity. a)* In order for $\mathbf{y}^{\nu-1}(k;\mathbf{y}_0^{\nu-1};\mathbf{0}_M)$ to be bounded on $\mathcal{N}_0$, it is necessary that every entry is bounded on $\mathcal{N}_0$. Hence, it is necessary that $\mathbf{y}(k;\mathbf{y}_0^{\nu-1};\mathbf{0}_M)$ is bounded on $\mathcal{N}_0$.

b) In order for $\mathbf{y}^{\nu-1}(k;\mathbf{y}_0^{\nu-1};\mathbf{0}_M)$ to vanish as $k \to \infty$, it is necessary that every entry of $\mathbf{y}^{\nu-1}(k;\mathbf{y}_0^{\nu-1};\mathbf{0}_M)$ vanishes as $k \to \infty$. Hence, it is necessary that $\mathbf{y}(k;\mathbf{y}_0^{\nu-1};\mathbf{0}_M)$ vanishes as $k \to \infty$.

Sufficiency. a) Let the condition of the statement of lemma under *a)* be valid. This, (3.56) in the free regime, the property of the $Z-$transform and the Generating Theorem 13.39 (in Subsection 13.4.1) yield

$$\mathbf{y}(k;\mathbf{y}_0^{\nu-1};\mathbf{0}_M) =$$

$$= \left\{ \delta_d(k)\boldsymbol{R}_0 + \sum_{i=1}^{i=\nu N} (z_i^*)^{k-1} \left[\sum_{r=1}^{r=\nu_i^*} \frac{1}{(z_i^*)^{r-1}\,(r-1)!} \frac{(k-1)!}{(k-r)!} \boldsymbol{R}_{ir} \right] \right\} \mathbf{y}_0^{\nu-1},$$

$$\boldsymbol{R}_0 \in \mathcal{R}^{N\times\nu N},\ \boldsymbol{R}_{i_r} \in \mathcal{C}^{N\times\nu N},\ |z_i^*| \leq 1,$$

$$\forall i = 1,2,\cdots,\nu N,\ |z_i^*| = 1 \Longrightarrow \nu_i^* = 1, \qquad (13.140)$$

where ν_i^* is the multiplicity of z_i^*. This shows that $\mathbf{y}(k;\mathbf{y}_0^{\nu-1};\mathbf{0}_M)$ is infinitely times shiftable at every $k \in \mathbb{Z}$. All its shifts rest bounded on $\mathcal{N}_0$. This, and the Generating Theorem 13.39 guarantee that $\mathbf{y}^{\nu-1}(k;\mathbf{y}_0^{\nu-1};\mathbf{0}_M)$ is also bounded on $\mathcal{N}_0$.

b) Let the condition of the statement of lemma under *b)* hold. This, (3.56) in the free regime, the properties of the $Z-$transform and (13.140) imply

$$\mathbf{y}(k;\mathbf{y}_0^{\nu-1};\mathbf{0}_M) =$$

$$= \left\{ \delta_d(k)\boldsymbol{R}_0 + \sum_{i=1}^{i=\nu N} (z_i^*)^{k-1} \left[\sum_{r=1}^{r=\nu_i^*} \frac{1}{(z_i^*)^{r-1}\,(r-1)!} \frac{(k-1)!}{(k-r)!} \boldsymbol{R}_{ir} \right] \right\} \mathbf{y}_0^{\nu-1}.$$

This shows that $\mathbf{y}(k;\mathbf{y}_0^{\nu-1};\mathbf{0}_M)$ is infinitely times shiftable at every $k \in \mathbb{Z}$. All its shifts rest bounded on $\mathcal{N}_0$ and vanish as $k \to \infty$ due to the Generating Theorem 13.39. Every entry of $\mathbf{y}^{\nu-1}(k;\mathbf{y}_0^{\nu-1};\mathbf{0}_M)$ is bounded and vanishes as $k \to \infty$. ■

Example 13.2 *Let us consider the second-order IO system described by*

$$E^2 y(k) - 0.8E^1 y(k) + 0.15E^0 y(k) = E^2 i(k) - 7.5E^1 i(k) + 9E^0 i(k).$$

The system discrete equation yields

$$\boldsymbol{G}_{IOy_0}(z) = \left[\begin{array}{cc} \frac{z(z-0.8)}{(z-0.3)(z-0.5)} & \frac{z}{(z-0.3)(z-0.5)} \end{array} \right] \Longrightarrow$$
$$Y(z) = Y\left(z;\mathbf{y}_0^1;0\right) = \boldsymbol{G}_{IOy_0}(z)\mathbf{y}_0^1,$$

$\boldsymbol{G}_{IOy_0}(z)$ *is proper, and so is* $\boldsymbol{F}_{IO}(z)$ *and* $\boldsymbol{G}_{IO}(z)$,

$$\boldsymbol{F}_{IO}(z) = \frac{\left[\begin{array}{ccccc} (z-1.5)\,(z-6) & -z\,(z-7.5) & -z & z\,(z-0.8) & z \end{array}\right]}{(z-0.3)(z-0.5)} \Longrightarrow$$

$$\boldsymbol{G}_{IO}(z) = \frac{(z-1.5)\,(z-6)}{(z-0.3)(z-0.5)} = 1 - \frac{34.2}{z-0.3} + \frac{27.5}{z-0.5}.$$

The unit discrete impulse response $y(k;\mathbf{0}_2;\delta_d)$ *is the inverse* $Z-$*transform of* $\boldsymbol{G}_{IO}(z)$,

$$y(k;\mathbf{0}_2;\delta_d) = \mathcal{Z}^{-1}\left\{\boldsymbol{G}_{IO}(z)\right\} = \delta_d(k) - 34.2\,(0.3)^{k-1} + 27.5\,(0.5)^{k-1} \Longrightarrow$$
$$\Longrightarrow y(0;\mathbf{0}_2;\delta_d) = 1.$$

It is bounded at the initial moment due to its discrete unit impulse component $\delta_d(k)$, so is at any other moment. We conclude about stability of the zero equilibrium state by testing $\mathbf{G}_{IOy_0}(z)$ which is proper and the modulus of all its poles are less than one, i.e., 0.3 and 0.5. Hence, the zero equilibrium state is globally asymptotically stable. Let us verify this in the time domain. The time evolutions of $y(k;\mathbf{y}^1;0)$ and $y^{(1)}(k;\mathbf{y}^1;0)$ describe completely the system output behavior in the free regime

$$y(k;\mathbf{y}_0^1;0) = \left[2.5\,(0.3)^k - 1.5\,(0.5)^k\right] y_0 \;+\; 5\left[-\,(0.3)^k + (0.5)^k\right] y_0^{(1)}\,,$$
$$y(0;\mathbf{y}_0^1;0) = y_0.$$

The output response $y(k;\mathbf{y}_0^1;0)$ is bounded on $\mathcal{N}_0$ for every bounded $\mathbf{y}_0^1$, and vanishes as $k \to \infty$. We determine easily its first shift as

$$y^{(1)}(k;\mathbf{y}_0^1;0) = \left[2.5\,(0.3)^{k+1} - 1.5\,(0.5)^{k+1}\right] y_0 \;+$$
$$+\,5\left[-\,(0.3)^{k+1} + (0.5)^{k+1}\right] y_0^{(1)},\; y^{(1)}(0;\mathbf{y}_0^1;0) = y_0^{(1)}.$$

Altogether, $\mathbf{y}^1(k;\mathbf{y}_0^1;0)$,

$$y^1(k;0;y_0^1) = \begin{bmatrix} 2.5\,(0.3)^k - 1.5\,(0.5)^k & -\,5\,(0.3)^k + 5\,(0.5)^k \\ 2.5\,(0.3)^{k+1} - 1.5\,(0.5)^{k+1} & -\,5\,(0.3)^{k+1} + 5\,(0.5)^{k+1} \end{bmatrix} y_0^1,$$
$$\mathbf{y}^1(0;\mathbf{y}_0^1;0) = \mathbf{y}_0^1,$$

is also bounded on $\mathcal{N}_0$ for every bounded $\mathbf{y}_0^1$, and vanishes as $k \to \infty$.

Note 13.13 *Equation (13.140), $|z_i^*| < 1$, $\forall k = 1, 2, ..., p = \nu N$, together with its shifts, and (13.14) (in the proof of sufficiency of Theorem 13.8, in Subsection 13.2.1) imply the existence of positive numbers α and β such that*

$$\left\|\mathbf{y}^{\nu-1}(k;\mathbf{y}_0^{\nu-1};\mathbf{0}_M)\right\| \leq$$
$$\leq \alpha \exp\left(-\beta k\right) \left\|\mathbf{y}_0^{\nu-1}\right\|,\; \forall k \in \mathcal{N}_0,\; \forall \mathbf{y}_0^{\nu-1} \in \mathcal{R}^{\nu N}. \tag{13.141}$$

Lemma 13.2 *As $\mathbf{G}_{IOy_0}(z)$ is proper, in order to test stability, asymptotic stability or exponential stability of the zero equilibrium vector $\mathbf{y}_e^{\nu-1} = \mathbf{0}_{\nu N}$ of the IO system (3.56) via the transfer function matrices of the system, it is necessary and sufficient to use the denominator polynomial $f_{IOy_0rnd}(z)$ of the row nondegenerate form $\mathbf{G}_{IOy_0rnd}(z)$ of the submatrix $\mathbf{G}_{IOy_0}(z)$ of $\mathbf{F}_{IO}(z)$.*

Proof. The output response of the IO system (3.56) in the free regime on $\mathcal{N}_0$ is determined, due to Lemma 6.1 (in Chapter 6) and Lemma 13.1, by

$$\mathbf{y}(k;\mathbf{y}_0^{\nu-1};\mathbf{0}_M) = \mathcal{Z}^{-1}\left\{\boldsymbol{G}_{IOy_{0rnd}}(z)\right\}\mathbf{y}_0^{\nu-1},\ \forall k \in \mathcal{N}_0. \tag{13.142}$$

This implies, since $\boldsymbol{G}_{IOy_0}(z)$, hence $\boldsymbol{G}_{IOy_0rnd}(z)$, is proper, that the zeros of the denominator polynomial $f_{IOy_0rnd}(z) = \det\left[\text{Den}\ \boldsymbol{G}_{IOy_0rnd}(z)\right]$ of the row nondegenerate form $\boldsymbol{G}_{IOy_0rnd}(z)$ of the submatrix $\boldsymbol{G}_{IOy_0}(z)$ of $\boldsymbol{F}_{IO}(z)$, which are the poles of $\boldsymbol{G}_{IOy_0rnd}(z)$, determine completely the character of $\mathbf{y}(k;\mathbf{y}_0^{\nu-1};\mathbf{0}_M)$. ■

The preceding results enable the simple form of the proof of the conditions for stability and for asymptotic stability of the equilibrium vector.

Theorem 13.40 *In order for the zero equilibrium vector $\mathbf{y}_e^{\nu-1} = \mathbf{0}_{\nu N}$ of the IO system (3.56) to be*

a) stable, it is necessary and sufficient that

1. *$\mathbf{G}_{IOy_0}(z)$ is either proper or strictly proper, real rational matrix function of z,*

   and that

2. *the modulus of all the poles of the row nondegenerate form $\mathbf{G}_{IOy_0rnd}(z)$ of $\mathbf{G}_{IOy_0}(z)$ are less or equal to one, and the multiplicity of its poles with modulus equal to one should be equal to one (i.e., the poles should be simple);*

b) asymptotically stable it is necessary and sufficient that

1. *$\mathbf{G}_{IOy_0}(z)$ is either proper or strictly proper, real rational matrix function of z,*

   and that

2. *the modulus of all poles of the row nondegenerate form $\mathbf{G}_{IOy_0rnd}(z)$ of $\mathbf{G}_{IOy_0}(z)$ are less than one.*

Proof. Lemma 6.1, the Generating Theorem 13.39, Lemma 13.1 and Lemma 13.2 prove this theorem. ■

Note 13.14 *This theorem discovers that the competent transfer function matrix of the IO system (3.56) for the test of a Lyapunov stability property is the row nondegenerate form $\mathbf{G}_{IOy_0rnd}(z)$ of $\mathbf{G}_{IOy_0}(z)$, and not the system transfer function matrix $\mathbf{G}_{IO}(z)$. This confirms the statement in Comment 13.6.*

Corollary 13.1 *In order for the cancellation of the same zero and pole of the equal order and with modulus grater or equal to one in the transfer function matrix $\boldsymbol{G}_{IO}(z)$ of the IO system (3.56) to be without any influence on the results on any Lyapunov stability property of the equilibrium vector, it is necessary and sufficient that their same cancellation is possible in all rows of the transfer function matrix $\boldsymbol{G}_{IOy0}(z)$ of the system with respect to $\mathbf{y}_0^{\nu-1}$.*

The matrix A_{IO} (13.117) permits the transformation of the IO system (13.111) in the free regime, i.e., (13.112), into its ISO equivalent system (13.118) (in Section 13.3).

Lemma 13.3 *The eigenvalues of the matrix A_{IO} (13.117) of the IO system (13.111) are the poles of the system transfer function matrix $\boldsymbol{G}_{IOy0}(z)$, and vice versa.*

Proof. Equation (8.10) (in Section 8.1), together with (2.2a) and (2.2b) (in Section 2.1), (8.4), and (8.5) (in Section 8.1), show that the denominator polynomial $f_{IOyo}(z)$ of $\boldsymbol{G}_{IOy0}(z)$ is given by

$$f_{IOyo}(z) = \det\left(\boldsymbol{A}^{(\nu)} S_N^{(\nu)}(z)\right) = \det\left(\sum_{r=0}^{r=\nu} z^r \boldsymbol{A}_r\right). \tag{13.143}$$

Its zeros are the poles of $\boldsymbol{G}_{IOy0}(z)$.

The characteristic polynomial $f_{IOA}(z)$ of the matrix $\boldsymbol{A}_{IO}$ reads [7, Th. 1-10-1, p. 10]

$$f_{IOA}(z) = \det\left(z\boldsymbol{I}_{\nu N} - \boldsymbol{A}_{IO}\right) =$$

$$= \det \begin{bmatrix} z\boldsymbol{I}_N & -\boldsymbol{I}_N & \cdots & \boldsymbol{O}_N & \boldsymbol{O}_N \\ \boldsymbol{O}_N & z\boldsymbol{I}_N & \cdots & \boldsymbol{O}_N & \boldsymbol{O}_N \\ \vdots & \vdots & \vdots & \vdots & \vdots \\ \boldsymbol{O}_N & \boldsymbol{O}_N & \cdots & z\boldsymbol{I}_N & -\boldsymbol{I}_N \\ \boldsymbol{A}_\nu^{-1}\boldsymbol{A}_0 & \boldsymbol{A}_\nu^{-1}\boldsymbol{A}_1 & \cdots & \boldsymbol{A}_\nu^{-1}\boldsymbol{A}_{\nu-2} & z\boldsymbol{I}_N + \boldsymbol{A}_\nu^{-1}\boldsymbol{A}_{\nu-1} \end{bmatrix} =$$

$$= \det\begin{bmatrix} \boldsymbol{L} & \boldsymbol{M} \\ \boldsymbol{N} & \boldsymbol{P} \end{bmatrix} = \det \boldsymbol{L} \det\left(\boldsymbol{P} - \boldsymbol{N}\boldsymbol{L}^{-1}\boldsymbol{M}\right),$$

$$\boldsymbol{L} = \begin{bmatrix} z\boldsymbol{I}_N & -\boldsymbol{I}_N & \cdots & \boldsymbol{O}_N \\ \boldsymbol{O}_N & z\boldsymbol{I}_N & \cdots & \boldsymbol{O}_N \\ \vdots & \vdots & \vdots & \vdots \\ \boldsymbol{O}_N & \boldsymbol{O}_N & \cdots & z\boldsymbol{I}_N \end{bmatrix} \in \mathcal{C}^{N(\nu-1)\times N(\nu-1)},$$

$$M = \begin{bmatrix} O_N \\ O_N \\ \vdots \\ -I_N \end{bmatrix} \in \mathcal{R}^{N(\nu-1)\times N}, \ N = \begin{bmatrix} \left(A_\nu^{-1}A_0\right)^T \\ \left(A_\nu^{-1}A_1\right)^T \\ \vdots \\ \left(A_\nu^{-1}A_{\nu-2}\right)^T \end{bmatrix}^T \in \mathcal{R}^{N\times N(\nu-1)},$$

$$P = zI_N + A_\nu^{-1}A_{\nu-1} \in \mathcal{C}^{N\times N}$$

$$f_{IOA}(z) = z^{(\nu-1)N}\cdot$$

$$\cdot \det\left\{\left(zI_N + A_\nu^{-1}A_{\nu-1}\right) + A_\nu^{-1}\left(\frac{1}{z^{\nu-1}}A_0 + \frac{1}{z^{\nu-2}}A_1 + \cdots + \frac{1}{z}A_{\nu-2}\right)\right\} =$$

$$= \det\left(z^\nu I_N + A_\nu^{-1}\sum_{r=0}^{r=\nu-1} z^r A_r\right) = \det A_\nu^{-1} \det\left(\sum_{r=0}^{r=\nu} z^r A_r\right).$$

The eigenvalues of the matrix A_{IO} are the zeros of $f_{IOA}(z)$, which are the zeros of

$$\det\left(\sum_{r=0}^{r=\nu} z^r A_r\right) = f_{IOyo}(z),$$

and vice versa. This, and (13.143) prove that the eigenvalues of the matrix A_{IO} (13.117) of the IO system (13.111) are the poles of the system transfer function matrix $G_{IOy_0}(z)$, and vice versa. ∎

Theorem 13.41 ***The Lyapunov matrix theorem for the IO system (3.56)***

In order for the zero equilibrium vector $\mathbf{y}_e^{\nu-1} = \mathbf{0}_{\nu N}$ *of the IO system (3.56) to be asymptotically stable, equivalently for the matrix* A_{IO} *(13.117) to be stable matrix, it is necessary and sufficient that for any positive definite symmetric matrix* G_{IO}, $G_{IO} = G_{IO}^T \in \mathcal{R}^{\nu N\times\nu N}$, *the matrix solution* H_{IO} *of the Lyapunov matrix equation*

$$A_{IO}^T H_{IO} A_{IO} - H_{IO} = -G_{IO} \tag{13.144}$$

is also positive definite symmetric matrix and the unique solution to (13.144).

Proof. Theorem 13.40, and Lemma 13.3 show the equivalence between asymptotic stability of the zero equilibrium vector $\mathbf{y}_e^{\nu-1} = \mathbf{0}_{\nu N}$ of the IO system (3.56) and stability of the matrix A_{IO}, (13.117) (in Section 13.3). We apply the replacements determined by (13.110) (in Section 13.3), to Theorem 13.31 that then becomes this theorem. ∎

Note 13.15 *This theorem, called* **the Lyapunov matrix theorem,** *is the fundamental matrix theorem for stability of the IO system (3.56). It enables the following estimate of* $\left\|\mathbf{y}^{\nu-1}(k;\mathbf{0}_M;\mathbf{y}_0^{\nu-1})\right\|$ *of* $\mathbf{y}^{\nu-1}(k;\mathbf{0}_M;\mathbf{y}_0^{\nu-1})$ *of the IO system (3.56) due to (13.110), (13.114), (13.117) and (13.118) (in Section 13.3) and (13.144):*

$$v(\mathbf{y}^{\nu-1}) = \left(\mathbf{y}^{\nu-1}\right)^T \boldsymbol{H}_{IO}\mathbf{y}^{\nu-1} \Longrightarrow \Delta v_{IO}\left(\mathbf{y}^{\nu-1}\right) = -\left(\mathbf{y}^{\nu-1}\right)^T \boldsymbol{G}_{IO}\mathbf{y}^{\nu-1} \Longrightarrow$$
$$\Delta v_{IO}\left(\mathbf{y}^{\nu-1}\right) \leq -\lambda_m(\boldsymbol{G}_{IO})\lambda_M^{-1}(\boldsymbol{H}_{IO})v_{IO}\left(\mathbf{y}^{\nu-1}\right) \Longrightarrow$$

$$v_{IO}\left(\mathbf{y}^{\nu-1}\right) \leq \exp\left(-\beta k\right) v_{IO}\left(\mathbf{y}_0^{\nu-1}\right),\ \beta = \ln\frac{\lambda_M\left(\boldsymbol{H}_{IO}\right)}{\lambda_M\left(\boldsymbol{H}_{IO}\right) - \lambda_m\left(\boldsymbol{G}_{IO}\right)},$$
$$\lambda_M\left(\boldsymbol{H}_{IO}\right) > \lambda_m\left(\boldsymbol{G}_{IO}\right) \Longrightarrow$$

$$\lambda_m(\boldsymbol{H}_{IO})\left\|\mathbf{y}^{\nu-1}(k)\right\|^2 \leq \lambda_M(\boldsymbol{H}_{IO})e^{-\beta k}\left\|\mathbf{y}_0^{\nu-1}\right\|^2,$$
$$\forall k \in \mathbb{Z},\ \forall \mathbf{y}_0^{\nu-1} \in \mathcal{R}^{\nu N} \Longrightarrow$$

$$\left\|\mathbf{y}^{\nu-1}(k;\mathbf{0}_M;\mathbf{y}_0^{\nu-1})\right\| \leq \alpha e^{-\gamma k}\left\|\mathbf{y}_0^{\nu-1}\right\|,\ \forall k \in \mathcal{N}_0,\ \forall \mathbf{y}_0^{\nu-1} \in \mathcal{R}^{\nu N},$$
$$\alpha = \left[\lambda_m^{-1/2}(\boldsymbol{H}_{IO})\lambda_M^{1/2}(\boldsymbol{H}_{IO})\right],\ \gamma = \frac{\beta}{2} = \frac{1}{2}\ln\frac{\lambda_M\left(\boldsymbol{H}_{IO}\right)}{\lambda_M\left(\boldsymbol{H}_{IO}\right) - \lambda_m\left(\boldsymbol{G}_{IO}\right)}.$$

Comment 13.7 ***The physical meaning of the Lyapunov matrix theorem for the IO system (3.56)***

Let the system power $p(\cdot)$ *be a negative definite quadratic form,* $p(\mathbf{w}) = \mathbf{w}^T\boldsymbol{P}\mathbf{w}$, $\boldsymbol{P} = \boldsymbol{P}^T < \boldsymbol{O}$. *It has the properties determined in Comment 13.4 (in Section 13.3). Equations (13.89) of Comment 13.4 imply*

$$\boldsymbol{A}_w^T\boldsymbol{E}\boldsymbol{A}_w - \boldsymbol{E} = \boldsymbol{P},\ \boldsymbol{P} = \boldsymbol{P}^T < \boldsymbol{O}.$$

In order for the zero equilibrium state of the system to be asymptotically (hence, exponentially) stable, it is both necessary and sufficient that the system energy $e(\cdot)$ *is positive definite quadratic form,* $e(\mathbf{w}) = \mathbf{w}^T\boldsymbol{E}\mathbf{w}$ *with* $\boldsymbol{E} = \boldsymbol{E}^T > \boldsymbol{O}$.

13.4.3 *ISO* systems

The motion $\mathbf{x}(\cdot;\mathbf{x}_0;\mathbf{0}_M)$ of the *ISO* system (3.60a) and (3.60b) (in Subsection 3.5.3), in the free regime is determined by the state transition matrix $\boldsymbol{A}$ and the initial state vector $\mathbf{x}_0$, (10.14) and (10.16) (in Section 10.2),

$$\mathbf{x}(k;\mathbf{x}_0;\mathbf{0}_M) = \boldsymbol{A}^k\mathbf{x}_0. \tag{13.145}$$

The well known Lyapunov stability conditions are naturally expressed in terms of the eigenvalues of the matrix $\boldsymbol{A}$. They are in the complex domain usually stated in terms of the poles of the system transfer function matrix $\boldsymbol{G}_{ISO}(z)$. If it is obtained from the state-space system description (3.60a) and (3.60b), then it reads

$$\boldsymbol{G}_{ISO}(z) = \left[\det(z\boldsymbol{I}_n - \boldsymbol{A})\right]^{-1} \left[\boldsymbol{C}\,\mathrm{adj}(z\boldsymbol{I}_n - \boldsymbol{A})\boldsymbol{B} + \boldsymbol{D}\det(z\boldsymbol{I}_n - \boldsymbol{A})\right].$$

Its denominator polynomial $f_{ISO}(z) = \det(z\boldsymbol{I}_n - \boldsymbol{A})$ is the characteristic polynomial of the matrix $\boldsymbol{A}$, if $\boldsymbol{G}_{ISO}(z)$ is nondegenerate matrix. However, if it is degenerate matrix and the same poles and zeros were cancelled, and if among them there were poles with modulus greater or equal to one, then we should not use such nondegenerate form of $\boldsymbol{G}_{ISO}(z)$ for the Lyapunov stability tests.

Example 13.3 *We analyze the ISO system similar to the one given in Example 8.9 in Section 8.2:*

$$E^1\mathbf{x} = \underbrace{\begin{bmatrix} 0 & 1 & 0 \\ 1.45 & 0 & 2 \\ 1.02 & 0 & 1.2 \end{bmatrix}}_{\boldsymbol{A}} \mathbf{x} + \begin{bmatrix} 0 \\ 0 \\ 0.5 \end{bmatrix} i,$$

$$y = \begin{bmatrix} -2 & 1 & 0 \end{bmatrix} \mathbf{x}.$$

The eigenvalues of the matrix $\boldsymbol{A}$ are

$$\lambda_1 = -0.5, \ \lambda_2 = 2, \ \lambda_3 = -0.3.$$

Since $\lambda_2 = 2$ is positive real eigenvalue of the matrix $\boldsymbol{A}$ with modulus greater than one, then we conclude that the equilibrium state $\mathbf{x}_e = \mathbf{0}_n$ of the system is unstable. Its transfer function $\boldsymbol{G}_{ISO}(z)$,

$$\boldsymbol{G}_{ISO}(z) = \frac{z-2}{z^3 - 1.2z^2 - 1.45z - 0.3} = \frac{z-2}{(z+0.5)(z-2)(z+0.5)},$$

is degenerate. If we used its nondegenerate form $\boldsymbol{G}_{ISOnd}(z)$,

$$\boldsymbol{G}_{ISOnd}(z) = \frac{1}{(z+0.5)(z+0.3)},$$

for Lyapunov stability test, then we would conclude wrongly that the equilibrium state $\mathbf{x}_e = \mathbf{0}_n$ of the system is asymptotically stable because both poles $z_1^* = -0.5$ and $z_2^* = -0.3$ of $\boldsymbol{G}_{ISOnd}(z)$ are with modulus less than one.

Besides, either the transfer function matrix in its original form $\boldsymbol{G}_{ISO}(z)$, or in its nondegenerate form $\boldsymbol{G}_{ISOnd}(z)$, does not determine the system motion in the free regime. They are not adequate for the Lyapunov stability tests.

The system transfer function matrix $\boldsymbol{G}_{ISOSS}(z)$ with respect to the initial state determines the $Z-$transform of the system motion in the free regime, (7.26) (in Section 7.3),

$$\mathbf{X}(z) = \boldsymbol{G}_{ISOSS}(z)\mathbf{x}_0, \ \mathbf{i}(t) \equiv \mathbf{0}_M, \tag{13.146}$$

where $\boldsymbol{G}_{ISOSS}(z)$ is in fact the system resolvent matrix multiplied by z, $z(z\boldsymbol{I}_n - \boldsymbol{A})^{-1}$, (8.37), and (8.2),

$$\boldsymbol{G}_{ISOSS}(z) = z(z\boldsymbol{I}_n - \boldsymbol{A})^{-1} = z\frac{\text{adj}(z\boldsymbol{I}_n - \boldsymbol{A})}{\det(z\boldsymbol{I}_n - \boldsymbol{A})}. \tag{13.147}$$

Lemma 13.4 *In order to test a Lyapunov stability property of the zero equilibrium vector $\mathbf{x}_e = \mathbf{0}_n$ of the ISO system (3.60a) and (3.60b) via the transfer function matrices of the system, it is necessary and sufficient to use the denominator polynomial of the row nondegenerate form of the system SS transfer function matrix $\boldsymbol{G}_{ISOSS}(z)$.*

Proof. Equation (13.147) shows that $\boldsymbol{G}_{ISOSS}(z)$ is proper. The dynamic behavior of the *ISO* system (3.60a) and (3.60b) in the free regime is determined by (13.146) in the complex domain. This equation shows that the same poles and zeros of any row of $\boldsymbol{G}_{ISOSS}(z)$ do not influence $\mathbf{x}(k;\mathbf{x}_0;\mathbf{0}_M)$,

$$\mathbf{x}(k;\mathbf{x}_0;\mathbf{0}_M) = \mathcal{Z}^{-1}\{\boldsymbol{G}_{ISOSS}(z)\mathbf{x}_0\}.$$

Hence, Lemma 6.1 (in Chapter 6), shows that we should use the row nondegenerate form of $\boldsymbol{G}_{ISOSS}(z)$ for any Lyapunov stability test, which leads to the use of its denominator polynomial. ■

Theorem 13.42 *In order for the zero equilibrium vector $\mathbf{x}_e = \mathbf{0}_n$ of the ISO system (3.60a) and (3.60b) to be*

a) stable, it is necessary and sufficient that the modulus of all poles of the row nondegenerate form of $\boldsymbol{G}_{ISOSS}(z)$ are less or equal to one, and the multiplicity of its poles with modulus equal to one should be equal to one (i.e., the poles should be simple),

b) asymptotically stable, it is necessary and sufficient that the modulus of all poles of the row nondegenerate form of $\boldsymbol{G}_{ISOSS}(z)$ are with modulus less than one.

Proof. The Generating Theorem 13.39 (in Subsection 13.4.1), and Lemma 13.4 imply directly the statement of the theorem. ■

Note 13.16 *The preceding theorem discovers that the competent transfer function matrix of the ISO system (3.60a) and (3.60b) for the test of a Lyapunov stability property is the row nondegenerate form* $\mathbf{G}_{ISOSSnd}(z)$ *of* $\mathbf{G}_{ISOSS}(z)$*, and not the system transfer function matrix* $\mathbf{G}_{ISO}(z)$*.*

Corollary 13.2 *In order for the cancellation of the same zero and pole of the equal order and with modulus greater or equal to one in the transfer function matrix* $\mathbf{G}_{ISO}(z)$ *of the ISO system (3.60a) and (3.60b) to be without any influence on the results on any Lyapunov stability property of the equilibrium vector, it is necessary and sufficient that their same cancellation is possible in all rows of the transfer function matrix* $\mathbf{G}_{ISOSS}(z)$ *of the system with respect to* $\mathbf{x}_0$*.*

Example 13.4 *We continue to use the system of Example 13.3. We compute the matrix* $z(z\boldsymbol{I}_n - \boldsymbol{A})^{-1}$ *that is* $\mathbf{G}_{ISOSS}(z)$*, (13.147),*

$$\mathbf{G}_{ISOSS}(z) = z(z\boldsymbol{I}_n - \boldsymbol{A})^{-1} =$$

$$= \begin{bmatrix} \frac{z^2(z-1.2)}{z^3-1.2z^2-1.45z-0.3} & \frac{z(z-1.2)}{z^3-1.2z^2-1.45z-0.3} & \frac{2z}{z^3-1.2z^2-1.45z-0.3} \\ \frac{z(1.45z+0.3)}{z^3-1.2z^2-1.45z-0.3} & \frac{z^2(z-1.2)}{z^3-1.2z^2-1.45z-0.3} & \frac{2z^2}{z^3-1.2z^2-1.45z-0.3} \\ \frac{1.02z^2}{z^3-1.2z^2-1.45z-0.3} & \frac{1.02z}{z^3-1.2z^2-1.45z-0.3} & \frac{z(z^2-1.45)}{z^3-1.2z^2-1.45z-0.3} \end{bmatrix}.$$

It is row nondegenerate. Its poles are all eigenvalues of the resolvent matrix $(sI_n - A)^{-1}$*,*

$$z_1^* = \lambda_1 = -0.5,\ z_2^* = \lambda_2 = 2,\ z_3^* = \lambda_3 = -0.3.$$

This shows that we should not cancel the pole whose modulus is greater than one with the same zero whose modulus is greater than one, $z_2^* = z_2^0 = 2$*, in* $\mathbf{G}_{ISO}(z)$ *if we use* $\mathbf{G}_{ISO}(z)$ *for the Lyapunov stability test.*

Lemma 13.5 *The eigenvalues of the matrix* $\boldsymbol{A}$ *of the ISO system (3.60a) and (3.60b) are the poles of the system transfer function matrix* $\mathbf{G}_{ISOSS}(z)$*, and vice versa.*

Proof. Equation (8.37) (in Section 8.2), shows that the denominator polynomial $f_{ISOSS}(z)$ of $\mathbf{G}_{ISOSS}(z)$ obeys

$$f_{ISOSS}(z) = \det(z\boldsymbol{I}_n - \boldsymbol{A}). \tag{13.148}$$

Its zeros are the poles of $\mathbf{G}_{ISOSS}(z)$.

The characteristic polynomial $f_{IOA}(z)$ of the matrix $\boldsymbol{A}$ is also $\det(z\boldsymbol{I}_n - \boldsymbol{A})$,

$$f_{ISOA}(z) = \det(z\boldsymbol{I}_n - \boldsymbol{A}) = f_{ISOSS}(z).$$

Hence, the eigenvalues of the matrix $\boldsymbol{A}$ are the zeros of $f_{ISOSS}(z)$ and the poles of $\boldsymbol{G}_{ISOSS}(z)$, and vice versa. ■

Theorem 13.43 ***The Lyapunov matrix theorem for the*** ISO ***system (3.60a) and (3.60b)***

In order for the zero equilibrium vector $\mathbf{x}_e = \mathbf{0}_n$ *of the* ISO *system (3.60a) and (3.60b) to be asymptotically stable, equivalently for its matrix* $\boldsymbol{A}$ *to be stable matrix, it is necessary and sufficient that for any positive definite symmetric matrix* $\mathbf{G}$, $\boldsymbol{G} = \boldsymbol{G}^T \in \mathcal{R}^{n \times n}$, *the matrix solution* $\boldsymbol{H}$ *of the Lyapunov matrix equation*

$$\boldsymbol{A}^T \boldsymbol{H}\boldsymbol{A} - \boldsymbol{H} = -\boldsymbol{G} \qquad (13.149)$$

is also positive definite symmetric matrix and the unique solution to (13.149).

Proof. Theorem 13.42, and Lemma 13.5 show the equivalence between asymptotic stability of the zero equilibrium vector $\mathbf{x}_e = \mathbf{0}_n$ of the ISO system (3.60a) and (3.60b) and stability of its matrix $\boldsymbol{A}$. We apply the replacements determined by (13.121) and (13.122) (in Section 13.3), to Theorem 13.31 that then becomes this theorem. ■

Note 13.17 *Theorem 13.43 is the fundamental matrix theorem for stability of the* ISO *system (3.60a) and (3.60b).*

Comment 13.8 ***The physical meaning of Lyapunov matrix theorem for the*** ISO ***system (3.60a) and (3.60b)***

Let the system energy $e(\cdot)$ *and power* $p(\cdot)$ *have the properties determined in Comment 13.4 (in Section 13.3). Equations (13.89) of Comment 13.4 imply*

$$\boldsymbol{A}^T \boldsymbol{E}\boldsymbol{A} - \boldsymbol{E} = \boldsymbol{P},\ \boldsymbol{P} = \boldsymbol{P}^T < \boldsymbol{O}.$$

The matrix $\boldsymbol{P}$ *of the system power* $p(\mathbf{w}) = \mathbf{w}^T \boldsymbol{P}\mathbf{w}$ *is negative definite and symmetric. In order for the zero equilibrium state of the* ISO *system (3.60a) and (3.60b) to be asymptotically (hence, exponentially) stable, it is both necessary and sufficient that the matrix* $\mathbf{E}$ *of the system energy* $e(\mathbf{w}) = \mathbf{w}^T \boldsymbol{E}\mathbf{w}$ *is positive definite and symmetric,* $\mathbf{E} = \boldsymbol{E}^T > \boldsymbol{O}$.

13.4.4 *IIO* systems

We will treat simultaneously the case $\nu \geq 1$ and $\nu = 0$ of the *IIO* systems (3.64a) and (3.64b) (in Subsection 3.5.4). The latter incorporates both the Rosenbrock systems (3.66a) and (3.66b) and the *GISO* systems (3.68a) and (3.68b) (in Subsection 3.5.4).

Lemma 13.6 *In order for*

$$\left[\ \left(\mathbf{r}^{\alpha-1}(k;\mathbf{r}_0^{\alpha-1};\mathbf{0}_M)\right)^T \quad \left(\mathbf{y}^{\nu-1}(k;\mathbf{r}_0^{\alpha-1};\mathbf{y}_0^{\nu-1};\mathbf{0}_M)\right)^T\ \right]^T,\ \nu \geq 1,$$
$$\mathbf{r}^{\alpha-1}(k;\mathbf{r}_0^{\alpha-1};\mathbf{0}_M),\ \nu = 0,$$

of the IIO system (3.64a) and (3.64b), in the free regime

a) to be bounded on $\mathcal{N}_0$,

$$\exists\zeta \in \mathcal{R}^+ \Longrightarrow \left\| \begin{bmatrix} \mathbf{r}^{\alpha-1}(k;\mathbf{r}_0^{\alpha-1};\mathbf{0}_M) \\ \mathbf{y}^{\nu-1}(k;\mathbf{r}_0^{\alpha-1};\mathbf{y}_0^{\nu-1};\mathbf{0}_M) \end{bmatrix} \right\| < \zeta,\ \forall k \in \mathcal{N}_0,\ \nu \geq 1,$$
$$\exists\theta \in \mathcal{R}^+ \Longrightarrow \left\|\mathbf{r}^{\alpha-1}(k;\mathbf{r}_0^{\alpha-1};\mathbf{0}_M)\right\| < \theta,\ \forall k \in \mathcal{N}_0,\ \nu = 0,$$

it is necessary and sufficient that

$$\left[\ \mathbf{r}^T(k;\mathbf{r}_0^{\alpha-1};\mathbf{0}_M) \quad \mathbf{y}^T(k;\mathbf{r}_0^{\alpha-1};\mathbf{y}_0^{\nu-1};\mathbf{0}_M)\ \right]^T,\ \nu \geq 1,$$
$$\mathbf{r}(k;\mathbf{r}_0^{\alpha-1};\mathbf{0}_M),\ \nu = 0,$$

is bounded on $\mathcal{N}_0$,

$$\exists\xi \in \mathcal{R}^+ \Longrightarrow \left\| \begin{bmatrix} \mathbf{r}(k;\mathbf{r}_0^{\alpha-1};\mathbf{0}_M) \\ \mathbf{y}(k;\mathbf{r}_0^{\alpha-1};\mathbf{y}_0^{\nu-1};\mathbf{0}_M) \end{bmatrix} \right\| < \xi,\ \forall k \in \mathcal{N}_0,\ \nu \geq 1,$$
$$\exists\sigma \in \mathcal{R}^+ \Longrightarrow \left\|\mathbf{r}(k;\mathbf{r}_0^{\alpha-1};\mathbf{0}_M)\right\| < \sigma,\ \forall k \in \mathcal{N}_0,\ \nu = 0,$$

b) to vanish as $k \to \infty$,

$$k \to \infty \Longrightarrow$$
$$\left[\ \left(\mathbf{r}^{\alpha-1}(k;\mathbf{r}_0^{\alpha-1};\mathbf{0}_M)\right)^T \quad \left(\mathbf{y}^{\nu-1}(k;\mathbf{r}_0^{\alpha-1};\mathbf{y}_0^{\nu-1};\mathbf{0}_M)\right)^T\ \right]^T \to$$
$$\to \mathbf{0}_{\alpha\rho+\nu N},\ \nu \geq 1, \mathbf{r}^{\alpha-1}(k;\mathbf{r}_0^{\alpha-1};\mathbf{0}_M) \to \mathbf{0}_{\alpha\rho},\ \nu = 0,$$

it is necessary and sufficient that

$$\left[\ \mathbf{r}^T(k;\mathbf{r}_0^{\alpha-1};\mathbf{0}_M) \quad \mathbf{y}^T(k;\mathbf{r}_0^{\alpha-1};\mathbf{y}_0^{\nu-1};\mathbf{0}_M)\ \right]^T,\ \nu \geq 1,$$
$$\mathbf{r}(k;\mathbf{r}_0^{\alpha-1};\mathbf{0}_M),\ \nu = 0,$$

vanishes as $k \to \infty$,

$$k \to \infty \Longrightarrow \begin{bmatrix} \mathbf{r}(k; \mathbf{r}_0^{\alpha-1}; \mathbf{0}_M) \\ \mathbf{y}(k; \mathbf{r}_0^{\alpha-1}; \mathbf{y}_0^{\nu-1}; \mathbf{0}_M) \end{bmatrix} \to \mathbf{0}_{\rho+N}, \ \nu \geq 1,$$
$$k \to \infty \Longrightarrow \mathbf{r}(k; \mathbf{r}_0^{\alpha-1}; \mathbf{0}_M) \to \mathbf{0}_\rho, \ \nu = 0.$$

Proof. *Necessity. a)* Let

$$\begin{bmatrix} \left(\mathbf{r}^{\alpha-1}(k; \mathbf{r}_0^{\alpha-1}; \mathbf{0}_M)\right)^T & \left(\mathbf{y}^{\nu-1}(k; \mathbf{r}_0^{\alpha-1}; \mathbf{y}_0^{\nu-1}; \mathbf{0}_M)\right)^T \end{bmatrix}^T, \ \nu \geq 1,$$
$$\mathbf{r}^{\alpha-1}(k; \mathbf{r}_0^{\alpha-1}; \mathbf{0}_M), \ \nu = 0,$$

be bounded on $\mathcal{N}_0$. Each of its vector entries is bounded on $\mathcal{N}_0$. Its subvector

$$\begin{bmatrix} \mathbf{r}^T(k; \mathbf{r}_0^{\alpha-1}; \mathbf{0}_M) & \mathbf{y}^T(k; \mathbf{r}_0^{\alpha-1}; \mathbf{y}_0^{\nu-1}; \mathbf{0}_M) \end{bmatrix}^T, \ \nu \geq 1,$$
$$\mathbf{r}(k; \mathbf{r}_0^{\alpha-1}; \mathbf{0}_M), \ \nu = 0,$$

is therefore bounded on $\mathcal{N}_0$.

b) Let

$$\begin{bmatrix} \left(\mathbf{r}^{\alpha-1}(k; \mathbf{r}_0^{\alpha-1}; \mathbf{0}_M)\right)^T & \left(\mathbf{y}^{\nu-1}(k; \mathbf{r}_0^{\alpha-1}; \mathbf{y}_0^{\nu-1}; \mathbf{0}_M)\right)^T \end{bmatrix}^T, \ \nu \geq 1,$$
$$\mathbf{r}^{\alpha-1}(k; \mathbf{r}_0^{\alpha-1}; \mathbf{0}_M), \ \nu = 0,$$

vanish as $k \to \infty$. Every subvector of it vanishes as $k \to \infty$, which holds also for its first subvector,

$$k \to \infty \Longrightarrow \begin{bmatrix} \mathbf{r}(k; \mathbf{r}_0^{\alpha-1}; \mathbf{0}_M) \\ \mathbf{y}(k; \mathbf{r}_0^{\alpha-1}; \mathbf{y}_0^{\nu-1}; \mathbf{0}_M) \end{bmatrix} \to \mathbf{0}_{\rho+N}, \ \nu \geq 1,$$
$$t \to \infty \Longrightarrow \mathbf{r}(t; \mathbf{r}_0^{\alpha-1}; \mathbf{0}_M) \to \mathbf{0}_\rho, \ \nu = 0.$$

Sufficiency. Let the conditions of the statement of lemma hold.

a) The properties of the $Z-$transform and the conditions of the Generating Theorem 13.39 (in Subsection 13.4.1) permit the following:

$$\begin{bmatrix} \begin{bmatrix} \mathbf{r}(k; \mathbf{r}_0^{\alpha-1}; \mathbf{0}_M) \\ \mathbf{y}(k; \mathbf{r}_0^{\alpha-1}; \mathbf{y}_0^{\nu-1}; \mathbf{0}_M) \end{bmatrix}, \ \nu \geq 1, \\ \mathbf{r}(k; \mathbf{r}_0^{\alpha-1}; \mathbf{0}_M), \ \nu = 0, \end{bmatrix} =$$

$$
= \begin{bmatrix} \begin{bmatrix} \left\{ \delta_d(k) \boldsymbol{R}_{0r} + \sum_{i=1}^{i=\alpha\rho} (z_i^*)^{k-1} \left[\sum_{r=1}^{r=\nu_i^*} \frac{1}{(z_i^*)^{r-1}(r-1)!} \frac{(k-1)!}{(k-r)!} \boldsymbol{R}_{irr} \right] \right\} \mathbf{r}_0^{\alpha-1} \\ \left\{ \delta_d(k) \boldsymbol{R}_{0y} + \sum_{i=1}^{i=\nu N} (z_i^*)^{k-1} \left[\sum_{r=1}^{r=\nu_i^*} \frac{1}{(z_i^*)^{r-1}(r-1)!} \frac{(k-1)!}{(k-r)!} \boldsymbol{R}_{iry} \right] \right\} \cdot \\ \cdot \begin{bmatrix} \mathbf{r}_0^{\alpha-1} \\ \mathbf{y}_0^{\nu-1} \end{bmatrix} \end{bmatrix}, \\ \left\{ \delta_d(k) \boldsymbol{R}_{0r} + \sum_{i=1}^{i=\alpha\rho} (z_i^*)^{k-1} \left[\sum_{r=1}^{r=\nu_i^*} \frac{1}{(z_i^*)^{r-1}(r-1)!} \frac{(k-1)!}{(k-r)!} \boldsymbol{R}_{irr} \right] \right\} \mathbf{r}_0^{\alpha-1}, \end{bmatrix}
$$

$$
|z_i^*| \leq 1, \ \forall i = 1, 2, \cdots, \nu N, \ |z_i^*| = 1 \Longrightarrow \nu_i^* = 1, \tag{13.150}
$$

where ν_i^* is the multiplicity of z_i^* and the constant matrices $\boldsymbol{R}_{0r}$, $\boldsymbol{R}_{irr}$, $\boldsymbol{R}_{0y}$, and $\boldsymbol{R}_{iry}$ are of the appropriate dimensions. All shifts of

$$
\begin{bmatrix} \begin{bmatrix} \mathbf{r}(k; \mathbf{r}_0^{\alpha-1}; \mathbf{0}_M) \\ \mathbf{y}(k; \mathbf{r}_0^{\alpha-1}; \mathbf{y}_0^{\nu-1}; \mathbf{0}_M) \end{bmatrix}, \ \nu \geq 1, \\ \mathbf{r}(k; \mathbf{r}_0^{\alpha-1}; \mathbf{0}_M), \ \nu = 0, \end{bmatrix}
$$

are bounded and infinitely times shiftable, which guarantees the same properties to

$$
\begin{bmatrix} \left(\mathbf{r}^{\alpha-1}(k; \mathbf{r}_0^{\alpha-1}; \mathbf{0}_M)\right)^T & \left(\mathbf{y}^{\nu-1}(k; \mathbf{r}_0^{\alpha-1}; \mathbf{y}_0^{\nu-1}; \mathbf{0}_M)\right)^T \end{bmatrix}^T, \ \nu \geq 1,
$$
$$
\mathbf{r}^{\alpha-1}(k; \mathbf{r}_0^{\alpha-1}; \mathbf{0}_M), \ \nu = 0.
$$

b) Since

$$
\begin{bmatrix} \begin{bmatrix} \mathbf{r}(k; \mathbf{r}_0^{\alpha-1}; \mathbf{0}_M) \\ \mathbf{y}(k; \mathbf{r}_0^{\alpha-1}; \mathbf{y}_0^{\nu-1}; \mathbf{0}_M) \end{bmatrix}, \ \nu \geq 1, \\ \mathbf{r}(k; \mathbf{r}_0^{\alpha-1}; \mathbf{0}_M), \ \nu = 0, \end{bmatrix}
$$

vanishes as $k \to \infty$ then, in view of (13.150), all its shifts also vanish as $k \to \infty$. This guarantees that

$$
\begin{bmatrix} \left(\mathbf{r}^{\alpha-1}(k; \mathbf{r}_0^{\alpha-1}; \mathbf{0}_M)\right)^T & \left(\mathbf{y}^{\nu-1}(k; \mathbf{r}_0^{\alpha-1}; \mathbf{y}_0^{\nu-1}; \mathbf{0}_M)\right)^T \end{bmatrix}^T, \ \nu \geq 1,
$$
$$
\mathbf{r}^{\alpha-1}(k; \mathbf{r}_0^{\alpha-1}; \mathbf{0}_M), \ \nu = 0,
$$

vanishes as $k \to \infty$. ■

Note 13.18 $|z_i^*| < 1$, $\forall i = 1, 2, ..., p = \alpha\rho + \nu N$, *(13.150), Generating Theorem 13.39 (in Subsection 13.4.1), and its shifts imply the existence of* *positive numbers* $\varsigma \geq 1$ *and* ξ *such that*

$$\nu \geq 1 \Longrightarrow \left\| \begin{bmatrix} \left(\mathbf{r}^{\alpha-1}(k;\mathbf{r}_0^{\alpha-1};\mathbf{0}_M)\right)^T & \left(\mathbf{y}^{\nu-1}(k;\mathbf{r}_0^{\alpha-1};\mathbf{y}_0^{\nu-1};\mathbf{0}_M)\right)^T \end{bmatrix}^T \right\| \leq$$

$$\leq \varsigma e^{-\xi k} \left\| \begin{bmatrix} \mathbf{r}_0^{\alpha-1} \\ \mathbf{y}_0^{\nu-1} \end{bmatrix} \right\|, \ \forall k \in \mathcal{N}_0, \ \forall \begin{bmatrix} \mathbf{r}_0^{\alpha-1} \\ \mathbf{y}_0^{\nu-1} \end{bmatrix} \in \mathcal{R}^{\alpha\rho+\nu N},$$

$$\nu = 0 \Longrightarrow \left\|\mathbf{r}^{\alpha-1}(k;\mathbf{r}_0^{\alpha-1};\mathbf{0}_M)\right\| \leq \varsigma e^{-\xi k} \left\|\left[\mathbf{r}_0^{\alpha-1}\right]\right\|,$$

$$\forall k \in \mathcal{N}_0, \ \forall \mathbf{r}_0^{\alpha-1} \in \mathcal{R}^{\alpha\rho}. \tag{13.151}$$

Lemma 13.7 *The IIO system overall transfer function matrix* $\boldsymbol{G}_{IIOr_0y_0}(z)$ *(13.152) with respect to the initial conditions,*

$$\boldsymbol{G}_{IIOr_0y_0}(z) = \left\{ \begin{array}{c} \begin{bmatrix} \boldsymbol{G}_{IIOr_0IS}(z) & \boldsymbol{O}_{\rho,\nu N} \\ \boldsymbol{G}_{IIOr_0}(z) & \boldsymbol{G}_{IIOy_0}(z) \end{bmatrix} \text{ iff } \nu \geq 1, \\ \boldsymbol{G}_{IIOr_0IS}(z) \text{ iff } \nu = 0, \end{array} \right. \tag{13.152}$$

is proper or strictly proper such that, then in order to test stability, asymptotic stability or exponential stability of the zero equilibrium vector

- $\begin{bmatrix} \left(\mathbf{r}_e^{\alpha-1}\right)^T & \left(\mathbf{y}_e^{\nu-1}\right)^T \end{bmatrix}^T = \mathbf{0}_{\alpha\rho+\nu N}$ *of the IIO system (3.64a) and (3.64b), iff* $\nu \geq 1$,

- $\mathbf{r}_e^{\alpha-1} = \mathbf{0}_{\alpha\rho}$ *of the IIO system (3.64a) and (3.64b), iff* $\nu = 0$, *or of the GISO systems (3.68a) and (3.68b),*

via the system transfer function matrices, it is necessary and sufficient to use the denominator polynomial $f_{IIOr_0y_0rnd}(z)$,

$$f_{IIOr_0y_0rnd}(z) = \left\{\det\left[\text{Den } \boldsymbol{G}_{IIOr_0ISrnd}(z)\right]\right\} \cdot$$

$$\cdot \left\{ \begin{array}{c} \left\{\det\left[\text{Den } \boldsymbol{G}_{IIOr_0rnd}(z)\right]\right\}\left\{\det\left[\text{Den } \boldsymbol{G}_{IIOy_0rnd}(s)\right]\right\}, \ \nu \geq 1, \\ 1, \ \nu = 0, \end{array} \right. , \tag{13.153}$$

of the row nondegenerate form $\boldsymbol{G}_{IIOr_0y_0rnd}(z)$,

$$\boldsymbol{G}_{IIOr_0y_0rnd}(z) = \left\{ \begin{array}{c} \begin{bmatrix} \boldsymbol{G}_{IIOr_0ISrnd}(z) & \boldsymbol{O}_{\rho,\nu N} \\ \boldsymbol{G}_{IIOr_0rnd}(z) & \boldsymbol{G}_{IIOy_0rnd}(z) \end{bmatrix} \text{ iff } \nu \geq 1, \\ \boldsymbol{G}_{IIOr_0ISrnd}(z) \text{ iff } \nu = 0 \end{array} \right. , \tag{13.154}$$

of $\boldsymbol{G}_{IIOr_0y_0}(z)$ *(13.152).*

Proof. The dynamic behavior of the IIO system (3.64a) and (3.64b) in the free regime is determined by (13.155) due to (10.33) (Definition 10.4), (10.35) (Theorem 10.4), i.e., (10.47), (10.45) (Definition 10.5), (10.48) (Theorem 10.5) (in Section 10.3), and Lemma 6.1 (in Chapter 6):

$$\begin{bmatrix} \mathbf{r}^{\alpha-1}(k;\mathbf{r}_0^{\alpha-1};\mathbf{0}_M) \\ \mathbf{y}^{\nu-1}(k;\mathbf{r}_0^{\alpha-1};\mathbf{y}_0^{\nu-1};\mathbf{0}_M) \end{bmatrix} =$$
$$= \mathcal{Z}^{-1}\left\{ \boldsymbol{G}_{IIOr_0y_0rnd}(z) \begin{bmatrix} \mathbf{r}_0^{\alpha-1} \\ \left\{ \begin{array}{c} \mathbf{y}_0^{\nu-1} \text{ iff } \nu \geq 1, \\ \mathbf{0}_{\nu N} \text{ iff } \nu = 0 \end{array} \right\} \end{bmatrix} \right\} \tag{13.155}$$

This, and Lemma 13.6 imply, because $\boldsymbol{G}_{IIOr_0y_0}(z)$, hence $\boldsymbol{G}_{IIOr_0y_0rnd}(z)$, is strictly proper, that the zeros of the denominator polynomial $f_{IIOr_0y_0rnd}(z)$ of the row nondegenerate form $\boldsymbol{G}_{IIOr_0y_0rnd}(z)$ (13.154) of the submatrix $\boldsymbol{G}_{IIOr_0y_0}(z)$ (13.152) of $\boldsymbol{F}_{IIO}(z)$, which are the poles of $\boldsymbol{G}_{IIOr_0y_0rnd}(z)$ (13.154), determine completely the character of

$$\begin{bmatrix} \left(\mathbf{r}^{\alpha-1}(k;\mathbf{r}_0^{\alpha-1};\mathbf{0}_M)\right)^T & \left(\mathbf{y}^{\nu-1}(k;\mathbf{r}_0^{\alpha-1};\mathbf{y}_0^{\nu-1};\mathbf{0}_M)\right)^T \end{bmatrix}^T \text{ iff } \nu \geq 1,$$
$$\mathbf{r}(k;\mathbf{r}_0^{\alpha-1};\mathbf{0}_M) \text{ iff } \nu = 0.$$

The proof is completed. ■

Theorem 13.44 *In order for the zero equilibrium vector*

- $\begin{bmatrix} (\mathbf{r}_e^{\alpha-1})^T & (\mathbf{y}_e^{\nu-1})^T \end{bmatrix}^T = \mathbf{0}_{\alpha\rho+\nu N}$ *of the IIO system (3.64a) and (3.64b) iff* $\nu \geq 1$,
- $\mathbf{r}_e^{\alpha-1} = \mathbf{0}_{\alpha\rho}$ *of the IIO system (3.64a) and (3.64b) iff* $\nu = 0$, *or of the GISO systems (3.68a) and (3.68b)*

  to be

a) *stable, it is necessary and sufficient that*

   1. $\boldsymbol{G}_{IIOr_0y_0}(z)$ *is either proper or strictly proper, real rational matrix function of* z,

      and that

   2. *the modulus of all poles of the row nondegenerate form* $\boldsymbol{G}_{IIOr_0y_0rnd}(z)$ *(13.154) of the IIO system combined transfer function matrix* $\boldsymbol{G}_{IIOr_0y_0}(z)$ *(13.152), i.e., that the modulus of*

all zeros of the characteristic polynomial $f_{IIOroyornd}(z)$ *(13.153), are less or equal to one, and the multiplicity of each of them with modulus equal to one should be equal to one (i.e., they should be simple),*

b) *asymptotically stable, it is necessary and sufficient that*

1. $\mathbf{G}_{IIOroyo}(z)$ *is either proper or strictly proper, real rational matrix function of* z,

   and that

2. *the modulus of all poles of the row nondegenerate form* $\mathbf{G}_{IIOroyornd}(z)$ *(13.154) of the IIO combined transfer function matrix* $\mathbf{G}_{IIOroyo}(z)$ *(13.152), i.e., that the modulus of all zeros of its characteristic polynomial* $f_{IIOroyornd}(z)$ *(13.153), are less than one.*

Proof. Lemma 6.1, the Generating Theorem 13.39, Lemma 13.6 and Lemma 13.7 prove this theorem ■

Note 13.19 *This theorem discovers that the competent transfer function matrix of the IIO system (3.64a) and (3.64b) for the test of a Lyapunov stability property is the row nondegenerate form* $\mathbf{G}_{IIOroyornd}(z)$ *(13.154) of* $\mathbf{G}_{IIOroyo}(z)$ *(13.152), but not the system transfer matrix* $\mathbf{G}_{IIO}(z)$.

Corollary 13.3 *In order for the cancellation of the same zero and pole of the equal order and with modulus greater or equal to one in the transfer function matrix* $\mathbf{G}_{IIO}(z)$ *of the IIO system (3.64a) and (3.64b) to be without any influence on the results on any Lyapunov stability property of the equilibrium vector, it is necessary and sufficient that their same cancellation is possible in the IIO system overall transfer function matrix* $\mathbf{G}_{IIOroyo}(z)$ *(13.152) with respect to the initial conditions.*

Theorem 13.45 ***Lyapunov matrix theorem for the IIO system (3.64a) and (3.64b)***

In order for the zero equilibrium vector

– $\left[\left(\mathbf{r}_e^{\alpha-1}\right)^T \ \left(\mathbf{y}_e^{\nu-1}\right)^T \right]^T = \mathbf{0}_{\alpha\rho+\nu N}$ *of the IIO system (3.64a) and (3.64b) iff* $\nu \geq 1$,

– $\mathbf{r}_e^{\alpha-1} = \mathbf{0}_{\alpha\rho}$ *of the IIO system (3.64a) and (3.64b) iff* $\nu = 0$, *or of the GISO systems (3.68a) and (3.68b)*

to be asymptotically stable, it is necessary and sufficient that for any positive definite symmetric matrix $\boldsymbol{G}_{IIO}$,

$$\boldsymbol{G}_{IIO} = \boldsymbol{G}_{IIO}^T \in \left\{ \begin{array}{c} \mathcal{R}^{(\alpha\rho+\nu N)\times(\alpha\rho+\nu N)},\ iff\ \nu \geq 1, \\ \mathcal{R}^{\alpha\rho\times\alpha\rho},\ iff\ \nu = 0 \end{array} \right\},$$

the matrix solution $\boldsymbol{H}_{IIO}$,

$$\boldsymbol{H}_{IIO} \in \left\{ \begin{array}{c} \mathcal{R}^{(\alpha\rho+\nu N)\times(\alpha\rho+\nu N)},\ iff\ \nu \geq 1, \\ \mathcal{R}^{\alpha\rho\times\alpha\rho},\ iff\ \nu = 0 \end{array} \right\},$$

of the Lyapunov matrix equation

$$\boldsymbol{A}_{IIO}^T \boldsymbol{H}_{IIO} \boldsymbol{A}_{IIO} - \boldsymbol{H}_{IIO} = -\boldsymbol{G}_{IIO} \tag{13.156}$$

is also positive definite symmetric matrix and the unique solution to (13.156).

Proof. The application of the matrix $\boldsymbol{A}_{IIO}$ defined by (13.39) through (13.42) (in Subsection 13.2.3), and by

$$\boldsymbol{A} = \boldsymbol{A}_{IIO} = \left\{ \begin{array}{c} \boldsymbol{A}_{IIOry} = \left[\begin{array}{cc} \boldsymbol{A}_{IIO1} & \boldsymbol{O}_{\alpha\rho,\nu N} \\ \boldsymbol{A}_{IIO2} & \boldsymbol{A}_{IIO3} \end{array} \right] iff\ \nu \geq 1, \\ \boldsymbol{A}_{IIOr} = \boldsymbol{A}_{IIO1}\ iff\ \nu = 0 \end{array} \right\}. \tag{13.157}$$

permits the transformation of the IIO system (3.64a) and (3.64b) in the free regime into the ISO system (13.58) (in Section 13.3) also in the free regime. This, and the replacements

$$\begin{aligned} p &= \alpha\rho + \nu N,\ \mathbf{w} = \left\{ \begin{array}{c} \left[\begin{array}{cc} \left(\mathbf{r}^{\alpha-1}\right)^T & \left(\mathbf{y}^{\nu-1}\right)^T \end{array} \right]^T iff\ \nu \geq 1, \\ \mathbf{r}^{\alpha-1},\ iff\ \nu = 0 \end{array} \right\}, \\ v(\mathbf{w}) &= v_{IIO}\left(\left\{ \begin{array}{c} \left[\begin{array}{cc} \left(\mathbf{r}^{\alpha-1}\right)^T & \left(\mathbf{y}^{\nu-1}\right)^T \end{array} \right]^T iff\ \nu \geq 1, \\ \mathbf{r}^{\alpha-1},\ iff\ \nu = 0 \end{array} \right\} \right), \end{aligned}$$

$$\boldsymbol{G}_{IIO} = \boldsymbol{G}_{IIO}^T \in \left\{ \begin{array}{c} \mathcal{R}^{(\alpha\rho+\nu N)\times(\alpha\rho+\nu N)},\ iff\ \nu \geq 1, \\ \mathcal{R}^{\alpha\rho\times\alpha\rho},\ iff\ \nu = 0 \end{array} \right\},$$

$$\boldsymbol{H}_{IIO} = \boldsymbol{H}_{IIO}^T \in \left\{ \begin{array}{c} \mathcal{R}^{(\alpha\rho+\nu N)\times(\alpha\rho+\nu N)},\ iff\ \nu \geq 1, \\ \mathcal{R}^{\alpha\rho\times\alpha\rho},\ iff\ \nu = 0 \end{array} \right\},$$

transform Theorem 13.43 (in Subsection 13.4.3) into this theorem. ■

Note 13.20 *This theorem is the fundamental matrix theorem for stability of the IIO system (3.64a) and (3.64b) and for stability of the GISO systems (3.68a) and (3.68b).*

Comment 13.9 ***The physical meaning of the Lyapunov matrix theorem for the IIO system (3.64a) and (3.64b)***

Let the system power $p(\cdot)$ *be negative definite quadratic form* $p(\mathbf{w}) = \mathbf{w}^T \boldsymbol{P}\mathbf{w}$, $\boldsymbol{P} = \boldsymbol{P}^T < \boldsymbol{O}$. *Theorem 13.31 (in Section 13.3), implies*

$$\boldsymbol{A}_{IIO}^T \boldsymbol{E} \boldsymbol{A}_{IIO} - \boldsymbol{E} = \boldsymbol{P},\ \boldsymbol{P} = \boldsymbol{P}^T < \boldsymbol{O}.$$

In order for the zero equilibrium state of the system to be asymptotically (hence, exponentially) stable, it is both necessary and sufficient that the system energy $e(\cdot)$ *is positive definite quadratic form,* $e(\mathbf{w}) = \mathbf{w}^T \boldsymbol{E}\mathbf{w}$ *with* $\boldsymbol{E} = \boldsymbol{E}^T > \boldsymbol{O}$.

Chapter 14

Bounded input stability

14.1 BI stability and initial conditions

The **Bounded Input** (BI) **stability** concept is essentially different from Lyapunov's (LY) stability concept for its following characteristics:

- The BI stability concept concerns the system behavior under actions of input variables, while the original LY stability concept treats the system behavior in the free regime considered in the framework of deviations, i.e., in the total nominal regime in terms of the total coordinates [93].

- The original BI stability concept in the framework of the linear continuous-time dynamical systems demands boundedness of the system behaviors for bounded input variables and *for all zero initial conditions* [117, p. 311], which was broaden to the boundedness of the system behaviors for bounded input variables *and any bounded initial conditions [68], [69]*, while the LY stability concept demands Lyapunov's ε-δ closeness of the system behaviors to the zero equilibrium vector in the framework of deviations under zero input deviations. This means that various new BI stability properties were introduced and defined, and complex domain criteria were discovered for them. E. D. Sontag et al. [118], [120], [121] established the theory of "**input to state stability**" that concerns both the bounded input vector and arbitrary initial state conditions. It is valid in the general setting of nonlinear systems. Their results, which hold in *time* domain, exploit the Lyapunov method in combination with the comparison functions of W. Hahn [83].

- The BI stability concept characterizes various stability properties of a dynamical system, while the LY stability concept determines various stability properties of a desired (nominal) system behavior in terms of total coordinates, i.e., of an equilibrium vector in terms of deviations.

- The BI stability concept does not demand asymptotic convergence (as $t \to \infty$) of the system behaviors to the equilibrium vector, while LY stability properties — attraction, asymptotic and exponential stability of the equilibrium vector — do.

All further we immediately broaden to discrete time systems and give related to them. Both the LY stability concept and the BI stability concept treat system behaviors over the infinite *discrete time* interval $[k = 0, k = 1, \cdots, k = \infty]$, i.e., over $\mathcal{N}_0$.

Since we consider only *time*-invariant systems, then the initial instant is zero moment, $k_0 = 0$, for both the BI stability concept and the LY stability concept.

The BI stability concept that demands boundedness of the system behaviors for bounded input variables and all zero or arbitrary initial conditions incorporates the following BI stability properties:

- **Bounded Input-Bounded Internal State** ($BIBIS$) **stability** demands the bounded system internal state, for all $k \in \mathcal{N}_0$ and for every bounded input vector function under all zero initial conditions.

- **Bounded Input-Bounded Output** (for short $BIBO$) **stability** demands the bounded system output response for all $k \in \mathcal{N}_0$ and for every bounded input vector function under all zero initial state [5], [18], [84].

- **Bounded Input-Bounded Output State** (for short $BIBOS$) **stability** demands the bounded system output state for all $k \in \mathcal{N}_0$ and for every bounded input vector function under all zero initial conditions.

- **Bounded Input-Bounded State** ($BIBS$) **stability** demands bounded the system state for all $k \in \mathcal{N}_0$, and for every bounded input vector function under all zero initial conditions.

- **Bounded Input-Bounded State and Output** ($BIBSO$) **stability** demands the bounded system state and its output response for all $k \in \mathcal{N}_0$, and for every bounded input vector function under all zero initial conditions.

These BI stability properties assume *all zero initial conditions*, and they are valid only under this assumption. In reality, at least one initial condition is, most often, different from zero. It is natural to consider system behaviors under the bounded input vector function and arbitrary initial conditions. This leads to the following generalizations of the BI stability concept and properties that hold for arbitrary initial conditions:

- **Bounded Input and Initial Internal State-Bounded Internal State** (for short $BIISBIS$) **stability** demands the bounded system internal state for all $k \in \mathcal{N}_0$ and for every bounded the input vector function and the initial internal state.

- **Bounded Input and Initial Internal State-Bounded Output** (for short $BIISBO$) **stability** demands the bounded system output response for all $k \in \mathcal{N}_0$ and for every bounded input vector function and the initial internal state.

- **Bounded Input and Initial Internal State-Bounded Output State** ($BIISBOS$) **stability** demands the bounded system output state for all $k \in \mathcal{N}_0$, and for every bounded input vector function and the initial internal state.

- **Bounded Input and Initial Internal State-Bounded Internal State and Output** ($BIISBISO$) **stability** demands the bounded system internal state and the system output response, for all $k \in \mathcal{N}_0$, and for every bounded input vector function and the initial internal state.

- **Bounded Input and Initial Internal State-Bounded State** ($BIISBS$) **stability** demands the bounded system state for all $k \in \mathcal{N}_0$ and for every bounded input vector function and the initial internal state.

- **Bounded Input and Initial Internal State-Bounded State and Output** ($BIISBSO$) **stability** demands the bounded system state and the system output response for all $k \in \mathcal{N}_0$ and for every bounded input vector function and the initial internal state.

- **Bounded Input and Initial State-Bounded Output** ($BISBO$) **stability** demands the bounded system output response, for all $k \in \mathcal{N}_0$ and for every bounded input vector function and the initial state.

- **Bounded Input and Initial State-Bounded Output State** $(BISBOS)$ **stability** demands the bounded system output state for all $k \in \mathcal{N}_0$ and for every bounded input vector function and the initial state.

- **Bounded Input and Initial State-Bounded State and Output** $(BISBSO)$ **stability** demands the bounded system output state for all $k \in \mathcal{N}_0$ and for every bounded input vector function and the initial state.

- **Bounded Input and Initial State-Bounded State** $(BISBS)$ **stability** demands the bounded system state for all $k \in \mathcal{N}_0$ and for every bounded input vector function and the initial state.

Comment 14.1 *Every BI stability property valid under arbitrary initial conditions implies the corresponding BI stability property valid only under all zero initial conditions.*

We will prove the conditions for each of the BI stability properties for input vector functions $\mathbf{i}(\cdot) : \mathcal{N}_0 \to \mathcal{R}^M$ belonging to the class $\mathcal{J}$ (2.11) (in Section 2.1).

The conditions will discover other relationships among BI stability properties. They will enable us also to compare them with those for Lyapunov stability properties.

14.2 Definitions

14.2.1 *IO* systems

It might seem at first glance that $\mathbf{y}(k;\mathbf{i}_0^{\mu-1};\mathbf{y}_0^{\nu-1};\mathbf{i})$ expresses both the internal dynamics behavior of the IO system (3.56) (in Subsection 3.5.2), and its output dynamics. It is not correct. The evolution of $\mathbf{y}^{\nu-1}(k;\mathbf{i}_0^{\mu-1};\mathbf{y}_0^{\nu-1};\mathbf{i})$ in *time* expresses both the internal dynamical behavior and the output dynamical behavior of the IO system. Hence, it represents the IO system dynamics. The *time* evolution $\mathbf{y}(k;\mathbf{i}_0^{\mu-1};\mathbf{y}_0^{\nu-1};\mathbf{i})$ is the ordinary temporal output response of the IO system and it expresses only the output variation in *time*. We can say that $\mathbf{y}^{\nu-1}(k;\mathbf{i}_0^{\mu-1};\mathbf{y}_0^{\nu-1};\mathbf{i})$ is also *the system output response of the order* $\nu - 1$, or the *system full output response*. It is also the system internal state, the system output state and the system state at the moment k.

Definition 14.1 *The IO system (3.56) is*

a) ***Bounded Input-Bounded State*** *(BIBS)* ***stable,*** *if and only if for every positive real number α there exists a positive real number μ, the value of which depends on α, $\mu = \mu(\alpha)$, such that for every input vector function obeying $\|\mathbf{i}(k)\| < \mu$ for all $k \in \mathcal{N}_0$, under all zero initial conditions the system state satisfies $\left\|\mathbf{y}^{\nu-1}(k;\mathbf{0}_{\mu M};\mathbf{0}_{\nu N};\mathbf{i})\right\| < \alpha$ for all $k \in \mathcal{N}_0$,*

$$\begin{gathered}\forall\alpha \in \mathcal{R}^{+},\ \exists\mu \in \mathcal{R}^{+},\ \mu = \mu(\alpha),\ \ \|\mathbf{i}(k)\| < \mu,\ \forall k \in \mathcal{N}_0, \Longrightarrow \\ \left\|\mathbf{y}^{\nu-1}(k;\mathbf{0}_{\mu M};\mathbf{0}_{\nu N};\mathbf{i})\right\| < \alpha,\ \forall k \in \mathcal{N}_0. \end{gathered} \tag{14.1}$$

b) ***Bounded Input-Bounded Output*** *(BIBO)* ***stable,*** *if and only if for every positive real number γ there exists a positive real number μ, the value of which depends on γ, $\mu = \mu(\gamma)$, such that for every input vector function obeying $\|\mathbf{i}(k)\| < \mu$ for all $k \in \mathcal{N}_0$, the output response under all zero initial conditions satisfies $\|\mathbf{y}(k;\mathbf{0}_{\mu M};\mathbf{0}_{\nu N};\mathbf{i})\| < \gamma$ for all $k \in \mathcal{N}_0$,*

$$\begin{gathered}\forall\gamma \in \mathcal{R}^{+},\ \exists\mu \in \mathcal{R}^{+},\ \mu = \mu(\gamma),\ \ \|\mathbf{i}(k)\| < \mu,\ \forall k \in \mathcal{N}_0, \Longrightarrow \\ \|\mathbf{y}(k;\mathbf{0}_{\mu M};\mathbf{0}_{\nu N};\mathbf{i})\| < \gamma,\ \forall k \in \mathcal{N}_0. \end{gathered} \tag{14.2}$$

c) ***Bounded Input and Initial State-Bounded State*** *(BISBS)* ***stable,*** *if and only if for every positive real number $\varkappa$ there exist positive real numbers μ and υ, the values of which depend on $\varkappa$, $\mu = \mu(\varkappa)$, $\upsilon = \upsilon(\varkappa)$, such that for every input vector function obeying $\|\mathbf{i}(k)\| < \mu$ for all $k \in \mathcal{N}_0$, and for every initial state satisfying $\left\|\mathbf{y}_0^{\nu-1}\right\| < \upsilon$ the state satisfies $\left\|\mathbf{y}^{\nu-1}(k;\mathbf{y}_0^{\nu-1};\mathbf{i})\right\| < \varkappa$ for all $k \in \mathcal{N}_0$,*

$$\begin{gathered}\forall\varkappa \in \mathcal{R}^{+},\ \exists\mu \in \mathcal{R}^{+},\ \mu = \mu(\varkappa),\ \exists\upsilon \in \mathcal{R}^{+},\ \upsilon = \upsilon(\varkappa), \\ \|\mathbf{i}(k)\| < \mu,\ \forall k \in \mathcal{N}_0, \left\|\mathbf{y}_0^{\nu-1}\right\| < \upsilon \Longrightarrow \\ \left\|\mathbf{y}^{\nu-1}(k;\mathbf{y}_0^{\nu-1};\mathbf{i})\right\| < \varkappa,\ \forall k \in \mathcal{N}_0. \end{gathered} \tag{14.3}$$

d) ***Bounded Input and Initial State-Bounded Output*** *(BISBO)* ***stable,*** *if and only if for every positive real numbers γ there exist positive real numbers μ and υ, the values of which depend on γ, $\mu = \mu(\gamma)$, $\upsilon = \upsilon(\gamma)$, such that for every input vector function obeying $\|\mathbf{i}(k)\| < \mu$ for all $k \in \mathcal{N}_0$, and for every initial state satisfying $\left\|\mathbf{y}_0^{\nu-1}\right\| < \upsilon$ the* *output response satisfies* $\left\|\mathbf{y}(k;\mathbf{y}_0^{\nu-1};\mathbf{i})\right\| < \gamma$ *for all* $k \in \mathcal{N}_0$,

$$\forall\gamma \in \mathcal{R}^+,\ \exists\mu \in \mathcal{R}^+,\ \mu = \mu(\gamma),\ \exists\upsilon \in \mathcal{R}^+,\ \upsilon = \upsilon(\gamma),$$
$$\|\mathbf{i}(k)\| < \mu,\ \forall k \in \mathcal{N}_0, \left\|\mathbf{y}_0^{\nu-1}\right\| < \upsilon \Longrightarrow$$
$$\left\|\mathbf{y}(k;\mathbf{y}_0^{\nu-1};\mathbf{i})\right\| < \gamma,\ \forall k \in \mathcal{N}_0. \tag{14.4}$$

Note 14.1 *The boundedness condition on the input vector that*

$$\|\mathbf{i}(k)\| < \mu \ \textit{ for all } k \in \mathcal{N}_0$$

implies the boundedness by μ *of the initial input vector,*

$$\|\mathbf{i}(0)\| = \|\mathbf{i}_0\| < \mu.$$

Note 14.2 *Since the initial state vector* $\mathbf{y}_0^{\nu-1}$ *of the IO system (3.56) is also the vector of all its initial conditions, then the following BI stability properties coincide: BIBIS, BIBOS, BIBS and BIBSO; BIISBIS, BIISBOS, BIISBSO and BIISBS; BISBOS and BISBS.*

Note 14.3 *These BI stability properties, like Lyapunov stability of the equilibrium vector, do not demand that system behaviors converge to the equilibrium vector. Moreover, the BI stability properties do not demand the existence of an equilibrium vector in general.*

14.2.2 *ISO* systems

A characteristic of the *ISO* system (3.60a) and (3.60b) (in Subsection 3.5.3), is the linear algebraic dependence of the system output vector on both the system state vector and on the input vector. This simplifies their study. This is clearer from the definition of the *BI* stability properties of these systems.

Definition 14.2 *The ISO system (3.60a) and (3.60b) is*

a) ***Bounded Input-Bounded State*** *(BIBS)* ***stable*** *if and only if for every positive real number* $\varkappa$ *there exists a positive real number* η*, the value of which depends on* $\varkappa$, $\eta = \eta(\varkappa)$*, such that for every input vector function obeying* $\|\mathbf{i}(k)\| < \eta$ *for all* $k \in \mathcal{N}_0$*, the system state vector under all zero initial conditions satisfies* $\|\mathbf{x}(k;\mathbf{0}_n;\mathbf{i})\| < \varkappa$ *for all* $k \in \mathcal{N}_0$,

$$\forall\varkappa \in \mathcal{R}^+,\ \exists\eta \in \mathcal{R}^+,\ \eta = \eta(\varkappa),\ \|\mathbf{i}(k)\| < \eta,\ \forall k \in \mathcal{N}_0, \Longrightarrow$$
$$\|\mathbf{x}(k;\mathbf{0}_n;\mathbf{i})\| < \varkappa,\ \forall k \in \mathcal{N}_0. \tag{14.5}$$

b) ***Bounded Input-Bounded Output** (BIBO) **stable** if and only if for every positive real number γ there exists a positive real number μ, the value of which depends on γ, $\mu = \mu(\gamma)$, such that for every input vector function obeying $\|\mathbf{i}(k)\| < \mu$ for all $k \in \mathcal{N}_0$, the system output vector under all zero initial conditions obeys $\|\mathbf{y}(k;\mathbf{0}_n;\mathbf{i})\| < \gamma$ for all $k \in \mathcal{N}_0$,*

$$\forall \gamma \in \mathcal{R}^+,\ \exists \mu \in \mathcal{R}^+,\ \mu = \mu(\gamma),\ \ \|\mathbf{i}(k)\| < \mu,\ \forall k \in \mathcal{N}_0, \Rightarrow \\ \|\mathbf{y}(k;\mathbf{0}_n;\mathbf{i})\| < \gamma,\ \forall k \in \mathcal{N}_0. \tag{14.6}$$

c) ***Bounded Input-Bounded State and Output** (BIBSO) **stable** if and only if for every positive real number $\varkappa$ there exist positive real numbers η and θ, the values of which depend on $\varkappa$, $\eta = \eta(\varkappa)$ and $\theta = \theta(\varkappa)$, such that for every input vector function obeying $\|\mathbf{i}(k)\| < \eta$ for all $k \in \mathcal{N}_0$, the system motion and the system output response under all zero initial conditions satisfy $\|\mathbf{x}(k;\mathbf{0}_n;\mathbf{i})\| < \varkappa$ and $\|\mathbf{y}(k;\mathbf{0}_n;\mathbf{i})\| < \theta$ for all $k \in \mathcal{N}_0$,*

$$\forall \varkappa \in \mathcal{R}^+,\ \exists \eta, \theta \in \mathcal{R}^+,\ \eta = \eta(\varkappa), \theta = \theta(\varkappa),\ \ \|\mathbf{i}(k)\| < \eta,\ \forall k \in \mathcal{N}_0, \Rightarrow \\ \|\mathbf{x}(k;\mathbf{0}_n;\mathbf{i})\| < \varkappa \ \ \text{and}\ \ \|\mathbf{y}(k;\mathbf{0}_n;\mathbf{i})\| < \theta,\ \forall k \in \mathcal{N}_0. \tag{14.7}$$

d) ***Bounded Input and Bounded Initial State-Bounded State** (BISBS) **stable** if and only if for every positive real number $\varkappa$ there exist positive real numbers η and θ, the values of which depend on $\varkappa$, $\eta = \eta(\varkappa)$ and $\theta = \theta(\varkappa)$, such that for every input vector function obeying $\|\mathbf{i}(k)\| < \eta$ for all $k \in \mathcal{N}_0$, and for every initial state $\mathbf{x}_0$ satisfying $\|\mathbf{x}_0\| < \theta$, the system motion obeys $\|\mathbf{x}(k;\mathbf{x}_0;\mathbf{i})\| < \varkappa$ for all $k \in \mathcal{N}_0$,*

$$\forall \varkappa \in \mathcal{R}^+,\ \exists \eta \in \mathcal{R}^+,\ \eta = \eta(\varkappa),\ \exists \theta \in \mathcal{R}^+,\ \theta = \theta(\varkappa), \\ \|\mathbf{i}(k)\| < \eta,\ \forall k \in \mathcal{N}_0,\ \text{and}\ \|\mathbf{x}_0\| < \theta \Longrightarrow \\ \|\mathbf{x}(k;\mathbf{x}_0;\mathbf{i})\| < \varkappa,\ \forall k \in \mathcal{N}_0. \tag{14.8}$$

e) ***Bounded Input and Initial State-Bounded Output** (BISBO) **stable** if and only if for every positive real number γ there exist positive real numbers μ and ν, the values of which depend on γ, $\mu = \mu(\gamma)$ and $\nu = \nu(\gamma)$, such that for every input vector function obeying $\|\mathbf{i}(k)\| < \mu$ for all $k \in \mathcal{N}_0$, and for every initial state $\mathbf{x}_0$ satisfying $\|\mathbf{x}_0\| < \nu$, the system output response obeys $\|\mathbf{y}(k;\mathbf{x}_0;\mathbf{i})\| < \gamma$ for* *all* $k \in \mathcal{N}_0$,

$$\forall \gamma \in \mathcal{R}^+, \ \exists \mu \in \mathcal{R}^+, \ \mu = \mu(\gamma), \ \exists \nu \in \mathcal{R}^+, \ \nu = \nu(\gamma),$$
$$\|\mathbf{i}(k)\| < \mu, \ \forall k \in \mathcal{N}_0, \ and \ \|\mathbf{x}_0\| < \nu \Longrightarrow$$
$$\|\mathbf{y}(k;\mathbf{x}_0;\mathbf{i})\| < \gamma, \ \forall k \in \mathcal{N}_0. \tag{14.9}$$

f) ***Bounded Input and Initial State-Bounded State and Output*** *(BISBSO)* **stable** *if and only if for every positive real number* $\varkappa$ *there exist positive real numbers* ν, μ *and* ρ, *the values of which depend on* $\varkappa$, $\nu = \nu(\varkappa)$, $\mu = \mu(\varkappa)$ *and* $\rho = \rho(\varkappa)$, *such that for every input vector function obeying* $\|\mathbf{i}(k)\| < \mu$ *for all* $k \in \mathcal{N}_0$, *and for every initial state* $\mathbf{x}_0$ *satisfying* $\|\mathbf{x}_0\| < \nu$, *the system motion and the system output response obey, respectively,* $\|\mathbf{x}(k;\mathbf{x}_0;\mathbf{i})\| < \varkappa$ *and* $\|\mathbf{y}(k;\mathbf{x}_0;\mathbf{i})\| < \rho$ *for all* $k \in \mathcal{N}_0$,

$$\forall \varkappa \in \mathcal{R}^+, \ \exists \nu \in \mathcal{R}^+, \nu = \nu(\varkappa), \ \exists \mu \in \mathcal{R}^+, \ \mu = \mu(\varkappa), \ \exists \rho \in \mathcal{R}^+,$$
$$\rho = \rho(\varkappa), \|\mathbf{i}(k)\| < \mu, \ \forall k \in \mathcal{N}_0, \ and \ \|\mathbf{x}_0\| < \nu \Longrightarrow$$
$$\|\mathbf{x}(k;\mathbf{x}_0;\mathbf{i})\| < \varkappa \ and \ \|\mathbf{y}(k;\mathbf{x}_0;\mathbf{i})\| < \rho, \ \forall k \in \mathcal{N}_0. \tag{14.10}$$

Note 14.4 *The condition that* $\|\mathbf{i}(k)\|$ *is bounded for all* $k \in \mathcal{N}_0$ *guarantees the boundedness of the norm* $\|\mathbf{i}_0\|$ *of the initial input vector* $\mathbf{i}_0$.

Note 14.5 *These BI stability properties do not demand that system motions converge to the equilibrium vector* $\mathbf{x}_e = \mathbf{0}_n$.

Claim 14.1 *The fact that the initial state vector* $\mathbf{x}_0$ *is the vector of all system initial conditions implies that the following BI stability properties are the same for the ISO system (3.60a) and (3.60b): BIBIS, BIBOS, BIBS and BIBSO; BIISBIS, BIISBOS, BIISBSO and BIISBS; BISBSO, BISBOS and BISBS.*

The following two Theorems and their proofs explain and verify Claim 14.1. Besides, they show how we can easily prove Claim 14.2 in Subsection 14.2.1.

Theorem 14.1 *In order for the ISO system (3.60a) and (3.60b) to be BIBSO stable, it is necessary and sufficient to be BIBS stable.*

Proof. *Necessity.* Let the *ISO* system (3.60a) and (3.60b) be *BIBSO* stable. Then, *c*) of Definition 14.2 shows that for every input vector function obeying $\|\mathbf{i}(k)\| < \eta$ for all $k \in \mathcal{N}_0$, the system motion and the system output response under all zero initial conditions satisfy $\|\mathbf{x}(t;\mathbf{0}_n;\mathbf{i})\| < \varkappa$ and $\|\mathbf{y}(t;\mathbf{0}_n;\mathbf{i})\| < \theta$ for all $k \in \mathcal{N}_0$. The system is $BIBS$ stable in view of a) of Definition 14.2.

Sufficiency. Let the ISO system (3.60a) and (3.60b) be $BIBS$ stable. From a) of Definition 14.2, i.e., (14.5), follows

$$\forall \varkappa \in \mathcal{R}^+,\ \exists \eta \in \mathcal{R}^+,\ \eta = \eta(\varkappa),$$
$$\|\mathbf{i}(k)\| < \eta,\ \forall k \in \mathcal{N}_0, \Longrightarrow \|\mathbf{x}(k;\mathbf{0}_n;\mathbf{i})\| < \varkappa,\ \forall k \in \mathcal{N}_0,$$

which, together with (3.60b), implies

$$\forall \varkappa \in \mathcal{R}^+,\ \eta = \eta(\varkappa),$$
$$\exists \theta \in \mathcal{R}^+,\ \theta = \theta(\varkappa) \Longrightarrow \|C\|\,\varkappa + \|D\|\,\eta < \theta,\ \theta = \theta\left[\varkappa, \eta(\varkappa)\right] = \theta\left(\varkappa\right),$$
$$\Longrightarrow$$

$$\|\mathbf{i}(k)\| < \eta,\ \forall k \in \mathcal{N}_0,\ \Longrightarrow$$
$$\|\mathbf{x}(k;\mathbf{0}_n;\mathbf{i})\| < \varkappa,\ \forall k \in \mathcal{N}_0 \Longrightarrow$$
$$\|\mathbf{y}(k;\mathbf{0}_n;\mathbf{i})\| \leq \|C\|\,\|\mathbf{x}(k;\mathbf{0}_n;\mathbf{i})\| + \|D\|\,\|\mathbf{i}(k)\| <$$
$$< \|C\|\,\varkappa + \|D\|\,\eta < \theta < \infty,\ \forall k \in \mathcal{N}_0.$$

The system is $BIBSO$ stable in view of c) of Definition 14.2. ■

By following the proof of Theorem 14.1 we can easily prove the following claim.

Claim 14.2 *Variations of the state vector and the input vector determine completely the output dynamical behavior of the ISO system (3.60a) and (3.60b). This comes out from the algebraic nature of the system output equation. The BI stability properties hold for bounded input vectors. They and various Bounded Input-Bounded State stability properties, which guarantee the boundedness of the system state, imply the boundedness of both the output behavior and the full output behavior because they coincide for the ISO system (3.60a) and (3.60b). This explains that BIBS stability guarantees BIBO stability, BIBOS stability and BIBSO stability.*

Theorem 14.2 *In order for the ISO system (3.60a) and (3.60b) to be BISBSO stable, it is necessary and sufficient to be BISBS stable.*

Proof. *Necessity.* Let the ISO system (3.60a) and (3.60b) be $BISBSO$ stable. Then, f) of Definition 14.2 shows that for every input vector function obeying $\|\mathbf{i}(k)\| < \mu$ for all $\forall k \in \mathcal{N}_0$, and for any initial state satisfying

$\|\mathbf{x}_0\| < \nu$, the system motion obeys $\|\mathbf{x}(k;\mathbf{x}_0;\mathbf{i})\| < \varkappa$ for all $\forall k \in \mathcal{N}_0$. The system is $BISBS$ stable in view of d) of Definition 14.2.

Sufficiency. Let the ISO system (3.60a) and (3.60b) be $BISBS$ stable. From d) of Definition 14.2, i.e., (14.8), follows

$$\forall \varkappa \in \mathcal{R}^+,\ \exists \eta \in \mathcal{R}^+,\ \eta = \eta(\varkappa),\ \exists \theta \in \mathcal{R}^+,\ \theta = \theta(\varkappa),$$
$$\|\mathbf{i}(k)\| < \eta,\ \forall k \in \mathcal{N}_0, \text{ and } \|\mathbf{x}_0\| < \theta \Longrightarrow$$
$$\|\mathbf{x}(k;\mathbf{x}_0;\mathbf{i})\| < \varkappa,\ \forall k \in \mathcal{N}_0,$$

which, together with (3.60b), implies

$$\forall \varkappa \in \mathcal{R}^+,\ \exists \psi \in \mathcal{R}^+,\ \psi = \psi(\varkappa) \leq \varkappa,$$
$$\exists \eta \in \mathcal{R}^+,\ \eta = \eta(\psi),\ \exists \theta \in \mathcal{R}^+,\ \theta = \theta(\psi) \Longrightarrow$$
$$\|C\|\,\psi + \|D\|\,\eta < \varkappa$$

$$\Longrightarrow$$

$$\exists \rho \in \mathcal{R}^+, \|\mathbf{i}(k)\| < \mu,\ \forall k \in \mathcal{N}_0, \text{ and } \|\mathbf{x}_0\| < \nu \Longrightarrow$$
$$\|\mathbf{x}(k;\mathbf{x}_0;\mathbf{i})\| < \varkappa,\ \forall k \in \mathcal{N}_0 \Longrightarrow$$
$$\|\mathbf{y}(k;\mathbf{x}_0;\mathbf{i})\| \leq \|C\|\,\|\mathbf{x}(k;\mathbf{x}_0;\mathbf{i})\| + \|D\|\,\|\mathbf{i}(k)\| <$$
$$< \|C\|\,\varkappa + \|D\|\,\mu < \rho < \infty,\ \forall k \in \mathcal{N}_0.$$

The system is $BISBSO$ stable in view of f) of Definition 14.2. ∎

With trivial appropriate adjustments of this proof we prove the following:

Claim 14.3 *Variations of the state vector and the input vector determine completely the output dynamical behavior of the ISO system (3.60a) and (3.60b). This comes out from the algebraic nature of the system output equation. The BI stability properties hold for bounded input vectors. They and various Bounded Input-Bounded State stability properties, which guarantee the boundedness of the system state, imply the boundedness of both the output behavior and the full output behavior because they coincide for the ISO system (3.60a) and (3.60b). This explains that $BISBS$ stability guarantees $BISBO$ stability and $BISBSO$ stability.*

We will establish the conditions for the BI stability properties of the ISO system (3.60a) and (3.60b) in Subsection 14.3.2.

14.2.3 *IIO* systems

The *IIO* system (3.64a), and (3.64b) (in Subsection 3.5.4), differently than the *IO* system (3.56) (in Subsection 3.5.2), and the *ISO* system (3.60a) and (3.60b) (in Subsection 3.5.3), has both *the internal dynamics* expressed by *its internal state behavior,* which is described by the left-hand side of its first discrete equation, and *the output dynamics* defined by the left-hand side of its second discrete equation if and only if $\nu > 0$ (in Section 2.3). The output dynamics determines the output state vector. This enriches the Bounded Input (*BI*) stability concept with new and more complex *BI* stability properties than those of the *IO* system (3.56) and of the *ISO* system (3.60a) and (3.60b). In this regard we recall Note 3.9 (in Subsection 3.5.4).

The fact that the output behavior depends on the output dynamics (if and only if $\nu > 0$) implies that the following *BI* stability properties are not sufficiently meaningful for the *IIO* system (3.64a) and (3.64b): *BIISBIS*, *BIISBO*, *BIISBOS*, *BIISBISO* and *BIISBSO*.

Definition 14.3 *The IIO system (3.64a) and (3.64b) is*

a) ***Bounded Input-Bounded Internal State*** *(BIBIS)* ***stable*** *if and only if for every positive real number $\varkappa$ there exists a positive real number η, the value of which depends on $\varkappa$, $\eta = \eta(\varkappa)$, such that for every input vector function obeying $\|\mathbf{i}(k)\| < \eta$ for all $k \in \mathcal{N}_0$, the system internal state vector under all zero initial conditions satisfies $\left\|\mathbf{r}^{\alpha-1}(k;\mathbf{0}_{\alpha\rho};\mathbf{i})\right\| < \varkappa$ for all $k \in \mathcal{N}_0$,*

$$\forall \varkappa \in \mathcal{R}^+,\ \exists \eta \in \mathcal{R}^+,\ \eta = \eta(\varkappa),$$
$$\|\mathbf{i}(k)\| < \eta,\ \forall k \in \mathcal{N}_0, \Longrightarrow \left\|\mathbf{r}^{\alpha-1}(k;\mathbf{0}_{\alpha\rho};\mathbf{i})\right\| < \varkappa,\ \forall k \in \mathcal{N}_0. \quad (14.11)$$

b) ***Bounded Input-Bounded Output*** *(BIBO)* ***stable*** *if and only if for every positive real number γ there exists a positive real number η, the value of which depends on γ, $\eta = \eta(\gamma)$, such that for every input vector function obeying $\|\mathbf{i}(k)\| < \eta$ for all $k \in \mathcal{N}_0$, the system output vector in the forced regime under all zero initial conditions satisfies $\|\mathbf{y}(k;\mathbf{0}_{\alpha\rho};\mathbf{0}_{\nu N};\mathbf{i})\| < \gamma$ for all $k \in \mathcal{N}_0$,*

$$\forall \gamma \in \mathcal{R}^+,\ \exists \eta \in \mathcal{R}^+,\ \eta = \eta(\gamma),\ \ \|\mathbf{i}(k)\| < \eta,\ \forall k \in \mathcal{N}_0,\ \Longrightarrow$$
$$\|\mathbf{y}(k;\mathbf{0}_{\alpha\rho};\mathbf{0}_{\nu N};\mathbf{i})\| < \gamma,\ \forall k \in \mathcal{N}_0. \quad (14.12)$$

c) ***Bounded Input-Bounded Output State*** *(BIBOS)* ***stable*** *if and only if for every positive real number* γ *there exists a positive real number* η*, the value of which depends on* γ, $\eta = \eta(\gamma)$*, such that for every input vector function obeying* $\|\mathbf{i}(k)\| < \eta$ *for all* $k \in \mathcal{N}_0$*, and for all zero initial conditions, the system output state fulfills* $\left\|\mathbf{y}^{\nu-1}(k;\mathbf{0}_{\alpha\rho};\mathbf{0}_{\nu N};\mathbf{i})\right\| < \gamma$ *for all* $k \in \mathcal{N}_0$,

$$\forall\gamma \in \mathcal{R}^+,\ \exists\eta \in \mathcal{R}^+,\ \eta = \eta(\gamma),\ \|\mathbf{i}(k)\| < \eta,\ \forall k \in \mathcal{N}_0,\ \Longrightarrow \\ \left\|\mathbf{y}^{\nu-1}(k;\mathbf{0}_{\alpha\rho};\mathbf{0}_{\nu N};\mathbf{i})\right\| < \gamma,\ \forall k \in \mathcal{N}_0. \tag{14.13}$$

d) ***Bounded Input-Bounded Internal State and Output*** *(BIBISO)* ***stable*** *if and only if for every positive real number* $\varkappa$ *there exist positive real numbers* η *and* θ*, the values of which depend on* $\varkappa$, $\eta = \eta(\varkappa)$ *and* $\theta = \theta(\varkappa)$*, such that for every input vector function obeying* $\|\mathbf{i}(k)\| < \eta$ *for all* $k \in \mathcal{N}_0$*, and for all zero initial conditions, the system internal state vector fulfills* $\left\|\mathbf{r}^{\alpha-1}(k;\mathbf{0}_{\alpha\rho};\mathbf{i})\right\| < \varkappa$ *and its output vector obeys* $\|\mathbf{y}(k;\mathbf{0}_{\alpha\rho};\mathbf{0}_{\nu N};\mathbf{i})\| < \theta$ *for all* $k \in \mathcal{N}_0$,

$$\forall\varkappa \in \mathcal{R}^+,\ \exists\eta,\theta \in \mathcal{R}^+,\ \eta = \eta(\varkappa),\ \theta = \theta(\varkappa), \\ \|\mathbf{i}(k)\| < \eta,\ \forall k \in \mathcal{N}_0\ \Longrightarrow \\ \left\|\mathbf{r}^{\alpha-1}(k;\mathbf{0}_{\alpha\rho};\mathbf{i})\right\| < \varkappa, \|\mathbf{y}(k;\mathbf{0}_{\alpha\rho};\mathbf{0}_{\nu N};\mathbf{i})\| < \theta,\ \forall k \in \mathcal{N}_0. \tag{14.14}$$

e) ***Bounded Input and Initial Internal State-Bounded Internal State*** *(BIISBIS)* ***stable*** *if and only if for every positive real number* $\varkappa$ *there exist positive real numbers* η *and* θ*, the values of which depend on* $\varkappa$, $\eta = \eta(\varkappa)$ *and* $\theta = \theta(\varkappa)$*, such that for every input vector function obeying* $\|\mathbf{i}(k)\| < \eta$ *for all* $k \in \mathcal{N}_0$*, and for every initial internal state vector satisfying* $\left\|\mathbf{r}_0^{\alpha-1}\right\| < \theta$*, the system internal state vector fulfills* $\left\|\mathbf{r}^{\alpha-1}(k;\mathbf{r}_0^{\alpha-1};\mathbf{i})\right\| < \varkappa$ *for all* $k \in \mathcal{N}_0$,

$$\forall\varkappa \in \mathcal{R}^+,\ \exists\eta \in \mathcal{R}^+,\ \eta = \eta(\varkappa),\ \exists\theta \in \mathcal{R}^+,\ \theta = \theta(\varkappa), \\ \|\mathbf{i}(k)\| < \eta,\ \forall k \in \mathcal{N}_0 \text{ and } \left\|\mathbf{r}_0^{\alpha-1}\right\| < \theta \Longrightarrow \\ \left\|\mathbf{r}^{\alpha-1}(k;\mathbf{r}_0^{\alpha-1};\mathbf{i})\right\| < \varkappa,\ \forall k \in \mathcal{N}_0. \tag{14.15}$$

f) ***Bounded Input and Initial State-Bounded Output*** *(BISBO)* ***stable*** *if and only if for every positive real number* γ *there exist positive real numbers* η *and* ξ*, the values of which depend on* γ, $\eta = \eta(\gamma)$ *and* $\xi = \xi(\gamma)$*, such that for every input vector function obeying* $\|\mathbf{i}(k)\| < \eta$ *for all* $k \in \mathcal{N}_0$*, and for every overall initial state vector* *satisfying* $\left\| \left[\begin{array}{cc} \left(\mathbf{r}_0^{\alpha-1}\right)^T & \left(\mathbf{y}_0^{\nu-1}\right)^T \end{array} \right]^T \right\| < \xi$, *the system response obeys* $\left\|\mathbf{y}(k;\mathbf{r}_0^{\alpha-1};\mathbf{y}_0^{\nu-1};\mathbf{i})\right\| < \gamma$ *for all* $k \in \mathcal{N}_0$,

$$\forall\gamma \in \mathcal{R}^+,\ \exists\eta \in \mathcal{R}^+,\ \eta = \eta(\gamma),\ \exists\xi \in \mathcal{R}^+,\ \xi = \xi(\gamma),$$
$$\|\mathbf{i}(k)\| < \eta,\ \forall k \in \mathcal{N}_0,\ and\ \left\| \left[\begin{array}{cc} \left(\mathbf{r}_0^{\alpha-1}\right)^T & \left(\mathbf{y}_0^{\nu-1}\right)^T \end{array} \right]^T \right\| < \xi \Longrightarrow$$
$$\left\|\mathbf{y}(k;\mathbf{r}_0^{\alpha-1};\mathbf{y}_0^{\nu-1};\mathbf{i})\right\| < \gamma,\ \forall k \in \mathcal{N}_0. \tag{14.16}$$

g) ***Bounded Input and Initial State-Bounded Output State (BISBOS) stable*** *if and only if for every positive real number* γ *there exist positive real numbers* η *and* ξ, *the values of which depend on* γ, $\eta = \eta(\gamma)$ *and* $\xi = \xi(\gamma)$, *such that for every input vector function obeying* $\|\mathbf{i}(k)\| < \eta$ *for all* $k \in \mathcal{N}_0$, *and for every overall initial state vector satisfying* $\left\| \left[\begin{array}{cc} \left(\mathbf{r}_0^{\alpha-1}\right)^T & \left(\mathbf{y}_0^{\nu-1}\right)^T \end{array} \right]^T \right\| < \xi$, *the system output state fulfills* $\left\|\mathbf{y}^{\nu-1}(k;\mathbf{r}_0^{\alpha-1};\mathbf{y}_0^{\nu-1};\mathbf{i})\right\| < \gamma$ *for all* $k \in \mathcal{N}_0$,

$$\forall\gamma \in \mathcal{R}^+,\ \exists\eta \in \mathcal{R}^+,\ \eta = \eta(\gamma),\ \exists\xi \in \mathcal{R}^+,\ \xi = \xi(\gamma),$$
$$\|\mathbf{i}(k)\| < \eta,\ \forall k \in \mathcal{N}_0,\ and\ \left\| \left[\begin{array}{cc} \left(\mathbf{r}_0^{\alpha-1}\right)^T & \left(\mathbf{y}_0^{\nu-1}\right)^T \end{array} \right]^T \right\| < \xi \Longrightarrow$$
$$\left\|\mathbf{y}^{\nu-1}(k;\mathbf{r}_0^{\alpha-1};\mathbf{y}_0^{\nu-1};\mathbf{i})\right\| < \gamma,\ \forall k \in \mathcal{N}_0. \tag{14.17}$$

h) ***Bounded Input and Initial State-Bounded Internal State and Output*** *(BISBISO)* ***stable*** *if and only if for every positive real number* $\varkappa$ *there exist positive real numbers* γ, η *and* ξ, *the values of which depend on* $\varkappa$, $\gamma = \gamma(\varkappa)$, $\eta = \eta(\varkappa)$ *and* $\xi = \xi(\varkappa)$, *such that for every input vector function obeying* $\|\mathbf{i}(k)\| < \eta$ *for all* $k \in \mathcal{N}_0$, *and for every overall initial state vector satisfying* $\left\| \left[\begin{array}{cc} \left(\mathbf{r}_0^{\alpha-1}\right)^T & \left(\mathbf{y}_0^{\nu-1}\right)^T \end{array} \right]^T \right\| < \xi$, *the system internal state vector fulfills* $\left\|\mathbf{r}^{\alpha-1}(k;\mathbf{r}_0^{\alpha-1};\mathbf{i})\right\| < \varkappa$ *and its output vector obeys* $\left\|\mathbf{y}(k;\mathbf{r}_0^{\alpha-1};\mathbf{y}_0^{\nu-1};\mathbf{i})\right\| < \gamma$ *for all* $k \in \mathcal{N}_0$,

$$\forall\varkappa \in \mathcal{R}^+, \exists\gamma \in \mathcal{R}^+, \gamma = \gamma(\varkappa), \exists\eta \in \mathcal{R}^+, \eta = \eta(\varkappa), \exists\xi \in \mathcal{R}^+, \xi = \xi(\varkappa),$$
$$\|\mathbf{i}(k)\| < \eta,\ \forall k \in \mathcal{N}_0,\ and\ \left\| \left[\begin{array}{cc} \left(\mathbf{r}_0^{\alpha-1}\right)^T & \left(\mathbf{y}_0^{\nu-1}\right)^T \end{array} \right]^T \right\| < \xi \Longrightarrow$$
$$\left\|\mathbf{r}^{\alpha-1}(k;\mathbf{r}_0^{\alpha-1};\mathbf{i})\right\| < \varkappa, \left\|\mathbf{y}(k;\mathbf{r}_0^{\alpha-1};\mathbf{y}_0^{\nu-1};\mathbf{i})\right\| < \gamma,\ \forall k \in \mathcal{N}_0. \tag{14.18}$$

i) ***Bounded Input and Initial State-Bounded State*** *(BISBS)* ***stable*** *if and only if for every positive real number* $\varkappa$ *and* γ *there exist* *positive real numbers η and ξ, the values of which depend on $\varkappa$ and γ, $\eta = \eta(\gamma, \varkappa)$ and $\xi = \xi(\gamma, \varkappa)$, such that for every input vector function obeying $\|\mathbf{i}(k)\| < \eta$ for all $k \in \mathcal{N}_0$, and for every overall initial state vector satisfying $\left\| \begin{bmatrix} (\mathbf{r}_0^{\alpha-1})^T & (\mathbf{y}_0^{\nu-1})^T \end{bmatrix}^T \right\| < \xi$, the system internal state vector fulfills $\left\|\mathbf{r}^{\alpha-1}(k;\mathbf{r}_0^{\alpha-1};\mathbf{i})\right\| < \varkappa$ and its output state obeys $\left\|\mathbf{y}^{\nu-1}(k;\mathbf{r}_0^{\alpha-1};\mathbf{y}_0^{\nu-1};\mathbf{i})\right\| < \gamma$ for all $k \in \mathcal{N}_0$,*

$$\forall \varkappa \in \mathcal{R}^+,\ \forall \gamma \in \mathcal{R}^+,\ \exists \eta \in \mathcal{R}^+,\ \eta = \eta(\gamma, \varkappa),\ \exists \xi \in \mathcal{R}^+,\ \xi = \xi(\gamma, \varkappa),$$
$$\|\mathbf{i}(k)\| < \eta,\ \forall k \in \mathcal{N}_0,\ and\ \left\| \begin{bmatrix} (\mathbf{r}_0^{\alpha-1})^T & (\mathbf{y}_0^{\nu-1})^T \end{bmatrix}^T \right\| < \xi \Longrightarrow$$
$$\left\|\mathbf{r}^{\alpha-1}(k;\mathbf{r}_0^{\alpha-1};\mathbf{i})\right\| < \varkappa, \left\|\mathbf{y}^{\nu-1}(k;\mathbf{r}_0^{\alpha-1};\mathbf{y}_0^{\nu-1};\mathbf{i})\right\| < \gamma, \forall k \in \mathcal{N}_0. \tag{14.19}$$

Note 14.6 *The norm $\|\mathbf{i}_0\|$ of the initial input vector $\mathbf{i}_0$ is bounded due to the boundedness of $\|\mathbf{i}(k)\|$ for all $k \in \mathcal{N}_0$.*

Note 14.7 *Positive real numbers $\varkappa$ and γ can be replaced by a positive real number ζ so that (14.19) takes a more compact equivalent form:*

$$\forall \zeta \in \mathcal{R}^+,\ \exists \eta \in \mathcal{R}^+,\ \eta = \eta(\zeta),\ \exists \xi \in \mathcal{R}^+,\ \xi = \xi(\zeta),$$
$$\|\mathbf{i}(k)\| < \eta,\ \forall k \in \mathcal{N}_0,\ and\ \left\| \begin{bmatrix} (\mathbf{r}_0^{\alpha-1})^T & (\mathbf{y}_0^{\nu-1})^T \end{bmatrix}^T \right\| < \xi \Longrightarrow$$
$$\left\| \begin{bmatrix} (\mathbf{r}^{\alpha-1}(k;\mathbf{r}_0^{\alpha-1};\mathbf{i}))^T & (\mathbf{y}^{\nu-1}(k;\mathbf{r}_0^{\alpha-1};\mathbf{y}_0^{\nu-1};\mathbf{i}))^T \end{bmatrix}^T \right\| < \zeta,\ \forall k \in \mathcal{N}_0. \tag{14.20}$$

Note 14.8 *Positive real numbers $\varkappa$ and γ can be replaced by a positive real number ζ so that (14.20) takes a more compact equivalent form:*

$$\forall \zeta \in \mathcal{R}^+,\ \exists \eta \in \mathcal{R}^+,\ \eta = \eta(\zeta),$$
$$\|\mathbf{i}(k)\| < \eta,\ \forall k \in \mathcal{N}_0,\ \Longrightarrow$$
$$\left\| \begin{bmatrix} (\mathbf{r}^{\alpha-1}(k;\mathbf{0}_{\alpha\rho};\mathbf{i}))^T & (\mathbf{y}^{\nu-1}(k;\mathbf{0}_{\alpha\rho};\mathbf{0}_{\nu N};\mathbf{i}))^T \end{bmatrix} \right\| < \zeta,\ \forall k \in \mathcal{N}_0. \tag{14.21}$$

Note 14.9 *These BI stability properties allow that the system (internal, output or full) state does not converge to the corresponding zero state vector.*

Comment 14.2 *Every BI stability property valid under nonzero initial conditions guarantees the corresponding BI stability property under all zero initial conditions.*

Subsection 14.3.3 contains the conditions for the main *BI* stability properties of the *IIO* system (3.64a) and (3.64b).

14.3 Conditions

14.3.1 *IO* systems

We will explore complex domain conditions for various BI stability properties of the IO system (3.56) under arbitrary all initial (internal and output) conditions (in Subsection 3.5.2). The system is repeated as

$$A^{(\nu)}\mathbf{y}^{\nu}(k) = B^{(\mu)}\mathbf{i}^{\mu}(k),\ \forall k \in \mathcal{N}_0,\ \nu \geq \mu. \tag{14.22}$$

Lemma 14.1 *Let the input vector function* $\mathbf{i}(\cdot)$ *belong to the family* $\mathcal{J}$, *(2.11) (in Section 2.1). In order for* $\mathbf{y}^{\nu-1}(k;\mathbf{i}_0^{\mu-1};\mathbf{y}_0^{\nu-1};\mathbf{i})$ *of the IO system (14.22) in the forced regime*

a) to be bounded on $\mathcal{N}_0$,

$$\exists \zeta \in \mathcal{R}^+, \zeta = \zeta\left(\mathbf{i}_0^{\mu-1}, \mathbf{y}_0^{\nu-1}, \mathbf{i},\right), \Rightarrow$$
$$\left\|\mathbf{y}^{\nu-1}(k;\mathbf{i}_0^{\mu-1};\mathbf{y}_0^{\nu-1};\mathbf{i})\right\| < \zeta, \forall k \in \mathcal{N}_0,$$

it is necessary and sufficient that $\mathbf{y}(k;\mathbf{i}_0^{\mu-1};\mathbf{y}_0^{\nu-1};\mathbf{i})$ *is bounded on* $\mathcal{N}_0$,

$$\exists \xi \in \mathcal{R}^+, \xi = \xi\left(\mathbf{i}_0^{\mu-1}, \mathbf{y}_0^{\nu-1}, \mathbf{i},\right), \Rightarrow \left\|\mathbf{y}(k;\mathbf{i}_0^{\mu-1};\mathbf{y}_0^{\nu-1};\mathbf{i})\right\| < \xi, \forall k \in \mathcal{N}_0,$$

b) to vanish as $k \to \infty$,

$$k \to \infty \Longrightarrow \mathbf{y}^{\nu-1}(k;\mathbf{i}_0^{\mu-1};\mathbf{y}_0^{\nu-1};\mathbf{i}) \to \mathbf{0}_{\nu N},$$

it is necessary and sufficient that $\mathbf{y}(k;\mathbf{i}_0^{\mu-1};\mathbf{y}_0^{\nu-1};\mathbf{i})$ *vanishes as* $k \to \infty$,

$$k \to \infty \Longrightarrow \mathbf{y}(k;\mathbf{i}_0^{\mu-1};\mathbf{y}_0^{\nu-1};\mathbf{i}) \to \mathbf{0}_N.$$

Proof. Let the input vector function $\mathbf{i}(\cdot)$ be arbitrarily chosen from the family $\mathcal{J}$, (2.11) (in Section 2.1) and be fixed. Let all the initial conditions be also arbitrarily chosen so that they are bounded and fixed.

Necessity. a) In order for $\mathbf{y}^{\nu-1}(k;\mathbf{i}_0^{\mu-1};\mathbf{y}_0^{\nu-1};\mathbf{i})$ to be bounded on $\mathcal{N}_0$ it is necessary its every entry is bounded on $\mathcal{N}_0$. Hence, it is necessary that $\mathbf{y}(k;\mathbf{i}_0^{\mu-1};\mathbf{y}_0^{\nu-1};\mathbf{i})$ is bounded on $\mathcal{N}_0$.

b) In order for $\mathbf{y}^{\nu-1}(k;\mathbf{i}_0^{\mu-1};\mathbf{y}_0^{\nu-1};\mathbf{i})$ to vanish as $k \to \infty$ it is necessary that every entry of $\mathbf{y}^{\nu-1}(k;\mathbf{i}_0^{\mu-1};\mathbf{y}_0^{\nu-1};\mathbf{i})$ vanishes as $k \to \infty$. Hence, it is necessary that $\mathbf{y}(k;\mathbf{i}_0^{\mu-1};\mathbf{y}_0^{\nu-1};\mathbf{i})$ vanishes as $k \to \infty$.

Sufficiency. The $Z-$transform of $\mathbf{y}(k;\mathbf{i}_0^{\mu-1};\mathbf{y}_0^{\nu-1};\mathbf{i})$ for $\mu \geq 1$ is determined by (8.6a) and (8.7a):

$$\mathbf{Y}(z) = \left[\begin{array}{ccc} \boldsymbol{G}_{IO}(z) & \boldsymbol{G}_{IOi_0}(z) & \boldsymbol{G}_{IOy_0}(z) \end{array}\right] \cdot$$
$$\cdot \left[\begin{array}{ccc} \mathbf{I}^T(z) & \left(\mathbf{i}_0^{\mu-1}\right)^T & \left(\mathbf{y}_0^{\nu-1}\right)^T \end{array}\right]^T.$$

Equations (2.10) and (2.11) transform the preceding equation into:

$$\mathbf{Y}(z) = \boldsymbol{H}_{IO}(z)\left[\begin{array}{ccc} \mathbf{1}_M^T & \left(\mathbf{i}_0^{\mu-1}\right)^T & \left(\mathbf{y}_0^{\nu-1}\right)^T \end{array}\right]^T,$$
$$\boldsymbol{H}_{IO}(z) = \left[\begin{array}{ccc} \boldsymbol{G}_{IO}(z)\boldsymbol{I}(z) & \boldsymbol{G}_{IOi_0}(z) & \boldsymbol{G}_{IOy_0}(z) \end{array}\right], \; \mathbf{I}(z) = \boldsymbol{I}(z)\,\mathbf{1}_M. \tag{14.23}$$

a) Let the conditions of the statement of lemma under *a*) be valid. This, $\mathbf{i}(\cdot) \in \mathcal{J}$, the boundedness of the initial conditions, the Heaviside expansion of (14.23) and its inverse $Z-$transform yield

$$\mathbf{y}(k;\mathbf{i}_0^{\mu-1};\mathbf{y}_0^{\nu-1};\mathbf{i}) =$$
$$= \left\{ \delta_d(k)\boldsymbol{R}_0 + \sum_{i=1}^{i=\nu N} (z_i^*)^{k-1} \left[\sum_{r=1}^{r=\nu_i^*} \frac{1}{(z_i^*)^{r-1}(r-1)!} \frac{(k-1)!}{(k-r)!} \boldsymbol{R}_{ir} \right] \right\} \cdot$$
$$\cdot \begin{bmatrix} \mathbf{1}_M \\ \mathbf{i}_0^{\mu-1} \\ \mathbf{y}_0^{\nu-1} \end{bmatrix}, \boldsymbol{R}_0 \in \mathcal{R}^{N \times ((\mu+1)M+\nu N)}, \; \boldsymbol{R}_{ir} \in \mathcal{C}^{N \times ((\mu+1)M+\nu N)},$$
$$\boldsymbol{R}_0 = \text{const.}, \; \boldsymbol{R}_{ir} = \text{const.},$$
$$|z_i^*| \leq 1, \; \forall i = 1, 2, .., \nu N, \; |z_i^*| = 1 \Longrightarrow \nu_i^* = 1, \tag{14.24}$$

where ν_i^* is the multiplicity of z_i^*. This shows that $\mathbf{y}(k;\mathbf{i}_0^{\mu-1};\mathbf{y}_0^{\nu-1};\mathbf{i})$ is infinitely times shiftable at every $k \in \mathbb{Z}$. All its shifts rest bounded on $\mathcal{N}_0$ due to the Generating Theorem 13.39, which guarantees boundedness also of $\mathbf{y}^{\nu-1}(k;\mathbf{i}_0^{\mu-1};\mathbf{y}_0^{\nu-1};\mathbf{i})$ on $\mathcal{N}_0$ if $\mu \geq 1$. In case $\mu = 0$ the submatrix $\boldsymbol{G}_{IOi_0}(z)$ is to be replaced by the zero matrix and the vector $\mathbf{i}_0^{\mu-1}$ by the zero vector so that (14.24) becomes

$$\mathbf{y}(k;\mathbf{y}_0^{\nu-1};\mathbf{i}) =$$

$$= \left\{ \delta_d(k)\boldsymbol{R}_0 + \sum_{i=1}^{i=\nu N} (z_i^*)^{k-1} \left[\sum_{r=1}^{r=\nu_i^*} \frac{1}{(z_i^*)^{r-1}(r-1)!} \frac{(k-1)!}{(k-r)!} \boldsymbol{R}_{ir} \right] \right\} \cdot$$

$$\cdot \begin{bmatrix} \mathbf{1}_M \\ \mathbf{y}_0^{\nu-1} \end{bmatrix}, \boldsymbol{R}_0 \in \mathcal{R}^{N\times(M+\nu N)}, \ \boldsymbol{R}_{ir} \in \mathcal{C}^{N\times(M+\nu N)},$$

$$\boldsymbol{R}_0 = \text{const.}, \ \boldsymbol{R}_{ir} = \text{const.},$$

$$|z_i^*| \leq 1, \ \forall i = 1, 2, .., \nu N, \ \ |z_i^*| = 1 \Longrightarrow \nu_i^* = 1, \tag{14.25}$$

The boundedness of the output response $\mathbf{y}(k;\mathbf{i}_0^{\mu-1};\mathbf{y}_0^{\nu-1};\mathbf{i})$, the condition under a), and Equations (14.24) and (14.25) imply that the output response $\mathbf{y}(k;\mathbf{i}_0^{\mu-1};\mathbf{y}_0^{\nu-1};\mathbf{i})$ is infinitely times shiftable and with each shift bounded on $\mathcal{N}_0$ due to the Generating Theorem 13.39), which ensures that $\mathbf{y}^{\nu-1}(k;\mathbf{i}_0^{\mu-1};\mathbf{y}_0^{\nu-1};\mathbf{i})$ is also bounded on $\mathcal{N}_0$.

b) Let the condition of the statement of lemma under b) hold. It, and Equations (14.24) and (14.25) imply that the output response $\mathbf{y}(k;\mathbf{i}_0^{\mu-1};\mathbf{y}_0^{\nu-1};\mathbf{i})$ is infinitely times shiftable at every $k \in \mathbb{Z}$. All its shifts rest bounded on $\mathcal{N}_0$ and vanishes as $k \to \infty$ due to the Generating Theorem 13.39. Every entry of $\mathbf{y}^{\nu-1}(k;\mathbf{i}_0^{\mu-1};\mathbf{y}_0^{\nu-1};\mathbf{i})$ is bounded and vanishes as $k \to \infty$. ■

This lemma reduces largely the study of the *BI* stability properties of the *IO* system (14.22).

In view of the fact that the internal state $\mathbf{S}_I$, the output state $\mathbf{S}_O$ and the (full) state $\mathbf{S}$ of the *IO* system (14.22) coincide, (2.6), the *BIISBIS*, *BIISBOS*, *BIISBISO*, and *BIISBSO* stability coincide with *BISBS*, and *BIISBO* with *BISBO*.

Theorem 14.3 *Let the input vector function* $\mathbf{i}(\cdot)$ *belong to the family* $\mathcal{J}$. *In order for the IO system (14.22) in the forced regime to be:*

a) ***Bounded Input-Bounded State** (BIBS) **stable,** it is necessary and sufficient that*

1. $\boldsymbol{G}_{IO}(\cdot)$ *is either proper or strictly proper, real rational matrix function of* z,

   and that

2. *the modulus* $|z_i^*(\mathbf{G}_{IOrnd})|$ *of all poles* $z_i^*(\mathbf{G}_{IOrnd})$ *of the row nondegenerate form* $\mathbf{G}_{IOrnd}(z)$ *of the system transfer function matrix* $\mathbf{G}_{IO}(z)$ *are less than one,*

$$|z_i^*(\mathbf{G}_{IOrnd})| < 1, \ \forall i \in \{1, 2, \cdots, \nu N\}. \tag{14.26}$$

b) ***Bounded Input-Bounded Output*** *(BIBO)* ***stable,*** *it is necessary and sufficient that the conditions under a) hold.*

c) ***Bounded Input and Initial State-Bounded State*** *(BISBS)* ***stable,*** *it is necessary and sufficient that*

1. $\mathbf{G}_{IO}(\cdot)$ *is either proper or strictly proper, real rational matrix function of* z, $\mathbf{G}_{IOi_0}(\cdot)$ *and* $\mathbf{G}_{IOy_0}(\cdot)$ *are also either proper or strictly proper real rational matrix functions of* z,
   and that
2. *the modulus* $|z_i^*(\mathbf{G}_{IOrnd})|$ *of all poles* $z_i^*(\mathbf{G}_{IOrnd})$ *of the row nondegenerate form* $\mathbf{G}_{IOrnd}(z)$ *of the system transfer function matrix* $\mathbf{G}_{IO}(z)$ *are less than one, the modulus* $|z_i^*(\mathbf{G}_{IOi_0rnd})|$ *and* $|z_i^*(\mathbf{G}_{IOy_0rnd})|$ *of all poles* $z_i^*(\mathbf{G}_{IOi_0rnd})$ *and* $z_i^*(\mathbf{G}_{IOy_0rnd})$ *of the row nondegenerate forms* $\mathbf{G}_{IOi_0rnd}(\cdot)$ *and* $\mathbf{G}_{IOy_0rnd}(\cdot)$ *of the system transfer function matrices* $\mathbf{G}_{IOi_0}(\cdot)$ *and* $\mathbf{G}_{IOy_0}(\cdot)$ *are less or equal to one and those with the modulus equal to one are simple,*

$$\begin{gathered} |z_i^*(\mathbf{G}_{IOrnd})| < 1, \ \forall i \in \{1, 2, \cdots, \nu N\}, \\ |z_i^*(\mathbf{G}_{IOi_0rnd})| \le 1 \text{ and } |z_i^*(\mathbf{G}_{IOy_0rnd})| \le 1, \forall i \in \{1, 2, \cdots, \nu N\}, \\ \left|z_j^*(\mathbf{G}_{IOi_0rnd})\right| = 1 \Longrightarrow \nu_j^* = 1, \\ |z_m^*(\mathbf{G}_{IOy_0rnd})| = 1 \Longrightarrow \nu_m^* = 1. \end{gathered} \tag{14.27}$$

d) ***Bounded Input and Initial State-Bounded Output*** *(BISBO)* ***stable,*** *it is necessary and sufficient that the conditions under c) hold.*

Proof. Let the input vector function $\mathbf{i}(\cdot)$ be arbitrarily chosen from the family $\mathcal{J}$, (2.11) (in Section 2.1) and be fixed. Let all the initial conditions be also arbitrarily chosen so that they are bounded and fixed.

Necessity. a) Let the IO system (14.22) be $BIBS$ stable. Hence, (14.1), (item a) of Definition 14.1, in Subsection 14.2.1), holds. The overall output response $\mathbf{y}^{\nu-1}(k; \mathbf{0}_{\mu M}; \mathbf{0}_{\nu N}; \mathbf{i})$ of the system may contain or not a discrete impulse component. The same holds for $\mathbf{y}(k; \mathbf{0}_{\mu M}; \mathbf{0}_{\nu N}; \mathbf{i})$ as a subvector of $\mathbf{y}^{\nu-1}(k; \mathbf{0}_{\mu M}; \mathbf{0}_{\nu N}; \mathbf{i})$. Its $Z-$transform

$$\mathcal{Z}\{\mathbf{y}(k; \mathbf{0}_{\mu M}; \mathbf{0}_{\nu N}; \mathbf{i})\} = \mathbf{G}_{IO}(z)\mathbf{I}(z),$$

is either proper or strictly proper real rational vector function due to the Generating Theorem 13.39 (in Subsection 13.4.1). Let

$$\mathbf{i}(k) = h(k)\mathbf{1}_M, \mathbf{1}_M = \left[\begin{array}{cccc} 1 & 1 & \cdots & 1 \end{array}\right]^T \in \mathcal{R}^M \Longrightarrow \mathbf{I}(z) = \frac{z}{z-1}\mathbf{1}_M \in \mathcal{C}^M.$$

Hence,

$$\mathcal{Z}\{\mathbf{y}(k; \mathbf{0}_{\mu M}; \mathbf{0}_{\nu N}; h(k)\mathbf{1}_M)\} = z\,(z-1)^{-1}\, \boldsymbol{G}_{IO}(z)\mathbf{1}_M$$

is either proper or strictly proper real rational vector function. This proves necessity of condition a-1. From this equation and Lemma 6.1 (in Chapter 6) follows that only the poles of the row nondegenerate form $\boldsymbol{G}_{IOrnd}(z)$ of $\boldsymbol{G}_{IO}(z)$ and the poles of $\mathbf{I}(z)$ determine the character of $\mathbf{y}^{\nu-1}(k; \mathbf{0}_{\mu M}; \mathbf{0}_{\nu N}; \mathbf{i})$ for every $\mathbf{i}(\cdot) \in \mathcal{J}$. The Generating Theorem 13.39 and (14.1) prove necessity of condition a-2, i.e., (14.26).

b) The preceding proof of the necessity of the conditions under a) applies to b).

c) Let the IO system (14.22) be $BISBS$ stable. Hence, (14.3), (item c) of Definition 14.1, in Subsection 14.2.1), holds. The overall output response $\mathbf{y}^{\nu-1}(k; \mathbf{i}_0^{\mu-1}; \mathbf{y}_0^{\nu-1}; \mathbf{i})$ of the system may contain or not a discrete impulse component. The same holds for $\mathbf{y}(k; \mathbf{i}_0^{\mu-1}; \mathbf{y}_0^{\nu-1}; \mathbf{i})$ as a subvector of $\mathbf{y}^{\nu-1}(k; \mathbf{i}_0^{\mu-1}; \mathbf{y}_0^{\nu-1}; \mathbf{i})$. Its $Z-$transform

$$\mathcal{Z}\{\mathbf{y}(k; \mathbf{i}_0^{\mu-1}; \mathbf{y}_0^{\nu-1}; \mathbf{i})\} = \boldsymbol{F}_{IO}(z) \begin{bmatrix} \mathbf{I}(z) \\ \mathbf{i}_0^{\mu-1} \\ \mathbf{y}_0^{\nu-1} \end{bmatrix}, \quad k = 1, 2, \ \dots \ , \nu - 1,$$

is either proper or strictly proper real rational vector function due to the Generating Theorem 13.39. Let

$$\mathbf{i}(k) = h(k)\mathbf{1}_M, \mathbf{1}_M = \left[\begin{array}{cccc} 1 & 1 & \cdots & 1 \end{array}\right]^T \in \mathcal{R}^M \Longrightarrow \mathbf{I}(z) = \frac{z}{z-1}\mathbf{1}_M \in \mathcal{C}^M.$$

This and (8.6a) (in Section 8.1) yield

$$\mathcal{Z}\{\mathbf{y}(k; \mathbf{i}_0^{\mu-1}; \mathbf{y}_0^{\nu-1}; h(k)\mathbf{1}_M)\} =$$
$$z\,(z-1)^{-1}\, \boldsymbol{G}_{IO}(z)\mathbf{1}_M + \boldsymbol{G}_{IOi_0}(z)\mathbf{i}_0^{\mu-1} + \boldsymbol{G}_{IOy_0}(z)\mathbf{y}_0^{\nu-1},$$

which is either proper or strictly proper real rational vector function. This proves necessity of condition c-1. From this equation and Lemma 6.1 (in Chapter 6) follows that only the poles of the row nondegenerate forms $\boldsymbol{G}_{IOrnd}(z)$, $\boldsymbol{G}_{IOi_0rnd}(z)$ and $\boldsymbol{G}_{IOy_0rnd}(z)$ of $\boldsymbol{G}_{IO}(z)$, $\boldsymbol{G}_{IOi_0}(z)$ and

$\boldsymbol{G}_{IOy_0}(z)$, as well as the poles of $\mathbf{I}(z)$, determine for every $\mathbf{i}(\cdot) \in \mathcal{J}$, the character of $\mathbf{y}^{\nu-1}(k;\mathbf{i}_0^{\mu-1};\mathbf{y}_0^{\nu-1};\mathbf{i})$. The Generating Theorem 13.39 (in Subsection 13.4.1), and (14.1) prove necessity of condition *c*-2, i.e., (14.27).

d) Let the *IO* system (14.22) be *BISBO* stable so that (14.4) (item *d)* of Definition 14.1, in Subsection 14.2.1) is valid. The output response $\mathbf{y}(k;\mathbf{i}_0^{\mu-1};\mathbf{y}_0^{\nu-1};\mathbf{i})$ of the system may contain or not a discrete impulse component. The repetition of the proof of the necessity under *c)* from this point on completes the proof of the necessity of *d)*.

Sufficiency. The boundedness of every $\mathbf{i}(\cdot)$ that belongs to the family $\mathcal{J}$ guarantees that all poles of $\mathbf{I}(z)$ are with modulus less or equal to one, that its poles with modulus equal to one are simple, and that $\mathbf{I}(z)$ is either proper or strictly proper real vector function.

a) Let the conditions under *a)* hold. The $Z-$transform of the system response $\mathbf{y}(k;\mathbf{0}_{\mu M};\mathbf{0}_{\nu N};\mathbf{i})$,

$$\mathcal{Z}\{\mathbf{y}(k;\mathbf{0}_{\mu M};\mathbf{0}_{\nu N};\mathbf{i})\} = \boldsymbol{G}_{IO}(z)\mathbf{I}(z),$$

is either proper or strictly proper and does not have either a pole with infinitely large modulus or a pole with modulus greater than one or a multiple pole with modulus equal to one for every $\mathbf{i}(\cdot)$ that belongs to the family $\mathcal{J}$. Lemma 6.1, and the Generating Theorem 13.39 imply that

$$\mathbf{y}(k;\mathbf{0}_{\mu M};\mathbf{0}_{\nu N};\mathbf{i}) = \mathcal{Z}^{-1}\left\{\boldsymbol{G}_{IO}(z)\mathbf{I}(z)\right\}$$

is bounded for every $\mathbf{i}(\cdot) \in \mathcal{J}$. Hence,

$$\mathbf{y}^{\nu-1}(k;\mathbf{0}_{\mu M};\mathbf{0}_{\nu N};\mathbf{i}) = \mathcal{Z}^{-1}\left\{\begin{bmatrix} \boldsymbol{I}_N & z\boldsymbol{I}_N & \cdots & z^{\nu-1}\boldsymbol{I}_N \end{bmatrix}^T \boldsymbol{G}_{IO}(z)\mathbf{I}(z)\right\}$$

is also bounded for every $\mathbf{i}(\cdot) \in \mathcal{J}$ due to Lemma 14.1.

b) Let the conditions under *b)* hold. The $Z-$transform

$$\mathcal{Z}\{\mathbf{y}(k;\mathbf{0}_{\mu M};\mathbf{0}_{\nu N};\mathbf{i})\} = \boldsymbol{G}_{IO}(z)\mathbf{I}(z)$$

of $\mathbf{y}(k;\mathbf{0}_{\mu M};\mathbf{0}_{\nu N};\mathbf{i})$ is either proper or strictly proper real rational function and it does not have either a pole with infinitely large modulus or a pole with modulus greater than one or a multiple pole with modulus equal to one for every $\mathbf{i}(\cdot)$ that belongs to the family $\mathcal{J}$. Lemma 6.1 and the Generating Theorem 13.39 imply that

$$\mathbf{y}(k;\mathbf{0}_{\mu M};\mathbf{0}_{\nu N};\mathbf{i}) = \mathcal{Z}^{-1}\left\{\boldsymbol{G}_{IO}(z)\mathbf{I}(z)\right\},$$

is bounded for every $\mathbf{i}(\cdot) \in \mathcal{J}$.

c) Let the conditions under *c)* hold. The system response $\mathbf{y}(k;\mathbf{i}_0^{\mu-1};\mathbf{y}_0^{\nu-1};\mathbf{i})$ obeys Equations (14.24) and (14.25). The matrix function $\boldsymbol{H}_{IO}(\cdot)$ is either proper or strictly proper real rational function for every $\mathbf{i}(\cdot) \in \mathcal{J}$, and it does not have either a pole with infinitely large modulus or a pole with modulus greater than one or a multiple pole with the modulus equal to one for every $\mathbf{i}(\cdot)$ that belongs to the family $\mathcal{J}$. The conditions under *c)*, Lemma 6.1, the Generating Theorem 13.39 imply that the system response $\mathbf{y}(k;\mathbf{i}_{0^-}^{\mu-1};\mathbf{y}_{0^-}^{\nu-1};\mathbf{i})\}$ is bounded for every $\mathbf{i}(\cdot) \in \mathcal{J}$. This and Lemma 14.1 prove that $\mathbf{y}^{(\nu-1)}(k;\mathbf{i}_0^{\mu-1};\mathbf{y}_0^{\nu-1};\mathbf{i})$ is also bounded for every $\mathbf{i}(\cdot) \in \mathcal{J}$.

d) The statement under *c)* proves the statement under *d)* as its special case. ■

The *BI* stability conditions, like the asymptotic stability conditions, demand that the modulus of all poles of the row nondegenerate form of the appropriate system transfer function matrix are at least less or equal to one, or less than one, depending on the transfer function matrix. However, the appropriate transfer function matrices are different for different *BI* stability properties and for the asymptotic stability.

Example 14.1 *For the IO system of Example 8.1 (in Section 8.1)*

$$E^2y(k) - E^1y(k) - 0.75E^0y(k) = E^2i(k) - 7.5E^1i(k) + 9E^0i(k),$$

we determined the following transfer function $\boldsymbol{G}_{IO}(z)$ *and the full transfer function matrix* $\boldsymbol{F}_{IO}(z)$*:*

$$\boldsymbol{G}_{IO}(z) = \frac{z^2 - 7.5z + 9}{z^2 - z - 0.75} = \frac{(z-1.5)(z-6)}{(z-1.5)(z+0.5)} \Longrightarrow G_{IOnd}(z) = \frac{z-6}{z+0.5},$$

$$\boldsymbol{F}_{IO}(z) = \begin{bmatrix} \underbrace{\dfrac{z^2 - 7.5z + 9}{(z-1.5)(z+0.5)}}_{G_{IO}(z)} \\ \left(\underbrace{\frac{-z(z-7.5)}{(z-1.5)(z+0.5)} \quad \frac{-z}{(z-1.5)(z+0.5)}}_{G_{IOi_0}(z)} \right)^T \\ \left(\underbrace{\frac{z(z-1)}{(z-1.5)(z+0.5)} \quad \frac{z}{(z-1.5)(z+0.5)}}_{G_{IOy_0}(z)} \right)^T \end{bmatrix}^T = \boldsymbol{F}_{IOnd}(z).$$

The system transfer function matrix $\mathbf{G}_{IO}(z)$ *has the same zero* $z^0 = 1.5$ *and pole* $z^* = 1.5$ *so that they cancel yielding the nondegenerate form* $\mathbf{G}_{IOnd}(z)$ *of* $\mathbf{G}_{IO}(z)$ *that has only one pole. The system is both* $BIBS$ *and* $BIBO$ *stable.*

Since the pole is with modulus less than one, $z^* = -0.5$*, then we could conclude wrongly from* $\mathbf{G}_{IOnd}(z)$ *that the system is* $BISBS$ *and* $BISBO$ *stable. However, the cancellation of the same zero* $z^0 = 1.5$ *and pole* $z^* = 1.5$ *is not possible in either the system full transfer function matrix* $\boldsymbol{F}_{IO}(z)$ *or in its submatrices* $\mathbf{G}_{IOi_0}(z)$ *and* $\mathbf{G}_{IOy_0}(z)$*. Since they have one pole with modulus greater than one* $z^* = 1.5$*, then it follows that the system is* $BISBS$ *and* $BISBO$ *unstable for some* $i(\cdot) \in \mathcal{J}$*. This shows that we should not apply the* BI *stability criteria established under all zero initial conditions to* BI *stability properties in general, i.e., under arbitrary initial conditions.*

Example 14.2 *We analyze further system presented under* 2) *in Example 8.7 (in Section 8.2), which is analogy to the* $s-$*complex case given in [5, Example 3.8].*

The state space model under 2) *in Example 8.7 yields the second-order* IO *discrete equation, hence the degenerate transfer function matrix* $\mathbf{G}_{IO2}(z)$ *and the full transfer function matrix* $\boldsymbol{F}_{IOISO2}(z)$ *that is nondegenerate,*

$$E^2 y_2(k) - y_2(k) = Ei(k) - i(k),$$
$$\mathbf{G}_{IO2}(z) = (z^2-1)^{-1}(z-1) \Longrightarrow \mathbf{G}_{IO2nd}(z) = (z+1)^{-1},$$
$$\boldsymbol{F}_{IOISO2}(z) = \left[\underbrace{\frac{z-1}{z^2-1}}_{G_{IO2}(z)} \quad \underbrace{\frac{-z}{z^2-1}}_{G_{IOi_02}(z)} \quad \underbrace{\frac{z^2}{z^2-1} \quad \frac{z}{z^2-1}}_{G_{IOy_02}(z)} \right].$$

The cancellation of the same zero $z^0 = 1$ *and pole* $z^* = 1$ *of the transfer functions* $\mathbf{G}_{IO2}(z)$ *is possible only in them, but it cannot be carried out either in the full transfer function matrix* $\boldsymbol{F}_{IOISO2}(z)$ *or in* $\mathbf{G}_{IOi_02}(z)$ *and in* $\mathbf{G}_{IOy_02}(z)$*.*

The systems are both $BIBS$ *and* $BIBO$ *unstable. They are both* $BISBS$ *and* $BISBO$ *unstable for some* $i(\cdot) \in \mathcal{J}$*.*

Example 14.3 *Let*

$$E^2y(k) + 0.9E^1y(k) + 0.2E^0y(k) = 0.36E^0i(k) + 1.2E^1i(k) + E^2i(k) \Longrightarrow$$
$$\nu = \mu = 2.$$

Its Z−transform yields

$$\boldsymbol{G}_{IO}(z) = \frac{z^2+1.2z+0.36}{z^2+0.9z+0.2} = \boldsymbol{G}_{IOnd}(z),$$

$$\boldsymbol{F}_{IO}(z) = \begin{bmatrix} \frac{z^2+1.2z+0.36}{z^2+0.9z+0.2} \\ -\frac{z(z+1.2)}{z^2+0.9z+0.2} \\ -\frac{z}{z^2+0.9z+0.2} \\ \frac{z(z+0.9)}{z^2+0.9z+0.2} \\ \frac{z}{z^2+0.9z+0.2} \end{bmatrix}^T = \boldsymbol{F}_{IOnd}(z).$$

Both $\boldsymbol{G}_{IO}(z)$ *and* $\boldsymbol{F}_{IO}(z)$ *is proper and nondegenerate. However,* $\boldsymbol{G}_{IOi_0}(z)$ *and* $\boldsymbol{G}_{IOy_0}(z)$ *are strictly proper.*

The system transfer function matrices obey all the conditions of Theorem 14.3. The system is $BIBS$*,* $BIBO$*,* $BISBS$ *and* $BISBO$ *stable for every* $i(\cdot) \in \mathcal{J}$*.*

For example, the unit step response of the system under all zero initial conditions

$$y(t;\mathbf{0}_2;\mathbf{0}_2;h) = \mathcal{Z}^{-1}\left\{\frac{z^2+1.2z+0.36}{z^2+0.9z+0.2}\frac{z}{z-1}\right\},$$

and the unit step response of the system under nonzero initial conditions,

$$y(t;\mathbf{i}_0^1;\mathbf{y}_0^1;h) = \mathcal{Z}^{-1}\left\{\frac{z^2+1.2z+0.36}{z^2+0.9z+0.2}\frac{z}{z-1} - \frac{z(z+1.2)}{z^2+0.9z+0.2}i_0 - \right.$$

$$\left. -\frac{z}{z^2+0.9z+0.2}i_0(1) + \frac{z(z+0.9)}{z^2+0.9z+0.2}y_0 + \frac{z}{z^2+0.9z+0.2}y_0(1)\right\}$$

are bounded for the input $i(k) = h(k)$ *and for bounded initial conditions (they contain discrete impulse components but without of disruption of boundedness),*

$$i_0 \in \mathcal{R},\ i_0(1) \in \mathcal{R},\ y_0 \in \mathcal{R},\ y_0(1) \in \mathcal{R}.$$

The modulus of the poles of $\boldsymbol{G}_{IO}(z)$ *and of* $\boldsymbol{F}_{IO}(z)$ *are less than one.*

Example 14.4 *We consider the following* IO *system,*

$$\begin{bmatrix} 1 & 1 \\ 1 & 2 \end{bmatrix} E^2\mathbf{y}(k) - 2\begin{bmatrix} 1 & 1 \\ 1 & 2 \end{bmatrix} E^1\mathbf{y}(k) + 0.75\begin{bmatrix} 1 & 1 \\ 1 & 2 \end{bmatrix} E^0\mathbf{y}(k) =$$

$$= \begin{bmatrix} -1.5 & 2 \\ 0 & -3 \end{bmatrix} E^0\mathbf{i}(k) + \begin{bmatrix} 1 & 1 \\ -1.5 & 1 \end{bmatrix} E^1\mathbf{i}(k) + \begin{bmatrix} 0 & 0 \\ 1 & 0 \end{bmatrix} E^2\mathbf{i}(k).$$

Its transfer function matrix

$$\boldsymbol{G}_{IO}(z) = \frac{(z-0.5)(z-1.5)}{[(z-0.5)(z-1.5)]^2}\begin{bmatrix} -(z-1.5)(s-2) & z+7 \\ (z-1.5)(z-1) & -5 \end{bmatrix}$$

is degenerate. The nondegenerate form $\boldsymbol{G}_{IOnd}(z)$ *of* $\boldsymbol{G}_{IO}(z)$ *reads*

$$\boldsymbol{G}_{IOnd}(z) = \frac{1}{(z-0.5)(z-1.5)} \begin{bmatrix} -(z-1.5)(z-2) & z+7 \\ (z-1.5)(z-1) & -5 \end{bmatrix}.$$

It is also the row nondegenerate form $\boldsymbol{G}_{IOrnd}(z)$ *of* $\boldsymbol{G}_{IO}(z)$,

$$\boldsymbol{G}_{IOnd}(z) = \boldsymbol{G}_{IOrnd}(z).$$

The column nondegenerate form $\boldsymbol{G}_{IOcnd}(z)$ *of* $\boldsymbol{G}_{IO}(z)$ *is different from them,*

$$\boldsymbol{G}_{IOcnd}(z) = \frac{1}{z-0.5} \begin{bmatrix} -(z-2) & z+7 \\ z-1 & -5 \end{bmatrix}.$$

If we wish to test whether the system is $BIBS$ *stable and/or* $BIBO$ *stable we should use the row nondegenerate form* $\boldsymbol{G}_{IOrnd}(z)$ *of* $\boldsymbol{G}_{IO}(z)$*, but we may not use its column nondegenerate form* $\boldsymbol{G}_{IOcnd}(z)$ *because the pole* $z^* = 1.5$ *cannot be cancelled in the rows of* $\boldsymbol{G}_{IOrnd}(z)$*. If we used the column nondegenerate form* $\boldsymbol{G}_{IOcnd}(z)$ *we would conclude wrongly that the system is* $BIBS$ *stable and/or* $BIBO$ *stable. However, it is* $BIBS$ *unstable and* $BIBO$ *unstable for some* $i(\cdot) \in \mathcal{J}$ *due to the pole* $z^* = 1.5$ *of the row nondegenerate form* $\boldsymbol{G}_{IOrnd}(z)$ *of* $\boldsymbol{G}_{IO}(z)$*, which is with modulus greater than one.*

The full system transfer function matrix

$$\boldsymbol{F}_{IO}(z) = \frac{(z-0.5)(z-1.5)}{[(z-0.5)(z-1.5)]^2} \cdot \begin{bmatrix} -(z-1.5)(z-2) & (z-1.5)(z-1) \\ z+7 & -5 \\ z(z-3.5) & -z(z-2.5) \\ -z & 0 \\ z & -z \\ 0 & 0 \\ z(z-2) & 0 \\ 0 & z(z-2) \\ z & 0 \\ 0 & z \end{bmatrix}^T$$

is degenerate. Its nondegenerate form

$$\boldsymbol{F}_{IOnd}(z) = \frac{1}{(z-0.5)(z-1.5)} \cdot$$
$$\cdot \begin{bmatrix} -(z-1.5)(z-2) & (z-1.5)(z-1) \\ z+7 & -5 \\ z(z-3.5) & -z(z-2.5) \\ -z & 0 \\ z & -z \\ 0 & 0 \\ z(z-2) & 0 \\ 0 & z(z-2) \\ z & 0 \\ 0 & z \end{bmatrix}^T$$

is also its row nondegenerate form $\boldsymbol{F}_{IOrnd}(z)$ *and its column nondegenerate form* $\boldsymbol{F}_{IOcnd}(z)$,

$$\boldsymbol{F}_{IOrnd}(z) = \boldsymbol{F}_{IOnd}(z) = \boldsymbol{F}_{IOcnd}(z).$$

They show that the system is both $BISBS$ *unstable for some* $i(\cdot) \in \mathcal{J}$ *and* $BISBO$ *unstable for some* $i(\cdot) \in \mathcal{J}$ *because* $\mathbf{G}_{IOi_0cnd}(z)$ *and* $\mathbf{G}_{IOy_0cnd}(z)$ *have pole* $z^* = 1.5$ *with modulus greater than one.*

14.3.2 *ISO* systems

For the ISO system (3.60a) and (3.60b) (in Subsection 3.5.3),

$$\mathbf{x}(k+1) = \boldsymbol{A}\mathbf{x}(k) + \boldsymbol{B}\mathbf{i}(k),\ \forall k \in \mathcal{N}_0, \tag{14.28}$$
$$\mathbf{y}(k) = \boldsymbol{C}\mathbf{x}(k) + \boldsymbol{D}\mathbf{i}(k),\ \forall k \in \mathcal{N}_0, \tag{14.29}$$

we recall the fact that its internal state $\mathbf{S}_I$, output state $\mathbf{S}_O$ and (full) state $\mathbf{S}$ coincide, (2.15). This explains that the $BIISBIS$, $BIISBOS$, $BIISBISO$, and $BIISBSO$ stability coincide with $BISBS$, and $BIISBO$ with $BISBO$.

Theorem 14.4 *Let the input vector function* $\mathbf{i}(\cdot)$ *belong to the family* $\mathcal{J}$, *(2.11) (in Section 2.1). In order for the ISO system (14.28) and (14.29) in the forced regime to be:*

a) ***Bounded Input-Bounded State*** *(BIBS)* ***stable****, it is necessary and sufficient that:*

1. *the system IS transfer function matrix* $\mathbf{G}_{ISOIS}(z)$ *is either proper or strictly proper, real rational matrix function of* z*,*
   and that
2. *the modulus* $|z_i^*(\mathbf{G}_{ISOISrnd})|$ *of all poles* $z_i^*(\mathbf{G}_{ISOISrnd})$ *of the row nondegenerate form* $\mathbf{G}_{ISOISrnd}(z)$ *of the system IS transfer function matrix* $\mathbf{G}_{ISOIS}(z)$ *are less than one,*

$$|z_i^*(\mathbf{G}_{ISOISrnd})| < 1, \ \forall i \in \{1, 2, \cdots, n\}. \tag{14.30}$$

b) ***Bounded Input-Bounded Output*** *(BIBO)* ***stable****, it is necessary and sufficient that:*

1. *the system IO transfer function matrix* $\mathbf{G}_{ISO}(z)$ *is either proper or strictly proper, real rational matrix function of* z*,*
   and that
2. *the modulus* $|z_i^*(\mathbf{G}_{ISOrnd})|$ *of all poles* $z_i^*(\mathbf{G}_{ISOrnd})$ *of the row nondegenerate form* $\mathbf{G}_{ISOrnd}(z)$ *of the system IO transfer function matrix* $\mathbf{G}_{ISO}(z)$ *are less than one,*

$$|z_i^*(\mathbf{G}_{ISOrnd})| < 1, \ \forall i \in \{1, 2, \cdots, n\}. \tag{14.31}$$

c) ***Bounded both Input and Initial State-Bounded State*** *(BISBS)* ***stable****, it is necessary and sufficient that:*

1. *the system IS transfer function matrix* $\mathbf{G}_{ISOIS}(z)$ *is either proper or strictly proper, real rational matrix function of* z*, and the system SS transfer function matrix* $\mathbf{G}_{ISOSS}(z)$ *is also either proper or strictly proper real rational matrix function of* z*,*
2. *the modulus* $|z_i^*(\mathbf{G}_{ISOISrnd})|$ *and* $|z_i^*(\mathbf{G}_{ISOSSrnd})|$ *of every pole* $z_i^*(\mathbf{G}_{ISOISrnd})$ *and* $z_i^*(\mathbf{G}_{ISOSSrnd})$ *of the row nondegenerate forms*

   $$\mathbf{G}_{ISOISrnd}(z) \text{ and } \mathbf{G}_{ISOSSrnd}(z)$$

   of the system IS transfer function matrix $\mathbf{G}_{ISOIS}(z)$ *are less than one, and of* $\mathbf{G}_{ISOSS}(z)$ *are less or equal to one and those with the modulus equal to one are simple, respectively,*

$$\begin{array}{c} |z_i^*(\mathbf{G}_{ISOISrnd})| < 1, \ \forall i \in \{1, 2, \cdots, n\} \\ |z_i^*(\mathbf{G}_{ISOSSrnd})| \leq 1, \ \forall i \in \{1, 2, \cdots, n\} \text{ and} \\ \left|z_j^*(\mathbf{G}_{ISOSSrnd})\right| = 1 \Longrightarrow \nu_j^* = 1. \end{array} \tag{14.32}$$

d) ***Bounded both Input and Initial State-Bounded Output*** *(BISBO)* ***stable****, it is necessary and sufficient that:*

1. *the system IO transfer function matrix* $\boldsymbol{G}_{ISO}(z)$ *is either proper or strictly proper, real rational matrix function of* z*, and the system IISO transfer function matrix* $\boldsymbol{G}_{ISOx_0}(z)$ *is also either proper or strictly proper real rational matrix function of* z*,*
2. *the modulus* $|z_i^*(\boldsymbol{G}_{ISOrnd})|$ *and* $|z_i^*(\boldsymbol{G}_{ISOxornd})|$ *of every pole* $z_i^*(\boldsymbol{G}_{ISOrnd})$ *and* $z_i^*(\boldsymbol{G}_{ISOxornd})$ *of the row nondegenerate forms* $\boldsymbol{G}_{ISOrnd}(z)$ *and* $\boldsymbol{G}_{ISOxornd}(z)$ *of the system IO transfer function matrix* $\boldsymbol{G}_{ISO}(z)$ *are less than one, and of* $\boldsymbol{G}_{ISOxo}(z)$ *are less or equal to one and those with the modulus equal to one are simple, respectively,*

$$\begin{gathered}|z_i^*(\boldsymbol{G}_{ISOrnd})| < 1,\ \forall i \in \{1, 2, \cdots, n\},\\ |z_i^*(\boldsymbol{G}_{ISOxornd})| \leq 1,\ \forall i \in \{1, 2, \cdots, n\}\\ \left|z_j^*(\boldsymbol{G}_{ISOxornd})\right| = 1 \Longrightarrow \nu_j^* = 1.\end{gathered} \tag{14.33}$$

Proof. Let the input vector function $\mathbf{i}(\cdot)$ be arbitrarily chosen from the family $\mathcal{J}$, (2.11) (in Section 2.1) and be fixed. Let all the initial conditions be also arbitrarily chosen so that they are bounded and fixed. Every $\mathbf{i}(\cdot) \in \mathcal{J}$ is bounded and guarantees that the modulus of all poles of $\mathbf{I}(z)$ are less or equal to one, that its poles with the modulus equal to one are simple, and that $\mathbf{I}(z)$ is either proper or strictly proper real vector function.

a) Since Equation (7.25) (in Section 7.3), implies

$$\mathbf{x}(k; \mathbf{0}_n; \mathbf{i}) = \mathcal{Z}^{-1}\left\{\boldsymbol{G}_{ISOIS}(z)\mathbf{I}(z)\right\},\ \mathbf{x}_0 = \mathbf{0}_n,$$

then *a)* of Definition 14.2 (in Subsection 14.2.2), $\mathbf{i}(\cdot) \in \mathcal{J}$, Lemma 6.1 (in Chapter 6), and the Generating Theorem 13.39 (in Subsection 13.4.1), prove necessity and sufficiency of the condition *a)*.

b) Equation (7.21) (in Section 7.3), yields

$$\mathbf{y}(k; \mathbf{0}_n; \mathbf{i}) = \mathcal{Z}^{-1}\left\{\boldsymbol{G}_{ISO}(z)\mathbf{I}(z)\right\},\ \mathbf{x}_0 = \mathbf{0}_n.$$

This, *b)* of Definition 14.2 (in Subsection 14.2.2), $\mathbf{i}(\cdot) \in \mathcal{J}$, Lemma 6.1, and the Generating Theorem 13.39 prove necessity and sufficiency of condition *b)*.

c) From (7.24) (in Section 7.3), we deduce

$$\mathbf{x}(k; \mathbf{x}_0; \mathbf{i}) = \mathcal{Z}^{-1}\left\{\boldsymbol{F}_{ISOIS}(z)\left[\begin{array}{cc}\mathbf{I}^T(z) & \mathbf{x}_0^T\end{array}\right]^T\right\}.$$

This, d) of Definition 14.2, $\mathbf{i}(\cdot) \in \mathcal{J}$, (7.24) and (7.27) (in Section 7.3), Lemma 6.1 (in Chapter 6), and the Generating Theorem 13.39 (in Subsection 13.4.1), prove necessity and sufficiency of condition c).

d) We write (7.20a) (in Section 7.3), in the time domain by applying the inverse $Z-$transform,

$$\mathbf{y}(k; \mathbf{x}_0; \mathbf{i}) = \mathcal{Z}^{-1}\left\{ \boldsymbol{F}_{ISO}(z) \left[\begin{array}{cc} \mathbf{I}^T(z) & \mathbf{x}_0^T \end{array} \right]^T \right\}.$$

This, e) of Definition 14.2, $\mathbf{i}(\cdot) \in \mathcal{J}$, (7.20a) and (7.23) (in Section 7.3), Lemma 6.1, and the Generating Theorem 13.39 prove necessity and sufficiency of condition d) ■

Note 14.10 *The conditions under d) are necessary and sufficient for BIBSO stability of the ISO system (14.28) and (14.29) due to Theorem 14.1 (in Subsection 14.2.2).*

Note 14.11 *The conditions under d) are necessary and sufficient for BISBSO stability of the ISO system (14.28) and (14.29) due to Theorem 14.2 (in Subsection 14.2.2).*

Example 14.5 *We analyze the ISO system similar to the one given in Example 8.9 in Section 8.2, and the same as in Example 13.3 in Subsection 13.4.3:*

$$E^1\mathbf{x} = \begin{bmatrix} 0 & 1 & 0 \\ 1.45 & 0 & 2 \\ 1.02 & 0 & 1.2 \end{bmatrix} \mathbf{x} + \begin{bmatrix} 0 \\ 0 \\ 0.5 \end{bmatrix} i,$$

$$y = \begin{bmatrix} -2 & 1 & 0 \end{bmatrix} \mathbf{x}.$$

We determined (8.38):

$$\boldsymbol{F}_{ISO}(z) =$$

$$= \frac{z-2}{(z+0.3)(z+0.5)(z-2)} \left[\begin{array}{cccc} 1 & -2z(z+0.075) & z(z-1.2) & 2z \end{array} \right]$$

$$\Longrightarrow$$

$$\boldsymbol{G}_{ISO}(z) = \frac{z-2}{z^3 - 1.2z^2 - 1.45z - 0.3} = \frac{z-2}{(z+0.3)(z+0.5)(z-2)}.$$

Both, $\boldsymbol{F}_{ISO}(z)$ and $\boldsymbol{G}_{ISO}(z)$ are degenerate. Their nondegenerate forms, which are also their row nondegenerate forms, read

$$\boldsymbol{F}_{ISOrnd}(z) = \boldsymbol{F}_{ISOnd}(z) = \frac{\left[\begin{array}{cccc} 1 & -2z(z+0.075) & z(z-1.2) & 2z \end{array} \right]}{(z+0.3)(z+0.5)},$$

$$\boldsymbol{G}_{ISOrnd}(z) = \boldsymbol{G}_{ISOnd}(z) = \frac{1}{(z+0.3)(z+0.5)}.$$

Their poles are with modulus less than one,

$$z_1^*(\boldsymbol{F}_{ISOrnd}) = z_1^*(\boldsymbol{G}_{ISOrnd}) = -0.3,\ z_2^*(\boldsymbol{F}_{ISOrnd}) = z_2^*(\boldsymbol{G}_{ISOrnd}) = -0.5.$$

The system is both $BIBO$ and $BISBO$ stable.

Besides,

$$\boldsymbol{G}_{ISOIS}(z) = (z\boldsymbol{I}_n - \boldsymbol{A})^{-1}\boldsymbol{B} =$$

$$= \begin{bmatrix} \frac{z(z-1.2)}{z^3-1.2z^2-1.45z-0.3} & \frac{z-1.2}{z^3-1.2z^2-1.45z-0.3} & \frac{2}{z^3-1.2z^2-1.45z-0.3} \\ \frac{1.45z+0.3}{z^3-1.2z^2-1.45z-0.3} & \frac{z(z-1.2)}{z^3-1.2z^2-1.45z-0.3} & \frac{2z}{z^3-1.2z^2-1.45z-0.3} \\ \frac{1.02z}{z^3-1.2z^2-1.45z-0.3} & \frac{1.02}{z^3-1.2z^2-1.45z-0.3} & \frac{z^2-1.45}{z^3-1.2z^2-1.45z-0.3} \end{bmatrix}.$$

$$\cdot \begin{bmatrix} 0 \\ 0 \\ 0.5 \end{bmatrix} = \begin{bmatrix} \frac{1}{z^3-1.2z^2-1.45z-0.3} \\ \frac{z}{z^3-1.2z^2-1.45z-0.3} \\ \frac{0.5(z^2-1.45)}{z^3-1.2z^2-1.45z-0.3} \end{bmatrix}.$$

It is row nondegenerate,

$$\boldsymbol{G}_{ISOISrnd}(z) = \boldsymbol{G}_{ISOIS}(z).$$

Its poles are

$$z_1^* = -0.3,\ z_2^* = -0.5,\ z_3^* = 2.$$

Since the third pole is with modulus greater than one, the system is $BIBS$ unstable.

Furthermore, $\boldsymbol{F}_{ISOIS}(z)$,

$$\boldsymbol{F}_{ISOIS}(z) = \begin{bmatrix} \underbrace{(z\boldsymbol{I}_n - \boldsymbol{A})^{-1}\boldsymbol{B}}_{G_{ISOIS}(z)} & \underbrace{(z\boldsymbol{I}_n - \boldsymbol{A})^{-1}}_{G_{ISOSS}(z)} \end{bmatrix} =$$

$$= \begin{bmatrix} \underbrace{\begin{matrix} \frac{1}{z^3-1.2z^2-1.45z-0.3} & \frac{z}{z^3-1.2z^2-1.45z-0.3} & \frac{0.5(z^2-1.45)}{z^3-1.2z^2-1.45z-0.3} \end{matrix}}_{(\boldsymbol{G}_{ISOIS}(z))^T} \\ \underbrace{\begin{matrix} \frac{z(z-1.2)}{z^3-1.2z^2-1.45z-0.3} & \frac{1.45z+0.3}{z^3-1.2z^2-1.45z-0.3} & \frac{1.02z}{z^3-1.2z^2-1.45z-0.3} \\ \frac{z-1.2}{z^3-1.2z^2-1.45z-0.3} & \frac{z(z-1.2)}{z^3-1.2z^2-1.45z-0.3} & \frac{1.02}{z^3-1.2z^2-1.45z-0.3} \\ \frac{2}{z^3-1.2z^2-1.45z-0.3} & \frac{2z}{z^3-1.2z^2-1.45z-0.3} & \frac{z^2-1.45}{z^3-1.2z^2-1.45z-0.3} \end{matrix}}_{(\boldsymbol{G}_{ISOSS}(z))^T} \end{bmatrix}^T,$$

is also row nondegenerate,

$$F_{ISOISrnd}(s) = F_{ISOIS}(s).$$

Its submatrices $\boldsymbol{G}_{ISOIS}(z)$ and $\boldsymbol{G}_{ISOSS}(z)$ are also row nondegenerate. They, all three, have the same poles. Their third pole, $z_3^* = 2$, is with modulus greater than one. The system is both $BIBS$ and $BISBS$ unstable.

Note 14.12 *This example illustrates the inadequacy of the system transfer function matrix $\boldsymbol{G}_{ISO}(z)$ for testing every BI stability property of the ISO system. We should use the full system transfer function matrix $\boldsymbol{F}_{ISO}(z)$ (or $\boldsymbol{F}_{ISOIS}(z)$) and its corresponding submatrix(es).*

This example illustrates the following Corollary to Theorem 14.4:

Corollary 14.1 *If the ISO system (14.28) and (14.29) is:*

a) Bounded Input and Initial State-Bounded State ($BISBS$) stable, then it is also Bounded Input-Bounded State ($BIBS$) stable.

b) Bounded Input and Initial State-Bounded Output ($BISBO$) stable, then it is also Bounded Input-Bounded Output ($BIBO$) stable.

This Corollary results also from Definition 14.2 and agrees with Comment 14.1.

14.3.3 *IIO* systems

We recall the following subclass of the *IIO* systems described by (3.64a) and (3.64b) (in Subsection 3.5.4):

$$\begin{gathered}\boldsymbol{Q}^{(\alpha)}\mathbf{r}^{\alpha}(k) = \boldsymbol{P}^{(\beta)}\mathbf{i}^{\beta}(k), \det \boldsymbol{Q}_{\alpha} \neq 0, \forall k \in \mathcal{N}_0,\ \alpha \geq 1, \alpha \geq \beta \geq 0, \\ \boldsymbol{E}^{(\nu)}\mathbf{y}^{\nu}(k) = \boldsymbol{R}^{(\alpha)}\mathbf{r}^{\alpha}(k) + \boldsymbol{T}^{(\mu)}\mathbf{i}^{\mu}(k), \det \boldsymbol{E}_{\nu} \neq 0,\ \forall k \in \mathcal{N}_0, \\ \alpha,\ \beta,\ \nu,\ \mu \in \mathcal{R}_+,\ \nu \geq \mu. \end{gathered} \tag{14.34}$$

The left-hand side of the first Equation of (14.34) expresses *the internal state (dynamical) behavior* of the system (in Section 2.3),

$$\boldsymbol{Q}^{(\alpha)}\mathbf{r}^{\alpha}(k) = \boldsymbol{P}^{(\beta)}\mathbf{i}^{\beta}(k),\ \forall k \in \mathcal{N}_0. \tag{14.35}$$

The left-hand side of the second Equation (14.34) determines *the output state (dynamical) behavior,* of the system (in Section 2.3),

$$\boldsymbol{E}^{(\nu)}\mathbf{y}^{\nu}(k) = \boldsymbol{R}^{(\alpha)}\mathbf{r}^{\alpha}(k) + \boldsymbol{T}^{(\mu)}\mathbf{i}^{\mu}(k),\ \forall k \in \mathcal{N}_0. \tag{14.36}$$

The left-hand sides of Equations (14.34), i.e., (14.35) and (14.36), or equivalently of the following equation, determine *the full dynamics* (the internal state and the output state dynamics) of the *IIO* system (14.34), which is also called *the full (the complete) system state dynamics*:

$$\begin{bmatrix} \boldsymbol{Q}^{(\alpha)} & \boldsymbol{O}_{\rho,(\nu+1)N} \\ -\boldsymbol{R}^{(\alpha)} & \boldsymbol{E}^{(\nu)} \end{bmatrix} \begin{bmatrix} \mathbf{r}^{\alpha}(k) \\ \mathbf{y}^{\nu}(k) \end{bmatrix} = \begin{bmatrix} \boldsymbol{P}^{(\gamma)} \\ \boldsymbol{T}^{(\gamma)} \end{bmatrix} \mathbf{i}^{\gamma}(k), \ \forall k \in \mathcal{N}_0,$$
$$\gamma = \max\{\beta, \mu\}, \ \nu \geq 1. \tag{14.37}$$

The *BI* stability properties of the *IIO* system (14.34) in the case $\nu = 0$ can be treated as those of the *ISO* system (3.60a) and (3.60b) (in Subsection 3.5.3). We will consider in the sequel only the case $\nu \geq 1$.

Lemma 14.2 *Let the input vector function* $\mathbf{i}(\cdot)$ *belong to the family* $\mathcal{J}$, *(2.11) (in Section 2.1), In order for*

$$\left[\left(\mathbf{r}^{\alpha-1}(k;\mathbf{r}_0^{\alpha-1};\mathbf{i})\right)^T \ \left(\mathbf{y}^{\nu-1}(k;\mathbf{r}_0^{\alpha-1};\mathbf{y}_0^{\nu-1};\mathbf{i})\right)^T \right]^T, \ \nu \geq 1,$$

of the IIO system (14.34) in the forced regime:

a) to be bounded on $\mathcal{N}_0$,

$$\exists \zeta = \zeta(\mathbf{r}_0^{\alpha-1};\mathbf{y}_0^{\nu-1};\mathbf{i}) \in \mathcal{R}^+ \Longrightarrow$$
$$\left\| \begin{bmatrix} \mathbf{r}^{\alpha-1}(k;\mathbf{r}_0^{\alpha-1};\mathbf{i}) \\ \mathbf{y}^{\nu-1}(k;\mathbf{r}_0^{\alpha-1};\mathbf{y}_0^{\nu-1};\mathbf{i}) \end{bmatrix} \right\| < \zeta, \forall k \in \mathcal{N}_0,$$

it is necessary and sufficient that

$$\left[\mathbf{r}^T(k;\mathbf{r}_0^{\alpha-1};\mathbf{i}) \ \ \mathbf{y}^T(k;\mathbf{r}_0^{\alpha-1};\mathbf{y}_0^{\nu-1};\mathbf{i}) \right]^T$$

is bounded on $\mathcal{N}_0$,

$$\exists \xi = \xi(\mathbf{r}_0^{\alpha-1};\mathbf{y}_0^{\nu-1};\mathbf{i}) \in \mathcal{R}^+ \Longrightarrow$$
$$\left\| \begin{bmatrix} \mathbf{r}(k;\mathbf{r}_0^{\alpha-1};\mathbf{i}) \\ \mathbf{y}(k;\mathbf{r}_0^{\alpha-1};\mathbf{y}_0^{\nu-1};\mathbf{i}) \end{bmatrix} \right\| < \xi, \ \forall k \in \mathcal{N}_0,$$

b) to vanish as $k \to \infty$,

$$k \to \infty \Longrightarrow \left[\left(\mathbf{r}^{\alpha-1}(k;\mathbf{r}_0^{\alpha-1};\mathbf{i})\right)^T \ \left(\mathbf{y}^{\nu-1}(k;\mathbf{r}_0^{\alpha-1};\mathbf{y}_0^{\nu-1};\mathbf{i})\right)^T \right]^T \to$$
$$\to \mathbf{0}_{\alpha\rho+\nu N},$$

it is necessary and sufficient that

$$\left[\mathbf{r}^T(k;\mathbf{r}_0^{\alpha-1};\mathbf{i}) \quad \mathbf{y}^T(k;\mathbf{r}_0^{\alpha-1};\mathbf{y}_0^{\nu-1};\mathbf{i}) \right]^T$$

vanishes as $k \to \infty$,

$$k \to \infty \Longrightarrow \left[\mathbf{r}^T(k;\mathbf{r}_0^{\alpha-1};\mathbf{i}) \quad \mathbf{y}^T(k;\mathbf{r}_0^{\alpha-1};\mathbf{y}_0^{\nu-1};\mathbf{i}) \right]^T \to \mathbf{0}_{\rho+N}.$$

Proof.

Let the input vector function $\mathbf{i}(\cdot)$ be arbitrarily chosen from the family $\mathcal{J}$, (2.10) and (2.11) (in Section 2.1) and be fixed. Let all the initial conditions be also arbitrarily chosen so that they are bounded and fixed.

Necessity. a) In order for

$$\left[\left(\mathbf{r}^{\alpha-1}(k;\mathbf{r}_0^{\alpha-1};\mathbf{i})\right)^T \quad \left(\mathbf{y}^{\nu-1}(k;\mathbf{r}_0^{\alpha-1};\mathbf{y}_0^{\nu-1};\mathbf{i})\right)^T \right]^T$$

to be bounded on $\mathcal{N}_0$, it is necessary that its every entry is bounded on $\mathcal{N}_0$. Hence, it is necessary that

$$\left[\mathbf{r}^T(k;\mathbf{r}_0^{\alpha-1};\mathbf{i}) \quad \mathbf{y}^T(k;\mathbf{r}_0^{\alpha-1};\mathbf{y}_0^{\nu-1};\mathbf{i}) \right]^T$$

is bounded on $\mathcal{N}_0$.

b) In order for

$$\left[\left(\mathbf{r}^{\alpha-1}(k;\mathbf{r}_0^{\alpha-1};\mathbf{i})\right)^T \quad \left(\mathbf{y}^{\nu-1}(k;\mathbf{r}_0^{\alpha-1};\mathbf{y}_0^{\nu-1};\mathbf{i})\right)^T \right]^T$$

to vanish as $k \to \infty$, it is necessary that its every entry vanishes as $k \to \infty$. Hence, it is necessary that

$$\left[\mathbf{r}^T(k;\mathbf{r}_0^{\alpha-1};\mathbf{i}) \quad \mathbf{y}^T(k;\mathbf{r}_0^{\alpha-1};\mathbf{y}_0^{\nu-1};\mathbf{i}) \right]^T$$

vanishes as $k \to \infty$.

Sufficiency. a) Let the conditions of the statement of lemma under a) be valid. By following the proof of Equations (14.24) and (14.25) in Subsection 14.3.1 and using $\mathbf{i}(\cdot) \in \mathcal{J}$, the properties of the $Z-$transform, the boundedness of the initial conditions and (2.36) (in Section 2.3) we derive

$$\mathbf{r}(k;\mathbf{i}_0^{\beta-1};\mathbf{r}_0^{\alpha-1};\mathbf{i}) =$$

$$= \left\{ \delta_d(k)\boldsymbol{R}_{0r} + \sum_{i=1}^{i=\alpha\rho} (z_i^*)^{k-1} \left[\sum_{r=1}^{r=\nu_i^*} \frac{1}{(z_i^*)^{r-1}(r-1)!} \frac{(k-1)!}{(k-r)!} \boldsymbol{R}_{ir} \right] \right\} \cdot$$

$$\cdot \begin{bmatrix} \mathbf{1}_M \\ \mathbf{i}_0^{\beta-1} \\ \mathbf{r}_0^{\alpha-1} \end{bmatrix}, \ \boldsymbol{R}_{0r} \in \mathcal{R}^{\rho\times((\beta+1)M+\alpha\rho)}, \ \boldsymbol{R}_{ir} \in \mathcal{C}^{\rho\times((\beta+1)M+\alpha\rho)},$$

$$\boldsymbol{R}_{0r} = \text{const.}, \ \boldsymbol{R}_{ir} = \text{const.}, \ \forall r \in \{1, 2, \cdots, \nu_i^*\},$$

$$\forall i \in \{1, 2, \cdots, \alpha\rho\}, \ |z_i^*| \leq 1, \ |z_i^*| = 1 \Longrightarrow \nu_i^* = 1,$$

and

$$\mathbf{y}(k;\mathbf{i}_0^{\gamma-1};\mathbf{r}_0^{\alpha-1};\mathbf{y}_0^{\nu-1};\mathbf{i}) =$$

$$= \left\{ \delta_d(k)\boldsymbol{R}_{0y} + \sum_{j=1}^{j=\nu N} \left(z_j^*\right)^{k-1} \left[\sum_{r=1}^{r=\nu_j^*} \frac{1}{(z_j^*)^{r-1}(r-1)!} \frac{(k-1)!}{(k-r)!} \boldsymbol{R}_{jry} \right] \right\} \cdot$$

$$\cdot \begin{bmatrix} \mathbf{1}_M \\ \mathbf{i}_0^{\gamma-1} \\ \mathbf{r}_0^{\alpha-1} \\ \mathbf{y}_0^{\nu-1} \end{bmatrix}, \ \boldsymbol{R}_{0y} \in \mathcal{R}^{N\times((\gamma+1)M+\alpha\rho+\nu N)}, \ \boldsymbol{R}_{jry} \in \mathcal{C}^{N\times((\gamma+1)M+\alpha\rho+\nu N)},$$

$$\left|z_j^*\right| \leq 1, \ \boldsymbol{R}_{0y} = \text{const.}, \ \boldsymbol{R}_{jry} = \text{const.},$$

$$\forall r \in \left\{1, 2, \cdots, \nu_j^*\right\}, \ \forall j \in \{1, 2, \cdots, \nu N\}, \left|z_j^*\right| = 1 \Longrightarrow \nu_j^* = 1,$$

where ν_j^* is the multiplicity of z_j^*. This shows, in view of the Generating Theorem (13.39) (in Subsection 13.4.1), that the system substate vector

$$\left[\ \mathbf{r}^T(k;\mathbf{r}_0^{\alpha-1};\mathbf{i}) \quad \mathbf{y}^T(k;\mathbf{r}_0^{\alpha-1};\mathbf{y}_0^{\nu-1};\mathbf{i}) \ \right]^T$$

is infinitely times shiftable at every $k \in \mathcal{N}_0$, and bounded on $\mathcal{N}_0$. All its shifts rest bounded on $\mathcal{N}_0$ due to the Generating Theorem (13.39), which guarantees that the system full state vector

$$\left[\ \left(\mathbf{r}^{\alpha-1}(k;\mathbf{i}_0^{\mu-1}\mathbf{i})\right)^T \quad \left(\mathbf{y}^{\nu-1}(k;\mathbf{r}_0^{\alpha-1};\mathbf{y}_0^{\nu-1};\mathbf{i})\right)^T \ \right]^T$$

is also bounded on $\mathcal{N}_0$.

b) Let the conditions of the statement of lemma under *b*) hold. This, $\mathbf{i}(\cdot) \in \mathcal{J}$ and the properties of the $Z-$transform and the boundedness of the initial conditions imply

$$\mathbf{r}(t; \mathbf{i}_0^{\beta-1}; \mathbf{r}_0^{\alpha-1}; \mathbf{i}) =$$

$$= \left\{ \delta_d(k) \boldsymbol{R}_{0r} + \sum_{i=1}^{i=\alpha\rho} (z_i^*)^{k-1} \left[\sum_{r=1}^{r=\nu_i^*} \frac{1}{(z_i^*)^{r-;1} (r-1)!} \frac{(k-1)!}{(k-r)!} \boldsymbol{R}_{ir} \right] \right\} \cdot$$

$$\cdot \begin{bmatrix} \mathbf{1}_M \\ \mathbf{i}_0^{\beta-1} \\ \mathbf{r}_0^{\alpha-1} \end{bmatrix}, \boldsymbol{R}_{0r} \in \mathcal{R}^{\rho \times ((\beta+1)M+\alpha\rho)}, \ \boldsymbol{R}_{ir} \in \mathcal{C}^{\rho \times ((\beta+1)M+\alpha\rho)}, \boldsymbol{R}_{0r} = \text{const.},$$

$$\boldsymbol{R}_{ir} = \text{const.}, \ |z_i^*| < 1,$$

$$\mathbf{y}(k; \mathbf{i}_0^{\gamma-1} \mathbf{r}_0^{\alpha-1}; \mathbf{y}_0^{\nu-1}; \mathbf{i}) =$$

$$= \left\{ \delta_d(k) \boldsymbol{R}_{0y} + \sum_{j=1}^{j=\nu N} (z_j^*)^{k-1} \left[\sum_{r=1}^{r=\nu_j^*} \frac{1}{(z_j^*)^{r-1} (r-1)!} \frac{(k-1)!}{(k-r)!} \boldsymbol{R}_{jry} \right] \right\} \cdot$$

$$\cdot \begin{bmatrix} \mathbf{1}_M \\ \mathbf{i}_0^{\gamma-1} \\ \mathbf{r}_0^{\alpha-1} \\ \mathbf{y}_0^{\nu-1} \end{bmatrix}, \boldsymbol{R}_{0y} \in \mathcal{R}^{N \times ((\gamma+1)M+\alpha\rho+\nu N)}, \ \boldsymbol{R}_{jry} \in \mathcal{C}^{N \times ((\gamma+1)M+\alpha\rho+\nu N)},$$

$$\boldsymbol{R}_{0y} = \text{const.}, \boldsymbol{R}_{jry} = \text{const.}, \ \left|z_j^*\right| < 1, \ \forall j \in \{1, 2, \cdots, \nu N\}.$$

This shows that the system substate vector

$$\left[\begin{array}{cc} \mathbf{r}^T(k; \mathbf{r}_0^{\alpha-1}; \mathbf{i}) & \mathbf{y}^T(k; \mathbf{r}_0^{\alpha-1}; \mathbf{y}_0^{\nu-1}; \mathbf{i}) \end{array} \right]^T$$

is infinitely times shiftable at every $k \in \mathbb{Z}$. All its shifts rest bounded on $\mathcal{N}_0$, and vanish as $k \to \infty$ in view of the Generating Theorem 13.39. Every entry of the system full state vector

$$\left[\left(\mathbf{r}^{\alpha-1}(k; \mathbf{i}_0^{\mu-1} \mathbf{i}) \right)^T \ \ (\mathbf{y}^{\nu-1}(k; \mathbf{r}_0^{\alpha-1}; \mathbf{y}_0^{\nu-1}; \mathbf{i}))^T \right]^T$$

is bounded and vanishes as $k \to \infty$, in view of the Generating Theorem 13.39, which holds also for the whole system full state vector. ■

Lemma 14.3 *Let the input vector function* $\mathbf{i}(\cdot)$ *belong to the family* $\mathcal{J}$*, (2.10) and (2.11) (in Section 2.1). In order for*

$$(\mathbf{r}^{\alpha-1}(k; \mathbf{r}_0^{\alpha-1}; \mathbf{i}))$$

of the IIO system (14.34) in the forced regime:

a) *to be bounded on* $\mathcal{N}_0$,

$$\exists\zeta = \zeta(\mathbf{r}_0^{\alpha-1};\mathbf{i}) \in \mathcal{R}^+ \Rightarrow \left\|\mathbf{r}^{\alpha-1}(k;\mathbf{r}_0^{\alpha-1};\mathbf{i})\right\| < \zeta,\ \forall k \in \mathcal{N}_0,$$

it is necessary and sufficient that

$$\mathbf{r}(k;\mathbf{r}_0^{\alpha-1};\mathbf{i})$$

is bounded on $\mathcal{N}_0$,

$$\exists\xi = \xi(\mathbf{r}_0^{\alpha-1};\mathbf{i}) \in \mathcal{R}^+ \Longrightarrow \left\|\mathbf{r}(k;\mathbf{r}_0^{\alpha-1};\mathbf{i})\right\| < \xi,\ \forall k \in \mathcal{N}_0,$$

b) *to vanish as* $k \to \infty$,

$$k \to \infty \Longrightarrow \mathbf{r}^{\alpha-1}(k;\mathbf{r}_0^{\alpha-1};\mathbf{i}) \to \mathbf{0}_{\alpha\rho},$$

it is necessary and sufficient that

$$\mathbf{r}(k;\mathbf{r}_0^{\alpha-1};\mathbf{i})$$

vanishes as $k \to \infty$,

$$k \to \infty \Longrightarrow \mathbf{r}(k;\mathbf{r}_0^{\alpha-1};\mathbf{i}) \to \mathbf{0}_{\rho}.$$

By following the proof of Lemma 14.2 we easily verify and prove this lemma.

Lemma 14.4 *Let the input vector function* $\mathbf{i}(\cdot)$ *belong to the family* $\mathcal{J}$, *(2.10) and (2.11) (in Section 2.1). In order for*

$$\mathbf{y}^{\nu-1}(k;\mathbf{r}_0^{\alpha-1};\mathbf{y}_0^{\nu-1};\mathbf{i}),\ \nu \geq 1,$$

of the IIO system (14.34) in the forced regime:

a) *to be bounded on* $\mathcal{N}_0$,

$$\exists\zeta = \zeta(\mathbf{r}_0^{\alpha-1};\mathbf{y}_0^{\nu-1};\mathbf{i}) \in \mathcal{R}^+ \Rightarrow \left\|\mathbf{y}^{\nu-1}(k;\mathbf{r}_0^{\alpha-1};\mathbf{y}_0^{\nu-1};\mathbf{i})\right\| < \zeta,\ \forall k \in \mathcal{N}_0,$$

it is necessary and sufficient that

$$\mathbf{y}(k;\mathbf{r}_0^{\alpha-1};\mathbf{y}_0^{\nu-1};\mathbf{i})$$

is bounded on $\mathcal{N}_0$,

$$\exists\xi = \xi(\mathbf{r}_0^{\alpha-1};\mathbf{y}_0^{\nu-1};\mathbf{i}) \in \mathcal{R}^+ \Longrightarrow \left\|\mathbf{y}(k;\mathbf{r}_0^{\alpha-1};\mathbf{y}_0^{\nu-1};\mathbf{i})\right\| < \xi,\ \forall k \in \mathcal{N}_0,$$

b) *to vanish as* $k \to \infty$,

$$k \to \infty \Longrightarrow \mathbf{y}^{\nu-1}(k; \mathbf{r}_0^{\alpha-1}; \mathbf{y}_0^{\nu-1}; \mathbf{i}) \to \mathbf{0}_{\nu N},$$

it is necessary and sufficient that

$$\mathbf{y}(k; \mathbf{r}_0^{\alpha-1}; \mathbf{y}_0^{\nu-1}; \mathbf{i})$$

vanishes as $k \to \infty$,

$$k \to \infty \Longrightarrow \mathbf{y}(k; \mathbf{r}_0^{\alpha-1}; \mathbf{y}_0^{\nu-1}; \mathbf{i}) \to \mathbf{0}_N.$$

The proof of this lemma is a special case of the proof of Lemma 14.2.

These lemmas are inherent for the BI stability conditions in the framework of the IIO systems (14.34) and for their proofs.

Theorem 14.5 *Let the input vector function* $\mathbf{i}(\cdot)$ *belong to the family* $\mathcal{J}$, *(2.10) and (2.11) (in Section 2.1). In order for the IIO system (14.34) in the forced regime to be:*

a) ***Bounded Input-Bounded Internal State*** *(BIBIS)* ***stable,*** *it is necessary and sufficient that* 1. *and* 2. *hold:*

1. *the system IRIS transfer function matrix* $\mathbf{G}_{IIOIS}(z)$ *is either proper or strictly proper, real rational matrix function of* z, *and*

2. *the modulus* $|z_i^*(\mathbf{G}_{IIOISrnd})|$ *of all poles* $z_i^*(\mathbf{G}_{IIOISrnd})$ *of the row nondegenerate form* $\mathbf{G}_{IIOISrnd}(z)$ *of* $\mathbf{G}_{IIOIS}(z)$ *are less than one,*

$$|z_i^*(\mathbf{G}_{IIOISrnd})| < 1, \ \forall i \in \{1, 2, \cdots, \ \alpha\rho\}. \tag{14.38}$$

b) ***Bounded Input-Bounded Output*** *(BIBO)* ***stable,*** *it is necessary and sufficient that* 1. *and* 2. *are valid:*

1. *the system IO transfer function matrix* $\mathbf{G}_{IIO}(z)$ *is either proper or strictly proper, real rational matrix function of* z, *and*

2. *the modulus* $|z_i^*(\mathbf{G}_{IIOrnd})|$ *of all poles* $z_i^*(\mathbf{G}_{IIOrnd})$ *of the row nondegenerate form* $\mathbf{G}_{IIOrnd}(z)$ *of* $\mathbf{G}_{IIO}(z)$ *are less than one,*

$$|z_i^*(\mathbf{G}_{IIOrnd})| < 1, \ \forall i \in \{1, 2, \cdots, \ \alpha\rho + \nu N\}. \tag{14.39}$$

c) ***Bounded Input-Bounded Output State** (BIBOS) **stable**, it is necessary and sufficient that the conditions under b) hold.*

d) ***Bounded Input-Bounded Internal State and Output** (BIBISO) **stable**, it is necessary and sufficient that the system is BIBIS stable and BIBO stable.*

e) ***Bounded Input and Initial Internal State-Bounded Internal State** (BIISBIS) **stable**, it is necessary and sufficient that 1 and 2 hold:*

1. *the system IRIS transfer function matrix $\mathbf{G}_{IIOIS}(z)$ is either proper or strictly proper, real rational matrix function of z, the system IRII transfer function matrix $\mathbf{G}_{IIOi_0IS}(z)$ and the system IRIR transfer function matrix $\mathbf{G}_{IIOr_0IS}(z)$ are also either proper or strictly proper real rational matrix functions of z,*
2. *the poles of the row nondegenerate form $\mathbf{G}_{IIOISrnd}(z)$ of the system IRIS transfer function matrix $\mathbf{G}_{IIOIS}(z)$ are with modulus less than one, the modulus of the poles of the row nondegenerate forms $\mathbf{G}_{IIOi_0ISrnd}(z)$ and $\mathbf{G}_{IIOr_0ISrnd}(z)$ of $\mathbf{G}_{IIOi_0IS}(z)$ and of $\mathbf{G}_{IIOr_0IS}(z)$, respectively, are less or equal to one and those with modulus equal to one are simple,*

$$\begin{gathered}
|z_i^*(\mathbf{G}_{IIOISrnd}(z)| < 1, \ \ |z_i^*(\mathbf{G}_{IIOi_0ISrnd}(z)| \leq 1, \\
|z_i^*(\mathbf{G}_{IIOr_0ISrnd}(z)| \leq 1, \ \forall i \in \{1, 2, \cdots, \alpha\rho\}, \\
\left|z_j^*(\mathbf{G}_{IIOi_0ISrnd})\right| = 1 \Longrightarrow \nu_j^* = 1, \\
\left|z_m^*(\mathbf{G}_{IIOr_0ISrnd})\right| = 1 \Longrightarrow \nu_m^* = 1.
\end{gathered} \tag{14.40}$$

f) ***Bounded Input and Initial State-Bounded Output** (BISBO) **stable**, it is necessary and sufficient that 1. and 2. hold:*

1. *the system IO transfer function matrix $\mathbf{G}_{IIO}(z)$ is either proper or strictly proper, real rational matrix function of z, its IICO, IIRO and IIYO transfer function matrices $\mathbf{G}_{IIOi_0}(z)$, $\mathbf{G}_{IIOr_0}(z)$ and $\mathbf{G}_{IIOy_0}(z)$ are also either proper or strictly proper real rational matrix functions of z,*
2. *the modulus $|z_i^*(\mathbf{G}_{IIOrnd})|$ of all poles $z_i^*(\mathbf{G}_{IIOrnd})$ of the row nondegenerate form $\mathbf{G}_{IIOrnd}(z)$ of $\mathbf{G}_{IIO}(z)$ are less than one, the modulus of the poles of the row nondegenerate forms $\mathbf{G}_{IIOi_0rnd}(z)$, $\mathbf{G}_{IIOr_0rnd}(z)$ and $\mathbf{G}_{IIOy_0rnd}(z)$ of $\mathbf{G}_{IIOi_0}(z)$,*

$\mathbf{G}_{IIOr_0}(z)$ *and of* $\mathbf{G}_{IIOy_0}(z)$ *are less or equal to one and those with modulus equal to one are simple,*

$$|z_i^*(\mathbf{G}_{IIOrnd}(z)| < 1, \ |z_i^*(\mathbf{G}_{IIOi_0rnd}(z)| \leq 1,$$
$$|z_i^*(\mathbf{G}_{IIOr_0rnd}(z)| \leq 1, \ |z_i^*(\mathbf{G}_{IIOy_0rnd}(z)| \leq 1,$$
$$\forall i \in \{1, 2, \cdots, \ \alpha\rho + \nu N\}, \left|z_j^*(\mathbf{G}_{IIOi_0rnd})\right| = 1 \Longrightarrow \nu_j^* = 1,$$
$$\left|z_l^*(\mathbf{G}_{IIOr_0rnd})\right| = 1 \Longrightarrow \nu_l^* = 1,$$
$$\left|z_m^*(\mathbf{G}_{IIOy_0rnd})\right| = 1 \Longrightarrow \nu_m^* = 1. \tag{14.41}$$

g) ***Bounded Input and Initial State-Bounded Output State*** *(BISBOS)* ***stable,*** *it is necessary and sufficient that f) holds.*

h) ***Bounded Input and Initial State-Bounded Internal State and Output*** *(BISBISO)* ***stable,*** *it is necessary and sufficient that e) and f) hold.*

i) ***Bounded Input and Initial State-Bounded State*** *(BISBS)* ***stable,*** *it is necessary and sufficient that g) and h) hold.*

Proof. Let $\mathbf{i}(\cdot) \in \mathcal{J}$, (2.10) and (2.11) (in Section 2.1).

Necessity. *a)* The condition $\mathbf{i}(\cdot) \in \mathcal{J}$, (2.10), (2.11), and $\alpha \geq \beta$ in (14.34) guarantee that the product $\mathbf{G}_{IIOIS}(z)\mathbf{I}(z) = \mathbf{Y}(z)$ is either proper or strictly proper real rational function of z. Let the IIO system (14.34) be $BIBIS$ stable, i.e., let *a)* of Definition 14.3 (in Subsection 14.2.3), hold. All initial conditions are equal to zero. The r-th shift $E^r\mathbf{r}(k;\mathbf{0}_{\beta M};\mathbf{0}_{\alpha\rho};\mathbf{i})$, $\forall r = 0, 1, 2, \cdots, \alpha - 1$, is bounded and it may contain or may not a discrete impulse component. The state behavior $\mathbf{r}(k;\mathbf{0}_{\beta M};\mathbf{0}_{\alpha\rho};\mathbf{i})$ of the system may or may not contain a discrete impulse component. This implies that the $Z-$transform $\mathcal{Z}\{\mathbf{r}(k;\mathbf{0}_{\beta M};\mathbf{0}_{\alpha\rho};\mathbf{i})\}$,

$$\mathcal{Z}\{\mathbf{r}(k;\mathbf{0}_{\beta M};\mathbf{0}_{\alpha\rho};\mathbf{i})\} = \mathbf{G}_{IIOIS}(s)\mathbf{I}(z),$$

is either proper or strictly proper real rational vector function, for which $\mathbf{G}_{IIOIS}(z)$ may be proper or strictly proper due to $\mathbf{i}(\cdot) \in \mathcal{J}$, (2.10) and (2.11). We prove this also as follows. Let

$$\mathbf{i}(\cdot) = h(\cdot)\mathbf{1}_M \in \mathcal{J}, \mathbf{1}_M = \begin{bmatrix} 1 \\ 1 \\ \vdots \\ 1 \end{bmatrix} \in \mathcal{R}^M \Rightarrow \mathbf{I}(z) = \frac{z}{z-1}\mathbf{1}_M \in \mathcal{C}^M.$$

Hence,

$$\mathcal{Z}\{\mathbf{r}^{\alpha-1}(k;\mathbf{0}_{\beta M};\mathbf{0}_{\alpha\rho};h(k)\mathbf{1}_M)\} =$$
$$= \left[\begin{array}{ccccc} \frac{z}{z-1}\boldsymbol{I}_N & \frac{z^2}{z-1}\boldsymbol{I}_N & \frac{z^3}{z-1}\boldsymbol{I}_N & \cdots & \frac{z^\alpha}{z-1}\boldsymbol{I}_N \end{array}\right]^T \boldsymbol{G}_{IIOIS}(z)\mathbf{1}_M$$

is either proper or strictly proper real rational vector function. This implies that the $Z-$transform $\mathcal{Z}\{\mathbf{r}(k;\mathbf{0}_{\beta M};\mathbf{0}_{\alpha\rho};\mathbf{i})\}$,

$$\mathcal{Z}\{\mathbf{r}(k;\mathbf{0}_{\beta M};\mathbf{0}_{\alpha\rho};\mathbf{i})\} = \boldsymbol{G}_{IIOII}(z)\frac{z}{z-1}\mathbf{1}_M,$$

is either proper or strictly proper real rational vector function, for which $\boldsymbol{G}_{IIOII}(z)$ may be proper or strictly proper. This proves necessity of condition *a*-1). Lemma 6.1 (in Chapter 6), $\mathbf{i}(\cdot) \in \mathcal{J}$, (2.10), (2.11), the Generating Theorem 13.39 (in Subsection 13.4.1), Lemma 14.3, and *a*) of Definition 14.3 (in Subsection 14.2.3), prove necessity of condition *a*-2), i.e., (14.38).

b) Let the *IIO* system (14.34) be *BIBO* stable so that *b*) of Definition 14.3 (in Subsection 14.2.3), is valid. Hence, $\boldsymbol{G}_{IIO}(z)\mathbf{I}(z)$ is either proper or strictly proper for every $\mathbf{i}(\cdot) \in \mathcal{J}$, (2.10) and (2.11), which, with $\mathbf{i}(\cdot) \in \mathcal{J}$, implies that $\boldsymbol{G}_{IIO}(z)$ is either proper or strictly proper. This proves the necessity of the condition *b*-1. Lemma 6.1, $\mathbf{i}(\cdot) \in \mathcal{J}$, the Generating Theorem 13.39 and

$$\mathbf{y}(k;\mathbf{0}_{\gamma M};\mathbf{0}_{\alpha\rho};\mathbf{0}_{\nu N};\mathbf{i}) = \mathcal{Z}^{-1}\left\{\boldsymbol{G}_{IIOrnd}(z)\mathbf{I}(z)\right\},$$

imply the necessity of condition (14.39), i.e., *b*-2).

c) Let the *IIO* system (14.34) be *BIBOS* stable, i.e., let *c*) of Definition 14.3, hold. Then it is also *BIBO* stable due to *b*) and *c*) of Definition 14.3. Hence, the conditions under *c*) are necessary due to Lemma 14.2, and the conditions under *b*).

d) Let the *IIO* system (14.34) be *BIBISO* stable. This, *a*), *b*), *c*) and *d*) of Definition 14.3 imply that the system is *BIBIS* stable and *BIBO* stable.

e) Let the *IIO* system (14.34) be *BIISBIS* stable, i.e., let (14.15) (in Subsection 14.2.3) hold. This guarantees, for every $\mathbf{i}(\cdot) \in \mathcal{J}$, the boundedness of $\left\|\mathbf{r}^{\alpha-1}(k;\mathbf{r}_0^{\alpha-1};\mathbf{i})\right\|$ on $\mathcal{N}_0$, so that $\left\|\mathbf{r}(k;\mathbf{r}_0^{\alpha-1};\mathbf{i})\right\|$ is bounded on $\mathcal{N}_0$ for every $\mathbf{i}(\cdot) \in \mathcal{J}$. This, (8.60) and (8.61) (in Section 8.3), Lemma 6.1, and the Generating Theorem 13.39 prove the necessity of the conditions under *e*).

f) Let the *IIO* system (14.34) be *BISBO* stable, i.e., let (14.16) (in Subsection 14.2.3), hold. Hence, $\boldsymbol{G}_{IIO}(z)\mathbf{I}(z)$ is either proper or strictly proper, which with $\mathbf{i}(\cdot) \in \mathcal{J}$, implies that $\boldsymbol{G}_{IIO}(z)$ is also proper or strictly proper, and together with (8.74) through (8.78) (in Section 8.3), imply the necessity of condition f-1.) $BISBO$ stability of the system,

$$\mathbf{y}(k; \mathbf{i}_0^{\gamma-1}; \mathbf{r}_0^{\alpha-1}; \mathbf{y}_0^{\nu-1}; \mathbf{i}) = \mathcal{Z}^{-1}\left\{ \boldsymbol{F}_{IIO}(z) \begin{bmatrix} \mathbf{I}(z) \\ \mathbf{i}_0^{\gamma-1} \\ \mathbf{r}_0^{\alpha-1} \\ \mathbf{y}_0^{\nu-1} \end{bmatrix} \right\},$$

(8.74), Lemma 6.1, and the Generating Theorem 13.39 imply the necessity of the condition (14.41), i.e., f-2).

g) Let the IIO system (14.34) be $BISBOS$ stable, i.e., let (14.17) (in Subsection 14.2.3), be valid. Then the system is $BISBO$ stable due to f) and g) of Definition 14.3, which proves the necessity of g).

h) The definition of $BISBISO$ stability under h) of Definition 14.3 (in Subsection 14.2.3) implies the validity of e) and f) of Definition 14.3 (in Subsection 14.2.3) due to Lemma 14.4, which prove the necessity of the condition under h).

i) The definition of $BISBS$ stability under i) of Definition 14.3 (in Subsection 14.2.3) and the fact that the state vector is composed of the internal state vector and the output state vector prove that $BISBS$ stability implies both $BISBOS$ and $BISBISO$ and due to g) and h) of Definition 14.3 (in Subsection 14.2.3).

Sufficiency. The boundedness of every $\mathbf{i}(\cdot)$ that belongs to the family $\mathcal{J}$, (2.10) and (2.11), guarantees that the modulus of all poles of $\mathbf{I}(z)$ are less or equal to one, that its poles with modulus equal to one are simple, and that $\mathbf{I}(z)$ is either proper or strictly proper real vector function.

a) Let the conditions under a) hold. Then, $\mathbf{i}(\cdot) \in \mathcal{J}$, (2.10) and (2.11), and the condition a-1.) guarantee that the $Z-$transform

$$\mathcal{Z}\{\mathbf{r}(k; \mathbf{0}_{\alpha\rho}; \mathbf{i})\} = \boldsymbol{G}_{IIOIS}(z)\mathbf{I}(z),$$

of $\mathbf{r}(k; \mathbf{0}_{\alpha\rho}; \mathbf{i})$ does not have a pole in infinity. This, the condition a-2), $\mathbf{i}(\cdot) \in \mathcal{J}$, Lemma 6.1, and the Generating Theorem 13.39 imply that

$$\mathbf{r}(k; \mathbf{0}_{\alpha\rho}; \mathbf{i}) = \mathcal{Z}^{-1}\left\{\boldsymbol{G}_{IIOIS}(z)\mathbf{I}(z)\right\},$$

is bounded for every $\mathbf{i}(\cdot) \in \mathcal{J}$. Hence, $\mathbf{r}^{\alpha-1}(k; \mathbf{0}_{\alpha\rho}; \mathbf{i})$ is also bounded for every $\mathbf{i}(\cdot) \in \mathcal{J}$ due to Lemma 14.3. The system is $BIBIS$ stable in view of a) of Definition 14.3 (in Subsection 14.2.3).

b) Let the conditions under *b*) hold. Then, condition *b*-1) and $\mathbf{i}(\cdot) \in \mathcal{J}$ guarantee that the $Z-$transform

$$\mathcal{Z}\{\mathbf{y}(k;\mathbf{0}_{\gamma M};\mathbf{0}_{\alpha\rho};\mathbf{0}_{\nu N};\mathbf{i})\} = \boldsymbol{G}_{IIO}(z)\mathbf{I}(z),$$

of $\mathbf{y}(k;\mathbf{0}_{\gamma M};\mathbf{0}_{\alpha\rho};\mathbf{0}_{\nu N};\mathbf{i})$ does not have a pole in infinity. This, the condition *b*-2), $\mathbf{i}(\cdot) \in \mathcal{J}$, Lemma 6.1, and the Generating Theorem 13.39 imply that

$$\mathbf{y}(k;\mathbf{0}_{\gamma M};\mathbf{0}_{\alpha\rho};\mathbf{0}_{\nu N};\mathbf{i}) = \mathcal{Z}^{-1}\left\{\boldsymbol{G}_{IIO}(z)\mathbf{I}(z)\right\},$$

is bounded for every $\mathbf{i}(\cdot) \in \mathcal{J}$. The system is *BIBO* stable in view of *b*) of Definition 14.3 (in Subsection 14.2.3).

c) Let the conditions under *c*) be satisfied. The conditions under *b*) hold. Therefore, $\mathbf{y}(k;\mathbf{0}_{\gamma M};\mathbf{0}_{\alpha\rho};\mathbf{0}_{\nu N};\mathbf{i})$ is bounded for every $\mathbf{i}(\cdot) \in \mathcal{J}$. This and Lemma 14.2 prove that the system is *BIBOS* stable in view of *c*) of Definition 14.3 (in Subsection 14.2.3).

d) Let the *IIO* system (14.34) be *BIBIS* stable and *BIBO* stable. This, *a*), and *c*) prove that *d*) of Definition 14.3 is satisfied, i.e., that the system is *BIBISO* stable.

e) Let the conditions under *e*) be valid. The condition *e*-1) and $\mathbf{i}(\cdot) \in \mathcal{J}$, (8.60) and (8.61) (in Section 8.3) show that the $Z-$transform,

$$\mathcal{Z}\{\mathbf{r}(k;\mathbf{r}_0^{\alpha-1};\mathbf{i})\} = \boldsymbol{F}_{IIOIS}(z)\begin{bmatrix} \mathbf{I}^T(z) & \left(\mathbf{i}_0^{\beta-1}\right)^T & \left(\mathbf{r}_0^{\alpha-1}\right)^T \end{bmatrix}^T,$$

$$\boldsymbol{F}_{IIOIS}(z) = \begin{bmatrix} \boldsymbol{G}_{IIOIS}(z) & \boldsymbol{G}_{IIOi_0IS}(z) & \boldsymbol{G}_{IIOr_0IS}(z) \end{bmatrix},$$

of $\mathbf{r}(k;\mathbf{r}_0^{\alpha-1};\mathbf{i})$ is either proper or strictly proper real rational function. This, condition *e*-2), $\mathbf{i}(\cdot) \in \mathcal{J}$, (8.61), Lemma 6.1, and the Generating Theorem 13.39 imply that $\mathbf{r}(k;\mathbf{r}_0^{\alpha-1};\mathbf{i})$ is bounded for every $\mathbf{i}(\cdot) \in \mathcal{J}$. Now, Lemma 14.2 proves the boundedness of $\mathbf{r}^{\alpha-1}(k;\mathbf{r}_0^{\alpha-1};\mathbf{i})$ for every $\mathbf{i}(\cdot\cdot) \in \mathcal{J}$. The system is *BIISBIS* stable in view of *e*) of Definition 14.3 (in Subsection 14.2.3).

f) Let the conditions under *f*) hold. The $Z-$transform,

$$\mathcal{Z}\left\{\mathbf{y}(k;\mathbf{i}_0^{\gamma-1};\mathbf{r}_0^{\alpha-1};\mathbf{y}_0^{\nu-1};\mathbf{i})\right\} = \boldsymbol{F}_{IIO}(z)\begin{bmatrix} \mathbf{I}(z) \\ \mathbf{i}_0^{\gamma-1} \\ \mathbf{r}_0^{\alpha-1} \\ \mathbf{y}_0^{\nu-1} \end{bmatrix},$$

$$\boldsymbol{F}_{IIO}(z) = \begin{bmatrix} \boldsymbol{G}_{IIO}(z) & \boldsymbol{G}_{IIOi_0}(z) & \boldsymbol{G}_{IIOr_0}(z) & \boldsymbol{G}_{IIOy_0}(z) \end{bmatrix},$$

of $\mathbf{y}(k;\mathbf{i}_0^{\gamma-1};\mathbf{r}_0^{\alpha-1};\mathbf{y}_0^{\nu-1};\mathbf{i})$ is either proper or strictly proper real rational function and it does not have either a pole in infinity or a pole with modulus greater than one or a multiple pole with modulus equal to one for every $\mathbf{i}(\cdot) \in \mathcal{J}$. Lemma 6.1, and the Generating Theorem 13.39 imply that

$$\mathbf{y}(k;\mathbf{i}_0^{\gamma-1};\mathbf{r}_0^{\alpha-1};\mathbf{y}_0^{\nu-1};\mathbf{i}) = \mathcal{Z}^{-1}\left\{ \boldsymbol{F}_{IIO}(z) \begin{bmatrix} \mathbf{I}(z) \\ \mathbf{i}_0^{\gamma-1} \\ \mathbf{r}_0^{\alpha-1} \\ \mathbf{y}_0^{\nu-1} \end{bmatrix} \right\},$$

is bounded for every $\mathbf{i}(\cdot) \in \mathcal{J}$. The system is $BISBO$ stable in view of $f)$ of Definition 14.3 (in Subsection 14.2.3).

$g)$ Let the conditions under $g)$ hold. We have just proved under $f)$ that the conditions 1) and 2) of $f)$ guarantee that $\mathbf{y}(k;\mathbf{i}_0^{\gamma-1};\mathbf{r}_0^{\alpha-1};\mathbf{y}_0^{\nu-1};\mathbf{i})$ is bounded for every $\mathbf{i}(\cdot) \in \mathcal{J}$. It implies boundedness of

$$\mathbf{y}^{\nu-1}(k;\mathbf{i}_0^{\gamma-1};\mathbf{r}_0^{\alpha-1};\mathbf{y}_0^{\nu-1};\mathbf{i})$$

for every $\mathbf{i}(\cdot) \in \mathcal{J}$ due to Lemma 14.2. The system is $BISBOS$ stable in view of $g)$ of Definition 14.3 (in Subsection 14.2.3).

$h)$ Let the condition under $h)$ hold, i.e., that the conditions under $e)$ and $f)$ are valid. The system is $BIISBIS$ stable and $BISBOS$ stable, which guarantee that the system is also $BISBISO$ due to $e)$, $g)$ and $h)$ of Definition 14.3 (in Subsection 14.2.3) having in mind that the output state does not influence the internal state.

$i)$ Let $i)$ hold. Hence, $g)$ and $h)$ are valid. The system is $BISBOS$ stable and $BISBISO$ stable, i.e., $g)$ and $h)$ of Definition 14.3 are satisfied, which implies that $i)$ of the same definition is fulfilled. The system is $BISBS$ stable. ■

Part IV

CONCLUSION

Chapter 15

Motivation for the book

Besides the well-studied linear dynamical systems via their classical mathematical models in the IO and ISO form, the authors have also introduced and studied the third class of the systems that are the *IIO systems*. The subclasses of the IIO systems are the *PMD systems*, *Rosenbrock systems*, and the $GISO$ systems.

A novel, unified and general approach has been developed to study in the same manner all three classes of the systems. The result is the unified general linear dynamical *time*-invariant discrete-*time* systems theory in the complex domain. Its applications are simple and straightforward.

To achieve clarity and accuracy in using different notions such as system regime, stationary vector, equilibrium vector, their definitions and procedures on how to determine them are presented. These issues are very simple, but make up the basis of the dynamical systems theory, which is often ignored in the literature on the linear dynamical systems. They assist in easier understanding of various qualitative properties of the systems (e.g., stability properties of the equilibrium vector).

Every dynamical physical system transfers and transmits simultaneously actions and influences of both the input vector and all initial conditions. The system transfer function matrix $\boldsymbol{G}(z)$ does not and cannot express and/or describe how the linear *time*-invariant (discrete-*time* or hybrid) system transforms all actions on itself into its internal and output behavior. This lack of $\boldsymbol{G}(z)$ is a consequence of its definition and validity only for zero initial conditions.

The $Z-$transform of the *n-th* order (scalar or vector) input-output discrete equation contains the double sums of the products of the system parameters and initial conditions. Such complex sums and products were the obstacle to treat, in the complex domain, the influence of nonzero (i.e., of arbitrary) initial conditions on the system behavior in the same effective manner as the input vector influence is treated. The existing theory has ignored the obstacle by unjustifiably assuming and accepting that all initial conditions are equal to zero. Such assumption is too harsh, a crucial simplification, unjustifiable and unreal.

The obstacle has been eliminated herein by following [68], which means by solving this mathematical problem of how to put the double sum in the form equivalent to $\boldsymbol{G}(z)\mathbf{I}(z)$ that characterizes the product of the system transfer function matrix $\boldsymbol{G}(z)$ and the $Z-$transform $\mathbf{I}(z)$ of the input vector $\mathbf{I}(k)$ under all zero initial conditions. Once this has been solved, it was possible to determine the system complex domain characteristic independent of the input vector and of all initial conditions, which completely expresses and describes how the system transfers, transmits and transforms influences of both the input vector and all initial conditions on the system state and output behavior. Such characteristic is *the system full transfer function matrix* $\boldsymbol{F}(z)$. It has the same features as the system transfer function matrix $\boldsymbol{G}(z)$:

- the independence of the input vector,
- the independence of all initial conditions,
- the invariance relative to the input vector and all initial conditions,
- the system order, dimension, structure and parameters completely determine it.

After presenting the definitions of *the system full transfer function matrix* $\boldsymbol{F}(z)$ and of its submatrices for every type of systems, the authors presented and proved how $\boldsymbol{F}(z)$ can be easily determined by using *the same mathematical knowledge that is applied to determine the system transfer function matrix* $\boldsymbol{G}(z)$. In addition, the book presents the physical meaning of the system full transfer function matrix $\boldsymbol{F}(z)$ and its submatrices.

The system full transfer function matrix $\boldsymbol{F}(z)$ **unifies** the complex domain theory of the linear *time*-invariant discrete-*time* systems.

In this context, the existence of row (non)degenerate, column (non)-degenerate, and (non)degenerate matrix functions were discovered.

Moreover, *the system full transfer function matrix* $\boldsymbol{F}(z)$ expresses the channeling information and the system structure through its submatrices.

This book discovers how the use of *the system full transfer function matrix* $\boldsymbol{F}(z)$ allows for new results on:

- Pole-zero cancellation, which is admissible if and only if it is possible in *the system full transfer function matrix* $\boldsymbol{F}(z)$ in general, or at least in its appropriate submatrix in a special case.

- The system realization and the minimal realization, which are different from the realization and the minimal realization of the transfer function matrix $\boldsymbol{G}(z)$, which is valid only under all zero initial conditions. The system realization and the minimal realization are the realization and the minimal realization of $\boldsymbol{F}(z)$ because they also hold for any initial conditions, but not of $\boldsymbol{G}(z)$.

- The generalization of the block diagram technique to *the full block diagram technique.*

Generalization of the notion of the system matrix has been applied to *the full system matrix* $\boldsymbol{P}(z)$, which produced new results on the system equivalence. In addition, the book establishes the link between *the system full transfer function matrix* $\boldsymbol{F}(z)$ and *the full system matrix* $\boldsymbol{P}(z)$.

The Lyapunov method has been established for the ISO systems. We made it directly applicable to both IO systems and IIO systems. This unifies the study of all three classes of the linear systems via the Lyapunov method.

After showing and explaining the inadequacy and the incompetence of the system transfer function matrix $\boldsymbol{G}(z)$ for the Lyapunov stability tests, it has been proved that only *the system full transfer function matrix* $\boldsymbol{F}(z)$, or at least its adequate submatrix, which is its submatrix related to the internal initial conditions, is competent for the Lyapunov stability tests.

Starting with the fact that the initial conditions are seldom equal to zero, the authors have extended the concept of the system stability under bounded inputs (BI stability) and zero initial conditions to the system stability under bounded inputs (BI stability) and *nonzero* initial conditions, which has led to new BI stability properties and the related criteria. They are expressed in terms of the submatrices of *the system full transfer function matrix* $\boldsymbol{F}(z)$, or directly in terms of *the system full transfer function matrix* $\boldsymbol{F}(z)$. The system transfer function matrix $\boldsymbol{G}(z)$ is inapplicable to BI stability tests in general.

Chapter 16

Summary of the contributions

This book completely solves the problems, induced by the missing parts of the theory, for all three classes of the systems. To be more specific:

- It discovers, defines and effectively exploits the state variables, i.e., the state vector, of both IO and IIO systems. They have the full physical sense. This allowed to extend the (Lyapunov and Bounded-Input, or BI) stability concepts and properties directly to these classes of the systems. This made provisions for their direct stability study rather than to study them formally mathematically, without any physical sense, in their formally mathematically (without a physical sense) transformed form of the ISO systems.

- It unifies the study and applications of all three classes of the systems. This is due to the following contributions that hold for all three classes of the systems.

- It discovers a complex domain fundamental dynamical characteristic of the systems, which is their full transfer function matrix $\boldsymbol{F}(z)$, and which in the domain of the complex variable z shows how the system in the course of *time* transmits and transfers the simultaneous influence of the system input vector and of all initial conditions on the system output (or on the system state).

- It provides the definition of $\boldsymbol{F}(z)$ and completely solves the problem of its determination so that it has the same properties as the well-known system transfer function matrix $\boldsymbol{G}(z)$: its compact matrix form, its independence of the external actions and initial conditions acting on the system, i.e., its dependence only on the system order, structure and parameters.

- It shows the physical meaning of $\boldsymbol{F}(z)$ and its link with the complete *time* response of the system.
- It establishes the full block diagram technique based on the use of $\boldsymbol{F}(z)$, which incorporates the $Z-$transform of the input vector and the vector of all initial conditions, and which generalizes the well-known block diagram technique.
- It exactly and completely solves the problem of the pole-zero cancellation.
- It introduces the concept of the system full matrix $\boldsymbol{P}(z)$ in the complex domain and establishes its link with the system full transfer function matrix $\boldsymbol{F}(z)$.
- It defines the system equivalence under nonzero initial conditions and proves the related conditions.
- It establishes the direct relationship between the system full transfer function matrix $\boldsymbol{F}(z)$ and the Lyapunov stability concept, definitions and conditions, which shows that the Lyapunov stability test via the system transfer function matrix $\boldsymbol{G}(z)$ can end with crucially wrong result. This refines inherently the complex domain criteria for the Lyapunov stability properties.
- It extends, broadens and generalizes the BI stability concept by introducing new BI stability properties that incorporate nonzero initial conditions, presents their exact definitions and proves the conditions for their validity in terms of the system full transfer function matrix $\boldsymbol{F}(z)$ and its submatrices.

Chapter 17

Future teaching and research

The basic notions, discoveries and results of this book are to be hopefully included, as the crucial parts, in the first course on linear dynamical *time*-invariant discrete-*time* systems. Other issues and results discovered herein should inherently enrich and/or refine the contents of the advanced courses in the linear dynamical *time*-invariant discrete-*time* systems.

The book contributions offer to engineers new, advanced, more powerful and complete complex domain theory of the linear discrete-*time* *time*-invariant systems.

The discoveries and new results presented herein open up new directions for further research, making up the basis for the observability, controllability, tracking and trackability theories of the linear control systems [70]. The optimization in general and the conditional optimization in particular (i.e., the system optimization by satisfying the stability demand) is another research direction.

The conditional optimization has been done in the parameter space for zero initial conditions first by Šiljak[1] in [113] for continuous-*time* systems and later in [32], [51] also for continuous-*time* systems and in [33], [55] for discrete-*time* systems. This book opens up the avenue to generalize their results related to the discrete-*time* systems, to the conditional optimization under simultaneous actions of the input vector and nonzero initial conditions on the system.

All that is presented herein is analogous to the *time*-invariant continuous-*time* linear dynamical systems, which is shown in the accompanying volume [69]. This book and *Linear Continuous-Time Systems* [69] constitute the entity.

[1]Pronounced Shilyak.

Part V

APPENDICES

Appendix A

Notation

The meaning of the notation is explained in the text at its first use.

A.1 Abbreviations

GISO system *General Input-State-Output system (2.35a) and (2.35b)*
iff *if and only if*
I *Input*
II *Input-Internal*
IIO *Input-Internal and Output dynamical*
IIO system *the Input-Internal and Output dynamical system (2.18a) and (2.18b)*
IO *Input-Output*
IO system *the Input-Output system (2.1)*
IS *Input-State*
ISO *Input-State-Output*
ISO system *the Input-State-Output system (3.60a) and (3.60b)*
LY *LY stability concept* means Lyapunov's stability concept
MIMO *Multiple-Input Multiple-Output*
OR *Output response*
PCUP *Physical Continuity and Uniqueness Principle*
PMD *Polynomial Matrix Description (2.32a) and (2.32b)*
RS system *Rosenbrock system (2.33a) and (2.33b)*
SISO *Single-Input Single-Output*
System *Discrete-time time-invariant linear dynamical system*
TCUP *Time Continuity and Uniqueness Principle*

A.2 Indexes

A.2.1 Subscripts

d the subscript d denotes "desired"

e *equilibrium*

i the subscript i denotes "the i-th"

j the subscript j denotes "the j-th"

nd *nondegenerate*

rd *reduced*

rnd *row nondegenerate*

$zero$ the subscript "*zero*" denotes "the zero value"

0 the subscript 0 (zero) associated with a variable $(\cdot)$ denotes its initial value $(\cdot)_0$; however, if $(\cdot) \subset \mathcal{T}_d$ then the subscript 0 (zero) associated with $(\cdot)$ denotes the *discrete time* set $\mathcal{T}_{d0}$, $(\cdot)_0 = \mathcal{T}_{d0}$, or if $(\cdot) \subset \mathbb{Z}$ then the subscript 0 (zero) associated with $(\cdot)$ denotes the extended set of natural numbers by zero $\mathcal{N}_0$, $(\cdot)_0 = \mathcal{N}_0$ which is equivalent (representative) of *discrete time* set $\mathcal{T}_{d0}$

A.2.2 Superscript

r the superscript r denotes "r-dimensional", $r \in \{1, 2, \cdots, n, \cdots\}$

A.3 Letters

Lowercase block or italic letters are used for scalars. Lowercase and uppercase bold block letters denote vectors. Uppercase block letters denote points. Uppercase bold italic letters denote matrices. Uppercase calligraphic letters designate sets or spaces.

A.3.1 Blackboard bold letters

$\mathbb{C}(\mathcal{R}^p)$ *the set of all defined and continuous functions on* $\mathcal{R}^p$

$\mathbb{Z}$ *the set of integers*

A.3.2 Calligraphic letters

Capital calligraphic letters are used for spaces or sets.

$\mathcal{A} \subseteq \mathcal{R}^n$ *a nonempty subset of* $\mathcal{R}^n$

$\mathcal{B} \subseteq \mathcal{R}^n$ *a nonempty subset of* $\mathcal{R}^n$

$\mathcal{B}_\xi(\mathbf{z})$ *an open hyperball with the radius* ξ *centered at the point* $\mathbf{z}$ in the corresponding space

$$\mathcal{B}_\xi(\mathbf{z}) = \{\mathbf{w} : \ \|\mathbf{w} - \mathbf{z}\| < \xi\}$$

$\mathcal{B}_\xi$ *an open hyperball with the radius* ξ *centered at the origin* of the corresponding space

$$\mathcal{B}_\xi = \mathcal{B}_\xi(\mathbf{0})$$

$\mathcal{C}$ *the set of complex numbers* z

$\mathcal{C}^i$ $i-$*dimensional complex vector space*

$\mathcal{D} = \mathcal{D}(\mathcal{T}_{d0})$ *the family of all functions defined and with a finite number of the first-order discontinuities on* $\mathcal{T}_{d0}$

$\mathcal{D}^i = \mathcal{D}(\mathcal{R}^i)$ *the family of all functions defined on* $\mathcal{R}^i$

$\mathcal{D}^i(\mathcal{S})$ *the family of all functions defined on the set* $\mathcal{S} \subseteq \mathcal{R}^i$, $\mathcal{D}^i(\mathcal{R}^i) = \mathcal{D}(\mathcal{R}^i) = \mathcal{D}^i$

$\mathcal{I}$ *the family of all* either proper or strictly proper real rational vector *complex functions, the originals of which are bounded discrete time-dependent functions* (2.10) (Section 2.1)

$$\mathcal{I} = \left\{ \begin{array}{c} \mathbf{I}(\cdot) : \exists\gamma(\mathbf{I}) \in \mathcal{R}^+ \Longrightarrow \|\mathbf{I}(k)\| < \gamma(\mathbf{I}), \ \forall k \in \mathcal{N}_0, \\ \mathcal{Z}\{\mathbf{I}(k)\} = \mathbf{I}(z) = \left[\begin{array}{cccc} I_1(z) & I_2(z) & \cdots & I_M(z) \end{array} \right]^T, \\ I_r(z) = \dfrac{\sum_{j=0}^{j=\zeta_r} a_{rj} z^j}{\sum_{j=0}^{j=\psi_r} b_{rj} z^j}, 0 \le \zeta_r \le \psi_r, \ \forall r = 1, 2, \cdots, M. \end{array} \right.$$

$\mathcal{J}$ a given, or to be determined, family of all bounded permitted input vector functions $\mathbf{I}(\cdot)$

$$\mathcal{J} \subset \mathcal{D} \cap \mathcal{I}.$$

$\mathcal{J}_-$ a subfamily of $\mathcal{J}$, $\mathcal{J}_- \subset \mathcal{J}$, such that the modulus of every pole of the $Z-$transform $\mathbf{I}(z)$ of every $\mathbf{I}(\cdot) \in \mathcal{J}_-$ is less than one

$\mathcal{N}$ the set of all natural numbers

$\mathcal{N}_0$ the extended set of natural numbers by zero,

$$\mathcal{N}_0 = \{k : k \in \mathbb{Z} \wedge k \in [0, +\infty[\} = \{0\} \cup \mathcal{N}$$

$\mathcal{R}$ *the set of all real numbers*

$\mathcal{R}^+$ *the set of all positive real numbers*

$\mathcal{R}_+$ *the set of all nonnegative real numbers*

$\mathcal{R}_d \subset \mathcal{R}$ *the set of some real numbers,*

$$\mathcal{R}_d = \{x : x = k \,\mathrm{num}\, T \in \mathcal{R}, k \in \mathbb{Z}\}$$

$\mathcal{R}^{\nu N}$ *the extended output space of the IO system*, which is simultaneously its *state space*

$\mathcal{R}^{n}$ an *n-dimensional real vector space, the state space of the ISO system*

$\mathcal{R}^{N\nu}\backslash\mathcal{B}_{\varepsilon}$ *the set of all vectors* $\mathbf{y}^{\nu-1}$ *in* $\mathcal{R}^{N\nu}$ *out of* $\mathcal{B}_{\varepsilon}$

$$\mathcal{R}^{N\nu}\backslash\mathcal{B}_{\varepsilon} = \left\{\mathbf{y}^{\nu-1} : \mathbf{y}^{\nu-1} \in \mathcal{R}^{N\nu},\ \mathbf{y}^{\nu-1} \notin \mathcal{B}_{\varepsilon}\right\}$$

$\mathcal{S}(\cdot)$ *system motion*

$\mathcal{T}$ the accepted reference time set, the arbitrary element of which is an arbitrary moment t and the *time* unit of which is second s, $1_t = s$, $t\langle s\rangle$,

$$\mathcal{T} = \{t : t\,[\mathrm{T}]\,\langle s\rangle, \mathrm{num}\, t \in \mathcal{R}, dt > 0\},\ \mathit{inf}\ \mathcal{T} = -\infty \notin \mathcal{T},\ \mathit{sup}\ \mathcal{T} = \infty \notin \mathcal{T}$$

$\mathcal{T}_d$ *the accepted reference discrete time set*, the arbitrary element of which is an arbitrary discrete moment t_d and the *discrete time* unit of which is second s, $1_t = s$, $t_d\langle s\rangle$

$$\mathcal{T}_d = \{t_d : \mathrm{num}\, t_d = k\, \mathrm{num}\, T \in \mathcal{R}_d \subset \mathcal{R},\ \Delta t_d = (k+1)T - kT = T > 0\},$$
$$\mathit{inf}\ \mathcal{T}_d = -\infty \notin \mathcal{T}_d,\ \ \mathit{sup}\ \mathcal{T}_d = \infty \notin \mathcal{T}_d$$

$\mathcal{T}_{d0}$ *the subset of* $\mathcal{T}_d$, which has the minimal element $min\mathcal{T}_{d0}$ that is the initial instant t_{d0}, $\mathrm{num}\, t_{d0} = 0$,

$$\mathcal{T}_{d0} = \{t_d : t_d \in \mathcal{T}_d,\ t_d \geq t_{d0},\ \ \mathrm{num}\, t_{d0} = 0\}, \mathcal{T}_{d0} \subset \mathcal{T}_d,$$
$$min\mathcal{T}_{d0} = t_{d0} \in \mathcal{T}_d,\ \ \mathit{sup}\ \mathcal{T}_{d0} = \infty \notin \mathcal{T}_{d0}$$

$\mathcal{T}_d^*$ the extended set of the set $\mathcal{T}_d$ by the set $\{-\infty, \infty\}$, $\mathcal{T}_d^* = \mathcal{T}_d \cup \{-\infty, \infty\infty\}$

$\mathcal{Y}_d \subset \mathcal{D}^N$ a given, or to be determined, *family of all bounded realizable desired total output vector functions* $\mathbf{Y}_d(\cdot)$, the $Z-$transforms of which are either proper or strictly proper real rational complex functions

$$\mathcal{Y}_d = \left\{\mathbf{Y}_d(\cdot) : \mathbf{Y}_d(k) \in \mathcal{D}^N,\ \exists\kappa \in \mathcal{R}^+ \Longrightarrow \|\mathbf{Y}_d(k)\| < \kappa,\ \forall k \in \mathcal{N}_0\right\}$$

$\mathcal{Z}\{\mathbf{i}(\cdot)\}$ *the* $Z-$*transform of a function* $\mathbf{i}(\cdot)$, Section B,

$$\mathcal{Z}\{\mathbf{i}(t)\} = \mathbf{I}(z) = \sum_{k=0}^{k=+\infty} \mathbf{i}(k) z^{-k}$$

A.3.3 Greek letters

α *a nonnegative integer*
β *a nonnegative integer*
$\gamma = \max\{\beta, \mu\}$
δ_{ij} *the Kronecker delta,* $\delta_{ij} = 1$ for $i = j$, and $\delta_{ij} = 0$ for $i \neq j$
ε *the output error vector* $\boldsymbol{\varepsilon} \in \mathcal{R}^N$, (3.46) (Subsection 3.5.1)

$$\boldsymbol{\varepsilon} = \mathbf{Y_d} - \mathbf{Y} = -\mathbf{y}, \boldsymbol{\varepsilon} = \begin{bmatrix} \varepsilon_1 & \varepsilon_2 & \cdots & \varepsilon_N \end{bmatrix}^T$$

θ *a nonnegative integer*
$\lambda_m(\boldsymbol{H})$ *the minimal eigenvalue of the symmetric matrix* $\boldsymbol{H} = \boldsymbol{H}^T$

$\lambda_M(\boldsymbol{H})$ *the maximal eigenvalue of the symmetric matrix* $\boldsymbol{H} = \boldsymbol{H}^T$
μ *a nonnegative integer*
ν *a nonnegative integer*
τ a subsidiary notation for *time* t
$\emptyset$ *the empty set*
ρ *a natural number*

A.3.4 Roman letters

$\boldsymbol{A} \in \mathcal{R}^{n \times n}$ *the matrix describing the internal dynamics of the ISO system*

$\boldsymbol{A}_r \in \mathcal{R}^{N \times N}$ *the matrix associated with the* $r-th$ *shift* $\mathbf{Y}(k+r)$ of the output vector $\mathbf{Y}$ of the *IO* system

$\boldsymbol{A}^{(\nu)} \in \mathcal{R}^{N \times (\nu+1)N}$ *the extended matrix describing the IO system internal dynamics,* $\boldsymbol{A}^{(\nu)} = \begin{bmatrix} \boldsymbol{A}_0 & \boldsymbol{A}_1 & \cdots & \boldsymbol{A}_\nu \end{bmatrix}$

$\boldsymbol{B}_r \in \mathcal{R}^{N \times M}$ *the matrix associated with the* $r-th$ *shift* $\mathbf{I}(k+r)$ of the input vector $\mathbf{I}$ of the *IO* system

$\boldsymbol{B}^{(\mu)} \in \mathcal{R}^{N \times (\mu+1)M}$ *the extended matrix describing the transmission of the influence of the input vector* $\mathbf{I}(k)$ *on the system dynamics,*

$$\boldsymbol{B}^{(\mu)} = \begin{bmatrix} \boldsymbol{B}_0 & \boldsymbol{B}_1 & \cdots & \boldsymbol{B}_\mu \end{bmatrix}$$

$\boldsymbol{C} \in \mathcal{R}^{N \times n}$ *the matrix of the ISO system, which describes the transmission of the state vector action on the system output vector* $\mathbf{Y}$

$\mathbf{C}_0$ *the vector of all initial conditions* acting on the system

$$\mathbf{C}_0 =$$

$$= \begin{bmatrix} \mathbf{I}_0^T & E^1\mathbf{I}_0^T & \cdots & E^{\mu-1}\mathbf{I}_0^T & \mathbf{X}_0^T & \mathbf{Y}_0^T & E^1\mathbf{Y}_0^T & \cdots & (E^{\nu-1}\mathbf{Y}_0^T)^T \end{bmatrix}^T \in$$
$$\in \mathcal{R}^{\mu M + n + \nu N}$$

d *a natural number*

$\mathbf{d} \in \mathcal{R}^d$ *the disturbance deviation vector*, (3.48) (Subsection 3.5.1)

$$\mathbf{d} = \mathbf{D} - \mathbf{D}_N$$

$\mathbf{D} \in \mathcal{R}^d$ *the total disturbance vector*

$\mathbf{D}_N \in \mathcal{R}^d$ *the nominal disturbance vector*

$\boldsymbol{D} \in \mathcal{R}^{N \times d}$ *the ISO system matrix describing the transmission of the influence of* $\mathbf{I}(k)$ *on the system output*

E *shifting operator; it shifts, i.e., translates, any time function along time axis in the negative direction for one, e.g.,* $E\mathbf{Y}(k) = \mathbf{Y}(k+1)$

$\boldsymbol{F}(\cdot) : \mathcal{N}_0 \longrightarrow \mathcal{R}^{N \times N}$ *a matrix function associated with* $\mathbf{f}(\cdot)$

$$\mathbf{f} = \left[\begin{array}{cccc} f_1 & f_2 & \cdots & f_N \end{array}\right]^T \Longrightarrow \boldsymbol{F} = \operatorname{diag}\left\{\begin{array}{cccc} f_1, & f_2, & \cdots & , f_N \end{array}\right\}$$

$\boldsymbol{F}(z)$ *the full (complete) transfer function matrix of a time-invariant discrete-time linear dynamical system*

$\boldsymbol{F}_{IIO}(z) \in \mathcal{C}^{N \times [(\gamma+1)M + \alpha\rho + \nu N]}$ *the full transfer function matrix of the IIO system (Definition 7.6)*

$\boldsymbol{F}_{IIOIS}(z) \in \mathcal{C}^{\rho \times [(\beta+1)M + \alpha\rho]}$ *the full (complete) IS transfer function matrix of the IIO system (Definition 7.7)*

$\boldsymbol{F}_{IO}(z) \in \mathcal{C}^{N \times [(\mu+1)M + \nu N]}$ *the full transfer function matrix of the IO system (Definition 7.3)*

$\boldsymbol{F}_{IOISO}(z)$ *the full transfer function matrix obtained from the IO mathematical model of the given ISO system* (Section 8.2, Note 8.5)

$\boldsymbol{F}_{ISO}(z) \in \mathcal{C}^{N \times (M+n)}$ *the full ISO transfer function matrix of the ISO system (Definition 7.4)*

$\boldsymbol{F}_{ISOIS}(z) \in \mathcal{C}^{n \times (M+n)}$ *the full (complete) IS transfer function matrix of the ISO system (Definition 7.5)*

$\boldsymbol{G} = \boldsymbol{G}^T \in \mathcal{R}^{p \times p}$ *the symmetric matrix of the quadratic form* $v(\mathbf{w}) = \mathbf{w}^T \boldsymbol{G} \mathbf{w}$

$\boldsymbol{G}(z)$ *the transfer function matrix of a time-invariant discrete-time linear dynamical system*

$\boldsymbol{G}_{IIO}(z) \in \mathcal{C}^{N \times M}$ *the transfer function matrix of the IIO system (Definition 7.6)*

$\boldsymbol{G}_{IIOIS}(z) \in \mathcal{C}^{\rho \times M}$ *the IS transfer function matrix of the IIO system (Definition 7.7)*

$\boldsymbol{G}_{IIOi_0}(z) \in \mathcal{C}^{N \times (\gamma+1)M}$ *the transfer function matrix relative to* $\mathbf{i}_0^{\gamma-1}$ *of the IIO system (Definition 7.6)*

$\boldsymbol{G}_{IIOi_0IS}(z) \in \mathcal{C}^{\rho \times (\beta+1)M}$ *the IS transfer function matrix relative to* $\mathbf{i}_0^{\beta-1}$ *of the IIO system (Definition 7.7)*

$\boldsymbol{G}_{IIOr_0}(z) \in \mathcal{C}^{N\times\alpha\rho}$ *the transfer function matrix relative to* $\mathbf{r}_0^{\alpha-1}$ *of the IIO system (Definition 7.6)*

$\boldsymbol{G}_{IIOr_0IS}(z) \in \mathcal{C}^{\rho\times\alpha\rho}$ *the IS transfer function matrix relative to* $\mathbf{r}_0^{\alpha-1}$ *of the IIO system (Definition 7.7)*

$\boldsymbol{G}_{IIOy_0}(z) \in \mathcal{C}^{N\times\nu N}$ *the transfer function matrix relative to* $\mathbf{y}_0^{\nu-1}$ *of the IIO system (Definition 7.6)*

$\boldsymbol{G}_{IO}(z) \in \mathcal{C}^{N\times M}$ *the transfer function matrix of the IO system (Definition 7.3)*

$\boldsymbol{G}_{IO_0}(z) \in \mathcal{C}^{N\times(\mu M+\nu N)}$ *the transfer function matrix relative to all initial conditions of the IO system (Definition 7.3)*

$\boldsymbol{G}_{IOi_0}(z) \in \mathcal{C}^{N\times\mu M}$ *the transfer function matrix relative to* $i_0^{\mu-1}$ *of the IO system (Definition 7.3)*

$\boldsymbol{G}_{IOy_0}(z) \in \mathcal{C}^{N\times\nu N}$ *the transfer function matrix relative to* $y_0^{\nu-1}$ *of the IO system (Definition 7.3)*

$\boldsymbol{G}_{IOISO}(z) \in \mathcal{C}^{N\times N}$ *the transfer function obtained from the IO mathematical model of the given ISO system* (Section 8.2, Note 8.5)

$\boldsymbol{G}_{ISO}(z) \in \mathcal{C}^{N\times M}$ *the ISO transfer function matrix of the ISO system (Definition 7.4)*

$\boldsymbol{G}_{ISOIS}(z) \in \mathcal{C}^{n\times M}$ *the IS transfer function matrix of the ISO system (Definition 7.5)*

$\boldsymbol{G}_{ISOSS}(z) \in \mathcal{C}^{n\times n}$ *the SS transfer function matrix of the ISO system (Definition 7.5)*

$\boldsymbol{G}_{ISOx_0}(z) \in \mathcal{C}^{N\times n}$ *the ISO transfer function matrix relative to* $\mathbf{x}_0$ *of the ISO system (Definition 7.4)*

$h(\cdot)$ *the Heaviside function,* i.e., *the unit step function (B.13) in Appendix B.1.2,*

$$h(\cdot):\mathcal{T}\to[0,1],\ h(t)\begin{cases} =0 \text{ for } t<0,\\ \in[0,1] \text{ for } t=0,\\ =1 \text{ for } t>0,\end{cases}$$

$h_d(\cdot)$ *the discrete Heaviside function,* i.e., *the unit discrete step function (B.12) in Appendix B.1.2,*

$$h_d(\cdot):\mathbb{Z}\to[0,1],\ h(k)\begin{cases} =0 \text{ for } k<0,\\ \in[0,1] \text{ for } k=0,\\ =1 \text{ for } k>0,\end{cases}$$

$\boldsymbol{H} = \boldsymbol{H}^T \in \mathcal{R}^{p\times p}$ *the symmetric matrix of the quadratic form* $v(\mathbf{w}) = \mathbf{w}^T\boldsymbol{H}\mathbf{w}$

i *an arbitrary natural number,* or *the input deviation variable*

$\mathbf{i} \in \mathcal{R}^M$ *the input deviation vector,* $\mathbf{i} = \begin{bmatrix} i_1 & i_2 & \cdots & i_M \end{bmatrix}^T$, (3.49) (in Subsection 3.5.1*)*

$$\mathbf{i} = \mathbf{I} - \mathbf{I}_N$$

$\mathbf{i}^\mu(k) \in \mathcal{R}^{(\mu+1)M}$ *the extended input vector at a moment k,*

$$\mathbf{i}^\mu(k) = \begin{bmatrix} \mathbf{i}^T(k) & E^1\mathbf{i}^T(k) & \cdots & E^\mu\mathbf{i}^T(k) \end{bmatrix}^T$$

$\mathbf{i}_0^{\mu-1} \in \mathcal{R}^{\mu M}$ *the initial extended input vector at the initial moment* $k_0 = 0$,

$$\mathbf{i}_0^{\mu-1} = \mathbf{i}^{\mu-1}(0) = \begin{bmatrix} \mathbf{i}_0^T & E^1\mathbf{i}_0^T & \cdots & E^{\mu-1}\mathbf{i}_0^T \end{bmatrix}^T \in \mathcal{R}^{\mu M}$$

I *total input variable*

$\boldsymbol{I}$ *the identity matrix of the n-th order,* $\boldsymbol{I} = \operatorname{diag}\left\{ 1 \quad 1 \quad \cdots \quad 1 \right\} \in \mathcal{R}^{n\times n}$

$\boldsymbol{I}_k$ *the identity matrix of the k-th order,* $\boldsymbol{I}_k = \operatorname{diag}\left\{ 1 \quad 1 \quad \cdots \quad 1 \right\} \in \mathcal{R}^{k\times k}$, $\boldsymbol{I}_k = \boldsymbol{I}$

$\mathbf{I} \in \mathcal{R}^M$ *the total input vector,* $\mathbf{I} = \begin{bmatrix} I_1 & I_2 & \cdots & I_M \end{bmatrix}^T$

$\mathbf{I}_N \in \mathcal{R}^M$ *the nominal total input vector,* $\mathbf{I}_N = \begin{bmatrix} I_{N1} & I_{N2} & \cdots & I_{NM} \end{bmatrix}^T$

$Int\mathcal{S}$ *the interior* of the set $\mathcal{S}$,

$Int\mathcal{T}_0$ *the interior* of the set $\mathcal{T}_0$, $Int\mathcal{T}_0 = \{t : \ t \in \mathcal{T}_0, \ t > 0\}$

$\operatorname{Im} z$ the imaginary part of $z = \sigma_z + j\omega_z$, $\operatorname{Im} z = \omega_z$

j *an arbitrary natural number,* or $j = \sqrt{-1}$ is the imaginary unit

k *an arbitrary integer*

m *a nonnegative integer*

n *a natural number*

N *a natural number,* if N is the dimension of the output vector and if n is the dimension of the state vector then $N \leq n$ in this book

$\boldsymbol{O}$ *the zero matrix of the appropriate order*

p *a natural number*

$\boldsymbol{P} \in \mathcal{R}^{n\times N}$ *a matrix,*

$\boldsymbol{P}_k \in \mathcal{R}^{\rho\times M}$ *a matrix*

$\boldsymbol{P}^{(\beta)} \in \mathcal{R}^{\rho\times M(\beta+1)}$ *an extended matrix describing the transmission of the influence of* $\mathbf{i}^\beta(k)$ *on the internal dynamics of the IIO system,*

$$\boldsymbol{P}^{(\beta)} = \begin{bmatrix} \boldsymbol{P}_0 & \boldsymbol{P}_1 & \cdots & \boldsymbol{P}_\beta \end{bmatrix}$$

q *a natural number*

$\boldsymbol{Q} \in \mathcal{R}^{N\times N}$ *a matrix*

$\boldsymbol{Q}_k \in \mathcal{R}^{\rho\times\rho}$ *a matrix*

$\boldsymbol{Q}^{(\alpha)} \in \mathcal{R}^{\rho\times\rho(\alpha+1)}$ *the extended matrix describing the internal dynamics of the IIO system,* $\boldsymbol{Q}^{(\alpha)} = \begin{bmatrix} \boldsymbol{Q}_0 & \boldsymbol{Q}_1 & \cdots & \boldsymbol{Q}_\alpha \end{bmatrix}$

$\mathbf{r} \in \mathcal{R}^\rho$ *a subsidiary deviation vector,* which is *the internal substate deviation vector of the IIO system,* (3.50) (Subsection 3.5.1*),*

$$\mathbf{r} = \mathbf{R} - \mathbf{R}_N$$

$\mathbf{R} \in \mathcal{R}^\rho$ *a subsidiary total vector,* which is *the total internal substate vector of the IIO system,* $\mathbf{R} = \begin{bmatrix} R_1 & R_2 & \cdots & R_\rho \end{bmatrix}$

$\mathbf{R}_N \in \mathcal{R}^\rho$ *a subsidiary nominal vector,* which is *the nominal total internal substate vector of the IIO system*

$\boldsymbol{R}_r \in \mathcal{R}^{N\times\rho}$ *a matrix*

$\boldsymbol{R}^{(\alpha)} \in \mathcal{R}^{N\times\rho(\alpha+1)}$ *the extended matrix describing the action of the extended internal dynamics vector* $\mathbf{r}^\alpha$ *on the output dynamics of the IIO system,*

$$\boldsymbol{R}^{(\alpha)} = \begin{bmatrix} \boldsymbol{R}_0 & \boldsymbol{R}_1 & \cdots & \boldsymbol{R}_\alpha \end{bmatrix}$$

$\operatorname{Re} z$ the real part of $z = \sigma_z + j\omega_z$, $\operatorname{Re} z = \sigma_z$

s *the basic time unit: second*

$\operatorname{sign}(\cdot) : \mathcal{R} \to \{-1, 0, 1\}$ *the scalar sign function,*

$$\operatorname{sign}(x) = |x|^{-1} x \text{ if } x \neq 0, \textit{ and } \operatorname{sign}(0) = 0$$

$\boldsymbol{S}_i^{(r)}(\cdot) : \mathcal{C} \longrightarrow \mathcal{C}^{i(r+1)\times i}$ *the matrix function of* z *defined by (8.4) in the Section 8.1:*

$$\boldsymbol{S}_i^{(r)}(z) = \begin{bmatrix} z^0\boldsymbol{I}_i & z^1\boldsymbol{I}_i & z^2\boldsymbol{I}_i & \cdots & z^r\boldsymbol{I}_i \end{bmatrix}^T \in \mathcal{C}^{i(r+1)\times i},$$
$$(r, i) \in \{(\mu, M), \ (\nu, N)\}$$

t *time* (*temporal variable*), or *an arbitrary time value* (*an arbitrary moment, an arbitrary instant*); and formally mathematically t denotes for short also the numerical *time* value $\operatorname{num} t$ if it does not create a confusion,

$$t[\mathrm{T}]\langle s\rangle, \ \operatorname{num} t \in \mathcal{R}, \ dt > 0 \ , \text{ or equivalently: } t \in \mathcal{T}.$$

It has been the common attitude to use the notation t of *time* and of its arbitrary temporal value also for its numerical value $\operatorname{num} t$, e.g., $t = 0$ is used in the sense $\operatorname{num} t = 0$. We do the same throughout the book if there is not any confusion because we can replace t everywhere by $t1_t^{-1}$, $(t1_t^{-1}) \in \mathcal{R}$, that we denote again by t, $\operatorname{num} t = \operatorname{num}(t1_t^{-1})$

t_0 a conventionally accepted *initial value of time* (*initial instant, initial moment*), $t_0 \in \mathcal{T}$, $\text{num}\, t_0 = 0$, i.e., simply $t_0 = 0$ in the sense $\text{num}\, t_0 = 0$

t_d *discrete time* (*discrete temporal variable*), or *an arbitrary discrete time value* (*an arbitrary discrete moment, an arbitrary discrete instant*); and formally mathematically t_d denotes for short also the numerical *discrete time* value $\text{num}\, t_d$ if it does not create a confusion,

$$t_d\, [\mathrm{T}]\, \langle s \rangle\, , \ \text{num}\, t_d \in \mathcal{R}_d \subset \mathcal{R},\ \Delta t_d = T > 0\ , \text{ or equivalently: } t_d \in \mathcal{T}_d.$$

$t_{d\inf}$ *the first discrete instant*, which has not happened, $t_{d\inf} = -\infty$

$t_{d\sup}$ *the last instant*, which will not occur, $t_{d\sup} = \infty$

$t_{dZeroTotal}$ *the total zero value of discrete time*, which has not existed and will not happen

t_{zero} *a conventionally accepted relative zero value of time*

t_{dzero} *a conventionally accepted relative zero value of discrete time*, $t_{dzero} = t_{zero} = 0$ $(\text{num}\, t_{dzero} = \text{num}\, t_{zero} = 0)$

T *the temporal dimension*, "the *time* dimension," which is the physical dimension of *time*

$T \in \mathcal{R}^+$ *sampling period; the period of a periodic behavior*

$\boldsymbol{T}_r \in \mathcal{R}^{N \times M}$ *a matrix*,

$\boldsymbol{T}^{(\mu)} \in \mathcal{R}^{N \times M(\mu+1)}$ *the extended matrix describing the action of the extended input vector* $\mathbf{i}^{\mu}$ *on the output dynamics of the IIO system*,

$$\boldsymbol{T}^{(\mu)} = \left[\begin{array}{cccc} \boldsymbol{T}_0 & \boldsymbol{T}_1 & \cdots & \boldsymbol{T}_\mu \end{array}\right]$$

$v(\cdot) : \mathcal{R}^p \to \mathcal{R}$ *a quadratic form*, $v(\mathbf{w}) = \mathbf{w}^T \boldsymbol{W} \mathbf{w}$,

$\mathbf{V}(z)$ the *Z−transform of all actions on the system*; it is composed of the *Z*−transform $\mathbf{I}(z)$ of the input vector $\mathbf{I}(k)$ and of all (input and output) initial conditions,

$$\mathbf{V}(z) = \left[\begin{array}{c} \mathbf{I}(z) \\ \mathbf{C}_0 \end{array}\right]$$

$\mathbf{w} \in \mathcal{R}^p$ *a subsidiary real-valued vector*,

$$\mathbf{w} = \left[\begin{array}{cccc} w_1 & w_2 & \cdots & w_p \end{array}\right]^T \in \left\{ \left[\begin{array}{cc} \left(\mathbf{r}^{\alpha-1}\right)^T & \left(\mathbf{y}^{\nu-1}\right)^T \end{array}\right]^T, \mathbf{x}, \mathbf{y}^{\nu-1} \right\},$$
$$p \in \{\alpha\rho + \nu N,\ n,\ \nu N\}$$

$\boldsymbol{W} = \boldsymbol{W}^T \in \mathcal{R}^{p \times p}$ *the symmetric matrix of the quadratic form* $v(\mathbf{w})$, $v(\mathbf{w}) = \mathbf{w}^T \boldsymbol{W} \mathbf{w}$, $\boldsymbol{W} \in \{\boldsymbol{G} = \boldsymbol{G}^T, \boldsymbol{H} = \boldsymbol{H}^T\}$

$x \in \mathcal{R}$ *a real-valued scalar state deviation variable*

$\mathbf{x} \in \mathcal{R}^n$ *the state vector deviation of the ISO system,* (3.51) (in Subsection 3.5.1*),*

$$\mathbf{x} = \begin{bmatrix} x_1 & x_2 & \cdots & x_n \end{bmatrix}^T, \ \mathbf{x} = \mathbf{X} - \mathbf{X}_N = \mathbf{X} - \mathbf{X}_d$$

$\mathbf{X} \in \mathcal{R}^n$ *the total state vector of the ISO system,*

$$\mathbf{X} = \begin{bmatrix} X_1 & X_2 & \cdots & X_n \end{bmatrix}^T$$

$\mathbf{X}_N \in \mathcal{R}^n$ *the total nominal state vector of the ISO system,*

$$\mathbf{X}_N = \begin{bmatrix} X_{N1} & X_{N2} & \cdots & X_{Nn} \end{bmatrix}^T$$

$y \in \mathcal{R}$ *a real-valued scalar output deviation variable*

$\mathbf{y} \in \mathcal{R}^N$ *a real-valued vector output deviation variable — the output deviation vector of the system,*

$$\mathbf{y} = \begin{bmatrix} y_1 & y_2 & \cdots & y_N \end{bmatrix}^T,$$

(3.45) (in Subsection 3.5.1*),*

$$\mathbf{y} = \mathbf{Y} - \mathbf{Y}_d = -\boldsymbol{\varepsilon}$$

$\mathbf{Y} \in \mathcal{R}^N$ *a real total valued vector output — the total output vector of the system,*

$$\mathbf{Y} = \begin{bmatrix} Y_1 & Y_2 & \cdots & Y_N \end{bmatrix}^T$$

$\mathbf{Y}_d \in \mathcal{R}^N$ *a desired (a nominal) total valued vector output — the desired total output vector of the system,*

$$\mathbf{Y}_d = \begin{bmatrix} Y_{d1} & Y_{d2} & \cdots & Y_{dN} \end{bmatrix}^T$$

$\mathbf{y}_0^{\nu-1} \in \mathcal{R}^{\nu N}$ *the initial extended output vector at the initial moment* $k_0 = 0$*,*

$$\mathbf{y}_0^{\nu-1} = \mathbf{y}^{\nu-1}(0) = \begin{bmatrix} \mathbf{y}_0^T & E^1\mathbf{y}_0^T & \cdots & E^{\nu-1}\mathbf{y}_0^T \end{bmatrix}^T,$$
$$\mathbf{y}_0^0 = \mathbf{y}^0(0) = \mathbf{y}_0 = \mathbf{y}(0)$$

z *a complex variable or a complex number* $z = \sigma_z + j\omega_z$

$\mathbf{Z}_r^{(\varsigma)}(\cdot) : \mathcal{C} \to \mathcal{C}^{(\varsigma+1)r\times\varsigma r}$ *the matrix function of z defined by (8.5) in Section 8.1:*

$$\mathbf{Z}_r^{(\varsigma)}(z) = \begin{bmatrix} \mathbf{O}_r & \mathbf{O}_r & \mathbf{O}_r & \cdots & \mathbf{O}_r \\ z^1\mathbf{I}_r & \mathbf{O}_r & \mathbf{O}_r & \cdots & \mathbf{O}_r \\ \vdots & \vdots & \vdots & \vdots & \vdots \\ z^{\varsigma-0}\mathbf{I}_r & z^{\varsigma-1}\mathbf{I}_r & z^{\varsigma-2}\mathbf{I}_r & \cdots & z^1\mathbf{I}_r \end{bmatrix}, \ \varsigma \geq 1,$$

$$\mathbf{Z}_r^{(\varsigma)}(z) \in \mathcal{C}^{(\varsigma+1)r\times\varsigma r}, \ (\varsigma, r) \in \{(\mu, M), \ (\nu, N)\}.$$

See Note 3.2 (in Subsection 3.4.2) on $\mathbf{Z}_r^{(\varsigma)}(\cdot)$ for $\zeta \leq 0$.

A.4 Names

General Input-State-Output Systems (GISO systems) are described in (2.35a) and (2.35b) (Section 2.3).

Input-Internal and Output dynamical systems (IIO systems) are described in (2.18a) and (2.18b) (Section 2.3).

Input-Output (IO) systems are described in (2.1) *(Section 2.1).*

Input-State-Output (ISO) systems are described by *the state space equation* (2.14a) and by *the output equation* (2.14b) *(Section 2.2).*

PMD systems are described by (2.32a) and (2.32b) (Section 2.3).

Rosenbrock systems (RS) are described in (2.33a) and (2.33b) (Section 2.3).

Stable (stability) matrix: a square matrix is *stable (stability) matrix* if and only if the moduli of all its eigenvalues are less than one.

A.5 Symbols and vectors

$(\cdot)$ *an arbitrary variable*, or *an index*

$|(\cdot)| : \mathcal{R} \to \mathcal{R}_+$ *the absolute value (module) of a (complex-valued) scalar variable* $(\cdot)$, respectively

$||\cdot|| : \mathcal{R}^n \to \mathcal{R}_+$ *an accepted norm on* $\mathcal{R}^n$, which is the *Euclidean norm on* $\mathcal{R}^n$ iff not stated otherwise:

$$||\mathbf{x}|| = ||\mathbf{x}||_2 = \sqrt{\mathbf{x}^T\mathbf{x}} = \sqrt{\sum_{i=1}^{i=n} x_i^2}$$

$\langle 1_{..} \rangle$ shows *the units 1.. of a physical variable*

$[\alpha, \beta] \subset \mathcal{R}$ *a compact interval,* $[\alpha, \beta] = \{x : x \in \mathcal{R}, \alpha \leq x \leq \beta\}$

$[\alpha, \beta[\subseteq \mathcal{R}$ *a left closed, right open interval,* $[\alpha, \beta[= \{x : x \in \mathcal{R}, \alpha \leq x < \beta\}$

$]\alpha, \beta] \subseteq \mathcal{R}$ *a left open, right closed interval,* $]\alpha, \beta] = \{x : x \in \mathcal{R}, \alpha < x \leq \beta\}$

$]\alpha, \beta[\subseteq \mathcal{R}$ *an open interval,* $]\alpha, \beta[= \{x : x \in \mathcal{R}, \alpha < x < \beta\}$,

$$(\sigma, \infty[\in \{]\sigma, \infty[, [\sigma, \infty[\}$$

$(\alpha, \beta) \subseteq \mathfrak{R}$ *a general interval,*

$$(\alpha, \beta) \in \{[\alpha, \beta], [\alpha, \beta[,]\alpha, \beta],]\alpha, \beta[\}$$

$\mathcal{A} \backslash \mathcal{B}$ is the set difference between the set $\mathcal{A}$ and the set $\mathcal{B}$,

$$\mathcal{A} \backslash \mathcal{B} = \{x : x \in \mathcal{A}, x \notin \mathcal{B}\}$$

$\lambda_i(\boldsymbol{A})$ *the i-th eigenvalue* $\lambda_i(\boldsymbol{A})$ *of the matrix* $\boldsymbol{A}$

$[A..]$ shows *the physical dimension A.. of a physical variable*

$\begin{bmatrix} \boldsymbol{A}_1 & \boldsymbol{A}_2 & \cdots & \boldsymbol{A}_\nu \end{bmatrix}$ *a structured matrix* composed of the submatrices $\boldsymbol{A}_1, \boldsymbol{A}_2, \cdots, \boldsymbol{A}_\nu$

$\mathbf{0}_r = \begin{bmatrix} 0 & 0 & \cdots & 0 \end{bmatrix}^T \in \mathcal{R}^r$ *the elementwise zero vector,* $\mathbf{0}_n = \mathbf{0}$

$\mathbf{1}_r = \begin{bmatrix} 1 & 1 & \cdots & 1 \end{bmatrix}^T \in \mathcal{R}^r$ *the elementwise unity vector,* $\mathbf{1}_n = \mathbf{1}$

$\forall$ *for every*

$\text{adj}\boldsymbol{A}$ *the adjoint matrix of the nonsingular square matrix* $\boldsymbol{A}$,

$$\det \boldsymbol{A} \neq 0 \Longrightarrow \boldsymbol{A} \,\text{adj}\, \boldsymbol{A} = (\det \boldsymbol{A}) \boldsymbol{I}$$

$\det\boldsymbol{A}$ *the determinant of the matrix* $\boldsymbol{A}$, $\det\boldsymbol{A} = |\boldsymbol{A}|$

$\boldsymbol{A}^{-1}$ *the inverse matrix of the nonsingular square matrix* $\boldsymbol{A}$,

$$\det \boldsymbol{A} \neq 0 \Longrightarrow \boldsymbol{A}^{-1} = \text{adj}\, \boldsymbol{A} / \det \boldsymbol{A}$$

$d(\mathbf{v}, \mathcal{S})$ *the scalar distance of a vector* $\mathbf{v}$ *from a set* $\mathcal{S}$,

$$d(\mathbf{v}, \mathcal{S}) = \inf[\|\mathbf{v} - \mathbf{w}\| : \mathbf{w} \in \mathcal{S}]$$

$\dim \mathbf{z}$ *the mathematical dimension of a vector* $\mathbf{z}, \mathbf{z} \in \mathcal{R}^n \Rightarrow \dim \mathbf{z} = n$

ddim *the dynamical dimension of a system* composed of the system order and the system dimension

Den$\boldsymbol{F}(z)$ *the denominator matrix polynomial of the real rational matrix* $\boldsymbol{F}(z)$,

$$\boldsymbol{F}(z) = [\mathrm{Den}\, \boldsymbol{F}(z)]^{-1}\, \mathrm{Num}\, \boldsymbol{F}(z), \mathrm{or}\ \boldsymbol{F}(z) = \mathrm{Num}\, \boldsymbol{F}(z)\, [\mathrm{Den}\, \boldsymbol{F}(z)]^{-1}$$

mddim *the minimal dynamical dimension of a system*

$\deg\left[\mathrm{adj}\left(\sum_{r=0}^{r=\nu} \boldsymbol{A}_r z^r\right)\right]$ denotes *the greatest power of* z *over all elements of* $\mathrm{adj}\left(\sum_{r=0}^{r=\nu} \boldsymbol{A}_r z^r\right)$

$\deg\left(\sum_{r=0}^{r=\mu} \boldsymbol{B}_r z^r\right)$ denotes *the greatest power of* z *over all elements of* $\sum_{r=0}^{r=\mu} \boldsymbol{B}_r z^r$

$\deg\left[\det\left(\sum_{r=0}^{r=\nu} \boldsymbol{A}_r z^r\right)\right]$ denotes *the greatest power of* z *in*

$$\det\left(\sum_{r=0}^{r=\nu} \boldsymbol{A}_r z^r\right)$$

$\mathrm{Im}\, \lambda_i(\boldsymbol{A})$ *the imaginary part* of the eigenvalue $\lambda_i(\boldsymbol{A})$ of the matrix $\boldsymbol{A}$

$\min(\delta, \Delta)$ denotes *the smaller between* δ *and* Δ,

$$\min(\delta, \Delta) = \begin{cases} \delta, & \delta \le \Delta, \\ \Delta, & \Delta \le \delta \end{cases}$$

Num$\boldsymbol{F}(z)$ *the numerator matrix polynomial of the real rational matrix* $\boldsymbol{F}(z)$

$$\boldsymbol{F}(z) = [\mathrm{Den}\, \boldsymbol{F}(z)]^{-1}\, \mathrm{Num}\, \boldsymbol{F}(z), \mathrm{or}\ \boldsymbol{F}(z) = \mathrm{Num}\, \boldsymbol{F}(z)\, [\mathrm{Den}\, \boldsymbol{F}(z)]^{-1}$$

$\mathrm{Re}\, \lambda_i(\boldsymbol{A})$ *the real part* of the eigenvalue $\lambda_i(\boldsymbol{A})$ of the matrix $\boldsymbol{A}$

$\exists$ *there exist(s)*

$\exists!$ *there exists exactly one*

$\in$ *belong(s) to, are (is) members (a member) of*, respectively

$\subset$ *a proper subset of* (it can not be equal to)

$\subseteq$ *a subset of* (it can be equal to)

$\simeq$ *equivalent*

$\sqrt{-1}$ *the imaginary unit* denoted by j, $j = \sqrt{-1}$

inf *infimum*

max *maximum*

min *minimum*

$\mathrm{num}\, x$ *the numerical value of* x, if $x = 50V$ then $\mathrm{num}\, x = 50$

phdim $x(\cdot)$ *the physical dimension of a variable* $x(\cdot)$,

$$x(\cdot) = t \Longrightarrow \mathit{phdim}\ x(\cdot) = \mathit{phdim}\ t = \mathrm{T},\ \text{but}\ \mathit{dim}\ t = 1$$

sup *supremum*

$\otimes$ the Kronecker matrix product, (13.92) (in Section 13.3)

A.6 Units

$1_{(\cdot)}$ *the unit of a physical variable* $(\cdot)$,

1_t *the time unit of the reference time axis* T, $1_t = s$

Appendix B

$Z-$transforms and unit impulses

B.1 $Z-$transforms

If and only if the following sum exists,

$$\exists \sum_{k=-\infty}^{k=+\infty} x(k)z^{-k} \tag{B.1}$$

then it is the *two-sided $Z-$transform of the scalar discrete-time function* $x(\cdot) : \mathbb{Z} \longrightarrow \mathcal{R}$, irrespective of the function origin. The most frequent way to generate discrete-time function $x(\cdot)$, in discrete-time systems, is by sampling its counterpart continuous-time function that should be continuous or piecewise continuous in sampling instants $t = kT$.

Assumption B.1 *Let the continuous-time functions $x(\cdot) : \mathcal{T} \to \mathcal{R}$ be such that:*

- *$x(t)$ is defined for all $t = kT$, i.e., $\forall t = kT$, $k \in \mathbb{Z}$, $T > 0$, while $x(kT) = 0$, $\forall k = \{\cdots, -2, -1\}$, or*

- *it is piecewise continuous for some k while then for such k, $x(kT) = x(kT^+)$ assuming that the following limit exists*

$$x\left(kT^+\right) = \lim_{t \to kT^+} x(t).$$

Anyway, $x(k)$ essentially represents the numerical sequence as the integer k increases. For such functions, the sum (B.1) becomes

$$\sum_{k=0}^{k=+\infty} x(k)z^{-k}. \tag{B.2}$$

If and only if it exists then it is the *one-sided $Z-$transform of the function* $x(\cdot)$,

$$\mathcal{Z}\{x(k)\} = X(z) = \sum_{k=0}^{k=+\infty} x(k)z^{-k}. \tag{B.3}$$

In this book this definition is exclusively used and called simply *the $Z-$transform of the function* $x(\cdot)$ that is called *the original.*

The crucial question related to the $Z-$transform (B.3) existence of the function $x(\cdot)$ is the negative power series convergence for some complex numbers z, $z \in \mathcal{C}$. Obviously the convergence depends on $x(k)$ and z.

It is possible that two different discrete time functions $x_1(\cdot)$ and $x_2(\cdot)$ give the same $Z-$transform, $X_1(z) = X_2(z)$, but for the $z-$complex numbers located in different regions of the $z-$complex plane, see examples in [101]. In that case, it is not possible to determine uniquely the original discrete time function based on its $Z-$transform, without taking into account the location of $z-$complex numbers. To conclude, knowing $X(z)$ of $x(k)$ and the location of $z-$complex numbers for which $X(z)$ converges, leads to the unique original $x(k)$. The region of $z-$complex numbers in the $z-$complex plane, for which $X(z)$ converges, is called ***region of convergence (ROC)** of $X(z)$*.

If, in the determination of the region of convergence, we use the absolute values of the series members, $\sum_{k=0}^{k=+\infty} \left|x(k) z^{-k}\right|$, instead of the original $Z-$transform series, the region of convergence in that case, is restricted to one side of a circle, which is convenient. Then, it is said that the original series is absolutely convergent. Generally, the absolute convergence of a series is sufficient condition for the series to be convergent, too, i.e., if a series is absolutely convergent, it is also convergent. It could be easily proved by using of Cauchy's criteria for a series convergence and an inequality relation. For more details see [98]. In such a way, it is allowed to use the absolute convergence of $X(z)$ in the determination of its region of convergence.

The next theorem and definition are taken from [41, p. 166, 167].

Theorem B.1 ***Conditions for the existence of the $Z-$transform***

If there exist two positive real numbers R and α, $R, \alpha \in \mathcal{R}^+$,

$$\mathcal{R}^+ = \{\beta : 0 < \beta < \infty\} =]0, +\infty[\,,$$

and a natural number N, $N \in \mathcal{N}$, such that the following is valid for a discrete time function $x(\cdot)$:

1. $|x(k)| < +\infty$, $\forall k = 0, 1, 2, \cdots$, *and*
2. $|x(k)| \leq \alpha R^k$, $\forall k = N, N+1, \cdots$,

then, the $Z-$transform of the function $x(\cdot)$,

$$X(z) = \sum_{k=0}^{k=+\infty} x(k) z^{-k},$$

absolutely converges for

$$|z| = \rho > R.$$

The proof of the Theorem B.1 is given in [41, p. 166, 167].

Definition B.1 ***Region of convergence of*** $X(z)$
The least positive real number R, $R \in \mathcal{R}^+$, denoted by r, such that $X(z)$ converges for all z whose modulus is greater then r, $\forall z : |z| > r$, is the radius of convergence of the $X(z)$. The $z-$complex plane region determined by $|z| > r$, Fig. B.1, is the region of the convergence of $X(z)$.

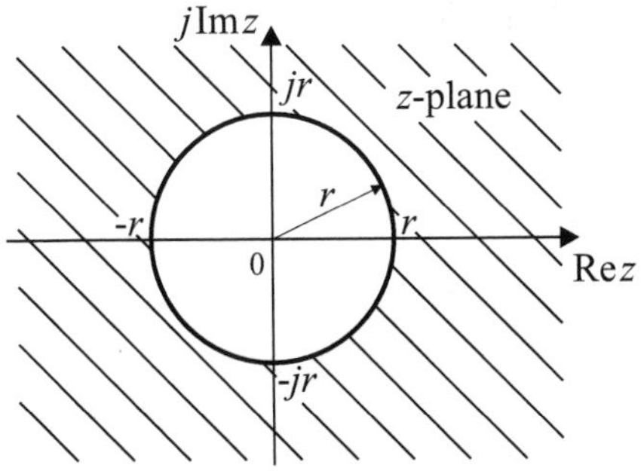

Figure B.1: Region of convergence of $X(z)$.

In the same figure, Fig. B.1, the region of the divergence of $X(z)$ is the interior of the circle of the radius r centered at the origin, $|z| < r$. On the boundary of the circle, i.e., for $|z| = r$, the $X(z)$ may either converge or diverge [14].

When the $Z-$transform $X(z)$ converges, it has a compact closed-form, which is generally the ratio of two polynomials in z:

$$X(z) = \frac{\sum\limits_{i=0}^{i=m} b_i z^i}{\sum\limits_{i=0}^{i=n} a_i z^i} = \frac{b_m z^m + b_{m-1} z^{m-1} + \cdots + b_1 z + b_0}{z^n + a_{n-1} z^{n-1} + \cdots + a_1 z + a_0}, \quad m \leq n.$$

After the factorization of the numerator and the denominator polynomials, $X(z)$ becomes:

$$X(z) = b_m \frac{\left(z - z_1^0\right)\left(z - z_2^0\right) \cdots \left(z - z_m^0\right)}{\left(z - z_1^*\right)\left(z - z_2^*\right) \cdots \left(z - z_n^*\right)},$$

where z_i^0, $i = 1, 2, \cdots, m$, are finite zeros of $X(z)$, and z_i^*, $i = 1, 2, \cdots, n$, are finite poles of $X(z)$. If $m < n$ then $X(z)$ has zero z^0 of $n - m$ order with infinite modulus, $\left|z^0\right| = \infty$, $\nu^0 = n - m$, where ν^0 is the multiplicity of the zero z^0. If $m > n$ then $X(z)$ has pole z^* of $m - n$ order with infinite modulus, $|z^*| = \infty$, $\nu^* = m - n$, where ν^* is the multiplicity of the pole z^*. Since we restricted the relationship between m and n to satisfy $m \leq n$ then $X(z)$ does not have poles of infinite modulus $|z^*| = \infty : \nexists |z^*| = \infty$.

The Inverse Z−transform $\mathcal{Z}^{-1}\{X(z)\}$ of $X(z)$ is defined by

$$\mathcal{Z}^{-1}\{X(z)\} = x(k) = \frac{1}{2\pi j} \oint_G X(z) z^{k-1} dz \tag{B.4}$$

where G is a circle within the region of convergence of $X(z)$ centered at the origin, and G encloses all poles of $X(z) z^{k-1}$. The circle integration in (B.4) is in the counterclockwise direction. The proof of (B.4) can be seen in [89]. The practical application of the inverse $Z-$transform in the determination of $x(k)$ is significantly simplified by using the well-known *Cauchy residue theorem:*

$$x(k) = \sum_i \operatorname{Res}\left[X(z) z^{k-1}\right]\Bigg|_{z = z_i^* - \text{pole of } X(z)}. \tag{B.5}$$

The $Z-$transform of a vector function $\mathbf{x}(\cdot) : \mathbb{Z} \longrightarrow \mathcal{R}^n$,

$$\mathbf{x}(\cdot) = \begin{bmatrix} x_1(\cdot) & x_2(\cdot) & \cdots & x_n(\cdot) \end{bmatrix}^T,$$

is the vector of the $Z-$transforms of the entries of $\mathbf{x}(\cdot)$, respectively,

$$\mathcal{Z}\{\mathbf{x}(k)\} = \mathbf{X}(z) = \begin{bmatrix} X_1(z) & X_2(z) & \cdots & X_n(z) \end{bmatrix}^T. \tag{B.6}$$

B.1.1 Z−transform properties

The following properties of the Z−transform are crucial:

- *linearity of the Z−transform*

$$\mathcal{Z}\left\{\sum_{r=1}^{r=p} M_r \mathbf{x}_r(k)\right\} = \sum_{r=1}^{k=p} M_r \mathcal{Z}\left\{\mathbf{x}_r(k)\right\} = \sum_{r=1}^{r=p} M_r \mathbf{X}_r(z), \tag{B.7}$$

- *the Z− transform of the function* $\mathbf{x}(\cdot)$ shifted for r discrete instants (sampling instants) along the discrete time axis reads:

  - in negative direction, which contains the influence of initial conditions

$$\mathcal{Z}\left\{\mathbf{x}(k+r)\right\} = z^r \left[\mathcal{Z}\left\{\mathbf{x}(k)\right\} - \sum_{k=0}^{r-1} \mathbf{x}(k) z^{-k}\right] =$$
$$= z^r \left[\mathbf{X}(z) - \sum_{k=0}^{r-1} \mathbf{x}(k) z^{-k}\right]; \tag{B.8}$$

  - in positive direction

$$\mathcal{Z}\left\{\mathbf{x}(k-r)\right\} = z^{-r}\mathcal{Z}\left\{\mathbf{x}(k)\right\} = z^{-r}\mathbf{X}(z). \tag{B.9}$$

- *the Z−transform $\mathcal{Z}\{x_1(k) * x_2(k)\}$ of the convolution sum $x_1(k) *$* $x_2(k)$ of the two functions $x_1(\cdot)$ and $x_2(\cdot)$,

$$x_1(k) * x_2(k) = \sum_{j=0}^{j=k} x_1(j)\, x_2(k-j) = \sum_{j=0}^{j=k} x_1(k-j)\, x_2(j)$$

  expressed in terms of the Z−transforms $X_1(z)$ and $X_2(z)$ of the functions $x_1(\cdot)$ and $x_2(\cdot)$, reads

$$\mathcal{Z}\left\{x_1(k) * x_2(k)\right\} = \mathcal{Z}\left\{x_1(k)\right\}\mathcal{Z}\left\{x_2(k)\right\} = X_1(z)\, X_2(z). \tag{B.10}$$

The following property of the Z−transform of a continuous-time function $x(\cdot)$ is also of the crucial importance:

- the $Z-$transform of the $n-th$ forward finite difference $\Delta^n x(t)$ of the continuous-time function $x(\cdot)$ expressed in terms of the $Z-$transform $X(z)$ and initial conditions reads

$$\mathcal{Z}\{\Delta^n x(t)\} = (z-1)^n X(z) - z \sum_{i=0}^{i=n-1} (z-1)^{n-i-1} \Delta^i x(0). \quad \text{(B.11)}$$

Let us remind of the definitions of the different order finite differences of a continuous-time function $x(\cdot)$:

- the first forward finite difference $\Delta x(t)$ of $x(t)$ is

$$\Delta x(t) = x(t+\Delta t) - x(t),$$

- the second forward finite difference $\Delta^2 x(t)$ of $x(t)$ is

$$\begin{gathered}\Delta^2 x(t) = \Delta[\Delta x(t)] = \Delta[x(t+\Delta t) - x(t)] = \Delta[x(t+\Delta t)] - \\ -\Delta[x(t)] = x(t+\Delta t+\Delta t) - x(t+\Delta t) - x(t+\Delta t) + x(t) = \\ = x(t+2\Delta t) - 2x(t+\Delta t) + x(t),\end{gathered}$$

- the $n-th$ forward finite difference $\Delta^n x(t)$ of $x(t)$ is

$$\begin{aligned}\Delta^n x(t) &= \Delta\left[\Delta^{n-1} x(t)\right] = \Delta\left[\Delta\left[\Delta^{n-2} x(t)\right]\right] = \\ &= \cdots = \underbrace{\Delta[\Delta[\cdots[\Delta x(t)]]]}_{n-\text{ times}},\end{aligned}$$

where Δt is finite difference of *time* t.

This property is used in approximate discretization of the mathematical model of the linear continuous-time time invariant systems by replacing derivatives with the appropriate finite differences. In general, the first derivative $\frac{dx(t)}{dt}$ of a variable $x(\cdot)$ at the instant t, by definition, reads

$$\frac{dx(t)}{dt} = \lim_{\Delta t \to 0} \frac{x(t+\Delta t) - x(t)}{\Delta t},$$

if the limit exists. Similarly, the second derivative $\frac{d^2x(t)}{dt^2}$ of a variable $x(\cdot)$ at the instant t, by definition, is as follows:

$$\frac{d^2 x(t)}{dt^2} = \lim_{\Delta t \to 0} \frac{x^{(1)}(t+\Delta t) - x^{(1)}(t)}{\Delta t},$$

if the limit exists. Finally, the $k-th$ derivative $\frac{d^k x(t)}{dt^k}$ of a variable $x(\cdot)$ at the instant t, by definition, is

$$\frac{d^k x(t)}{dt^k} = \lim_{\Delta t \to 0} \frac{x^{(k-1)}(t+\Delta t) - x^{(k-1)}(t)}{\Delta t},$$

if the limit exists. If Δt is sufficiently small and different from zero, then

$$\frac{dx(t)}{dt} \approx \frac{\Delta x(t)}{\Delta t} = \frac{x(t+\Delta t) - x(t)}{\Delta t},$$

$$\frac{d^2 x(t)}{dt^2} \approx \frac{\Delta x^{(1)}(t)}{\Delta t} \approx \frac{\Delta\left(\frac{\Delta x(t)}{\Delta t}\right)}{\Delta t} = \frac{\Delta^2 x(t)}{\Delta t^2} = \\ = \frac{x(t+2\Delta t) - 2x(t+\Delta t) + x(t)}{\Delta t^2},$$

and

$$\frac{d^k x(t)}{dt^k} \approx \frac{\Delta x^{(k-1)}(t)}{\Delta t} \approx \frac{\Delta\left(\frac{\Delta x^{(k-2)}(t)}{\Delta t}\right)}{\Delta t} = \\ = \frac{\Delta^2 x^{(k-2)}(t)}{\Delta t^2} \approx \cdots \approx \frac{\Delta^k x(t)}{\Delta t^k}.$$

Since all these approximations are carried out in context of discretization in time by sampling, the finite differences are observed at the sampling instants $t = kT$ and if sampling period T is enough small, Δt can be replaced by T. Then the approximations are as follows:

$$\frac{dx(t)}{dt} \approx \frac{\Delta x(kT)}{T} = \frac{x[(k+1)T] - x(kT)}{T},$$

$$\frac{d^2 x(t)}{dt^2} \approx \frac{\Delta^2 x(kT)}{T^2} = \frac{x[(k+2)T] - 2x[(k+1)T] + x(kT)}{T^2},$$

$$\frac{d^k x(t)}{dt^k} \approx \frac{\Delta^k x(kT)}{T^k},$$

or for the reason of denotation simplicity

$$\frac{dx(t)}{dt} \approx \frac{x(k+1) - x(k)}{T},$$

$$\frac{d^2 x(t)}{dt^2} \approx \frac{x(k+2) - 2x(k+1) + x(k)}{T^2},$$

$$\frac{d^k x(t)}{dt^k} \approx \frac{\Delta^k x(k)}{T^k},$$

where T as known and fixed is omitted in $\Delta^k x(k)$, but to be aware that it actually exists.

These features of the $Z-$transform permit us to transfer time domain studies of properties of linear time-invariant discrete-time systems to their complex domain studies.

Notice the complete analogy between the Laplace transform and the $Z-$transform.

B.1.2 $Z-$transforms of the basic functions

$Z-$transforms of the basic functions are as follows:

- **Unite step (Heaviside) function**

  We will use discrete *Heaviside function*, called also *the unit discrete step function*, $h_d(\cdot) : \mathbb{Z} \longrightarrow [0, 1]$,

$$h_d(k) \begin{cases} = 0,\ k < 0, \\ \in [0,1],\ k = 0, \\ = 1,\ k > 0, \end{cases} \tag{B.12}$$

  It is the counterpart function of the continuous Heaviside function (the unit step function), $h(\cdot) : \mathcal{T} \to \mathcal{R}$,

$$h(t) \begin{cases} = 0,\ t < 0, \\ \in [0,1],\ t = 0, \\ = 1,\ t > 0. \end{cases} \tag{B.13}$$

  Taking into account Assumption B.1, the $Z-$transform $\mathcal{Z}\{h_d(k)\}$ of $h_d(k)$ reads

$$\mathcal{Z}\{h_d(k)\} = H_d(z) = \sum_{k=0}^{k=+\infty} h_d(k) z^{-k} =$$
$$= \left[h_d(0) = h(0^+)\right] z^{-0} + h_d(1) z^{-1} + h_d(2) z^{-2} + \cdots =$$
$$= 1 \cdot z^{-0} + 1 \cdot z^{-1} + 1 \cdot z^{-2} + \cdots = 1 + z^{-1} + z^{-2} + \cdots =$$
$$= \frac{1}{1 - z^{-1}},\ |z| > 1 \Longrightarrow H_d(z) = \frac{z}{z-1},\ |z| > 1.$$

- **Exponential functions** $a^k h_d(k)$, $e^{\alpha k} h_d(k)$

$$\mathcal{Z}\left\{a^k h_d(k)\right\} = \sum_{k=0}^{k=+\infty} a^k h_d(k) z^{-k} =$$
$$= a^0 \left[h_d(0) = h\left(0^+\right)\right] z^{-0} + a^1 h_d(1) z^{-1} + a^2 h_d(2) z^{-2} + \cdots =$$
$$= 1 + az^{-1} + \left(az^{-1}\right)^2 + \cdots = \frac{z}{z-a}, \quad |z| > |a|.$$

Similarly,

$$\mathcal{Z}\left\{e^{\alpha k} h_d(k)\right\} = \sum_{k=0}^{k=+\infty} e^{\alpha k} h_d(k) z^{-k} =$$
$$= e^{\alpha 0} \left[h_d(0) = h\left(0^+\right)\right] z^{-0} + e^{\alpha 1} h_d(1) z^{-1} + e^{\alpha 2} h_d(2) z^{-2} + \cdots =$$
$$= 1 + e^{\alpha} z^{-1} + \left(e^{\alpha} z^{-1}\right)^2 + \cdots = \frac{z}{z-e^{\alpha}}, \quad |z| > |e^{\alpha}|. \tag{B.14}$$

- **Sine and Cosine functions** $\alpha(\sin\omega k) h_d(k)$, $\alpha(\cos\omega k) h_d(k)$

We will use here the following two identities

$$\sin\omega k = \frac{e^{j\omega k} - e^{-j\omega k}}{2j},$$

$$\cos\omega k = \frac{e^{j\omega k} + e^{-j\omega k}}{2},$$

which are obtained by using the well-known Euler identity:

$$e^{j\omega k} = \cos\omega k + j\sin\omega k.$$

$$\mathcal{Z}\left\{\alpha(\sin\omega k) h_d(k)\right\} = \alpha\mathcal{Z}\left\{(\sin\omega k) h_d(k)\right\} =$$
$$= \alpha\sum_{k=0}^{k=+\infty} (\sin\omega k) h_d(k) z^{-k} = \alpha\sum_{k=0}^{k=+\infty} \frac{e^{j\omega k} - e^{-j\omega k}}{2j} h_d(k) z^{-k} =$$
$$= \frac{\alpha}{2j}\left[\sum_{k=0}^{k=+\infty} e^{j\omega k} h_d(k) z^{-k} - \sum_{k=0}^{k=+\infty} e^{-j\omega k} h_d(k) z^{-k}\right] =$$
$$= \frac{\alpha}{2j}\left[\mathcal{Z}\left\{e^{j\omega k} h_d(k)\right\} - \mathcal{Z}\left\{e^{-j\omega k} h_d(k)\right\}\right]. \tag{B.15}$$

The expression (B.15) is further transformed by using (B.14) in:

$$\mathcal{Z}\left\{\alpha\left(\sin\omega k\right)h_d\left(k\right)\right\} = \frac{\alpha}{2j}\left[\frac{z}{z-e^{j\omega}} - \frac{z}{z-e^{-j\omega}}\right] =$$
$$= \frac{\alpha}{2j}\frac{z\left(e^{j\omega}-e^{-j\omega}\right)}{z^2 - z\left(e^{j\omega}+e^{-j\omega}\right)+1} = \frac{\alpha z\sin\omega}{z^2-2z\cos\omega+1}, \quad |z|>1.$$

Analogically obtained the $Z-$transform of the $\alpha\left(\cos\omega k\right)h_d\left(k\right)$ reads

$$\mathcal{Z}\left\{\alpha\left(\cos\omega k\right)h_d\left(k\right)\right\} = \frac{\alpha\left(z^2-z\cos\omega\right)}{z^2-2z\cos\omega+1}, \quad |z|>1.$$

Appendix B.2 presents separately the $Z-$transforms of unit impulses as the more complex issues.

B.2 Unit impulses

Continuous-time unit impulse function $\delta(\cdot) : t \to \{0, +\infty\}$ (unit Dirac impulse, unit Dirac delta distribution) is of the enormous theoretical and practical significance for the linear continuous-time time-invariant systems. According to its standard original definition in literature, see for example [99], the unit impulse function has got the following properties (B.16):

$$\int_{0^-}^{0^+}\delta(t)dt = 1;\ \delta\left(0\right) = +\infty;\ \delta\left(t\right) = 0,\ \forall t\in\left]-\infty,+\infty\right[,\ t\neq 0, \tag{B.16}$$

i.e., it is of, the infinitely short duration, of the infinitely great height, at $t = 0$, of the area equal to 1, and of the values equal to 0 for all $t \neq 0$. Much more about unit impulse functions was given in [68], [69]: definitions of the left/the right unit impulses $\delta^{\mp}\left(\cdot\right)$ and their Laplace transforms $\mathcal{L}^{\mp}\left\{\delta^{\mp}\left(t\right)\right\}$, including unit impulse $\delta\left(\cdot\right)$ and the Laplace transforms of it, $\mathcal{L}^{\mp}\left\{\delta\left(t\right)\right\}$, as well about the unit step functions: definitions of the left/the right unit step functions $h^{\mp}\left(t\right)$ and their use in the calculus of $\delta^{\mp}\left(t\right)$. For the different Laplace transforms, proofs in details were given in [68], [69].

Similarly, the unit impulse function has got very important theoretical and practical role and place in the linear discrete-time time-invariant systems, too. Especially it is the case when the ideal sampling process is mathematically modeled, i.e., mathematically described. The mathematical description of the ideal sample output is carried out through the weighting the delayed unit impulses for kT by the values of the ideal sample input variable at the same sampling instants kT. These weighting in literature is done in two different ways: through the area of the weighted impulse [41], [89], and through the height of the weighted impulse [8], [14], [101]. The latter approach requires to redefine the unit impulse as so called *discrete unit impulse* of infinitely small duration but of the unit height at $t_d = t_{dzero}$, $\text{num}\, t_{dzero} = 0$, i.e., for $k = 0$, and denoted by $\delta_d(\cdot)$ to avoid confusion with the unit impulse $\delta(\cdot)$.

Let the *discrete unit impulse* $\delta_d(\cdot) : \mathcal{T} \rightarrow \{0, 1\}$ be such that

$$\delta_d(t) = \begin{cases} 0, & t \neq 0 \\ 1, & t = 0 \end{cases}. \tag{B.17}$$

Then, a continuous-time variable $x(\cdot)$ discretized in time, by means of the ideal sample, is modeled as follows,

$$x^*(t) = x(t) \sum_{k=0}^{k=+\infty} \delta_d(t - kT) = x(t)\, \delta_d^*(t) = \sum_{k=0}^{k=+\infty} x(kT)\, \delta_d(t - kT),$$

$$k \in \mathbb{Z},\ T > 0,\ \delta_d^*(t) = \sum_{k=0}^{k=+\infty} \delta_d(t - kT), \tag{B.18}$$

where $x^*(\cdot)$ is variable $x(\cdot)$ discretized in time, T is sampling period, $T \in \mathcal{R}^+$, $\delta_d(\cdot - kT)$ is discrete unit impulse delayed for kT, and $\delta_d^*(\cdot)$ is the sequence of delayed discrete unit impulses for kT. It is obvious from (B.17) that:

$$\delta_d(j) = \begin{cases} 1, & j = 0 \\ 0, & j \neq 0 \end{cases},$$

or for the shifted discrete unit impulse,

$$\delta_d(j - k) = \begin{cases} 1, & j = k \\ 0, & j \neq k \end{cases}.$$

In the discrete-time systems we operate with the sequence of numbers, which are the values of the discrete-time variables and the system variables exactly at discrete instants. Because of that the latter approach is much more natural and it is accepted in this book. The previously discussed two approaches influence the determination of the $Z-$transform of the unit impulse and the discrete unit impulse. In the former approach, it is difficult to find in literature explicit determination of the $Z-$transform of the unit impulse, by its definition, although its value 1 is present in the $Z-$transform tables. This is controversial because the correct value of the $Z-$transform of the continuous-time unit impulse function is as follows:

$$\mathcal{Z}\{\delta(t)\} = \sum_{k=0}^{k=+\infty} \delta(kT)\, z^{-k} = +\infty \cdot z^{-0} = +\infty \cdot 1 = +\infty, \forall z \in \mathcal{C}. \quad \text{(B.19)}$$

Since the former weighting approach is carried out through the area of the continuous-time unit impulse then

$$\mathcal{Z}\left\{\int_{-\infty}^{+\infty} \delta(t)\, dt\right\} = \sum_{k=0}^{k=+\infty} \left[\int_{-\infty}^{+\infty} \delta(t)\, dt\right]_{t=kT} z^{-k} =$$

$$= \sum_{k=0}^{k=+\infty} \left[\int_{0^-}^{0^+} \delta(t)\, dt\right]_{t=kT} z^{-k} = 1 \cdot z^{-0} = 1 \cdot 1 = 1, \forall z \in \mathcal{C}, \quad \text{(B.20)}$$

or for the delayed unit impulse

$$\mathcal{Z}\left\{\int_{-\infty}^{+\infty} \delta(t - jT)\, dt\right\} = \sum_{k=0}^{k=+\infty} \left[\int_{-\infty}^{+\infty} \delta(t - jT)\, dt\right]_{t=kT} z^{-k} =$$

$$= \sum_{k=0}^{k=+\infty} \left[\int_{jT^-}^{jT^+} \delta(t - jT)\, dt\right]_{t=kT} z^{-k} = 1 \cdot z^{-j} = z^{-j}, \ \forall z \in \mathcal{C}. \quad \text{(B.21)}$$

In the latter approach, which we accepted, the determination of the $Z-$transform of the discrete unit impulse, is as follows,

$$\mathcal{Z}\{\delta_d(t)\} = \sum_{k=0}^{k=+\infty} \delta_d(k)\, z^{-k} = \delta_d(0)\, z^{-0} = 1 \cdot z^{-0} = 1 \cdot 1 = 1, \ \forall z \in \mathcal{C}, \quad \text{(B.22)}$$

or

$$\mathcal{Z}\{\delta_d(t - jT)\} = \sum_{k=0}^{k=+\infty} \delta[(k - j)T]\, z^{-k} = \delta(0)\, z^{-j} =$$

$$= 1 \cdot z^{-j} = z^{-j}, \ \forall z \in \mathcal{C}. \quad \text{(B.23)}$$

To determine the inverse $Z-$transform of 1, $\mathcal{Z}^{-1}\{1\}$, we represent the 1 as

$$1 = \delta_d(0)\, z^{-0} + \delta_d(1)\, z^{-1} + \delta_d(2)\, z^{-2} + \cdots = \sum_{k=0}^{k=+\infty} \delta_d(k)\, z^{-k}, \quad \text{(B.24)}$$

what is by definition the $Z-$transform of the discrete unit impulse $\delta_d(t)$. This implies that

$$\mathcal{Z}^{-1}\{1\} = \delta_d(t), \ \forall z \in \mathcal{C}. \quad \text{(B.25)}$$

Part VI

REFERENCES

References

[1] J. K. Acka, "Finding the Transfer Function of Time Invariant Linear System without Computing $(s\boldsymbol{I} - \boldsymbol{A})^{-1}$," *Int. J. Control*, Vol. 62, No. 6, pp. 1517 - 1522, 1995.

[2] J. Ackermann, *Sampled-Data Control Systems: Analysis and Synthesis, Robust System Design*, Springer-Verlag, Berlin, 1985., 1990., 2012.

[3] B. D. O. Anderson, "A Note on Transmission Zeros of a Transfer Function Matrix," *IEEE Transactions on Automatic Control*, Vol. AC-21, No. 4, pp. 589 - 591, August 1976.

[4] B. D. O. Anderson and J. B. Moore, *Linear Optimal Control*, Englewood Cliffs: Prentice Hall, 1971.

[5] P. J. Antsaklis and A. N. Michel, *Linear Systems*, New York: The McGraw-Hill Companies, Inc., 1997.

[6] K. J. Astrom, B. J. Wittenmark, *Computer Controlled Systems-Theory and Design,* Prentice-Hall, Englewood Cliffs, New Jersey, 07632, 1990., 1997., USA.

[7] S. Barnett, C. Storey, *Matrix Methods in Stability Theory,* Barnes & Noble, Inc., New York, 1970.

[8] A. B. Bishop, *Introduction to Discrete Linear Controls: Theory and Application,* Academic Press, New York, 1975., USA.

[9] J. G. Bollinger, N. A. Duffie, *Computer Control of Machines and Processes,* Addison-Wesley Publishing Company, Reading, Massachusetts, 1988., USA.

[10] P. Borne, G. Dauphin-Tanguy, J.-P. Richard, F. Rotella and I. Zambettakis, *Commande et Optimisation des Processus*, Paris: Éditions TECHNIP, 1990.

[11] R. D. Braatz, "On Internal Stability and Unstable Pole-Zero Cancellations," *IEEE Control Systems Magazine*, vol. 32, No. 5, October 2012, pp. 15, 16.

[12] W. L. Brogan, *Modern Control Theory*, New York: Quantum Publishers, Inc., 1974.

[13] G. S. Brown and D. P. Campbell, *Principles of Servomechanisms*, New York: Wiley, 1948.

[14] J. A. Cadzow, H. R. Martens, *Discrete-Time and Computer Control Systems,* Prentice-Hall, Inc., Englewood Cliffs, New Jersey, 1970., USA.

[15] F. M. Callier and C. A. Desoer, *Multivariable Feedback Systems*, New York: Springer-Verlag, 1982.

[16] F. M. Callier and C. A. Desoer, *Linear System Theory*, New York: Springer-Verlag, 1991.

[17] G. E. Carlson, *Signal and Linear Systems Analysis and MATLAB®*, second edition, New York: Wiley, 1998.

[18] C.-T. Chen, *Linear System Theory and Design*, New York: Holt, Rinehart and Winston, Inc., 1984.

[19] Chestnut and R. W. Mayer, *Servomechanisms and Regulating System Design*, New York: Wiley, 1955.

[20] R. I. Damper, *Introduction to Discrete-Time Signals and Systems,* Chapman & Hall, London, 1995.

[21] J. J. D'Azzo and C. H. Houpis, *Linear Control System Analysis & Design*, New York: McGraw-Hill Book Company, 1988.

[22] C. A. Desoer, *Notes for A Second Course on Linear Systems*, New York: Van Nostrand Reinhold Company, 1970.

[23] C. A. Desoer and M. Vidyasagar, *Feedback Systems: Input-Output Properties*, New York: Academic Press, 1975.

[24] O. I. Elgerd, *Control Systems Theory*, New York: McGraw-Hill Book Company, 1967.

[25] D. R. Fannin, W. H. Tranter, and R. E. Ziemer, *Signals & Systems Continuous and Discrete*, fourth edition, Englewood Cliffs: Prentice-Hall, 1998.

[26] T. E. Fortmann and K. L. Hitz, *An Introduction to Linear Control Systems*, New York: Marcel Dekker, Inc., 1977.

[27] F. R. Gantmacher, *The Theory of Matrices*, Vol. 1, New York: Chelsea Publishing Co., 1974.

[28] F. R. Gantmacher, *The Theory of Matrices*, Vol. 2, New York: Chelsea Publishing Co., 1974.

[29] Ly. T. Grouyitch, *Automatique: Dynamique Linéaire*, Notes de cours, Belfort: Ecole Nationale d'Ingénieurs de Belfort, 1997.

[30] Ly. T. Grouyitch, *Automatique: Dynamique Linéaire*, Lecture Notes, Belfort: University of Technology Belfort-Montbeliard, 1999-2000.

[31] Lj. T. Grujić, "Natural Trackability and Control: Multiple Time Scale Systems," *Preprints of the IFAC-IFIP-IMACS Conference: Control of Industrial Systems*, **2**, Pergamon, Elsevier, London, pp. 111 - 116, 1997.

[32] Lj. T. Grujić, "Possibilities of Linear System Design on the Basis of Conditional Optimization in Parameter Plane," Part I, *Automatika: Theoretical supplement*, Zagreb (Yugoslavia - Croatia), Vol. 2, No. 1-2, pp. 49 - 60, 1966.

[33] Lj. T. Grujić, "Possibilities of Linear System Design on the Basis of Conditional Optimization in Parameter Plane," Part II, Linear Sampled Data Systems with Constant Parameters, *Automatika: Theoretical Supplement*, Zagreb (Yugoslavia - Croatia), Vol. 2, No. 1-2, pp. 61 - 71, 1966.

[34] Lj. T. Grujić (Ly. T. Gruyitch), *Continuous Time Control Systems*, Lecture notes for the course "DNEL4CN2: Control Systems," Durban: Department of Electrical Engineering, University of Natal, South Africa, 1993.

[35] Lj. T. Grujić, *Automatique - Dynamique Linéaire*, Lecture Notes, Belfort: Ecole Nationale d'Ingénieurs de Belfort, 1994 - 1996.

[36] Lj. T. Grujić, *Automatique - Dynamique Linéaire*, Lecture Notes, Belfort: Université de Technologie de Belfort - Montbéliard, 2000 - 2003.

[37] Lj. T. Grujić, "Natural Trackability and Tracking Control of Robots," *IMACS-IEEE-SMC Multiconference CESA'96: Symposium on Control, Optimization and Supervision*, Vol. 1, Lille, France, pp. 38 - 43, 1996.

[38] Lj. T. Grujić, "Natural Trackability and Control: Multiple Time Scale Systems," *Proceedings of the IFAC Conference: Control of Industrial Systems*, (Ed's. Lj. T. Grujić, P. Borne, A. El Moudni and M. Ferney), Vol. 2, Pergamon, Elsevier, London, pp. 669 - 674, 1997.

[39] Lj. T. Grujić, "Natural Trackability and Control: Perturbed Robots," *Proceedings of the IFAC Conference: Control of Industrial Systems*, (Ed's. Lj. T. Grujić, P. Borne, A. El Moudni and M. Ferney), Vol. 3, Pergamon, Elsevier, London, pp. 1641 - 1646, 1997.

[40] Lj. T. Grujić and W. P. Mounfield, Jr., "Natural Tracking Control of Linear Systems," *Proceedings of the 13th IMACS World Congress on Computation and Applied Mathematics*, Eds. R. Vichnevetsky and J. J. H. Miller, Trinity College, Dublin, Ireland, Vol. 3, pp. 1269 - 1270, July 22 - 26, 1991.

[41] Lj. T. Grujić, *Discrete Systems (in Serbian)*, Mechanical Engineering Faculty, Belgrade, 1985., 1991., Serbia.

[42] Lj. T. Grujić and W. P. Mounfield, Jr., "Tracking Control of Time-Invariant Linear Systems Described by IO Differential Equations," *Proceedings of the 30th IMACS Conference on Decision and Control*, Brighton, England, Vol. 3, pp. 2441 - 2446, December 11 - 13, 1991.

[43] Lj. T. Grujić and W. P. Mounfield, "Natural Tracking Control of Linear Systems," in *Mathematics of the Analysis and Design of Process Control*, Ed. P. Borne, S.G. Tzafestas and N.E. Radhy, Elsevier Science Publishers B. V., IMACS, pp. 53 - 64, 1992.

[44] Lj. T. Grujić and W. P. Mounfield, "Stablewise Tracking with Finite Reachability Time: Linear Time-Invariant Continuous-Time MIMO

Systems," *Proc. of the 31st IEEE Conference on Decision and Control*, Tucson, Arizona, pp. 834 - 839, 1992.

[45] Lj. T. Grujić and W. P. Mounfield, "Natural Tracking PID Process Control for Exponential Tracking," *American Institute of Chemical Engineers Journal*, **38**, No. 4, pp. 555 - 562, 1992.

[46] Lj. T. Grujić and W. P. Mounfield, "PD Natural Tracking Control of an Unstable Chemical Reaction," *Proc. 1993 IEEE International Conference on Systems, Man and Cybernetics*, Le Touquet, Vol. 2, pp. 730 - 735, 1993.

[47] Lj. T. Grujić and Mounfield W. P., "PID Natural Tracking Control of a Robot: Theory," *Proc. 1993 IEEE International Conference on Systems, Man and Cybernetics*, Le Touquet, Vol. 4, pp. 323 - 327, 1993.

[48] Lj. T. Grujić and W. P. Mounfield, "PD-Control for Stablewise Tracking with Finite Reachability Time: Linear Continuous Time MIMO Systems with State-Space Description," *International Journal of Robust and Nonlinear Control*, England, Vol. 3, pp. 341 - 360, 1993.

[49] Lj. T. Grujić and W. P. Mounfield, *Natural Tracking Controller*, US Patent No 5,379,210, Jan. 3, 1995.

[50] Lj. T. Grujić and W. P. Mounfield, "Ship Roll Stabilization by Natural Tracking Control: Stablewise Tracking with Finite Reachability Time," *Proc. 3rd IFAC Workshop on Control Applications in Marine Systems*, Trondheim, Norway, pp. 202-207, 10-12 May, 1995.

[51] Lj. T. Grujić and Z. Novaković, "Conditional Optimization of Multi-Input Multi-Output Continuous-Time Automatic Control Systems (A.C.S.) with Application to a Pneumatic A.C.S.," in Serbo-Croatian, *Proceedings of HIPNEF*, Belgrade (Yugoslavia-Serbia), pp. 55 - 61, November 1980.

[52] Ly. T. Gruyitch, "Exponential stabilizing natural tracking control of robots: theory," *Proceedings of the Third ASCE Specialty Conference on Robotics for Challenging Environments*, held in Albuquerque, New Mexico, USA, (Ed's. Laura A. Demsetz, Raymond H. Bryne and John P. Wetzel), Reston, Virginia, USA: American Society of Civil Engineers (ASCE), pp. 286 - 292, April 26 - 30, 1998.

[53] Ly. T. Gruyitch, "Natural Tracking Control Synthesis for Lagrangian Systems," *V International Seminar on Stability and Oscillations of Nonlinear Control Systems*, Russian Academy of Sciences, Moscow, pp. 115 - 120, June 3 - 5, 1998.

[54] Ly. T. Gruyitch, "Robust Prespecified Quality Tracking Control Synthesis for 2D Systems," *Proc. International Conference on Advances in Systems, Signals, Control and Computers*, **3**, Durban, South Africa, pp. 171 - 175, 1998.

[55] Lj. T. Grujić and D. Obradović, "Conditional Optimization of Multi-Input Multi-Output Discrete-Time Automatic Control Systams (A.C.S.) with Application to a Hydraulic A.C.S.," in Serbo-Croatian, *Proceedings of HIPNEF*, Belgrade (Yugoslavia-Serbia), pp. 341 - 350, November 1980.

[56] Ly. T. Gruyitch, "Physical Continuity and Uniqueness Principle. Exponential Natural Tracking Control," *Neural, Parallel & Scientific Computations*, 6, pp. 143 - 170, 1998.

[57] Ly. T. Gruyitch, "Natural Control of Robots for Fine Tracking," *Proceedings of the 38th Conference on Decision and Control*, Phoenix, Arizona, USA, pp. 5102 - 5107, December 1999.

[58] Ly. T. Gruyitch, "Robot Global Tracking with Finite Vector Reachability Time," *Proceedings of the European Control Conference*, Karlsruhe, Germany, Paper # 132, pp. 1 - 6, 31 August - 3 September 1999.

[59] Ly. T. Gruyitch, *Conduite des systèmes*, Lecture Notes: Notes de cours SY 98, Belfort: University of Technology Belfort - Montbéliard, 2000, 2001.

[60] Ly. T. Gruyitch, *Systèmes d'asservissement industriels*, Lecture Notes: Notes de cours SY 40, Belfort: Université de Technologie de Belfort - Montbéliard, 2001.

[61] Ly. T. Gruyitch, *Contrôle commande des processus industriels*, Lecture Notes: Notes de cours SY 51, Belfort: University of Technology Belfort - Montbéliard, 2002, 2003.

[62] Ly. T. Gruyitch, "Time, Systems, and Control: Qualitative Properties and Methods," Chapter 2 in *Stability and Control of Dynamical*

Systems with Applications, Editors D. Liu and P. J. Antsaklis, pp. 23 - 46, Boston: Birkhâuser, pp. 23 - 46, 2003.

[63] Ly. T. Gruyitch, "Consistent Lyapunov Methodology for Exponential Stability: PCUP Approach," in *Advances in Stability Theory at the End of the 20th Century*, Ed. A. A. Martynyuk, London: Taylor & Francis, pp. 107 - 120, 2003.

[64] Ly. T. Gruyitch, *TIME. Fields, Relativity, and Systems*, ISBN 1-59526-671-2, LCCN 2006909437, Llumina, Coral Springs, Florida, USA, http://www.llumina.com/store/timefieldsrelativity.htm, 2006.

[65] Ly. T. Gruyitch, *TIME and TIME FIELDS. Modeling, Relativity, and Systems Control*, ISBN 1-4251-0726-5, Trafford, Victoria, Canada, http://www.trafford.com/06-2484, 2006.

[66] Ly. T. Gruyitch, *Time and Consistent Relativity. Physical and Mathematical Fundamentals,* Apple Academic Press, Inc., Waretown N.J. and Oakville ON, 2015.

[67] Ly. T. Gruyitch, *Advances in the Linear Dynamic Systems Theory,* Llumina Press, Tamarac, FL, USA, 2013.

[68] Ly. T. Gruyitch, *Advances in the Linear Dynamic Systems Theory: Time-Invariant Continuous-Time Systems*, ISBN: 978-1-60594-988-8, 978-1-60594-994-9, Llumina, Plantation, Florida, USA, 2013.

[69] Ly. T. Gruyitch, *Linear Dynamical Systems. Continuous-Time Time-Invariant Systems,* Apple Academic Press, Waretown, USA, 2017.

[70] Ly. T. Gruyitch, *Tracking Control of Linear Systems*, CRC Press/Taylor & Francis Group, Boca Raton, Florida, USA, 2013.

[71] Ly. T. Gruyitch, *Nonlinear Systems Tracking*, CRC Press/Taylor & Francis Group, Boca Raton, Florida, USA, 2016.

[72] Ly. T. Gruyitch and W. P. Mounfield, Jr., "Robust Elementwise Exponential Tracking Control: IO Linear Systems," *Proceedings of the 36th IEEE Conference on Decision and Control*, San Diego, California, USA, pp. 3836 - 3841, December 1997.

[73] Ly. T. Gruyitch and W. Pratt Mounfield, Jr., "Absolute Output Natural Tracking Control: MIMO Lurie Systems," *Proceedings of the*

14th Triennial World Congress, Beijing, P. R. China, Pergamon - Elsevier Science, Vol. C, pp. 389 - 394, July 5 - 9, 1999.

[74] Ly. T. Gruyitch and W. Pratt Mounfield, Jr., "Constrained Natural Tracking Control Algorithms for Bilinear DC Shunt Wound Motors," *Proceedings of the 40th IEEE Conference on Decision and Control*, Orlando, Florida USA, pp. 4433 - 4438, December 2001.

[75] Ly. T. Gruyitch and W. Pratt Mounfield, Jr., "Elementwise Stablewise Tracking with Finite Reachability Time: Linear Time-Invariant Continuous-Time MIMO Systems," *International Journal of Systems Science*, Vol. 33, No. 4, pp. 277 - 299, 2002.

[76] Ly. T. Gruyitch and W. Pratt Mounfield, Jr., "Stablewise Absolute Output Natural Tracking Control with Finite Reachability Time: MIMO Lurie Systems," *CD ROM Proceedings of the 17th IMACS World Congress, Invited session IS-2: Tracking theory and control of nonlinear systems*, Paris, France, pp. 1 - 17, July 11-15, 2005; *Mathematics and computers in simulation*, Vol. 76, pp. 330 - 344, 2008.

[77] W. P. Mounfield, Jr. and Lj. T. Grujić, "High-Gain Natural Tracking Control of Linear Systems," *Proceedings of the 13th IMACS World Congress on Computation and Applied Mathematics*, Eds. R. Vichnevetsky and J. J. H. Miller, Trinity College, Dublin, Ireland, Vol. 3, pp. 1271 - 1272, 1991.

[78] Y. Hasegawa, *Control Problems of Discrete-Time Dynamical Systems,* Springer International Publishing, 2015.

[79] R. A. Horn and C. R. Johnson, *Topics in Matrix Analysis*, Cambridge University Press, Cambridge, UK, 1991.

[80] C. H. Houpis, G.B. Lamont, *Digital Control Systems: Theory, Hardware, Software*, McGraw-Hill Book Company, New York, 1985, 1987, 1992, USA, New Dehli, 2014, India.

[81] R. Iserman, *Digital Control Systems*, Springer-Verlag, Berlin, 1981, 2013, 2014.

[82] J. Sarangapani, *Neural Network Control of Nonlinear Discrete-Time Systems,* CRC Press/Taylor & Francis Group, Boca Raton, Florida, USA, 2006.

[83] W. Hahn, *Stability of Motion*, New York: Springer-Verlag, 1967.

[84] T. Kailath, *Linear Systems*, Englewood Cliffs: Prentice-Hall, Inc., 1980.

[85] R. E. Kalman, "Mathematical Description of Linear Dynamical Systems," *J. S. I. A. M. Control*, Ser. A, Vol. 1, No. 2, pp. 152 - 192, 1963.

[86] N. N. Krasovskii, *Some Problems of the Theory of Stability of Motion*, in Russian, Moscow: FIZMATGIZ, 1959.

[87] N. N. Krasovskii, *Stability of Motion*, Stanford: Stanford University Press, 1963.

[88] B. C. Kuo, *Automatic Control Systems*, Englewood Cliffs: Prentice-Hall, Inc., 1967.

[89] B. C. Kuo, *Digital Control Systems*, Holt, Rinehart and Winston, New York, 1980, USA, Oxford University Press, 1992. UK.

[90] B. C. Kuo, *Automatic Control Systems*, Englewood Cliffs: Prentice-Hall, Inc., 1987.

[91] H. Kwakernaak and R. Sivan, *Linear Optimal Control Systems*, New York: Wiley Interscience, 1972.

[92] J. P. La Salle, *The Stability of Dynamical Systems,* SAIM, Philadelphia, 1976.

[93] A. M. Lyapunov, *The General Problem of Stability of Motion*, (in Russian), Kharkov Mathematical Society, Kharkov, 1892; in Academician A. M. Lyapunov: "Collected Papers," U.S.S.R. Academy of Science, Moscow, II, pp. 5 - 263, 1956. French translation: "Problème général de la stabilité du mouvement," *Ann. Fac. Toulouse*, 9, pp. 203 - 474; also in: *Annals of Mathematics Study*, No. 17, Princeton University Press, 1949. English translation: *Intern. J. Control*, 55, pp. 531 - 773, 1992; also published as the book in English, Taylor & Francis, London, 1992.

[94] L. A. MacColl, *Fundamental Theory of Servomechanisms*, New York: D. Van Nostrand Company, Inc., 1945.

[95] J. M. Maciejowski, *Multivariable Feedback Systems, Wokingham: Addison-Wesley Publishing Company, 1989.*

[96] M. Mandal, A. Asif, *Continuous and Discrete Time Signals and Systems,* Cambridge University Press, New York, 2007.

[97] J. L. Melsa and D. G. Schultz, *Linear Control Systems*, New York: McGraw-Hill Book Company, 1969.

[98] M. Merkle, *Mathematical Analysis: Theory (in Serbian),* Milan Merkle, Belgrade, 1996., Serbia.

[99] B. R. Milojković and Lj. T. Grujić, *Automatic Control,* in Serbo-Croatian, Belgrade: Faculty of Mechanical Engineering, University of Belgrade, 1977.

[100] B. R. Milojković and L. T. Grujić, *Automatic Control,* in Serbo-Croatian, Belgrade: Faculty of Mechanical Engineering, 1981.

[101] K. M. Moudgalya, *Digital control,* John Wiley & Sons Ltd, The Atrium, Southern Gate, Chichester, West Sussex PO19 8SQ, England, 2007.

[102] W. P. Mounfield, Jr. and Lj. T. Grujić, "High-Gain Natural Tracking Control of Time-Invariant Systems Described by IO Differential Equations," *Proceedings of the 30th Conference on Decision and Control,* Brighton, England, pp. 2447 - 2452, 1991.

[103] W. P. Mounfield and Lj. T. Grujić, "Natural Tracking Control for Exponential Tracking: Lateral High-Gain PI Control of an Aircraft System with State-Space Description," *Neural, Parallel & Scientific Computations*, Vol. 1, No. 3, pp. 357-370, 1993.

[104] W. P. Mounfield, Jr. and Ly. T. Gruyitch, "Control of Aircrafts with Redundant Control Surfaces: Stablewise Tracking Control with Finite Reachability Time," *Proceedings of the Second International Conference on Nonlinear Problems in Aviation and Aerospace*, European Conference Publishers, Cambridge, Vol. 2, 1999, pp. 547 - 554, 1999.

[105] K. Ogata, *State Space Analysis of Control Systems*, Englewood Cliffs: Prentice-Hall, 1967.

[106] K. Ogata, *Modern Control Engineering*, Englewood Cliffs: Prentice-Hall, 1970.

[107] D. H. Owens, *Feedback and Multivariable Systems*, Stevenage, Herts: Peter Peregrinus Ltd., 1978.

[108] C.L. Phillips, H.T. Nagle, *Digital Control Systems: Analysis and Design*, Prentice-Hall, Englewood Cliffs, New Jersey, 1984., 1995., 2002., USA.

[109] H. M. Power and R. J. Simpson, *Introduction to Dynamics and Control*, London: McGraw-Hill Book Company (UK) Limited, 1978.

[110] C. A. Rabbath, N. Léchevin, *Discrete-Time Control System Design with Application*, Springer Science + Business Media, New York, 2014.

[111] H. H. Rosenbrock, *State-Space and Multivariable Theory*, London: Thomas Nelson and Sons Ltd., 1970.

[112] E. N. Sanchez, F. Ornelas-Tellez, *Discrete-Time Inverse Optimal Control for Nonlinear Systems,* CRC Press/Taylor & Francis Group, Boca Raton, Florida, USA, 2013.

[113] D. D. Šiljak, *Feedback Systems Synthesis via Squared Error Conditional Optimization,* D. Sc. Dissertation (in Serbo-Croatian), Department of Electrical Engineering, University of Belgrade, Belgrade (Yugoslavia - Serbia), 1963.

[114] D. D. Šiljak, *Nonlinear Systems*, John Wiley & Sons, New York, 1969.

[115] D. D. Šiljak, *Large-Scale Dynamic Systems: Stability and Structure*, New York: North Holland, 1978.

[116] V. Simoncini, "Computational Methods for Linear Matrix Equations," *SIAM REVIEW*, Vol. 58, No. 3, pp. 377–441, 2016.

[117] R. E. Skelton, *DYNAMIC SYSTEMS CONTROL. Linear Systems Analysis and Synthesis*, New York: John Wiley & Sons, 1988.

[118] E. D. Sontag, "Further Facts About Input to State Stabilization," *IEEE Transactions on Automatic Control*, Vol. 35, No. 3, pp. 473 - 476, 1990.

[119] E. D. Sontag, *Mathematical Control Theory: Deterministic Finite Dimensional Systems*, New York: Springer, 1990.

[120] E. D. Sontag, "Input to State Stability: Basic Concepts and Results," pages 163-220 in P. Nistri and G. Stefani, editors, *Nonlinear and Optimal Control Theory*, Berlin: Springer-Verlag, 2007, available also at http://www.mit.edu/~esontag/FTP_DIR/04cetraro.pdf.

[121] E. D. Sontag, "Publications about 'input to state stability,'" http://www.math.rutgers.edu/~sontag/Keyword/INPUT-TO-STATE-STABILITY.html.

[122] H.F. Vanlandingham, *Introduction to Digital Control Systems*, Macmillan Publishing Company, New York, 1985., U.S.A.

[123] M. Vidyasagar, *Nonlinear Systems Analysis*, Englewood Cliffs: Prentice-Hall, 1978.

[124] J. C. West, *Textbook of Servomechanisms*, London: English Universities Press, 1953.

[125] D. M. Wiberg, *State Space and Linear Systems*, New York: McGraw-Hill Book Company, 1971.

[126] W. A. Wolovich, *Linear Multivariable Systems*, New York: Springer-Verlag, 1974.

[127] W. M. Wonham, *Linear Multivariable Control. A Geometric Approach*, Berlin: Springer-Verlag, 1974.

Part VII

INDEX

Author Index

Subject Index